QUANTUM TRANSPORT CALCULATIONS FOR NANOSYSTEMS

QUANTUM TRANSPORT CALCULATIONS FOR NANOSYSTEMS

Kenji Hirose & Nobuhiko Kobayashi

Pan Stanford Publishing

Published by

Pan Stanford Publishing Pte. Ltd.
Penthouse Level, Suntec Tower 3
8 Temasek Boulevard
Singapore 038988

Email: editorial@panstanford.com
Web: www.panstanford.com

British Library Cataloguing-in-Publication Data
A catalogue record for this book is available from the British Library.

Quantum Transport Calculations for Nanosystems

ISBN 978-981-4267-32-8 (Hardcover)
ISBN 978-981-4267-59-5 (eBook)

Printed in the USA

Contents

Preface

This book describes quantum transport calculations for nanosystems. Recently, as electric devices have become smaller and smaller to the nanometer scale, electron transport behaviors in such a tiny region have attracted much attention. To understand electron transport in nanosystems is important not only for the fundamental research but also for the practical applications to nanometer-scale devices. In this region, we need to consider electron transport from atomistic viewpoints, taking into account all the atomistic effects in the region on transport. Since atomistic behaviors are governed by the quantum mechanics, we have to treat electron transport purely on the basis of quantum mechanics, solving the Schrödinger equation for the wavefunctions and so on.

Then we might ask several questions about the problems. Is quantum electron transport different from classical electron transport such as Ohm's law? What is the size that changes electron transport from classical to quantum? Can we treat the behaviors of quantum transport numerically from atomic-scale calculations? Although to answer all of these questions is a difficult task, we will try to show the clues of methods to access these questions.

In this book, we will first give brief reviews of the basic concepts of electron transport. These are the classical Drude model, semi-classical Boltzmann equation, and transport theories based on the quantum mechanics such as linear-response theory of the Kubo formula and the scattering approach of the Landauer formula. Then we provide the methods to use them, Green's function and numerical ways to obtain wavefunctions directly. The objective of this book is to show the fundamental concepts of electron transport and provide various numerical calculation methods in detail, showing their applications to various nanosystems.

The invention of the scanning tunneling microscope (STM) has opened a way to directly access atoms and molecules, where the STM tip is utilized for atom manipulations. Recently, it has become possible to create nanosystems using a single atom and molecule. This opens the possibility for constructing ultra-small electron devices with these components as conducting electron channels. The studies of electron transport for the atomistic nanosystems from the *bottom-up* approach become more and more important for the future electron devices. Since electronic states for these materials have been studied very well based on the density-functional theory (DFT), we will show the computational methods for electron transport with use of the DFT and present their applications from atomistic viewpoints.

On the other hand, the *top-down* approach to reduce the size of electron devices has made rapid progress. At present, the size of silicon transistors has become much less than 30 nm and is being reduced further. Also, there is rapid progress of making electron devices using nano-carbon and organic materials. Correspondingly, various effects on electron transport have to be taken into account from atomistic viewpoints. So far, transistor devices have provided many fundamental transport characteristics from a number of transport experiments. Especially, transport properties in artificially constructed low-dimensional systems are very important for various applications. We present computational approaches to study quantum electron transport in the low-dimensional systems, such as quantum dots and quantum point contacts.

In this book, in addition to the descriptions of the basic formulas of electron transport, we present practical computational methods for transport, including recursion-transfer-matrix method, Lippmann-Schwinger method, nonequilibrium Green's function method, time-dependent wave-packet diffusion method, and quantum master equation method. With these tools, we show various transport properties to apply to the atomistic nanosystems from the bottom-up approach and to the artificial nanosystems from the top-down approach. We also show the way to extend the DFT formalism to more accurate calculations for transport. At the end of chapters, we provide several problems for the basic formulas together with the keys to solutions, which complement the contents of the book.

For acknowledgements, first of all, we would like to express sincere gratitude to Prof. Masaru Tsukada, who guided the authors to the quantum transport calculation world. Since we started to study and to develop an *ab initio* calculation method of open and nonequilibrium systems for quantum transport calculations, we are always in mind to his impressive and intuitive foresight on the importance of the forthcoming science and technologies. There are many other people that we would like to thank, from whom we learned so much on the subjects of the present book. Among them, especially, we acknowledge Prof. Boris Altshuler, Prof. Mads Brandbyge, Prof. Roberto Car, Dr. Hiroyuki Ishii, Prof. Hiroshi Kamimura, Prof. Yigal Meir, Dr. Stephan Roche, and Prof. Ned Wingreen for a number of valuable suggestions, comments, and supports. The research in this book has been partially supported by several national projects, for which we are grateful to acknowledge. These include the "Nano-Link" project from the MEXT of Japan, the CREST project on "Contact Effects and Transport Properties of Single Molecule" by the JST, and the Next Generation Supercomputer Project.

We hope that this book will be useful for readers to study quantum transport calculations and will stimulate the research and development for transport properties of nanometer-scale systems.

Kenji Hirose
Nobuhiko Kobayashi

Chapter 1

Prologue

Here we will give a brief introduction to the chapters in this book. We hope that this will help readers understand the concepts of quantum transport and realize the difficult points.

1.1 Nanometer-Scale Electron Systems

First of all, let us specify the nanosystems in view of the nanometer-scale length, $1 \text{ nm} = 10^{-9}$ m. In this book, we treat the nanosystems with the length of approximately $1 \text{ nm} \sim 1 \text{ µm} (= 1,000 \text{ nm})$. Since the size of the atom is about $1 \text{ Å} = 10^{-10} \text{ m} = 0.1 \text{ nm}$, this corresponds to the length of 10 to 10,000 atoms in one direction. Currently, there are two trends in the construction of nanometer-scale systems in experiments.

One is the bottom-up approach. In this approach, atoms and molecules are used as building blocks to create the atomistic nanosystems. The powerful experimental tool for making this approach possible is the scanning tunneling microscope (STM). The invention of the STM has enabled us to directly access atoms and molecules. The utilities of STMs not only are limited to

Quantum Transport Calculations for Nanosystems
Kenji Hirose and Nobuhiko Kobayashi

ISBN 978-981-4267-32-8 (Hardcover), 978-981-4267-59-5 (eBook)
www.panstanford.com

the observations of atomic structures on the surfaces but also are extended to the application of the manipulation of a single atom and molecule to create artificial atomistic structures on the surfaces. The appearance of this surprising experimental tool has stimulated research from an atomistic point of view. Nanoscience and nanotechnologies have become one of the most important research fields.

Recently, the research fields concerned with the construction of electric devices using molecules have attracted much attention because of the fascinating molecular electronics. The experimental technologies using the STM tip are indispensable for this purpose. The system lengths for the devices range from 1 nm to several tens of nanometer, depending on the molecules to use. A number of experiments have been performed to measure the electric current flowing between electrodes through molecules. However, since we still do not have the technologies to control the precise atomic positions of molecules to bridge the electrodes, the measurement of electric current depends sensitively on the atomic structures. The creation of a precise nanogap between electrodes for the molecules to bridge requires difficult techniques in experiments. Therefore, the analyses from the theoretical treatments become more and more important for the understanding of electric current in the molecular devices. As for larger molecular systems, nano-carbon materials such as carbon nanotubes and graphene become very important. The electric devices using nano-carbon materials have attracted much attention, since these materials have anomalous band structures and the devices are expected to show an extremely high-speed performance. The analyses from the atomic-scale computations for the transport properties of these devices have become more and more important (Fig. 1.1).

The other trend is the top-down approach. The size of electron devices becomes smaller and smaller due to the improvement of the lithography techniques. The rapid progress of the technologies has made the electron device size less than 50 nm for the channel length in the silicon metal–oxide–semiconductor (MOS) transistors. In such regimes, it becomes important to understand the transport from atomistic viewpoints for the applications. One of the important fundamental concepts that transistors have produced is that

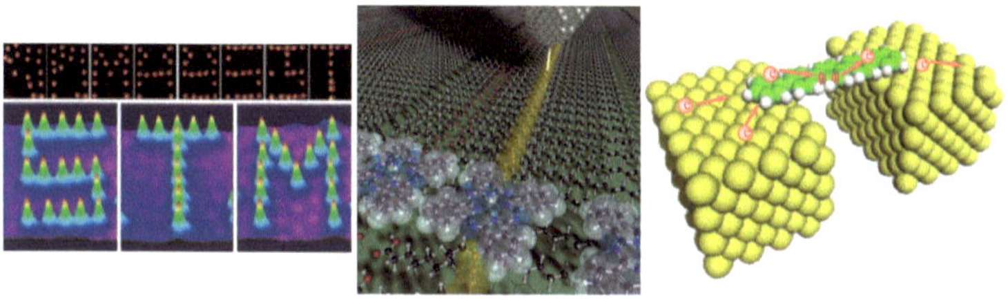

Figure 1.1 Examples of experiments for nanosystems from the bottom-up approach. (Left) Creation of the atomic characters by the atom manipulation using the tip of scanning tunneling microscopy (STM). From Surface and Interface Science labs of RIKEN with permission. (Center) Creation of atomistic molecular wire induced by the tip of STM. From Nano-system Organization Group of NIMS with permission. (Right) Schematic illustration of the atomistic nanosystem. A single molecule bridged between metallic electrodes forming a molecular device.

transport behaviors change due to the confinement of electrons into a narrow region to construct low-dimensional systems. Electrons are confined at the interface to construct a two-dimensional electron system. Electron transport is controlled by applying a bias voltage to the gate electrode on top of the electron device, creating the transistor. By the deposition of various shapes of electrodes on top of the device and by applying bias voltages, electrons beneath them are electrostatically confined and low-dimensional electron devices are created. They are called quantum wire for a one-dimensional system, quantum dot for a zero-dimensional system, and quantum point contact for connecting various parts of the system. Electron transport through such artificial nanosystems shows fascinating behaviors, such as single-electron transport, quantization of conductance, interference, and Kondo effect and may become the sources for a number of applications. These devices are called quantum devices. We will treat electron transport of these artificial nanosystems (Fig. 1.2). Figure 1.3 illustrates the schematic view of the various transport simulations treated in this book in view of multiscale calculations. We will see briefly the basic concepts and difficult points of the electron transport problems in the following sections.

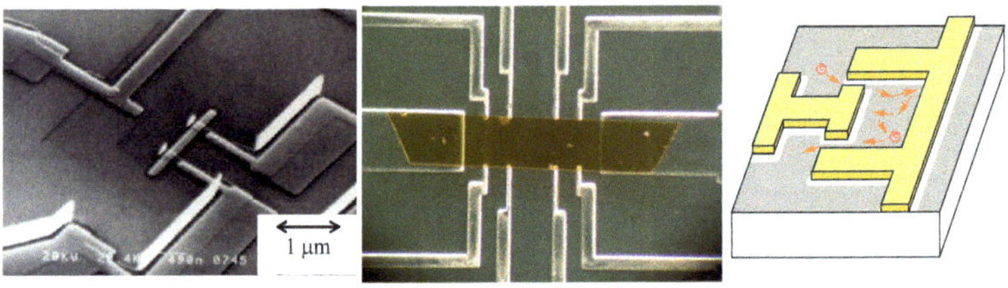

Figure 1.2 Examples of experiments for nanosystems from the top-down approach. (Left) SEM image of the quantum dot device. From Ootuka and Kanda labs of University of Tsukuba with permission. (Center) SEM image of organic semiconductor device. From Takeya labs of University of Tokyo with permission. (Right) Schematic illustration of the artificial nanosystem. An electron inside the nanosystem moves quantum-mechanically with the phase coherence to preserve.

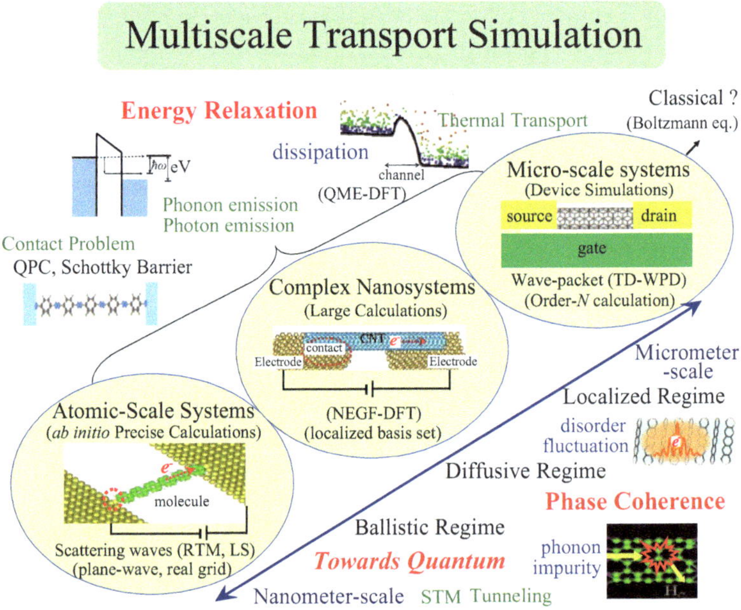

Figure 1.3 Schematic illustration of the multiscale computations for nanosystems.

1.2 Electric Current: What Pushes Electron to Flow?

From here, we consider several basic concepts. The first is why electron transport occurs. One may say that when we attach a battery, electric current appears. That is true for one possibility. There are other ways to make electrons flow. Then how can we construct the effective way to simulate electric current? There are mainly two different ways for the formulation of electron transport.

One is to consider that electron flow occurs as a response of an external electric field (Fig. 1.4 (left)). Electrons are accelerated by the electric field and reduce energies by the scatterings with other electrons, impurities, vibrations of atoms, etc. The steady state of electric current is achieved when these two contributions become comparable. In this formalism, the electric field or equivalently the vector potential is the main source of the electromotive force for electron transport, and corresponding bias voltage created at the electrodes is assumed to be the bias due to the battery. We describe electron transport in this picture, beginning from the classical Drude model where electrons are treated as point particles, then the semi-classical Boltzmann equation where the distribution function of electrons is introduced. The extension to quantum transport is possible by an introduction of the density matrix instead of the distribution function. The Kubo formula is derived as a linear-response theory for the density matrix. We also present the

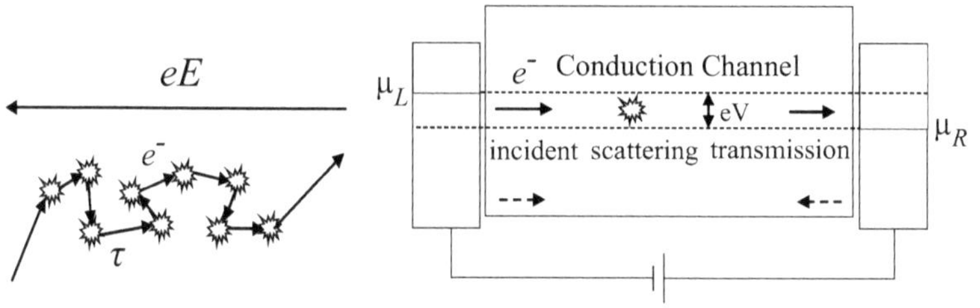

Figure 1.4 Schematic views of electron flow. (Left) Electron path due to scattering at different points in space with scattering time τ. Electron flow occurs as a response of an external electric field E. (Right) Electron flow between electrodes. Electron transport occurs at the energy between two different chemical potentials μ_L and μ_R in the electrodes.

quantum master equation method, which numerically solves the time-dependent density matrix equation for the current.

The other way to formulate electron transport is the scattering-wave approach (Fig. 1.4 (right)). Different chemical potentials are assigned to the electrodes so that their difference corresponds to an applied bias voltage. Electron transport as a scattering wave occurs at the energy between these two chemical potentials through nanosystems. As the region where electrons flow becomes smaller and smaller to nanometer scale, the quantum nature of electrons, such as tunneling and resonant behavior, becomes significant. Thus, description from the wavefunctions of electrons becomes more appropriate for electron transport. The Landauer formula is derived in this direction based on the scattering approach. This is mathematically formulated using the nonequilibrium Green's function (NEGF). These two formulations are shown to provide the same results when an appropriate boundary condition is imposed.

1.3 Phase Coherence: Particle or Wave?

The next concept is that an electron has two faces. One is the particle nature with the electron mass m and the electric charge $-e$. If we confine electrons to a narrow region, the number of electrons must be an integer value (Fig. 1.5 (left)). However, there is a finite

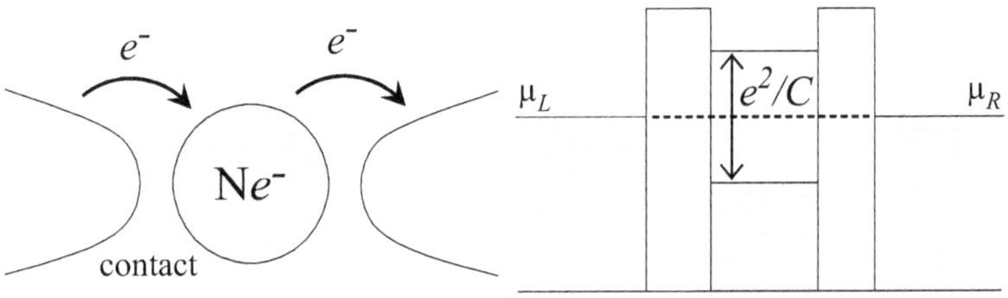

Figure 1.5 (Left) Schematic illustration of electron flow through a tiny island. Since the number of electrons in the island is fixed to an integer value, one by one electron hopping to the island shows a classical particle nature. Due to the coherent nature of electron, electron can go through the island via a virtual quantum level at low temperature. (Right) Schematic view of the energy diagram of the device with the Coulomb energy of e^2/C.

probability that an electron can hop to that region from electrodes. This should be accompanied by the tunneling phenomenon, which shows another face of the wave nature of an electron characterized by the phase θ of the wavefunction. Since the phase and the number have an uncertain relation of $[\theta, N] = i\hbar$, we cannot determine both quantities at the same time completely. Here we consider two cases such that this uncertain relation becomes apparent.

One example is the hopping transport in the confined system with strong Coulomb interaction. Each electron has its electric charge $-e$ and interacts with other electrons. When the confined region is so small that the capacitance energy e^2/C exceeds the temperature, the Coulomb interaction stops the other electron from hopping to the region (Fig. 1.5 (right)). Thus, the number of electrons N in that region is fixed. In this situation, electron behaves as a particle and this is called the Coulomb blockade. However, in the quantum mechanics, when the energy of the quantum level formed in the narrow region corresponds to the Fermi level of electrodes, the probability of hopping to the narrow region becomes finite. Correspondingly, the number of electrons is not fixed and fluctuates between N and $N+1$. This is an example in which the particle nature and the wave nature of electrons appear clearly. More interesting phenomena appear for the Coulomb interaction at low temperature, where the coherent wave nature of electron appears even in the blockade condition above. This is due to the hopping of electrons through a virtual quantum level formed at Fermi energy, called the Kondo effect.

Another example that an electron shows a particle and the wave nature is the hopping transport in the strongly disordered system. In this case, an electron is trapped in the conducting channel and thus behaves as a particle. Electron loses its phase memory and behaves as if it is a classical particle. The transport of an electron is determined mostly by the thermal hopping between the localized regions (Fig. 1.6 (left)). These transport mechanisms are considered to be realized in the organic crystals where the atomic configurations are random and electron mobility is very small. The wavefunction of an electron is localized, and as an electron hops to the other site, it loses its memory of the phase. As the crystalline structures become ordered, the wavefunction spreads

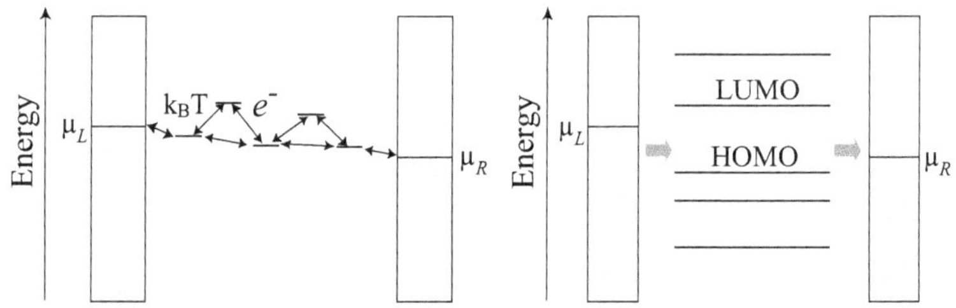

Figure 1.6 (Left) Schematic illustration of the transport between electrodes in dirty crystals. Electron transport is due to hopping through the localized regions as a classical particle. (Right) Schematic view of the transport between electrodes in ordered crystals. The phase is preserved and electron transport becomes coherent through quantum states characterized by HOMO and LUMO.

out and quantum states characterized by HOMO and LUMO are constructed (Fig. 1.6 (right)). The transport behavior from a particle hopping to a wave nature is an interesting and difficult problem.

1.4 Characteristic Length: Quantum vs. Classical?

Here, we argue about the several characteristic length scales to distinguish an electron transport into different regimes (Fig. 1.7).

The first is the Fermi wavelength, λ_F, which characterizes the electron wavelength itself at the Fermi energy for the transport, varying between a few Å in metals to a few hundred Å for the conducting electron in the semiconductor heterostructures.

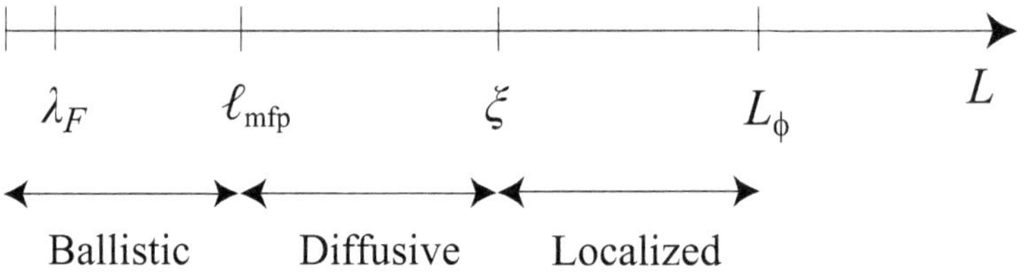

Figure 1.7 Typical relationship between characteristic lengths for the coherent transport in nanosystems.

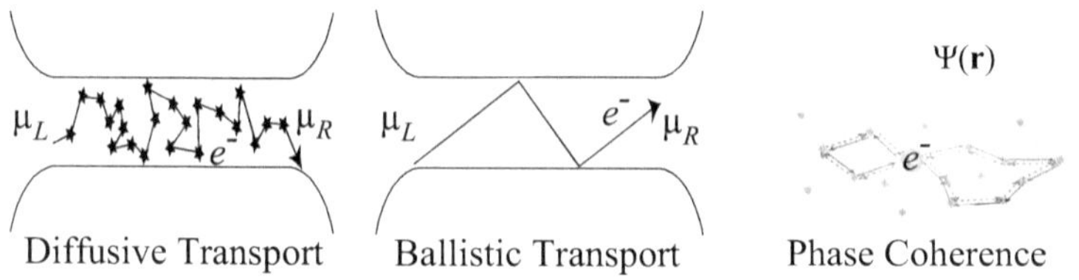

Figure 1.8 (Left) Schematic view of the diffusive transport. (Center) Schematic view of the ballistic transport. The mean free path distinguishes these two regimes. (Right) Schematic view of the phase coherence of an electron.

The second is the mean free path, ℓ_{mfp}, which characterizes the average length that an electron moves without scattering with other impurities, phonons, etc. It varies from a few hundred Å in amorphous alloys to an order of 10 μm in GaAs/AlAs heterostructures. Electron transport behavior changes from diffusive (Fig. 1.8 (left)) to ballistic (Fig. 1.8 (center)) beyond the mean free path, $\ell_{\text{mfp}} = v_F\tau$. Here v_F is the Fermi velocity and τ is the elastic scattering time for electron transport.

The third is the localization length, ξ, of electronic wavefunction. Due to disorders, the wavefunction shows a localized nature, which affects transport significantly. The localization character is strongly dependent on the dimensionality of the system.

The fourth is the phase coherence length, L_ϕ, which characterizes the length of an interference of the wavefunction. This is a typical length that an electron wave packet can travel without losing its phase coherence (Fig. 1.8 (right)). L_ϕ has a strong dependence on temperature, increasing as the temperature decreases, and might be larger than the mean free path ℓ_{mfp} in low-temperature regime. These characteristic length scales distinguish quantum transport from a classical one. We will describe the characteristic properties of electron transport in each regime in Chapter 2.

1.5 Where Does Electron Lose Energy for Dissipation?

Finally, we consider where the dissipation occurs in nanosystems. When an electron scatters with impurities, it changes its momentum

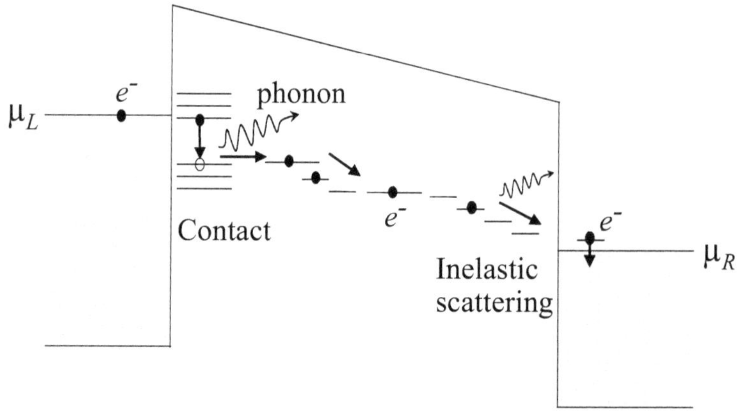

Figure 1.9 Schematic illustration of the energy diagram between electrodes. Electron injected from the left electrode loses its energy by the dissipation due to sequential inelastic phonon scatterings to reach the right electrode. The created phonon becomes the source of local heating.

within an elastic scattering, which characterizes the mean free path. When an electron encounters an inelastic scattering, it loses its energy, by emitting the phonon, photon, etc. (Fig. 1.9). When the phonon is emitted, it serves as a source of heating, which has a significant effect on nanoelectronic devices. Where an electron loses its energy for dissipation is, therefore, a very important problem for nanometer-scale device applications.

In nanosystems, we cannot neglect the effect of the contact on electrodes. At contacts, due to the geometrical structure and due to the mixing of wavefunctions between nanosystems and electrodes, the probability that inelastic scattering occurs becomes significant. The competition of the dissipation in the channel of nanosystems and at contacts to electrodes will offer practically important problems for nanoelectronic devices.

1.6 Scope of This Book

Following the introduction in this chapter, we describe the basic formulas to treat the transport properties in Chapter 2. Beginning from the classical treatment by the Drude model, we describe the semi-classical formula based on the Boltzmann equation where electron distribution function is introduced and is determined from the scatterings of impurities, phonons, etc., which are treated quantum mechanically. Also, we describe the basic formulas of classical nonequilibrium statistical mechanics such as the Langevin equation, Fokker–Planck equation, and stochastic master equation for treating the diffusive nature, fluctuation–dissipation theorem, and relaxation to equilibrium. These concepts lead to the quantum master equation in Chapter 4. Then, introducing the density matrix to replace the classical distribution function, we derive the Kubo formula within the linear-response theory. These formulas are based on the concepts that an applied electric field accelerates the electron motion and the electric current is the response to that field. There is another method to describe electron transport, which is the scattering approach. We describe the Landauer formula for an expression of the electric current and extend it to the multiterminal case. Based on this formalism, we show how the scattering approach describes the classical Ohm's law and the transport properties in the mesoscopic regime between microscopic and macroscopic regimes.

In Chapter 3, we introduce the mathematical method based on the Green's function technique in the framework of the quantum field approach, which describes both the particle and wave natures in the second quantization operators and derive the formulas for the electric current using Green's function. Linear-response Kubo formula is described by the equilibrium Green's function and the conductivity in the presence of impurity scatterings is presented. The nonequilibrium Green's functions are introduced to describe the transport with a finite bias voltage. Several examples to calculate the electric current including electron–electron and electron–phonon interactions are treated. The extension of the formalism to time-dependent current for the electron dynamics is also presented.

Chapter 4 is the central part of the present book. We show several calculation methods to obtain the electric current from an atomistic point of view, which is indispensable to nanometer-scale devices. Beginning from electronic state calculations based on the density-functional method, we explain the basic concepts of the ab initio calculations such as pseudopotential and exchange-correlation energy. Then, we present specific computational methods for the transport from atomistic viewpoints. The recursion-transfer-matrix method (RTM) and the Lippmann–Schwinger (LS) method are based on the scattering approach. The nonequilibrium Green's function (NEGF) approach using the atomically localized basis sets is illustrated. In addition to electron transport, we present the thermal transport, which becomes important in nanometer-scale electronics. The phonon transport, the local heating problem due to electron–phonon interaction, and the thermoelectric relations are treated by the NEGF method. The time-dependent wave-packet diffusion (TD-WPD) method uses the linear-response theory of the Kubo formula, and the quantum master equation (QME) method based on the density-matrix formalism is a generalization of the linear-response theory to an electric field with a finite bias and relaxation and is the quantum version of the nonequilibrium statistical mechanics. We also discuss an extension of the density-functional method to the time-dependent density-functional theory by using the NEGF method and the QME method, aiming at the rigorous treatment of electron transport for an application to realistic materials.

In Chapter 5, we give various applications of the computational methods in presented in Chapter 4, mostly for the electron transport from the bottom-up systems. These include the tunneling current of STM, atom manipulations by the tip of STM, electron transport through atomic wires, and transport properties of nano-carbon materials such as carbon nanotubes and graphene nanoribbons. We provide the characteristic lengths such as mean free path and phase coherence length to distinguish the ballistic regime from the diffusive regime and introduce their applications to field-effect transistors. We also present electron transport through a single molecule and molecular wires, which is expected to lead to the molecular electronics. Finally, we treat carrier transport properties of single-crystal organic semiconductors in which organic molecules are weakly bonded by the van der Waals interaction having flexible structures. We show the importance of thermal fluctuation, polaron formation, and static disorder for the understanding of transport properties, for which both thermal activated hopping transport behaviors and band-like transport behaviors are observed experimentally.

Chapter 6 describes applications to electron transport for the artificial nanostructures constructed in the two-dimensional (2D) electron systems, such as quantum wires, quantum dots, and quantum point contacts. We treat various characteristics of the electron transport, e.g., single-electron transport, Kondo effects, artificial atom, and the conductance fluctuation in view of the competition of electronic state energies and the Coulomb interaction. Anomalous spin-dependent transport observed at the lowest channel of quantum point contacts is also presented.

Chapter 7 is a brief summary of the presented computational methods and the future direction of the quantum transport calculations of nanosystems. The keys to solutions for the problems given at the end of chapters are provided separately, which are important to complement the basic concepts in the book.

Chapter 2

Basic Formulas for Electron Transport

In this chapter, we show the derivations of the basic formulas for electron transport. We begin from classical treatment for the electric current. Then we move to treat electron transport using quantum mechanics. We show the relationship between the quantum version and the classical version.

2.1 From Particle Description

2.1.1 *Drude Model*

The first is the Drude model [1]. Here we consider that an electron is a classical point charge. Then electric current is expressed by

$$I = \int_S \mathbf{j} \cdot d\mathbf{S}, \tag{2.1}$$

where $\mathbf{j}$ is the current density. It is caused by the response of an applied electric field E as

$$\mathbf{j} = \sigma \mathbf{E}, \tag{2.2}$$

where σ is called conductivity. Here we assume that the current is parallel to the direction of the field for simplicity.

Quantum Transport Calculations for Nanosystems
Kenji Hirose and Nobuhiko Kobayashi

ISBN 978-981-4267-32-8 (Hardcover), 978-981-4267-59-5 (eBook)
www.panstanford.com

The voltage drop V is obtained using the line integral

$$V = \int \mathbf{E} \cdot d\ell = \int \frac{\mathbf{j}}{\sigma} \cdot d\ell. \tag{2.3}$$

Here we assume the conductor is uniform and has a constant cross section S and length L, then we obtain the classical Ohm's law

$$V = RI, \tag{2.4}$$

where R is the resistance. In this situation, since Eq. (2.3) becomes

$$V = \frac{1}{\sigma}\frac{I}{S}L, \tag{2.5}$$

we have an expression for the resistance as

$$R = \frac{1}{\sigma}\frac{L}{S}, \tag{2.6}$$

or equivalently the conductance G is

$$G \equiv \frac{1}{R} = \sigma\frac{S}{L}. \tag{2.7}$$

This shows a well-known result that the resistance increases with the length L of the conductor and the conductance increases with the cross section S for the *Ohmic* conductors.

Now we derive the formula for conductivity σ, using the fundamental quantities of electrons and the classical treatment from the equation of motion of electrons. The current density is defined by

$$\mathbf{j} = en\mathbf{v}, \tag{2.8}$$

where $e(< 0)$ is the electron charge, n is the carrier density, and $\mathbf{v}$ is the velocity of an electron.[a]

Electrons are moving in an electric field and have a number of collisions, which will change their momentum $\mathbf{p}$. These collisions may be due to the scattering with impurities, or due to the scattering with the vibrations of atoms. Here we assume that these collisions are independent events and are not dependent on the previous collisions.[b] Suppose that the electrons have an average velocity $\mathbf{v}$ and we define the relaxation time τ as the average time between successive collisions. As a consequence of the collisions, electrons

[a] For a hole motion, we change the charge as $e > 0$.

[b] This is called the Markov assumption.

on average change their momentum $\mathbf{p}$ by the amount of $m\mathbf{v}/\tau$. Since electrons are accelerated by an electric field $\mathbf{E}$, the equation of motions of electrons is

$$m\frac{d\mathbf{v}}{dt} = e\mathbf{E} - \frac{m\mathbf{v}}{\tau}. \tag{2.9}$$

In a steady state, since the change of momentum of electrons is equal to the collision term, we have

$$\frac{m\mathbf{v}}{\tau} = e\mathbf{E}. \tag{2.10}$$

From this relation, the conductivity σ becomes

$$\sigma = \frac{\mathbf{j}}{\mathbf{E}} = \frac{en\mathbf{v}}{m\mathbf{v}/e\tau} = \frac{ne^2\tau}{m}. \tag{2.11}$$

This is called "Drude conductivity," derived completely from the classical treatment.

The carrier density n is determined from the Hall effect. In the Hall experiment, a magnetic field B_z is applied in the z direction in addition to an applied electric field E_x to the x direction. Then an electron, flowing in the x direction with a current density j_x, feels the Lorentz force $e\mathbf{v} \times \mathbf{B}/c$ in the y direction and there appears an electric field E_y in the y direction because electrons are accumulated at the edges of the sample. In equilibrium, the force from an electric field eE_y balances the Lorentz force.[a]

The equation of motion of an electron in this case becomes

$$m\frac{d\mathbf{v}}{dt} = e\left(\mathbf{E} + \frac{1}{c}\mathbf{v} \times \mathbf{B}\right) - \frac{m\mathbf{v}}{\tau}. \tag{2.12}$$

---

[a] In the two-dimensional (2D) system, the current density is determined from

$$j_x = \sigma E_x = \sigma_{xx}E_x + \sigma_{xy}E_y, \quad j_y = 0 = \sigma_{yx}E_x + \sigma_{yy}E_y.$$

Taking $\sigma_{xx} = \sigma_{yy}$ and $\sigma_{xy} = -\sigma_{yx}$ from the reciprocity relation, we have

$$\sigma_{xx} = \frac{1}{1+(\omega_c\tau)^2}\sigma, \quad \sigma_{xy} = \frac{\omega_c\tau}{1+(\omega_c\tau)^2}\sigma = \frac{nec}{B} - \frac{\sigma_{xx}}{\omega_c\tau}, \quad R_H = \frac{1}{B}\frac{\sigma_{xy}}{\sigma_{xx}^2+\sigma_{xy}^2}.$$

Here $\omega_c = eB/mc$ is the cyclotron frequency. Since the resistivity is the inverse of conductivity

$$\rho_{xx}(=\rho_{yy}) = \frac{\sigma_{xx}}{\sigma_{xx}^2+\sigma_{xy}^2}, \quad \rho_{xy}(=-\rho_{yx}) = \frac{\sigma_{xy}}{\sigma_{xx}^2+\sigma_{xy}^2}, \quad \text{thus } R_H = \frac{\rho_{xy}}{B}.$$

The Hall coefficient R_H is obtained from the measurements of B vs. ρ_{xy}.

In a steady state $d\mathbf{p}/dt = 0$, we have

$$eE_x + \frac{e}{c}v_y B - \frac{mv_x}{\tau} = 0$$
$$eE_y - \frac{e}{c}v_x B - \frac{mv_y}{\tau} = 0. \tag{2.13}$$

The Hall coefficient R_H is defined by

$$R_H = \frac{E_y}{j_x B}. \tag{2.14}$$

Since the electric field E_y appearing due to the Lorentz force is proportional to the applied magnetic field B and to the current density along the sample j_x, we expect this coefficient gives us information on the carrier density n. From the second equation of Eq. (2.13),[a] setting $v_y = 0$, we have

$$E_y = \frac{B}{nec} j_x. \tag{2.15}$$

Then the Hall coefficient becomes

$$R_H = \frac{1}{nec}, \tag{2.16}$$

which does not depend on any parameters of materials but on the carrier density only. In the usual case, the Hall coefficient R_H depends on the magnetic field, temperature, and the condition of the sample. However, at very low temperature and in pure sample conditions, the measured Hall coefficient approach a constant value. The measurement of the Hall coefficient is very important because it determines not only the magnitude of the carrier density n but also its sign, since the electric field E_y changes its sign according to the sign of the carrier. So we can find that the carrier for the transport is due to an electron or a hole in the material.

We note that in experiments the mobility μ is obtained from the measurements of conductivity σ. The Hall mobility

$$\mu_H = \sigma R_H c \tag{2.17}$$

corresponds to $\mu = e\tau/m$ Eq. (2.24) for the isotropic material.[b]

---

[a]Note that the first equation of Eq. (2.13) gives the magnetoresistance defined by $\rho(B) = E_x/j_x$.

[b]In the 2D systems for high magnetic fields $\omega_c\tau \gg 1$, we have $\sigma_{xx} = 0$ and $\sigma_{xy} = nec/B$. Quantum mechanics shows that the Landau states are formed in high

2.1.2 *Diffusive Motion and Einstein Relation*

Equation (2.11) due to the relaxation time τ reflects the scatterings of electrons driven by an external electric field. On the other hand, electrons move by the difference of the carrier concentration, which is described by the diffusion equation. In an equilibrium situation, these two factors for an electron flow should be balanced. This is the idea for the description of the diffusive motions of electrons.

Let us derive the relationship between τ and the diffusive nature, which is symbolically described by the diffusion constant D and obeys the diffusion equation

$$\frac{\partial n}{\partial t} = D\nabla^2 n. \tag{2.18}$$

Applying the charge conservation law

$$e\frac{\partial n}{\partial t} + \nabla \cdot \mathbf{j} = 0, \tag{2.19}$$

we can express the diffusive current (Fick's law) as

$$\mathbf{j} = -eD\nabla n. \tag{2.20}$$

This says that the diffusive current flows due to the spatial non-uniform carrier density n in directions opposite to the direction of changes in densities. Since this should be balanced by the difference of chemical potentials, which occurs with the distance L

$$\mathbf{j} = -eD\frac{\partial n}{\partial \epsilon}\frac{e\mathbf{E}L}{L} = -e^2\left[\frac{\partial n}{\partial \epsilon}\right]D\mathbf{E}, \tag{2.21}$$

the conductivity close to equilibrium is expressed by

$$\sigma = e^2\left[\frac{\partial n}{\partial \epsilon}\right]D = e^2\nu D \tag{2.22}$$

where $\nu = \partial n/\partial \epsilon$ is the density of states (DOS). We note that the diffusion constant D is regarded as a type of dissipative coefficient.

---

magnetic fields having $n = eB/hc$ electrons in each level with an energy separation of $\hbar\omega_c$. When the states are fully occupied up to the N-th Landau level at low temperature $k_B T \ll \hbar\omega_c$, we have $\sigma_{xy} = (ec/B)(NeB/hc) = (e^2/h)N$. Amazingly, this relation is satisfied when the Landau states are partially occupied, since most of the states in the disordered 2D systems are localized and not contribute to the transport. This leads to the *Integer Quantum Hall Effect* such that σ_{xy} are quantized with plateau structures at multiple integers of e^2/h for $\sigma_{xx} = 0$ when we change the Fermi energy by the gate voltage V_g.

Now we define the mobility μ to be a coefficient of the response of an electric field as

$$\mathbf{v} \equiv \mu \cdot \mathbf{E}, \tag{2.23}$$

or using Eq. (2.10)

$$\mu = \frac{e\tau}{m}. \tag{2.24}$$

Then we have

$$\frac{eD}{\mu} = \frac{n}{\partial n/\partial \epsilon} = k_B T. \tag{2.25}$$

Here we use the Boltzmann distribution $n = n_0 e^{-\epsilon/k_B T}$ for the nondegenerate classical particles. Correspondingly, the mobility is expressed by

$$\mu = \frac{eD}{k_B T}. \tag{2.26}$$

This is the famous "*Einstein relation*" for the Brownian motion, which relates the diffusion constant D with the mobility μ and shows that the diffusion constant is proportional to the mobility.[a]

Since each thermalized electron motion at temperature T has an average energy of

$$\frac{1}{2}m\langle v^2\rangle = \frac{1}{2}k_B T, \tag{2.27}$$

due to fluctuations (equipartition theorem),[b] we have the relation between the diffusion constant D and the relaxation time τ as

$$D = \langle v^2\rangle\tau = \frac{1}{d}v^2\tau, \tag{2.28}$$

---

[a]The idea for the Brownian motion by Einstein is to balance the osmotic pressure force and the frictional force in continuum fluid mechanics on the particles at steady state. Since the latter force described by the Stokes law depends on the particle velocity, dynamical viscosity, and the particle diameter, when the osmotic pressure is measured, the size of atoms is estimated and then the Avogadro number N_A and the Boltzmann constant k_B can be determined.

[b]The average value of a randomly fluctuating variable $x(t)$ is defined by

$$\langle x\rangle = \lim_{T\to\infty}\frac{1}{T}\sum_{i=1}^{T}x(t_i) = \lim_{T\to\infty}\frac{1}{T}\int_{-T/2}^{T/2}x(t)dt = \int_{-\infty}^{\infty}x(t)P(x,t)dt$$

where $P(x,t)$ is the probability distribution normalized by 1. When $P(x,t)$ is independent of time t, then the distribution is said to be stationary. Here we use the Maxwell–Boltzmann distribution $p(v) = \sqrt{\frac{m}{2\pi k_B T}}e^{-mv^2/2k_B T}$.

where d is the dimension of the conductor. Then the conductivity is obtained from

$$\sigma = ne\mu = e^2 \nu D = \frac{ne^2\tau}{m}, \tag{2.29}$$

which is frequently called the *Einstein relation* for conductivity.

Let us consider the diffusion equation of Eq. (2.18). Since the density at $\mathbf{r}$ and t is related to the position $\mathbf{r}'$ and time t', we define Green's function $G(\mathbf{r}, t)$ as

$$n(\mathbf{r}, t) = \int G(\mathbf{r} - \mathbf{r}', t - t')n(\mathbf{r}', t')d\mathbf{r}'dt' \tag{2.30}$$

with the boundary condition

$$G(\mathbf{r}, t = 0) = \delta(\mathbf{r} - \mathbf{r}'), \tag{2.31}$$

which assumes that carrier density starts from the origin at the initial time $t = 0$ as $n(\mathbf{r}, 0) = \delta(\mathbf{r})$. Then the equation for $G(\mathbf{r}, t)$

$$\frac{\partial G(\mathbf{r}, t)}{\partial t} - D\nabla^2 G(\mathbf{r}, t) = 0 \tag{2.32}$$

is solved using the Fourier transform as

$$G(\mathbf{q}, \omega) = \frac{1}{-i\omega + D\mathbf{q}2} \tag{2.33}$$

with

$$G(\mathbf{q}, t) = \frac{1}{2\pi}\int_{-\infty}^{\infty} G(\mathbf{q}, \omega)e^{-i\omega t}d\omega = e^{-D\mathbf{q}^2 t}. \tag{2.34}$$

This shows that there is a pole at $\omega = -iD\mathbf{q}^2$ and the density has a spatial modulation with the wave number of $q \sim 1/\lambda$, decaying of order $1/Dq^2 \sim \lambda^2/D$.

For a one-dimensional case, Green's function is obtained as

$$G(x, t) = \frac{1}{2\pi}\int_{-\infty}^{\infty} G(q_x, t)e^{iq_x x}dq_x = \frac{1}{\sqrt{4\pi Dt}}e^{-x^2/4Dt}. \tag{2.35}$$

From this expression, we see that electron density initially at the origin spreads out with time t and the average position and the mean-square radius[a] become

$$\langle x \rangle = 0, \qquad \langle x^2 \rangle = 2Dt. \tag{2.36}$$

This shows that the mean displacement $\sqrt{\langle x^2 \rangle}$ is not proportional to the time, but rather its square root for the diffusive motion.[b]

---

[a]Here we use $G(x, t)$ for the probability distribution $P(x, t)$ as we show in the proceeding section such as $\langle x \rangle = \int_{-\infty}^{\infty} xG(r, t)dx$ and $\langle x^2 \rangle = \int_{-\infty}^{\infty} x^2 G(r, t)dx$.

[b]For a d-dimensional case, $G(\mathbf{r}, t) = \frac{1}{(4\pi Dt)^{d/2}}e^{-\mathbf{r}^2/4Dt}$ and $\langle r^2 \rangle = 2dDt$.

2.1.3 *Fluctuation–Dissipation Theorem*

Here we consider the Einstein relation for the diffusive behaviors of electrons more [2]. When an external field $\mathbf{E}$ is absent in Eq. (2.9), the electron motion becomes $\mathbf{v}(t) = e^{-t/\tau}\mathbf{v}(0)$, which decays zero at long time. This is not true, since at equilibrium $\langle v^2 \rangle = k_B T/m \neq 0$ in Eq. (2.27). This means that only the average velocity is not sufficient for the diffusive motion of electrons [3] and we need the other force other than the frictional force $m\mathbf{v}/\tau$.

2.1.3.1 Brownian motion and Langevin equation

Since electron motions are random due to the scatterings from successive collisions, it would be appropriate to introduce the (randomly fluctuating) "stochastic" random force $\eta(t)$. Then the equation of motion for the stochastic process of an electron Eq. (2.9) is generalized into

$$m\frac{d\mathbf{v}}{dt} = -m\gamma\mathbf{v} + \eta(t) + \mathbf{F}. \tag{2.37}$$

This is the Langevin equation. The total force is the sum of the friction force $-m\gamma\mathbf{v}$ with damping rate of $\gamma = 1/\tau$ and the fluctuating force $\eta(t)$, both of which come from the interaction of an electron with its environment (called "heat bath"), and an external force $\mathbf{F} = e\mathbf{E}$. We note that the local friction γ proportional to velocity produces the Ohmic behavior and the system approaches and maintains thermal equilibrium by interactions with $\eta(t)$.[a]

First we consider the case where the external force is absent. The Langevin equation Eq. (2.37) is solved as

$$v(t) = v(0)e^{-\gamma t} + \frac{1}{m}\int_0^t \eta(t')e^{-\gamma(t-t')}dt'. \tag{2.38}$$

When we assume that the force $\eta(t)$ during the collisions vary rapidly as an independent random event whose average is zero, the effects of $\eta(t)$ become

$$\langle \eta(t) \rangle = 0, \qquad \langle \eta(t)\eta(t') \rangle = S_\eta \delta(t - t') \tag{2.39}$$

where S_η represents the strength of the random force.

---

[a]When the friction depends on the velocity in the past, the Langevin equation is generalized as $m\dfrac{d\mathbf{v}(t)}{dt} = -m\displaystyle\int_{-\infty}^{t} \gamma(t - t')\mathbf{v}(t')dt' + \eta(t) + \mathbf{F}$.

Then, using the relation of $\langle v(t)\eta(t')\rangle = 0$, its fluctuation becomes

$$\langle v(t)^2\rangle = e^{-2\gamma t}\langle v(0)^2\rangle + \frac{1}{m^2}\int_0^t dt' \int_0^t dt'' e^{-\gamma(t'+t'')}\langle \eta(t')\eta(t'')\rangle$$
$$= e^{-2\gamma t}\langle v(0)^2\rangle + \frac{S_\eta}{2m^2\gamma}\left(1 - e^{-2\gamma t}\right). \quad (2.40)$$

This approaches $\langle v^2\rangle = S_\eta/2m^2\gamma$ in the long time limit, which should be equal to $k_B T/m$. Thus, we have

$$S_\eta = 2mk_B T\gamma. \quad (2.41)$$

This is known as the fluctuation–dissipation theorem, which relates the strength S_η of the fluctuating force to the magnitude of the friction or dissipation γ. It expresses the balance between the dissipation that drives the system in diversity and the fluctuation that keeps the system in order at a thermal equilibrium state.

Next we consider the case where the external force is present. Since $\langle \eta(t)\rangle = 0$, the solution for the Langevin equation is

$$\langle v\rangle = v(0)e^{-\gamma t} + \frac{eE}{m\gamma}\left(1 - e^{-\gamma t}\right) \quad (2.42)$$

and the mobility as defined by $\langle v\rangle = \mu E$ becomes $\mu = e/m\gamma$. On the other hand, since we obtain the velocity correlation from Eq. (2.38) as

$$\int_0^\infty \langle v(t)v(0)\rangle dt = \int_0^\infty e^{-\gamma t}\frac{k_B T}{m}dt = \frac{k_B T}{m\gamma}, \quad (2.43)$$

the conductivity $\sigma = ne\mu$ is expressed by the current–current correlation function as

$$\sigma = \frac{1}{k_B T}\int_0^\infty \langle j(t)j(0)\rangle dt. \quad (2.44)$$

This expression by the autocorrelation functions is a special case for the transport parameter and can be generally derived from the linear-response theory.[a]

Then diffusion constant as obtained from $\langle x^2\rangle = 2Dt$ becomes

$$D = \lim_{t\to\infty}\frac{\langle (x(t) - x(0))^2\rangle}{2t}$$
$$= \lim_{t\to\infty}\frac{1}{2t}\int_0^t dt' \int_0^t dt'' \langle v(t')v(t'')\rangle dt'dt'' = \frac{S_\eta}{2(m\gamma)^2}. \quad (2.45)$$

---

[a] Two sets of the expressions are obtained from the autocorrelation functions for the mobility $\mu = \frac{e}{k_B T}\int_0^\infty \langle v(t)v(0)\rangle dt$ and $\gamma = \frac{1}{mk_B T}\int_0^\infty \langle \eta(t)\eta(0)\rangle dt$.

From the fluctuation–dissipation theorem in Eq. (2.41), we have $D = k_B T/m\gamma$ and the Einstein relation for the mobility and the diffusion constant $\mu = eD/k_B T$. This shows that the Einstein relation represents one example of realization of the fluctuation–dissipation theorem.[a]

2.1.3.2 Power spectrum and noise

Here let us consider the power spectrum [4] defined by

$$S_X(\omega) = \lim_{T\to\infty} \frac{1}{T}\langle |X(\omega)|^2\rangle \tag{2.46}$$

where $X(\omega)$ is the Fourier transform of $X(t)$. Since $|X(\omega)|^2$ represents the energy for the oscillator, $S_X(\omega)$ shows the strength of power in the spectrum ω. For example, the power spectrum of $\langle \eta(t)\eta(t')\rangle = S_\eta \delta(t-t')$ is a constant $S_\eta(\omega) = S_\eta$ and is independent of ω. Thus, random force $\eta(t)$ in the previous section is a *white noise*. The Fourier transform of $S_X(\omega)$ becomes

$$\begin{aligned}\kappa(\tau) &= \frac{1}{2\pi}\int_{-\infty}^{\infty} S_X(\omega)e^{-i\omega\tau}d\omega = \lim_{T\to\infty}\frac{1}{2\pi T}\int_{-\infty}^{\infty}\langle |X(\omega)|^2\rangle e^{-i\omega\tau}d\omega \\ &= \lim_{T\to\infty}\frac{1}{2\pi T^2}\int_{-\infty}^{\infty}\int_{-T/2}^{T/2}\int_{-T/2}^{T/2} X(t)X(t')e^{-i\omega(t-t'+\tau)}dtdt'd\omega \\ &= \lim_{T\to\infty}\frac{1}{T}\int_{-T/2}^{T/2} X(t)X(t+\tau)dt = \langle X(t)X(t+\tau)\rangle. \end{aligned} \tag{2.47}$$

This shows that the autocorrelation function $\kappa(\tau)$ appears as the Fourier transform of the power spectrum $S_X(\omega)$ known as the Wiener–Khinchin theorem. Since, as we see in the previous section, the fluctuation is related to the autocorrelation function $\langle X(t)X(0)\rangle$, the knowledge of power spectrum $S_X(\omega)$ enables us to understand the behavior of fluctuations produced in the system.

As an example, we consider the thermal noise in electronics. Small voltage fluctuations appear in the resistance due to the fluctuations of thermal electrons, which are back to the conductance. Such noise is called Johnson–Nyquist noise and is expressed by

$$S_V(\omega) = 4k_B T R \tag{2.48}$$

---

[a] The diffusion constant in the d dimension is in general expressed by the autocorrelation function as $D = \frac{1}{d}\int_0^\infty \langle v(t)v(0)\rangle dt$ (Green–Kubo formula), which is reduced to $D = (1/d)\langle v^2\rangle\tau$ in the relaxation time approximation.

where R is the resistance $R = V/I$ and T is the temperature. For the performance of electronics, the voltage noise by thermal fluctuations is very important. Let us show that Johnson–Nyquist noise is related to the fluctuation–dissipation relationship. Consider the resistor composed of fluctuating voltage source $V(t)$, resistance R, and inductance L connected in series. From Ohm's law, we have

$$L\frac{dI}{dt} + RI = V(t), \tag{2.49}$$

where we assume $V(t)$ is white as $\langle V(t)\rangle = 0$ and $\langle V(t)V(t')\rangle = S_V\delta(t - t')$. We see that this is the same form as the Langevin equation by changing $v \to I$, $m \to L$, and $\gamma \to R/L$. Then we have $\langle I^2\rangle = S_V/2LR$ in the long time limit. Since fluctuating thermal energy is balanced with storage of the current as $\langle LI^2/2\rangle = k_B T/2$, we have the power spectrum $S_V(\omega)$ of $\langle V(t)V(t')\rangle$ for the Johnson–Nyquist noise due to the fluctuating voltage $V(t)$ as Eq. (2.48).[a] This enables us to determine the temperature T from measurements of the Johnson–Nyquist noise. When the autocorrelation has a form of $\langle V(t)V(t')\rangle = \langle V^2\rangle e^{-t/\tau}$ with relaxation time τ, its power spectrum becomes

$$S_V(\omega) = \langle V^2\rangle \frac{\tau}{1 + (\omega\tau)^2}. \tag{2.50}$$

This type of noise is called Lorentzian noise.[b]

2.1.4 *Master Equation*

We see the diffusive motion of an electron from the stochastic viewpoints [5] where its motion is controlled by random processes.

Let us consider the random walks of an electron on an infinite one-dimensional lattice. When an electron is at the origin ($x_0 = 0$)

---

[a] Taking the symmetry $S(-\omega) = S(\omega)$ into account, here we define the power spectrum $S_V(\omega)$ as $2\langle V^2\rangle$ for $\omega \geq 0$ and 0 for $\omega < 0$.

[b] The noise for the current I due to the discrete arrival of many electrons is known as shot noise or Schottky noise. Since $\langle I\rangle = e\langle N\rangle$ for N electrons and $I(t) = e\sum_{i=1}^{N}\delta(t - t_i)$ where t_i is the arrival time for the i-th electron, the power spectrum is $S_I(\omega) = 2\langle I^2\rangle = 2e^2\langle N\rangle = 2e\langle I\rangle$ (classical Poisson limit). Shot noise becomes important when the number of electrons for the arrival is small with correlation $\langle(\Delta n)^2\rangle \approx \mathcal{T}(1 - \mathcal{T})$ where $\mathcal{T}$ is transmission. Then $S_I(\omega) \approx (2e^2/h)eV\sum_i \mathcal{T}_i(1 - \mathcal{T}_i)$ (quantum shot noise). There is another type of noise proportional to I whose power spectrum behaves as $S_I(\omega) \propto \omega^{-1}$, diverging at low frequencies. This is called $1/f$ noise.

initially ($t_0 = 0$), the probability $P(m, N)$ that an electron is on the location $x = m\ell$ after the N steps at time $t = N\tau$ becomes

$$P(m, N) = \left(\frac{1}{2}\right)^N \frac{N!}{\left[\frac{1}{2}(N+m)\right]!\left[\frac{1}{2}(N-m)\right]!}, \tag{2.51}$$

since an electron moves $(N + m)/2$ steps to the left and $(N - m)/2$ steps to the right to reach the location m after N steps. Here we take the lattice unit length ℓ and the time for the movement τ. Using Stirling's formula

$$N! = \int_0^\infty t^N e^{-t} dt \approx \left(\frac{N}{e}\right)^N \sqrt{2\pi N} \qquad N \gg 1, \tag{2.52}$$

we have [a]

$$P(m, N) = \sqrt{\frac{2}{\pi N}} e^{-m^2/2N}. \tag{2.53}$$

Then the probability $P(x, t|x_0, t_0)$ that an electron is located at x at time t becomes

$$P(x, t|x_0, t_0) = \frac{1}{2\ell} P(m, N) = \frac{1}{\sqrt{4\pi Dt}} e^{-x^2/4Dt}, \tag{2.54}$$

where $D = \ell^2/2\tau$ and 1/2 comes from even or odd lattice points. This Gaussian probability distribution shows that the present random walk model represents the diffusive motion of an electron on the lattice. Note that the conditional probability $\langle \delta(x - x^i(t)) \rangle_{x_0^i, t_0}$ to find an i-th electron at x at time t when it was at x_0^i at t_0 is $P(x, t|x_0^i, t_0)$. Since the electron density is given by $n(x, t) = \sum_i \delta(x - x^i(t))$, the average density becomes

$$\begin{aligned} \langle n(x, t) \rangle &= \int P(x, t|x_0, t_0) \sum_i \delta(x_0 - x_0^i) dx_0 \\ &= \int P(x, t|x_0, t_0) n(x_0, t_0) dx_0. \end{aligned} \tag{2.55}$$

From Eq. (2.30), this means that the probability $P(x, t|x_0, t_0)$[b] is identical to Green's function in the dilute case as

$$P(x, t|x_0, t_0) = G(x - x_0, t - t_0). \tag{2.56}$$

---

[a] $P(m, N) \approx \frac{2\sqrt{N}}{\sqrt{2\pi(N^2 - m^2)}} \cdot \frac{1}{\left(1+\frac{m}{N}\right)^{\frac{N+m}{2}} \left(1-\frac{m}{N}\right)^{\frac{N-m}{2}}}$. Using $\ln(1+\epsilon)^{1/\epsilon} \approx 1 - \epsilon/2$, we have $\left(1 \pm \frac{m}{N}\right)^{\frac{N \pm m}{2}} = e^{\pm \frac{m}{2} + \frac{m^2}{4N} + \cdots}$, which results in Eq. (2.53).

[b] Here we show the position probability distribution. The equation for the velocity probability $P(p, t)$ is called the Fokker–Planck equation, which is used to show how distribution decays to thermal equilibrium at long times. For the Langevin

The formula for the probability of random walks can be derived from the following master equation with stochastic processes:

$$\frac{dP_m(t)}{dt} = w(P_{m+1}(t) - P_m(t)) + w(P_{m-1}(t) - P_m(t)). \tag{2.57}$$

Here $P_m(t)$ is the probability that an electron is at m site at time t and the rate of change of $P_m(t)$ is the rate w of arrival from $m+1$ or $m-1$ sites and the rate w of departure from m site. With an initial condition of $P_m(t=0) = \delta_{m,0}$, this master equation is solved using the Fourier transform of $P_m(t)$ as

$$g(\theta, t) = \sum_{-\infty}^{\infty} e^{i\theta m} P_m(t) \tag{2.58}$$

with

$$\frac{d}{dt} g(\theta, t) = -2w(1 - \cos\theta) g(\theta, t), \tag{2.59}$$

to obtain

$$g(\theta, t) = e^{-2wt(1-\cos\theta)} g(\theta, 0). \tag{2.60}$$

Inverting the Fourier transform with $g(\theta, 0) = 1$, we have

$$P_m(t) = \frac{1}{2\pi} e^{-2wt} \int_{-\pi}^{\pi} e^{i\theta m} e^{2wt\cos\theta} d\theta = I_m(2wt) e^{-2wt} \tag{2.61}$$

with $I_m(z)$ a modified Bessel function. Expanding in θ, we have

$$P(x, t) \simeq \frac{1}{2\pi} \int_{-\infty}^{\infty} e^{i\theta m - wt\theta^2} d\theta = \frac{1}{\sqrt{4\pi wt}} e^{-m^2/4wt} \tag{2.62}$$

which reduces to Eq. (2.54) for $D = w\ell^2$ and $x = m\ell$.

In general, the dynamics of the probability $P_m(t)$ is expressed by the master equation as a gain-loss rate $W_{m,n}$ equation

$$\begin{aligned} \frac{dP_m(t)}{dt} &= \sum_n W_{m,n} P_n(t) - \sum_n W_{n,m} P_m(t) \\ &= \sum_n R_{m,n} P_n(t). \end{aligned} \tag{2.63}$$

---

equation $\frac{dp}{dt} = -\frac{\partial V}{\partial x} - \gamma p + \eta(t)$, $\langle \eta(t)\eta(t')\rangle = 2m\gamma k_B T \delta(t - t')$, the Fokker–Planck equation is $\frac{\partial P}{\partial t} = \frac{1}{\gamma m} \frac{\partial}{\partial x} \left[\frac{\partial V}{\partial x} + k_B T \frac{\partial}{\partial x} \right] P$, which decays to thermal equilibrium $P_{eq} \sim e^{-V/k_B T}$. The Fokker–Planck equation may be regarded as a continuous version of master equation and master equation is a discretized version of the Fokker–Plank equation (problem 5).

The matrix R is defined by

$$R_{m,n} = (1 - \delta_{m,n}) W_{m,n} - \delta_{m,n} \sum_k W_{k,n} \tag{2.64}$$

with the sum rule $\sum_m R_{m,n} = 0$ for the probability conservation. Here we note that the master equation is based on the Markov approximation without memory effects. For the requirements, the off-diagonal elements must be positive or zero, since the rates are positive. Then, R has at least one zero eigenvalue and other eigenvalues have negative real parts, which describe an approach to equilibrium. The eigenstate with a zero eigenvalue corresponds to an equilibrium state. If there are more than one zero eigenvalues or eigenvalues with purely imaginary parts, the process is not ergodic, leading to different stationary states for different initial conditions.

The Pauli master equation is an example of the quantum mechanical dynamics of the occupation probability of a state $|m\rangle$

$$\frac{dP_m(t)}{dt} = \sum_n \left[W_{m,n} P_n(t) - W_{n,m} P_m(t) \right] \tag{2.65}$$

where the transition rates $W_{m,n}$ between states $|m\rangle$, $|n\rangle$ are given by Fermi golden rule as

$$W_{m,n} = \frac{2\pi}{\hbar} |\langle m| \hat{V}_{\text{int}} |n\rangle|^2 \delta(\epsilon_n - \epsilon_m) \tag{2.66}$$

for the coupling interaction $\hat{V}_{\text{int}}$. The transition rate $W_{m,n}$ satisfies detailed balance, which approaches to thermal equilibrium distribution.

In Chapter 4, we will derive an approximate formula using the Liouville equation for the dynamics of density matrix $\hat{\rho}(t)$

$$\frac{d\hat{\rho}(t)}{dt} = -\frac{i}{\hbar} \left[\hat{H}, \hat{\rho}(t) \right] + C\left[\hat{\rho}(t) \right], \tag{2.67}$$

which is another example of the master equation called the quantum (Markovian) master equation. Here $\hat{H}$ is the Hamiltonian of the system and $C[\rho\hat{(t)}]$ describes the dissipation of the dynamics, which eliminates memory effects by the system-bath interaction.

2.1.5 *Semi-classical Approach*

In the previous sections, we considered that electrons are classical point particles or the probability distribution $P_n(t)$. Since the position $\mathbf{r}(t)$ and the momentum $\mathbf{p}(t)$ are both determined in classical mechanics, we can treat many-electron system as an N-electron problem with the treatment of electron density distribution of $f_N(\mathbf{r}, \mathbf{p}, t)$. In the nonequilibrium situations, this distribution function is changed due to interactions among electrons and other collisions and scatterings.

2.1.5.1 Phase space and Liouville equation

We consider a system composed of N electrons. Each electron has a generalized coordinate $q_i(t)$ and a momentum $p_i(t)$, which constructs a q-p phase space and whose time evolution is described by a trajectory. Although the electrons are interacting, it is important to note that the ensemble of systems described by the continuum distribution function $f_N(\mathbf{q}, \mathbf{p}, t)dq_1dq_2 \cdots dq_N dp_1 dp_2 \cdots dp_N$ in the phase space is a conservative quantity for time t and is uniquely determined once the initial condition is given with a constraint of

$$\int f_N(\mathbf{q}, \mathbf{p}, t)dq_1 \cdots dp_N = 1. \tag{2.68}$$

This Liouville theorem in classical mechanics is based on the fact that the divergence of the q-p phase space is zero as

$$\sum_{i=1}^{N}\left[\frac{\partial}{\partial q_i}\left(\frac{dq_i}{dt}\right) + \frac{\partial}{\partial p_i}\left(\frac{dp_i}{dt}\right)\right] = 0, \tag{2.69}$$

which is shown by using the Hamilton equation

$$\frac{dq_i(t)}{dt} = \frac{\partial H}{\partial p_i}, \quad \frac{dp_i(t)}{dt} = -\frac{\partial H}{\partial q_i} \tag{2.70}$$

for the trajectory of each electron where H is the Hamiltonian of the system. We note that $f_N dq_1 \cdots dp_N$ is the probability to find an electron at time t within the phase space of $d\mathbf{q}d\mathbf{p} = dq_1 \cdots dp_N$. Then since $df_N/dt = 0$, we have an equation for the distribution

$$\frac{\partial f_N}{\partial t} + \sum_{i=1}^{N}\left(\frac{\partial f_N}{\partial q_i}\frac{\partial H}{\partial p_i} - \frac{\partial f_N}{\partial p_i}\frac{\partial H}{\partial q_i}\right) = 0. \tag{2.71}$$

This is the Liouville equation, which governs the time evolution of the N-electron distribution function f_N.

Introducing the Poisson bracket defined by

$$\{A, B\} = \sum_{i=1}^{N} \left(\frac{\partial A}{\partial q_i} \frac{\partial B}{\partial p_i} - \frac{\partial A}{\partial p_i} \frac{\partial B}{\partial q_i} \right), \tag{2.72}$$

the Liouville equation is written as

$$\frac{\partial f_N}{\partial t} = \left\{ H, f_N \right\}. \tag{2.73}$$

Then time evolution of the physical quantity X becomes

$$\begin{aligned} \frac{\partial X}{\partial t} &= \sum_{i=1}^{N} \left(\frac{\partial X}{\partial q_i} \frac{\partial q_i(t)}{\partial t} + \frac{\partial X}{\partial p_i} \frac{\partial p_i(t)}{\partial t} \right) \\ &= \sum_{i=1}^{N} \left(\frac{\partial X}{\partial q_i} \frac{\partial H}{\partial p_i} - \frac{\partial X}{\partial p_i} \frac{\partial H}{\partial q_i} \right) = \{X, H\} \end{aligned} \tag{2.74}$$

and the (ensemble) average value of $X(t)$[a] is obtained from

$$\langle X(t) \rangle = \int X(\mathbf{q}, \mathbf{p}, t) f_N(\mathbf{q}, \mathbf{p}, t) d\mathbf{q} d\mathbf{p}. \tag{2.75}$$

Various conservation laws are obtained from these equations.[b]

For the mathematical treatments compared to the quantum mechanics ones, it is convenient to introduce the Liouville operator L such as

$$L = i \sum_{i=1}^{N} \left(\frac{\partial H}{\partial p_i} \frac{\partial}{\partial q_i} - \frac{\partial H}{\partial q_i} \frac{\partial}{\partial p_i} \right) \equiv -i \left\{ H, \right\}. \tag{2.76}$$

Then the Liouville equation becomes

$$\frac{\partial f_N}{\partial t} = i L f_N, \tag{2.77}$$

and its solution is formally written as

$$f_N(t) = e^{iLt} f_N(0). \tag{2.78}$$

Since $\partial X/\partial t = \{X, H\} = -\{H, X\} = -iLX$, we can write formally the time evolution of a physical quantity X as

---

[a] This is based on the assumption that $\langle X \rangle = (1/T) \int_{-T/2}^{T/2} X(t) dt$ as the time average is equal to the ensemble average, called the ergodic hypothesis.

[b] For example, since $\partial H/\partial t = \{H, H\} = 0$, this is the energy conservation. Also for the density $n(\mathbf{r}) = \sum_i m\delta(\mathbf{q}_i - \mathbf{r})$, it becomes $\partial n(\mathbf{r})/\partial t = \{n(\mathbf{r}), H\} = -\partial/\partial\mathbf{r} \sum_i \mathbf{p}_i \delta(\mathbf{q}_i - \mathbf{r}) = -\nabla \cdot \mathbf{p}(\mathbf{r})$, which represents the mass conservation law.

$$X(t) = e^{-iLt} X(0). \tag{2.79}$$

These relations are used to derive the response function later.

The direct solution to the Liouville equation for the N-electron nonequilibrium distribution f_N is impossible, although this equation should be valid for all classical systems. We will next show the Boltzmann equation to simplify this equation. Then, we develop the linear-response theory based on the perturbation approach of the Liouville equation, which will be extended to treat the quantum systems using the density matrix formalism.

2.1.5.2 Boltzmann equation

The Boltzmann equation is a simplification of the Liouville equation to consider the behavior of one electron having coordinate $\mathbf{r}$ and momentum $\mathbf{p}$ in a system. The distribution of an electron is constructed from the rest of the $(N-1)$ electrons as

$$f(\mathbf{r}, \mathbf{p}, t) = N \int f_N(r_2, \cdots, r_N, p_2, \cdots, p_N) dr_2 \cdots dr_N dp_2 \cdots dp_N \tag{2.80}$$

where we note that f is a one-particle distribution and $f d\mathbf{r} d\mathbf{p}$ represents to find an electron in the coordinate $d\mathbf{r}$ and in the momentum $d\mathbf{p}$ with electron density and electron number of

$$\int f(\mathbf{r}, \mathbf{p}, t) d\mathbf{p} = n(\mathbf{r}, t), \quad \int \int f(\mathbf{r}, \mathbf{p}, t) d\mathbf{r} d\mathbf{p} = N. \tag{2.81}$$

This approximation reduces the variables significantly. The reduced phase space should be considered as the projection of that of N-electron system.

Then from the Liouville equation Eq. (2.71), we have

$$\left(\frac{\partial}{\partial t} + \frac{\mathbf{p}}{m} \cdot \frac{\partial}{\partial \mathbf{r}} - \frac{\partial V(\mathbf{r}, t)}{\partial \mathbf{r}} \cdot \frac{\partial}{\partial \mathbf{p}} \right) f(\mathbf{r}, \mathbf{p}, t) = \left[\frac{\partial f(\mathbf{r}, \mathbf{p}, t)}{\partial t} \right]_{\text{coll}}. \tag{2.82}$$

This is the Boltzmann equation. Here $[\partial f / \partial t]_{\text{coll}}$, called the collision term, appears to represent the interaction of one-particle distribution $f(\mathbf{r}, \mathbf{p}, t)$ with the rest of electrons, showing that the one-particle phase space is not conserved. The average value over

$\mathbf{p}$ (or $\mathbf{k}$) of the physical quantity X is obtained from[a]

$$\langle X(\mathbf{r})\rangle = \int X(\mathbf{r},\mathbf{p}) f d\mathbf{p} \left(= \int X(\mathbf{r},\mathbf{k}) f \frac{d\mathbf{k}}{(2\pi)^3} \right). \tag{2.84}$$

Relaxation Time Approximation

Here we consider the carrier scattering effects in the right hand of the Boltzmann equation. This term includes all the types of processes that drive the system to local equilibrium. These processes, usually contain two-body potentials, are approximated as

$$\left[\frac{\partial f(\mathbf{r},\mathbf{p},t)}{\partial t}\right]_{\text{coll}} = -\int \left\{ W_{\mathbf{p},\mathbf{p}'} f(\mathbf{r},\mathbf{p},t)\left[1 - f(\mathbf{r},\mathbf{p}',t)\right] \right. \\ \left. - W_{\mathbf{p}',\mathbf{p}} f(\mathbf{r},\mathbf{p}',t)\left[1 - f(\mathbf{r},\mathbf{p},t)\right] \right\} d\mathbf{p}' \tag{2.85}$$

where $W_{\mathbf{p},\mathbf{p}'}$ represents the scattering probability that an electron with momentum $\mathbf{p}$ is scattered into any one of the unoccupied states contained with momentum $\mathbf{p}'$. Thus, the term $f(\mathbf{r},\mathbf{p},t)$ represents the number of electrons in the occupied state $\mathbf{p}$ and the term $1 - f(\mathbf{r},\mathbf{p}',t)$ represent the unoccupied states with the momentum $\mathbf{p}'$ for which electrons are scattered. This shows the increase and decrease of electrons due to scattering from other states. We note that this equation is similar (but not equal) to the master equation Eq. (2.65) and the collision term is frequently treated by the quantum mechanics using Fermi golden rule, the lowest-order perturbation to evaluate the scattering probability

$$W_{\mathbf{p},\mathbf{p}'} = \frac{2\pi}{\hbar}|\langle \mathbf{p}|\hat{V}_{\text{int}}|\mathbf{p}'\rangle|^2 \delta\left(\epsilon_{\mathbf{p}} - \epsilon_{\mathbf{p}'}\right). \tag{2.86}$$

---

[a] Using $\mathbf{v} = \mathbf{p}/m$ or $\mathbf{k} = \mathbf{p}/\hbar$, the Boltzmann equation is also written as

$$\left(\frac{\partial}{\partial t} + \mathbf{v}\cdot\frac{\partial}{\partial \mathbf{r}} + \frac{\mathbf{F}}{m}\cdot\frac{\partial}{\partial \mathbf{v}}\right) f(\mathbf{r},\mathbf{v},t) = \left[\frac{\partial f(\mathbf{r},\mathbf{v},t)}{\partial t}\right]_{\text{coll}}$$

$$\left(\frac{\partial}{\partial t} + \mathbf{v}\cdot\frac{\partial}{\partial \mathbf{r}} + \frac{\mathbf{F}}{\hbar}\cdot\frac{\partial}{\partial \mathbf{k}}\right) f(\mathbf{r},\mathbf{k},t) = \left[\frac{\partial f(\mathbf{r},\mathbf{k},t)}{\partial t}\right]_{\text{coll}} \tag{2.83}$$

where $\mathbf{F} = d\mathbf{p}/dt = -\partial V(\mathbf{r},t)/\partial \mathbf{r}$ is the external force acting on the electron. We note that the Boltzmann equation is used for quantum electron transport in view of a particle as long as the phase coherence is not taken into account. It is possible to consider the two-particle distribution

$$f_2(\mathbf{r}_1,\mathbf{p}_1,\mathbf{r}_2,\mathbf{p}_2,t) = \frac{N(N-1)}{2}\int f_N(r_3,\cdots,p_N)dr_3\cdots dr_N dp_3\cdots dp_N,$$

which is approximated as $f_2 \approx f_1 f_1$. We can construct f_n from f_{n-1}.

The two-electron scattering rates $W_{\mathbf{p},\mathbf{p}'}$ satisfy detailed balance, which approaches to thermal equilibrium distribution.[a]

To solve the Boltzmann equation, the relaxation time approximation simplifies the scattering term as follows:

$$\left[\frac{\partial f(\mathbf{r}, \mathbf{p}, t)}{\partial t}\right]_{\text{coll}} = -\frac{f(\mathbf{r}, \mathbf{p}, t) - f_0(\mathbf{r}, \mathbf{p})}{\tau}, \tag{2.87}$$

where $f_0(\mathbf{r}, \mathbf{p})$ is the equilibrium distribution. When we neglect the spatial non-uniformity, the Boltzmann equation becomes

$$\frac{\partial f}{\partial t} = -\frac{f - f_0}{\tau}, \tag{2.88}$$

which represents approach to the equilibrium distribution f_0 as

$$f - f_0 = Ce^{-t/\tau} \tag{2.89}$$

with the relaxation time τ. The successive scatterings with different processes are treated frequently as

$$\frac{1}{\tau} = \sum_i \frac{1}{\tau_i}, \tag{2.90}$$

which assumes the independent scattering mechanics and this approximation is called the Matthiessen's rule.

---

[a] Approach to local thermal equilibrium is shown by using the Boltzmann H theorem. First we note that Eq. (2.85) is written generally as a function of $\partial f/\partial t = \int W_{\mathbf{p}\mathbf{p}_1,\mathbf{p}'\mathbf{p}'_1}(ff_1 - f'f'_1)d\mathbf{p}_1 d\mathbf{p}' d\mathbf{p}'_1$. When we define the H function as $H = \int f(\mathbf{r}, \mathbf{p}, t)\ln f(\mathbf{r}, \mathbf{p}, t)d\mathbf{p}$, then we have $dH/dt = \int(\partial f/\partial t)\ln f d\mathbf{p} = \cdots = (1/4)\int\int \ln(ff_1/f'f'_1)(f'f'_1 - ff_1)d\mathbf{p}d\mathbf{p}_1 d\mathbf{p}' d\mathbf{p}_1$, which means $dH/dt \leq 0$. This shows that the H function decreases as t and becomes zero when $ff_1 = f'f'_1$, which corresponds to thermal equilibrium in the Boltzmann equation. Thus, the H function decreases as the system approaches to thermal equilibrium.

On the other hand, the entropy S in thermodynamics is known to increase in these processes. This suggests that the H function is related to the entropy S, randomness in the system. Since the number of accessible states W with N electron in a system becomes $W = (N!/n_1!n_2!\cdots)g_1^{n_1}g_2^{n_2}\cdots \propto -f\ln f$ with $f_i = (n_i/g_i)$, the entropy S defined by $S = k_B \ln W = -k_B f \ln f = -k_B H$ is shown to satisfy the thermodynamic relationships and increases as $dS/dt \geq 0$ for the scattering processes to approach thermal equilibrium. The relationship of $S = k_B \ln W$ is called the Boltzmann's principle. From the condition $ff_1 = f'f'_1$ or $\ln f(\mathbf{p}) + \ln f(\mathbf{p}_1) = \ln f(\mathbf{p}') + \ln f(\mathbf{p}'_1)$, we have $\ln f = 1 + c\mathbf{p}^2$, which leads to the Maxwell–Boltzmann distribution at equilibrium. The Boltzmann's principle $S = k_B \ln W$ is important also in the information theory.

Boltzmann Transport Equation for Electron

Next, we consider electron transport in a uniform electric field $\mathbf{E}$ with $V(\mathbf{r}, t) = -e\mathbf{E} \cdot \mathbf{r}$ within the relaxation time approximation. For the steady state $\partial f/\partial t = 0$, the Boltzmann equation becomes

$$\frac{\mathbf{p}}{m}\frac{\partial f}{\partial \mathbf{r}} + e\mathbf{E}\frac{\partial f}{\partial \mathbf{p}} = -\frac{f - f_0}{\tau} \tag{2.91}$$

or equivalently

$$f = f_0 - \tau\left(\mathbf{v}\frac{\partial f}{\partial \mathbf{r}} + \frac{e\mathbf{E}}{m}\frac{\partial f}{\partial \mathbf{v}}\right). \tag{2.92}$$

The current density defined by

$$\mathbf{j}(\mathbf{r}) = e\int \mathbf{v} f(\mathbf{r}, \mathbf{v})d\mathbf{v}, \tag{2.93}$$

is obtained using Eq. (2.28) as

$$\begin{aligned}\mathbf{j}(\mathbf{r}) &= -e\tau\langle \mathbf{v}^2\rangle \nabla n(\mathbf{r}) - e^2\tau\langle \mathbf{v}^2\rangle \left[\frac{\partial n(\mathbf{r})}{\partial \epsilon}\right]\mathbf{E} \\ &= -eD\nabla n(\mathbf{r}) - \sigma\mathbf{E}\end{aligned} \tag{2.94}$$

with the conductivity of

$$\sigma = e^2\nu D = ne\mu. \tag{2.95}$$

This shows that the Boltzmann equation from the distribution function f includes the Drude model based on the point charge in the classical mechanics treatment.

Let us use the quantum mechanics as $\mathbf{p} = \hbar\mathbf{k}$. The relaxation time τ is generally dependent on $\mathbf{k}$ as $\tau(\mathbf{k})$. Consider the metallic system[a] where the distribution at equilibrium is expressed by the Fermi–Dirac function

$$f_0(\mathbf{k}) = \frac{1}{e^{(\epsilon_\mathbf{k}-\mu)/k_BT} + 1} \tag{2.96}$$

and electrons are accelerated by electric fields by

$$\hbar\frac{\partial \mathbf{k}}{\partial t} = e\mathbf{E}. \tag{2.97}$$

In the metallic systems where many electrons are present, the spatial distribution is not perturbed by the electric fields due to

[a] In the semiconducting system, since carriers are non-degenerate and dilute, we use the Maxwell–Boltzmann distribution as $f_0(\mathbf{k}) = \sqrt{\frac{m}{2\pi k_BT}}e^{-\epsilon_\mathbf{k}/k_BT}$.

screening effects as $\partial f/\partial r = 0$. Thus, the distribution $f(\mathbf{k})$ changes as

$$f(\mathbf{k}) = f_0(\mathbf{k}) - e\tau(\mathbf{k})\frac{\hbar\mathbf{k}}{m}\frac{\partial f(\mathbf{k})}{\partial\mathbf{k}} \tag{2.98}$$

where we assume that the system is isotropic $\epsilon_{\mathbf{k}} = \hbar^2\mathbf{k}^2/2m$. Here we approximate $f(\mathbf{k})$ in the right term as $f_0(\mathbf{k})$

$$f(\mathbf{k}) \simeq f_0(\mathbf{k}) - e\tau(\mathbf{k})\frac{\hbar\mathbf{k}}{m}\mathbf{E}\frac{\partial f_0(\mathbf{k})}{\partial\epsilon_{\mathbf{k}}}, \tag{2.99}$$

then the current becomes (2 comes from spin degeneracy)

$$\begin{aligned}\mathbf{j} &= 2e\int\frac{\hbar\mathbf{k}}{m}f(\mathbf{k})\frac{d\mathbf{k}}{(2\pi)^3}\\ &\simeq \frac{2e^2}{(2\pi)^3}\int\left(-\frac{\partial f_0(\mathbf{k})}{\partial\epsilon_{\mathbf{k}}}\right)\tau(\mathbf{k})\mathbf{v}_{\mathbf{k}}\left(\mathbf{v}_{\mathbf{k}}\cdot\mathbf{E}\right)d\mathbf{k}\end{aligned} \tag{2.100}$$

and the conductivity as $\mathbf{j} = \sigma(T)\mathbf{E}$ is obtained from

$$\sigma(T) = \frac{2e^2}{(2\pi)^3}\int\left(-\frac{\partial f_0(\mathbf{k})}{\partial\epsilon_{\mathbf{k}}}\right)\tau(\mathbf{k}){\mathbf{v}_{\mathbf{k}}}^2 d\mathbf{k}. \tag{2.101}$$

At $T = 0$, since the Fermi–Dirac distribution becomes a step function as $-\partial f_0(\mathbf{k})/\partial\epsilon_{\mathbf{k}} = \delta(\epsilon_{\mathbf{k}} - \epsilon_F)$, we have

$$\begin{aligned}\sigma(0) &= \frac{2e^2}{3(2\pi)^3}\frac{\tau(k_F)v_F^2}{\hbar v_F}4\pi k_F^2\\ &= \frac{ne^2\tau(k_F)}{m},\end{aligned} \tag{2.102}$$

where $n = k_F^3/3\pi^2$ is the electron density of the metal. This shows that τ in Eq. (2.11) in the Drude model is the scattering time of electrons at the Fermi energy. At finite T, we obtain the conductivity $\sigma(T)$ from Eq. (2.98) once we have $\tau(\mathbf{k})$.

To calculate $\tau(\mathbf{k})$ and the current $\mathbf{j}$, the rate of change of $f(\mathbf{k})$ due to collisions from Eq. (2.85) and Eq. (2.98) is evaluated from

$$\begin{aligned}-e\tau(\mathbf{k})\frac{\hbar\mathbf{k}}{m}\cdot\mathbf{E}\frac{\partial f(\mathbf{k})}{\partial\epsilon_{\mathbf{k}}} &= \int\frac{2\pi}{\hbar}|\langle\mathbf{k}|V_{\mathrm{int}}|\mathbf{k}'\rangle|^2\delta\left(\epsilon_k - \epsilon_{k'}\right)\\ &\quad\times\left(f(\mathbf{k}') - f(\mathbf{k})\right)d\mathbf{k}'\end{aligned} \tag{2.103}$$

for various potentials V_{int} such as Coulomb scattering, phonon scattering, impurity and roughness scatterings, and others. This is called the Boltzmann transport equation (BTE). Here we note that

BTE is used for the dissipative (non-energy conserving) scatterings in the high-field transport.[a] This equation is frequently solved numerically using the Monte Carlo method [6], taking various scatterings as random events.

When the electron density is spatially inhomogeneous such as the junction systems, we need to solve the Poisson equation

$$\nabla \mathbf{E}(\mathbf{r}) = \frac{4\pi e}{\epsilon}\left(n_+ - n(\mathbf{r})\right) \tag{2.104}$$

with dielectric function ϵ, the positive charge n_+, and electron density $n(\mathbf{r}) = \int f(\mathbf{r},\mathbf{k})d\mathbf{k}$.[b] Then this should be inserted in the BTE with the inclusion of an additional spatial term $\frac{\hbar\mathbf{k}}{m}\frac{\partial f(\mathbf{r},\mathbf{k})}{\partial \mathbf{r}}$ in the left hand of Eq. (2.103).

2.1.5.3 Dimensionality

Up to now, we do not use quantum mechanical equations for the wavefunctions and the relationship is derived mostly from the classical statistical mechanics, except for the last part of Boltzmann equation. In the quantum mechanics, we know that electrons occupy the single-particle states from the bottom of energy levels and the highest-occupied single-particle energy level is called the Fermi energy ϵ_F.

Due to the progress of the fabrication technologies, we can construct various structures of nanosystems, where the lengths are controlled with nanometer-scale precision. When one length is less than the Fermi wavelength λ_F, electrons are confined to that direction and we have the low-dimensional systems. According to confinement, they are called quantum well for the two-dimensional (2D) system with ($L_z < \lambda_F < L_x, L_y$), quantum wire for the 1D system with ($L_y, L_z < \lambda_F < L_x$), and quantum dot for the 0D system with ($L_x, L_y, L_z < \lambda_F$) (Fig. 2.1 (top)).

[a] This comes from characteristics that the Boltzmann equation has for symmetry. In the Boltzmann equation Eq. (2.82), the term in the left hand changes its sign by $(\mathbf{p}, t) \to (-\mathbf{p}, -t)$, while scattering term in the right Eq. (2.85) is unchanged. This term, not present in the Liouville equation, enables the Boltzmann equation to treat nonequilibrium processes with dissipation.

[b] We note that the electron density $n(\mathbf{r})$ is a classical one in the BTE, not obtained from the quantum wavefunctions $\psi_\mathbf{k}(\mathbf{r})$ as $n(\mathbf{r}) = \int f(\mathbf{k})|\psi_\mathbf{k}(\mathbf{r})|^2 d\mathbf{k}$.

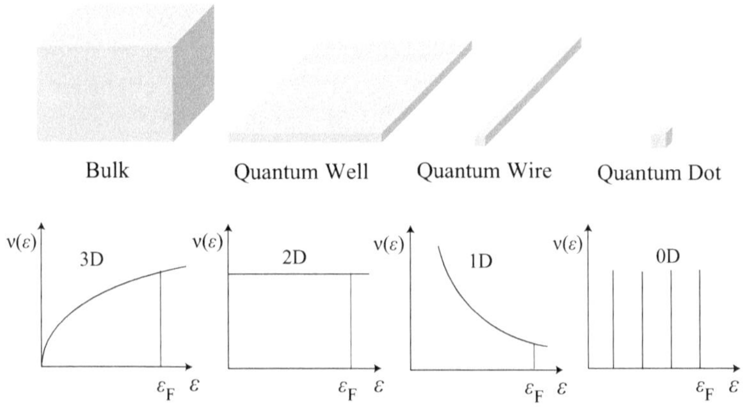

Figure 2.1 (Top) Schematic illustrations of low-dimensional systems: 3D bulk, 2D quantum well, 1D quantum wire, and 0D quantum dot. When one length is less than the Fermi wavelength, electrons are confined to that direction to create low-dimensional structures. (Bottom) Corresponding density of states $\nu(\epsilon)$ of the low-dimensional systems.

The density of states (DOS) $\nu(\epsilon)$ show characteristic features for each low-dimensional system. For the 3D electron gas system, since electron density is

$$n = 2\int \frac{d\mathbf{k}}{(2\pi)^3} = \frac{2}{(2\pi)^3}\frac{4}{3}\pi k^3 \left(= \frac{k^3}{3\pi^2}\right), \tag{2.105}$$

the density of states becomes

$$\begin{aligned} \nu(\epsilon) &= \frac{\partial n}{\partial \epsilon} = \frac{\partial n}{\partial k}\cdot\frac{\partial k}{\partial \epsilon} = \frac{2}{(2\pi)^3}4\pi k^2 \cdot \frac{m}{\hbar^2 k} \\ &= n\frac{3m}{\hbar^2 k^2} = \frac{\sqrt{2}m^{3/2}}{\pi^2\hbar^3}\sqrt{\epsilon}, \end{aligned} \tag{2.106}$$

where the index 2 comes from the spin degeneracy. This result can also be obtained directly from

$$\begin{aligned} \nu(\epsilon) &= 2\int \frac{d\mathbf{k}}{(2\pi)^3}\delta\left(\epsilon - \frac{\hbar^2 k^2}{2m}\right) \\ &= 2\int \frac{4\pi k^2}{(2\pi)^3}\delta\left(\epsilon - \frac{\hbar^2 k^2}{2m}\right) dk \\ &= \frac{1}{\pi^2}\frac{mk}{\hbar^2} = \frac{\sqrt{2}m^{3/2}}{\pi^2\hbar^3}\sqrt{\epsilon}. \end{aligned} \tag{2.107}$$

These results show that $\nu(\epsilon) \propto \sqrt{\epsilon}$ for 3D case. In the low-dimensional systems, $\nu(\epsilon) \propto$ const for 2D case, $\nu(\epsilon) \propto 1/\sqrt{\epsilon}$ for 1D case, and $\nu(\epsilon)$ take spiky peaks for 0D case. These characteristics are schematically shown in Fig. 2.1 (bottom).

From this we can show that the relationship of Eq. (2.29) is satisfied for the 3D electron gas system

$$\sigma = e^2 \left[\frac{\partial n}{\partial \epsilon}\right]_{\epsilon=\epsilon_F} D$$
$$= e^2 n \frac{3m}{\hbar^2 k_F^2} \frac{1}{3} v_F^2 \tau = \frac{ne^2\tau}{m}. \tag{2.108}$$

This shows that the Einstein relation is satisfied for the 3D gas electron system when we take the quantum physics into account.

The mean free path (MFP) is defined as an average distance where the electrons travel between two successive collisions. If the average electron velocity at Fermi energy ϵ_F is $|\mathbf{v_F}|$, the MFP $\ell_{\rm mfp}$ is[a]

$$\ell_{\rm mfp} = |\mathbf{v_F}|\tau. \tag{2.109}$$

In the usual metallic materials, the typical Fermi velocity v_F, which corresponds to the highest-occupied single-particle energy level, is of the order of 10^8 cm/s. The relaxation time due to the collisions with phonons at room temperature is $\tau = 10^{-14}$–10^{-15} s. Thus, we get $\ell_{\rm mfp} \sim 1$–10 nm. The relaxation time τ increases as a decrease of temperature since the scattering with ionic vibrations decreases, therefore the MFP increases accordingly. If there are no other contributions like impurity scatterings for τ, it should be in principle divergent as $T \to 0$.

When the MFP exceeds the length of the sample $\ell_{\rm mfp} > L$, the transport is in a ballistic regime. It is difficult to realize the ballistic transport regime in the metallic systems. However, in the semiconductor transistor systems, since the impurities are extremely reduced and also the effective mass of an electron is quite small, very high mobility μ is realized, especially at low temperature. The transport regime of such devices is close to the ballistic (quasi-ballistic) regime even at room temperature. This enables the semiconductor transistor devices to operate with very

[a]We note that the conductivity σ in d-dimension system is expressed by $\sigma \simeq (e^2/\hbar)k_F^{d-1}\ell_{\rm mfp}$. Here $k_F\ell_{\rm mfp} \gg 1$ represents the metallic states.

high speed. We will discuss the ballistic versus diffusive transport in nano-carbon materials in Chapter 5 and the semiconducting GaAs systems in Chapter 6.

2.1.6 *Linear Response Theory: Kubo Formula*

In this section, we derive the transport properties based on the linear-response theory both classical and quantum mechanically. When an electric field E is applied, the electric current is given by

$$\mathbf{j}(\mathbf{r},\omega) = \int \sigma(\mathbf{r},\mathbf{r}',\omega)\mathbf{E}(\mathbf{r}',\omega)d\mathbf{r}' \tag{2.110}$$

where we perform the Fourier transform for the time t. Taking the zero frequency limit where the electric field is assumed to be uniform, we obtain the conductivity $\sigma(0)$ from $\mathbf{j}(\omega) = \sigma(\omega)\mathbf{E}(\omega)$.

2.1.6.1 Response function and Green–Kubo formula

First we describe the electric transport from the distribution function f_N of the classical mechanics. In the next section, we describe it from the density matrix $\hat{\rho}$ of the quantum mechanics.

The time dependence of the electron density is determined from the Liouville equation

$$\frac{\partial f_N}{\partial t} = \{H, f_N\}, \tag{2.111}$$

where $\{\cdots, \cdots\}$ is the Poisson bracket defined in Eq. (2.72). Here let us consider the response of the density to the perturbation of $H_1(t) = -e\sum_{i=1}^{N} \mathbf{r}_i \cdot \mathbf{E}(t)$ (thus $\mathbf{F}_i(t) = -dH_1(t)/d\mathbf{r}_i = e\mathbf{E}(t)$),

$$\begin{aligned} H &= H_0 + H_1(t) = H_0 - e\sum_{i=1}^{N} \mathbf{r}_i \cdot \mathbf{E}(t) \\ f_N &= f_N^0 + f_N^1(t), \end{aligned} \tag{2.112}$$

where H_0 and f_N^0 are the Hamiltonian and the distribution function for the equilibrium situation. Within the linear-response regime, the Liouville equation becomes

$$\frac{\partial f_N^1(t)}{\partial t} \approx \{H_0, f_N^1(t)\} - \left\{e\sum_{i=1}^{N} \mathbf{r}_i, f_N^0\right\} \cdot \mathbf{E}(t). \tag{2.113}$$

Introducing the Liouville operator $L = -i\{H_0, \}$ as in Eq. (2.76), this equation is solved as[a]

$$f_N^1(t) = -\int_{-\infty}^{t} e^{iL(t-t')}\left\{e\sum_{i=1}^{N}\mathbf{r}_i,\ f_N^0\right\}\cdot\mathbf{E}(t')dt'. \tag{2.114}$$

The electric current is defined by

$$\mathbf{j}(t) = e\sum_{i=1}^{N}\mathbf{v}_i = e\sum_{i=1}^{N}\dot{\mathbf{r}}_i. \tag{2.115}$$

The ensemble average of the current becomes

$$\langle\mathbf{j}(t)\rangle = \int\mathbf{j}(t)f_N(t)d\mathbf{r}d\mathbf{p} \equiv \int_{-\infty}^{t}\phi(t-t')\mathbf{E}(t')dt', \tag{2.116}$$

where $\phi(t)$ is called the response function. Its Fourier transform

$$\chi(\omega)(=\sigma(\omega)) = \int_0^{\infty}\phi(t)e^{i\omega t}dt \tag{2.117}$$

as obtained from

$$\langle\mathbf{j}(\omega)\rangle = \chi(\omega)\mathbf{E}(\omega)\ (=\sigma(\omega)\mathbf{E}(\omega)) \tag{2.118}$$

is called the dynamical susceptibility (conductivity).

Let us obtain the response function $\phi(t)$ to electric field $\mathbf{E}(t)$. Putting Eq. (2.114) in Eq. (2.116) and using $\int\mathbf{j}(t)f_N^0 = 0$, we have

$$\begin{aligned}\langle\mathbf{j}(t)\rangle &= -\int_{-\infty}^{t}\left[\int\mathbf{j}(t)e^{iL(t-t')}\left\{e\sum_{i=1}^{N}\mathbf{r}_i,\ f_N^0\right\}d\mathbf{r}d\mathbf{p}\right]\cdot\mathbf{E}(t')dt' \\ &= -\int_{-\infty}^{t}\left[\int\left\{e\sum_{i=1}^{N}\mathbf{r}_i,\ f_N^0\right\}\mathbf{j}(t-t')d\mathbf{r}d\mathbf{p}\right]\cdot\mathbf{E}(t')dt',\end{aligned} \tag{2.119}$$

where we use $\int Be^{iL(t-t')}Cd\mathbf{r}d\mathbf{p} = \int BC(t-t')d\mathbf{r}d\mathbf{p} = \int CB\left(-(t-t')\right)d\mathbf{r}d\mathbf{p} = \int Ce^{-iL(t-t')}Bd\mathbf{r}d\mathbf{p}$ since the volume in the phase space $d\mathbf{r}d\mathbf{p}$ is a conservative quantity for time t. Thus, the response function is expressed by

$$\phi(t) = \int\left\{f_N^0,\ e\sum_{i=1}^{N}\mathbf{r}_i\right\}\mathbf{j}(t)d\mathbf{r}d\mathbf{p}. \tag{2.120}$$

[a]From the direct differentiation of $f_N^1(t)$ in Eq. (2.114), we have $\partial f_N^1(t)/\partial t = -\left\{e\sum_{i=1}^{N}\mathbf{r}_i,\ f_N^0\right\}\cdot\mathbf{E}(t) + iLf_N^1(t) = \left\{H_0,\ f_N^1(t)\right\} - \left\{e\sum_{i=1}^{N}\mathbf{r}_i,\ f_N^0\right\}\cdot\mathbf{E}(t)$.

On the other hand, when we assume that the system in equilibrium has the Boltzmann distribution as $f_N^0 \propto e^{-H_0/k_B T}$, we have the relation of

$$\left\{ f_N^0, e\sum_{i=1}^{N} \mathbf{r}_i \right\} = -\frac{f_N^0}{k_B T}\left\{ H_0, e\sum_{i=1}^{N} \mathbf{r}_i \right\} = \frac{f_N^0}{k_B T}\mathbf{j}(0), \qquad (2.121)$$

where we use $\left\{ H_0, e\sum_{i=1}^{N} \mathbf{r}_i \right\} = -e\sum_{i=1}^{N} \dot{\mathbf{r}}_i|_{t=0} = -\mathbf{j}(0)$. Since the average value of A in the phase space means $\langle A \rangle = \int f_N^0 A d\mathbf{r} d\mathbf{p}$, we obtain the expression for the response function as

$$\phi(t) = \frac{1}{k_B T}\langle \mathbf{j}(t)\mathbf{j}(0)\rangle \qquad (2.122)$$

and the conductivity $\sigma(\omega)$ is given by

$$\sigma(\omega) = \frac{1}{k_B T}\int_0^\infty \langle \mathbf{j}(t)\mathbf{j}(0)\rangle e^{i\omega t} dt. \qquad (2.123)$$

This expression is called the Green–Kubo formula for transport properties. We note that this is a generalized form of[a]

$$\sigma = \frac{1}{k_B T}\int_0^\infty \langle \mathbf{j}(t)\mathbf{j}(0)\rangle dt \qquad (2.129)$$

---

[a] In general, we have the response function [7] $\phi_{BA}(t)$ to the perturbation defined by $H_1(t) = -AF(t)$ as

$$\phi_{BA}(t) = \frac{1}{k_B T}\langle B(t)\dot{A}(0)\rangle \qquad (2.124)$$

in the linear-response theory. Here the average of $B(t)$ is expressed by

$$\langle B(t)\rangle = \int_{-\infty}^{t} \phi_{BA}(t-t')F(t')dt'. \qquad (2.125)$$

For the dynamical susceptibility defined by

$$\chi_{BA}(\omega) = \int_0^\infty \phi_{BA}(t)e^{i\omega t}dt = \chi'_{BA}(\omega) + i\chi''_{BA}(\omega), \qquad (2.126)$$

the Kramers–Kronig relations are applied due to a requirement of causality as

$$\chi''_{BA}(\omega) = -\mathrm{P}\int_{-\infty}^{\infty} \frac{\chi'_{BA}(\omega')}{\omega'-\omega}d\omega', \quad \chi'_{BA}(\omega) = \mathrm{P}\int_{-\infty}^{\infty} \frac{\chi''_{BA}(\omega')}{\omega'-\omega}d\omega' \qquad (2.127)$$

and this relates the dynamical susceptibility to the correlation function to produce the Green–Kubo formula [8] based on the classical mechanics

$$\chi_{BA}(\omega) = \frac{1}{k_B T}\int_0^\infty \langle B(t)\dot{A}(0)\rangle e^{i\omega t}dt. \qquad (2.128)$$

For the electric transport, we apply $A = \sum_{i=1}^{N} e\mathbf{r}_i$, $F(t) = \mathbf{E}(t)$, $\chi_{BA}(\omega) = \sigma(\omega)$, $B(t) = \mathbf{j}(t) = \sum_{i=1}^{N} e\dot{\mathbf{r}}_i$, and thus $\dot{A}(0) = \sum_{i=1}^{N} e\dot{\mathbf{r}}_i|_{t=0} = \mathbf{j}(0)$.

as we derive Eq. (2.44) in the fluctuation–dissipation section. This formula can be used for obtaining the conductivity σ due to the fluctuating ionized particles since we have the velocities $\mathbf{v}_i = \dot{\mathbf{r}}_i(t)$ from the classical molecular dynamics for $\mathbf{j}(t) = e\sum_{i=1}^{N} \dot{\mathbf{r}}_i(t)$.

2.1.6.2 Derivation of Kubo formula from density matrix

Next we derive the linear-response theory formally from the quantum mechanics and apply it to the transport properties of conductivity σ [9]. For that purpose, the density matrix $\hat{\rho}$ plays a central role, which corresponds to the distribution function $f_N(\mathbf{r}, \mathbf{p})$ in the classical mechanics.

The density matrix (for the mixed states) is defined by

$$\hat{\rho} = \sum_i p_i |i\rangle\langle i|, \tag{2.130}$$

where the states are orthogonal as $\langle i|j\rangle = \delta_{i,j}$ and p_i is the probability that a subsystem labeled as i emerges from among all the possible states; thus $\sum_i p_i = 1$. For example we can take p_i to be a statistical distribution for the grand-canonical ensemble $p_i = e^{-(\epsilon_i - \mu N)/k_B T} / \sum_i e^{-(\epsilon_i - \mu N)/k_B T}$.

The average value for $\hat{X}$ becomes

$$\begin{aligned} \langle \hat{X} \rangle &= \sum_i p_i \langle i|\hat{X}|i\rangle = \sum_{ij} p_i \langle i|\hat{X}|j\rangle\langle j|i\rangle \\ &= \sum_{ij} \langle j|i\rangle p_i \langle i|\hat{X}|j\rangle = \sum_j \langle j|\hat{\rho}\hat{X}|j\rangle = \mathrm{Tr}\left(\hat{\rho}\hat{X}\right), \end{aligned} \tag{2.131}$$

where the density matrix $\hat{\rho}$ in the grand-canonical ensemble is written as

$$\hat{\rho} = e^{-\beta(\hat{H} - \mu\hat{N})} / \mathrm{Tr}\left(e^{-\beta(\hat{H} - \mu\hat{N})}\right) = e^{\beta\left(\hat{\Omega} - \hat{H}\right)}. \tag{2.132}$$

Here $\hat{\Omega}$ is the thermodynamics potential. The time evolution of the density matrix $\hat{\rho}(t)$ follows the quantum version of the Liouville equation (called the Liouville–von Neumann equation),

$$i\hbar \frac{\partial \hat{\rho}(t)}{\partial t} = \left[\hat{H}, \hat{\rho}(t)\right], \tag{2.133}$$

which is derived directly from the Schrödinger equation $i\hbar\partial|i\rangle/\partial t = \hat{H}|i\rangle$ and its complex conjugate $-i\hbar\partial\langle i|\partial t = \langle i|\hat{H}$.

As in the classical mechanics case using the Poisson bracket, it is convenient to introduce the operator $\hat{L}$ defined by

$$\hat{L} \equiv -\frac{1}{\hbar}\left[\hat{H},\right]. \tag{2.134}$$

Then the Liouville equation becomes

$$\frac{\partial \hat{\rho}(t)}{\partial t} = i\hat{L}\hat{\rho}(t), \tag{2.135}$$

and its solution is formally written as

$$\hat{\rho}(t) = e^{i\hat{L}t}\hat{\rho}(0). \tag{2.136}$$

From the Heisenberg equation of $i\hbar\partial\hat{X}/\partial t = \left[\hat{X}, \hat{H}\right] = \hbar\hat{L}\hat{X}$, we can write formally time evolution of a physical quantity $\hat{X}$ as

$$\hat{X}(t) = e^{-i\hat{L}t}\hat{X}(0). \tag{2.137}$$

In the linear-response theory, we take the Hamiltonian as $\hat{H} = \hat{H}_0 + \hat{H}_1(t)$ and correspondingly the density matrix as $\hat{\rho}(t) = \hat{\rho}_0 + \hat{\rho}_1(t)$. As in the classical mechanics case in the previous section, the Liouville equation becomes

$$i\hbar\frac{\partial \hat{\rho}_1(t)}{\partial t} \approx \left[\hat{H}_0, \hat{\rho}_1(t)\right] + \left[\hat{H}_1(t), \hat{\rho}_0\right]. \tag{2.138}$$

Let us derive the Kubo formula based on the density matrix $\hat{\rho}(t)$. We assume electrons in an electromagnetic field with vector potential $\hat{\mathbf{A}}(t)$. The current density is expressed by

$$\hat{\mathbf{j}}(t) = e\sum_i \hat{\mathbf{v}}_i. \tag{2.139}$$

Since the velocity is expressed in the presence of $\hat{\mathbf{A}}(t)$ by

$$\hat{\mathbf{v}}_i = \frac{\hat{\mathbf{p}}_i - (e/c)\hat{\mathbf{A}}(t)}{m}, \tag{2.140}$$

the total current is the sum of two contributions

$$\hat{\mathbf{j}}(t) = \hat{\mathbf{j}}_p(t) - \frac{ne^2}{mc}\hat{\mathbf{A}}(t). \tag{2.141}$$

Here the first term $\hat{\mathbf{j}}_p(t) = (e/m)\sum_i \hat{\mathbf{p}}_i$ is called the paramagnetic current density and the second term $-(ne^2/mc)\hat{\mathbf{A}}(t)$ is called the diamagnetic current density.

The total Hamiltonian of the system is written as

$$\hat{H} = \frac{1}{2m}\sum_i \left(\hat{\mathbf{p}}_i - \frac{e}{c}\hat{\mathbf{A}}(t)\right)^2 + \hat{V}_{\text{ext}} \tag{2.142}$$

and we suppose that the field is small and take the linear term as

$$\hat{H} \approx \hat{H}_0 - \frac{1}{c}\hat{\mathbf{j}}_p(t)\cdot\hat{\mathbf{A}}(t) \equiv \hat{H}_0 + \hat{H}_1(t). \tag{2.143}$$

Now we consider the current defined by

$$\langle\hat{\mathbf{j}}(t)\rangle = \mathrm{Tr}\big(\hat{\rho}(t)\hat{\mathbf{j}}(t)\big), \tag{2.144}$$

where $\hat{\rho}(t)$ is the density matrix.

Using the interaction picture, its solution is written as[a,b]

$$\hat{\rho}_1(t) = -\frac{i}{\hbar}\int_{-\infty}^{t} e^{-i\hat{H}_0(t-t')/\hbar}\left[\hat{H}_1(t'), \hat{\rho}(t')\right] e^{i\hat{H}_0(t-t')/\hbar}dt'. \tag{2.145}$$

Since the field is small, we can replace $\hat{\rho}(t)$ to the equilibrium $\hat{\rho}_0$ in the right hand and have

$$\langle\hat{\mathbf{j}}(t)\rangle = -\frac{i}{\hbar}\mathrm{Tr}\left\{\int_{-\infty}^{t} e^{-i\hat{H}_0(t-t')/\hbar}\left[\hat{H}_1(t'), \hat{\rho}_0\right] e^{i\hat{H}_0(t-t')/\hbar}\hat{\mathbf{j}}_p dt'\right\}$$
$$-\mathrm{Tr}\left\{\frac{ne^2}{mc}\hat{\mathbf{A}}(t)\hat{\rho}_0\right\}. \tag{2.146}$$

Using the cyclic nature of the trace, it becomes

$$\langle\hat{\mathbf{j}}(t)\rangle = -\mathrm{Tr}\left\{\left[\int_{-\infty}^{\infty}\Pi_{j_p,j_p}(t-t')\frac{1}{c}\hat{\mathbf{A}}(t')dt' + \frac{ne^2}{m}\frac{1}{c}\hat{\mathbf{A}}(t)\right]\hat{\rho}_0\right\}, \tag{2.147}$$

where we introduce the retarded current–current correlation function by

$$\Pi_{j_p,j_p}(t-t') = -\frac{i}{\hbar}\theta(t-t')\langle\left[\hat{\mathbf{j}}_p(t), \hat{\mathbf{j}}_p(t')\right]\rangle. \tag{2.148}$$

Now we derive the formula for the conductivity. The electric field is related to the vector potential as

$$\hat{\mathbf{E}}(t) = -\frac{1}{c}\frac{d\hat{\mathbf{A}}(t)}{dt}. \tag{2.149}$$

---

[a] In the interaction picture, $\tilde{\rho}(t) = \tilde{\rho}(-\infty) - (i/\hbar)\int_{-\infty}^{t}\left[\tilde{H}_1(t'), \tilde{\rho}(t')\right]dt'$.

[b] From the direct differentiation of Eq. (2.145), we have $\partial\hat{\rho}_1(t)/\partial t = -(i/\hbar)\times[\hat{H}_1(t), \hat{\rho}_0] - (i/\hbar)\hat{H}_0\hat{\rho}_1(t) + (i/\hbar)\hat{\rho}_1(t)\hat{H}_0 = -(i/\hbar)[\hat{H}_1(t), \hat{\rho}_0] - (i/\hbar)[\hat{H}_0, \hat{\rho}_1(t)]$, which reduces to $i\hbar\partial\hat{\rho}_1(t)/\partial t \approx [\hat{H}_0, \hat{\rho}_1(t)] + [\hat{H}_1(t), \hat{\rho}_0]$. This is the formula for the density matrix $\hat{\rho}(t) = \hat{\rho}_0 + \hat{\rho}_1(t)$ with $\hat{\rho}_1(-\infty) = 0$ within the linear-response approximation in Eq. (2.138).

Assuming the periodic perturbation, we have

$$\hat{\mathbf{A}}(\omega) = \frac{c}{i\omega}\hat{\mathbf{E}}(\omega) \tag{2.150}$$

and from the transport of the current $\bar{\mathbf{j}}(\omega) = \sigma(\omega)\hat{\mathbf{E}}(\omega)$, the conductivity $\sigma(\omega)$ is obtained from[a]

$$\sigma(\omega) = \frac{i}{\omega}\left[\Pi_{j_p,j_p}(\omega) + \frac{ne^2}{m}\right]. \tag{2.151}$$

Here

$$\Pi_{j_p,j_p}(\omega) = -\frac{i}{\hbar}\int_0^\infty \langle\left[\hat{\mathbf{j}}_p(t), \hat{\mathbf{j}}_p(0)\right]\rangle e^{i\omega t}dt \tag{2.152}$$

is the current–current correlation function, which corresponds to the autocorrelation function of $\langle \mathbf{j}(t)\mathbf{j}(0)\rangle$ in the classical mechanics case.[b] $\Pi_{j_p,j_p}(\omega)$ is calculated using

$$\langle\left[\hat{\mathbf{j}}_p(t), \hat{\mathbf{j}}_p(0)\right]\rangle = \left(\frac{e}{m}\right)^2 \sum_{\alpha,\beta} f(\epsilon_\alpha)\langle\alpha|\hat{p}_z|\beta\rangle\langle\beta|\hat{p}_z|\alpha\rangle \times\{e^{i(\epsilon_\alpha-\epsilon_\beta)t/\hbar} - e^{-i(\epsilon_\alpha-\epsilon_\beta)t/\hbar}\} \tag{2.153}$$

as

$$\Pi_{j_p,j_p}(\omega) = 2\left(\frac{e}{m}\right)^2 \sum_{\alpha,\beta}\langle\alpha|\hat{p}_z|\beta\rangle\langle\beta|\hat{p}_z|\alpha\rangle\frac{f(\epsilon_\alpha) - f(\epsilon_\beta)}{\epsilon_\alpha - \epsilon_\beta + \hbar\omega + i0^+}. \tag{2.154}$$

With use of the relation

$$\frac{1}{x + i\delta} = \frac{P}{x} - i\pi\delta(x), \tag{2.155}$$

we have

$$\Pi_{j_p,j_p}(\omega) = -2\pi i\left(\frac{e}{m}\right)^2 \sum_{\alpha,\beta} |\langle\alpha|\hat{p}_z|\beta\rangle|^2 \left(f(\epsilon_\alpha) - f(\epsilon_\beta)\right) \times\delta\left(\epsilon_\alpha - \epsilon_\beta + \hbar\omega\right). \tag{2.156}$$

---

[a] This is one example for the general formulation of the linear-response theory of $X(\omega) = -\Pi(\omega)/i\omega$ with $\Pi(\omega) = -(i/\hbar)\int_0^\infty\langle\left[X(t), \dot{H}_1(0)\right]\rangle e^{i\omega t}dt$, where $X = \mathbf{j}_p = \sigma E$ and $\dot{H}_1(t) = -(1/c)\mathbf{j}_p \cdot dA/dt = \mathbf{j}_p\dot{E}$.

[b] Using the identity for an equilibrium average of two operators $\langle\hat{j}_p(t)\,\hat{j}_p(0)\rangle = \mathrm{Tr}\left(\hat{\rho}\hat{j}_p(t)j_p(0)\right) = \langle\hat{j}_p(0)\,\hat{j}_p(t + i\hbar/k_BT)\rangle$, the fluctuation–dissipation theorem is expressed as $S(\omega) = \hbar\mathrm{Im}\sigma(\omega)\coth(\hbar\omega/2k_BT)$ in the linear-response formula. We note $S_V(\omega) = 4k_BTR$, reducing to Johnson–Nyquist noise for $k_BT \gg \hbar\omega$.

Then taking $\omega \to 0$,

$$\sigma(0) = 2\pi \left(\frac{e}{m}\right)^2 \lim_{\omega \to 0} \frac{1}{\omega} \sum_{\alpha,\beta} |\langle \alpha | \hat{p}_z | \beta \rangle|^2 \left(f(\epsilon_\alpha) - f(\epsilon_\beta)\right)$$
$$\times \delta\left(\epsilon_\alpha - \epsilon_\beta + \hbar\omega\right)$$
$$= 2\pi\hbar \left(\frac{e}{m}\right)^2 \int \sum_{\alpha,\beta} |\langle \alpha | \hat{p}_z | \beta \rangle|^2 \left(-\frac{df}{d\epsilon}\right) \delta(\epsilon - \epsilon_\alpha)\delta(\epsilon - \epsilon_\beta) d\epsilon.$$

We note that the same formula is derived when we consider the master equation for energy transfer (problem 2).

Several applications using the Kubo formula, rewriting it in Green's function form, will be presented in Chapter 3. We show that the Kubo formula will give the classical Drude conductivity in the lowest approximation. Also we will show a sophisticated numerical method for the transport of huge systems by using the Kubo formula in Chapter 4 and its applications in Chapter 5.

2.1.6.3 Thermal conductivity as heat flow

Here we consider the heat flow as the thermal current transport [10]. The phenomenological concept of heat flow $\mathbf{j}_{\text{th}}$ in macroscopic materials is based on the assumption of Fourier's law

$$\mathbf{j}_{\text{th}} = -\kappa \nabla T, \qquad (2.157)$$

which corresponds to the Fick's law of an electron flow we described as $\mathbf{j} = -eD\nabla n$. Here κ is called the thermal conductivity and T is the local temperature of the position $\mathbf{r}$. We note that the heat flow occurs due to both electrons $\mathbf{j}_q$ and lattice vibrations $\mathbf{j}_{ph}$ known as phonons. Usually the values of κ differs from 10^{-2} [W/mK] for gas to 10^5 [W/mK] for solids.

Due to the energy conservation law as

$$\frac{\partial T}{\partial t} + \kappa \nabla \cdot \mathbf{j}_{\text{th}} = 0, \qquad (2.158)$$

we obtain the equation for heat flow (heat diffusion equation)

$$\frac{\partial T}{\partial t} = \kappa \nabla^2 T. \qquad (2.159)$$

Thus, Fourier's law assumes the diffusive motion of heat flow based on the particle pictures. In usual cases, Fourier's law is satisfied in the macroscopic materials, much larger than the MFP[a] due to phonon-impurity scatterings and phonon-phonon scatterings. To obtain the temperature distribution of T, we solve the heat diffusion equation Eq. (2.159) with appropriate boundary conditions numerically, for example by the finite element method (FEM). For a smaller system, the assumption of Fourier's law becomes questionable and we need different calculation methods, similar to the electron transport problems.

The Boltzmann equation method is used also for the thermal transport of $\mathbf{j}_{\mathrm{th}}$. For electron contributions $\mathbf{j}_q$, within the relaxation time approximation,

$$f \simeq f_0 - \tau \left(\mathbf{v}\frac{\partial f_0}{\partial \mathbf{r}} + \frac{e\mathbf{E}}{m}\frac{\partial f_0}{\partial \mathbf{v}} \right) \tag{2.160}$$

the thermal current density is obtained from

$$j_q(\mathbf{r}) = \int \epsilon \mathbf{v} f(\mathbf{r}, \mathbf{v}) d\mathbf{v}, \tag{2.161}$$

which should be compared with the electric current as

$$\mathbf{j}(\mathbf{r}) = e \int \mathbf{v} f(\mathbf{r}, \mathbf{v}) d\mathbf{v}. \tag{2.162}$$

The detailed calculations of these are described in the problem 2. We note that in the (free electron) metallic system where heat flow is mostly due to electrons $\mathbf{j}_q$ only, the ratio of κ and the electric conductivity σ for an electron transport takes a constant value of

$$\frac{\kappa}{\sigma T} = \frac{\pi^2}{3}\left(\frac{k_B}{e}\right)^2 = 2.44 \times 10^{-8}\ (W\Omega/K^2), \tag{2.163}$$

known as the Wiedemann–Franz law and the value is called the Lorentz number [1]. For phonon contributions $\mathbf{j}_{ph}$, we have

$$j_{ph}(\mathbf{r}) = \int \epsilon \mathbf{v} \eta(\mathbf{r}, \mathbf{v}) d\mathbf{v}, \tag{2.164}$$

where η is the Bose–Einstein distribution since phonons are bosons. Within the relaxation time approximation, it is given by

---

[a]The thermal conductivity κ is expressed using the MFP $\Lambda = \tau v$ approximately as $\kappa = (1/3) C v \Lambda$, where C is the specific heat.

$$\eta \simeq \eta_0 - \tau_{ph} \mathbf{v} \nabla T \frac{\partial \eta_0}{\partial T} \tag{2.165}$$

where the second term contributes to Fourier's law.[a]

An alternative calculation method is to use the molecular dynamics (MD) based on the linear-response theory. As in the electrical conductivity, the thermal conductivity is obtained from the autocorrelation function[b] of the thermal currents as

$$\kappa = \frac{1}{3k_B T^2} \int_0^\infty \langle \mathbf{j}_{\mathrm{th}}(t) \mathbf{j}_{\mathrm{th}}(0) \rangle dt. \tag{2.166}$$

Here $\mathbf{j}_{\mathrm{th}}(t)$ is obtained in the MD for a particle $(\mathbf{R}_i, \mathbf{V}_i)$ from

$$\mathbf{j}_{\mathrm{th}}(t) = \sum_{i=1}^{N} \left\{ \frac{1}{2} M_i \mathbf{V}_i^2 + \frac{1}{2} \sum_{j=1(i \neq j)}^{N} U_{ij}(R_{ij}) \right\} \mathbf{V}_i - \frac{1}{2} \sum_{i=1}^{N} \sum_{j=1(i \neq j)}^{N} \left[\mathbf{V}_i \cdot \frac{\partial U_{ij}(R_{ij})}{\partial \mathbf{R}_i} \right] \mathbf{R}_{ij} \tag{2.167}$$

where $R_{ij} = |\mathbf{R}_i - \mathbf{R}_j|$ and $U_{ij}(R_{ij})$ are the distance and the potential between particles i and j [11]. We note that direct calculation to assign different temperatures as the boundary conditions for the MD in the nonequilibrium situation is possible. This is called the nonequilibrium molecular dynamics (NEMD).[c]

2.2 From Wave Description

2.2.1 *Scattering Approach: Landauer Formula*

In the Kubo formula, we consider that an electric field drives the electron to move for the transport. Here we consider that electron transmission through nanostructures determines the transport. The Landauer formula is based on the concept that an electron has a wave nature [12] and thus it matches, in itself, to the quantum mechanics very well.

---

[a] Here we assume j_q and j_{ph} are independent and neglect τ_{e-ph} due to the electron-phonon (e–ph) interaction, which produces phonon drag effects.

[b] The average is taken as $\langle \mathbf{j}_{\mathrm{th}}(t) \mathbf{j}_{\mathrm{th}}(0) \rangle = (1/\tau) \int_0^\tau \mathbf{j}_{\mathrm{th}}(t + t') \mathbf{j}_{\mathrm{th}}(t') dt'$.

[c] We describe the numerical method for the molecular dynamics and phonon transport by the nonequilibrium Green's function method in Chapter 4.

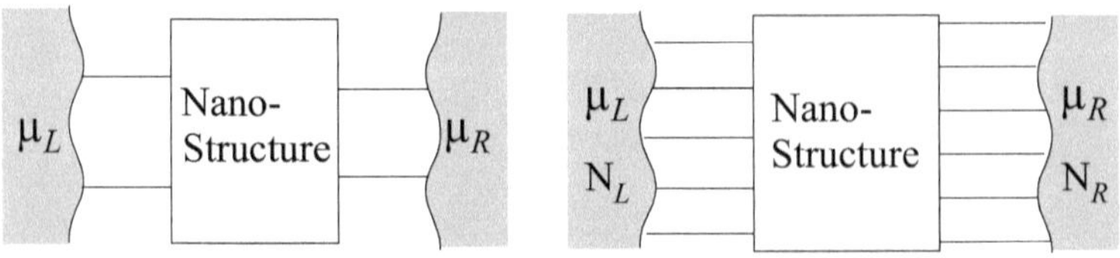

Figure 2.2 (Left) Schematic representation for a ballistic two-terminal conductance. The gray region shows the scattering region. Electrons move between electrodes with different chemical potentials of $\mu_{L,R}$ through scattering region. (Right) Schematic representation for multichannel conductions. The number of channels or modes in the left and right electrodes are N_L and N_R, respectively.

We first consider the strictly 1D case. We set a system that has a scattering region coupled with two semi-infinite electrodes as shown in the left panel of Fig. 2.2. The applied bias voltage corresponds to the difference of chemical potentials for left and right electrodes

$$\mu_L - \mu_R = eV. \tag{2.168}$$

The quantum state well inside the electrode is a plane wave and thus the number of electrons per energy (density of states ν) is equal to

$$dn = \nu d\epsilon = \frac{dk}{2\pi}. \tag{2.169}$$

Since $d\epsilon/dk = \hbar v_F$, the density of states per spin is

$$\nu = \frac{1}{2\pi\hbar v_F} \tag{2.170}$$

and we calculate the current flowing between electrodes as

$$I = 2e\frac{\mathcal{T}}{2\pi\hbar v_F}v_F(\mu_L - \mu_R) = \frac{2e}{h}\mathcal{T}(\mu_L - \mu_R). \tag{2.171}$$

Here $\mathcal{T}$ is the transmission probability and the index 2 comes from the spin degeneracy.

The conductance $G = I/V$ becomes

$$G = \frac{I}{(\mu_L - \mu_R)/e} = \frac{2e^2}{h}\mathcal{T}. \tag{2.172}$$

The coefficient $2e^2/h$ is a very important constant value comprising only the fundamental value of electron charge and the Planck constant. It is equal, in the resistance unit, to

$$\left[\frac{2e^2}{h}\right]^{-1} = 12.9\ \text{k}\Omega \tag{2.173}$$

or equivalently, in the conductance unit, as

$$\left[\frac{2e^2}{h}\right] = 77.5\ \mu\text{S}. \tag{2.174}$$

We expand this formula to the systems with multichannel cases in higher dimensions as shown in the right panel of Fig. 2.2. The important concept is the channel mode. If we assume that the leads have a finite width, the electronic states are separated with the direction in which electrons flow and their perpendicular directions. For the perpendicular directions, electrons form discrete quantum states due to confinement from the finite width. These states constitute the subband or transverse modes and we can label them as the channel number. It is apparent that each subband has a different k-vector in the current flow direction. When a bias voltage is applied to electrodes, electrons come from the left electrode and go out to the right electrode. The total current is the sum of electrons that transmit through the region. Therefore, the current is expressed by

$$I = \frac{2e}{h} \sum_{i,j} \mathcal{T}_{i,j}(\mu_L - \mu_R) \tag{2.175}$$

and the corresponding conductance becomes

$$G = \frac{2e^2}{h} \sum_{i,j} \mathcal{T}_{i,j}. \tag{2.176}$$

This is the formula for the multichannel conductance.

To obtain the Landauer formula, we assume ideal steady states where electrons in the reservoirs are always in thermal equilibrium. However, in experiments, we cannot put the probes to measure the current in such idealized situations. Usually, we use the four-terminal scheme for the measurement where the probes are set *locally* close to the samples so that they adjust the voltage to keep the current flow between probes zero. The measured conductance is so defined as the total current divided the measured voltage.

We consider the local chemical potentials $\bar{\mu}_L$ and $\bar{\mu}_R$ in the four-terminal situations. The number of states occupied above $\bar{\mu}_R$ on the right side due to injection from the left is $\mathcal{T}(\mu_L - \bar{\mu}_R)/\pi\hbar v_F$

and the number of unoccupied states below $\bar{\mu}_R$ is given by $2(\bar{\mu}_R - \mu_R)/\pi\hbar v_F - \mathcal{T}(\bar{\mu}_R - \mu_R)/\pi\hbar v_F$. Equating these two equations gives

$$\bar{\mu}_R = \frac{1}{2}\left[\mathcal{T}\mu_L + (2-\mathcal{T})\mu_R\right]. \tag{2.177}$$

Similarly, on the left side, $\bar{\mu}_L$ is defined as

$$\bar{\mu}_L = \frac{1}{2}\left[(2-\mathcal{T})\mu_L + \mathcal{T}\mu_R\right], \tag{2.178}$$

which gives

$$\bar{\mu}_L - \bar{\mu}_R = (1-\mathcal{T})(\mu_L - \mu_R). \tag{2.179}$$

Therefore, the measured conductance in the four-terminal case is given by

$$G = \frac{I}{(\bar{\mu}_L - \bar{\mu}_R)/e} = 2e\frac{\mathcal{T}}{2\pi\hbar v_F}v_F\frac{(\mu_L - \mu_R)}{(\bar{\mu}_L - \bar{\mu}_R)/e} = \frac{2e^2}{h}\frac{\mathcal{T}}{1-\mathcal{T}}. \tag{2.180}$$

So the conductance becomes infinite or equivalently the resistance becomes zero when we have the perfect conductor. This is a desired result. From these observations, we can say that the total current flowing between electrodes is obtained from the Landauer formula, and the conductance in the four-terminal scheme needs to multiply the coefficient of $1/(1-\mathcal{T})$.

2.2.2 *Multiterminal Transport*

We can expand the Landauer formula for the multiterminal case [13]. We assume that each terminal i has a chemical potential μ_i and take the minimum one as μ_0. Since the fraction of current $I_i = (2e/h)(\mu_i - \mu_0)$ is not dependent on the channel number, defining the transmission and reflection coefficients as

$$\mathcal{T}_{ij} = \sum_{mn}\mathcal{T}_{ij,mn}, \qquad \mathcal{R}_{ij} = \sum_{mn}\mathcal{R}_{ij,mn}, \tag{2.181}$$

the total current in terminal i is obtained as follows (Fig. 2.3).

If there are N_i channels at the terminal i with chemical potential μ_i, the incoming current is $(2e/h)N_i(\mu_i - \mu_0)$ and the outgoing current is $(2e/h)(\mathcal{R}_{ii}(\mu_i - \mu_0) + \sum_{j\neq i}\mathcal{T}_{i,j}(\mu_i - \mu_0))$. Using the

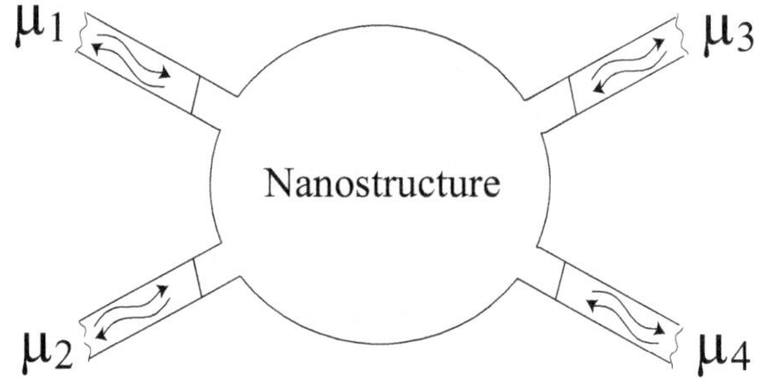

Figure 2.3 Schematic illustration of the multiterminal transport.

current conservation law of $N_i = \mathcal{R}_{ii} + \sum_{j \neq i} \mathcal{T}_{ij}$, which eliminates μ_0, we obtain the total current flowing at the terminal i as

$$I_i = \frac{2e}{h} \left[(N_i - \mathcal{R}_{ii})\mu_i - \sum_{j \neq i} \mathcal{T}_{ij} \mu_j \right]. \tag{2.182}$$

This is Landauer–Büttiker formula for multiterminal conductor.

Let us evaluate this formula for several cases. For the two-terminal conductor, we restrict the sum over the terminal indices i and j to 2. We can set $I = I_1 = -I_2$. By the conservation of the current $N_1 = \mathcal{R}_{11} + \mathcal{T}_{12}$ in terminal 1 and thus we get

$$I = \frac{2e}{h} [(N_1 - \mathcal{R}_{11})\mu_1 - \mathcal{T}_{12}\mu_2] = \frac{2e}{h} \mathcal{T}_{12}(\mu_1 - \mu_2), \tag{2.183}$$

which is the two-probe current in the previous section.

For the three-terminal conductor, we consider that the contact 3 represents an ideal voltage probe that draws no current and put $I_3 = 0$. Then the condition of zero current flowing in contact 3 is

$$0 = \frac{2e}{h} [(N_3 - \mathcal{R}_{33})\mu_3 - \mathcal{T}_{31}\mu_1 - \mathcal{T}_{32}\mu_2]. \tag{2.184}$$

By the conservation of the current, $N_3 = \mathcal{R}_{33} + \mathcal{T}_{31} + \mathcal{T}_{32}$ in terminal 3, we have

$$\mu_3 = \frac{\mathcal{T}_{31}\mu_1 + \mathcal{T}_{32}\mu_2}{\mathcal{T}_{31} + \mathcal{T}_{32}}. \tag{2.185}$$

Since the sum of all currents in all probes must be zero, we have $I = I_1 = -I_2$ and I is evaluated using the formula for μ_3 and $N_2 = \mathcal{R}_{22} + \mathcal{T}_{21} + \mathcal{T}_{23}$ as

$$I = \frac{2e}{h}\left(\mathcal{T}_{21} + \frac{\mathcal{T}_{31}\mathcal{T}_{23}}{\mathcal{T}_{31} + \mathcal{T}_{32}}\right)(\mu_1 - \mu_2). \tag{2.186}$$

For the four-terminal conductor, the same procedures are applied and the total current is obtained. This will be left for the exercise of the reader.

2.2.3 *S-Matrix Theory*

A coherent conductor can generally be characterized at each energy by the scattering matrix, usually called the S-matrix. The definition of the S-matrix is

$$S = \begin{bmatrix} r & \tilde{t} \\ t & \tilde{r} \end{bmatrix}, \tag{2.187}$$

where the matrices r and $\tilde{r}$ describe the reflection amplitudes and the matrices t and $\tilde{t}$ describe the transmission amplitudes. The S-matrix relates the outgoing wave amplitudes to the incoming wave amplitudes at the different electrodes L, R as

$$\begin{pmatrix} b_L \\ b_R \end{pmatrix} = S \begin{pmatrix} a_L \\ a_R \end{pmatrix}, \tag{2.188}$$

where a_L is a column vector representing the incoming wave amplitudes in all the various modes in lead L, b_L is a column vector representing the outgoing wave amplitudes in all the various modes in lead R, etc. (Fig. 2.4 (left)).

Due to the current conservation,

$$|a_L|^2 + |a_R|^2 = |b_L|^2 + |b_R|^2, \tag{2.189}$$

which implies the relation of

$$\begin{pmatrix} b_L^* & b_R^* \end{pmatrix}\begin{pmatrix} b_L \\ b_R \end{pmatrix} = \begin{pmatrix} a_L^* & a_R^* \end{pmatrix} S^\dagger S \begin{pmatrix} a_L \\ a_R \end{pmatrix} = \begin{pmatrix} a_L^* & a_R^* \end{pmatrix}\begin{pmatrix} a_L \\ a_R \end{pmatrix}. \tag{2.190}$$

From this we can see that the S-matrix is unitary

$$S^\dagger S = I. \tag{2.191}$$

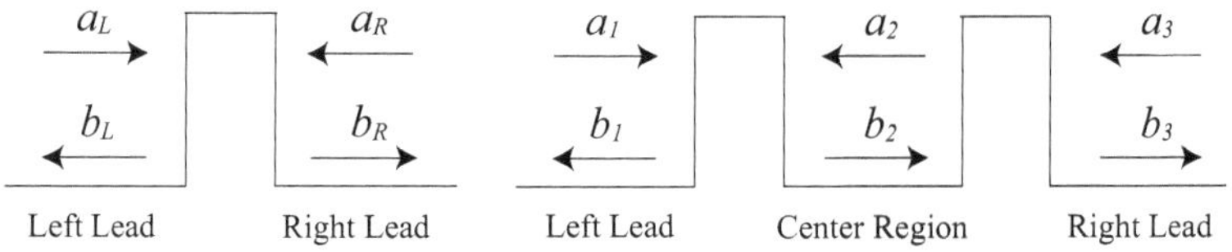

Figure 2.4 (Left) Schematic view of the energy diagram for the single barrier system. (Right) Schematic view of energy diagram for junction system connected to leads with double barriers.

The transmission probability $\mathcal{T}$ is obtained by taking the squared magnitude of the corresponding elements of the S-matrix

$$\mathcal{T} = \sum_{i,j} \mathcal{T}_{i,j} = \sum_{i,j} tt^* = \mathrm{Tr}\left\{tt^\dagger\right\}, \tag{2.192}$$

so the two-terminal conductance is obtained using the S-matrix as

$$G = \frac{2e^2}{h}\mathrm{Tr}\left\{tt^\dagger\right\}. \tag{2.193}$$

The overall S-matrix can be evaluated to divide the conductor into several sections (Fig. 2.4 (right)). We compute the individual S-matrices S_1, S_2, $\cdots$ and then combine them to obtain the overall S-matrix. Here we show the rules for combining S-matrices.

$$\begin{pmatrix} b_1 \\ b_2 \end{pmatrix} = \begin{bmatrix} r_1 & \tilde{t}_1 \\ t_1 & \tilde{r}_1 \end{bmatrix} \begin{pmatrix} a_1 \\ a_2 \end{pmatrix} \quad \text{and} \quad \begin{pmatrix} a_2 \\ b_3 \end{pmatrix} = \begin{bmatrix} r_2 & \tilde{t}_2 \\ t_2 & \tilde{r}_2 \end{bmatrix} \begin{pmatrix} b_2 \\ a_3 \end{pmatrix} \tag{2.194}$$

to obtain the combined S-matrix,

$$\begin{pmatrix} b_1 \\ b_3 \end{pmatrix} = \begin{bmatrix} r & \tilde{t} \\ t & \tilde{r} \end{bmatrix} \begin{pmatrix} a_1 \\ a_3 \end{pmatrix}. \tag{2.195}$$

It is straightforward to eliminate a_2 and b_2 from the above equations to obtain the composite structures of combined S-matrix,

$$\begin{aligned} r &= r_1 + \tilde{t}_1 r_2 \left[I - \tilde{r}_1 r_2\right]^{-1} t_1 \\ \tilde{r} &= \tilde{r}_2 + t_2 \left[I - \tilde{r}_1 r_2\right]^{-1} \tilde{r}_1 \tilde{t}_2 \\ t &= t_2 \left[I - \tilde{r}_1 r_2\right]^{-1} t_1 \\ \tilde{t} &= \tilde{t}_1 \left[I - r_2 \tilde{r}_1\right]^{-1} \tilde{t}_2. \end{aligned} \tag{2.196}$$

It is interesting to see that the transmission is expressed by expanding

$$t = t_2 t_1 + t_2 \left[\tilde{r}_1 r_2\right] t_1 + t_2 \left[\tilde{r}_1 r_2\right]\left[\tilde{r}_1 r_2\right] t_1 + \cdots. \tag{2.197}$$

These successive terms show the trajectories that the incoming electron behaves, the first term is the direct transmission, the second is with two reflections, the third is with four reflections, etc. We note that this behavior appears if the coherent nature of an electron is preserved.

For the calculations, the transfer matrix M defined by

$$\begin{pmatrix} b_R \\ a_R \end{pmatrix} = M \begin{pmatrix} a_L \\ b_L \end{pmatrix} \tag{2.198}$$

is also effective. By comparing the S-matrix with the definition of transfer matrix, we have

$$M = \begin{bmatrix} t^{-1} & \tilde{r}\tilde{t}^{-1} \\ -\tilde{t}^{-1}r & \tilde{t}^{-1} \end{bmatrix}. \tag{2.199}$$

The advantage of using the M is that it is *multiplicative*, that is, we can divide the scattering regions into N regions in space, each described by M_i matrix, then the total transfer matrix is the product of these single matrices

$$M = M_N M_{N-1} \cdots M_1. \tag{2.200}$$

2.2.4 *Coherent Transport*

As an example of the above property, let us consider the single-particle scattering through two junctions, separated by a region where electrons have no scatterings due to double tunneling barriers (Fig. 2.5). For simplicity, we assume there is only one incoming and one outgoing channel; thus the matrix $t,r,\tilde{t},\tilde{r}$ comprises numbers. From Eq. (2.196), the total transmission amplitude is obtained from

$$t = \frac{t_1 t_2}{1 - \tilde{r}_1 r_2} \tag{2.201}$$

and the transmission probability from the left electrode to the right electrode is

$$\mathcal{T} = tt^* = \frac{\mathcal{T}_1 \mathcal{T}_2}{1 + \mathcal{R}_1 \mathcal{R}_2 - 2\sqrt{\mathcal{R}_1 \mathcal{R}_2}\cos\theta} \tag{2.202}$$

where $\theta = \text{phase}(\tilde{r}_1) + \text{phase}(r_2)$ is the phase shift acquired from the scattering, which shows the coherence of waves inside the region

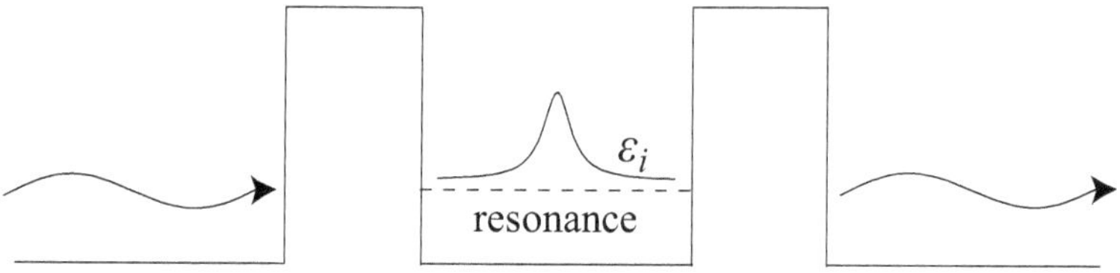

Figure 2.5 Schematic representation of energy diagram of junction system. Resonant tunneling occurs at a certain energy ϵ_i.

separated by two electrodes. We can rewrite this expression as

$$\begin{aligned}\mathcal{T} &= \frac{\mathcal{T}_1\mathcal{T}_2}{\left(1-\sqrt{\mathcal{R}_1\mathcal{R}_2}\right)^2 + 2\sqrt{\mathcal{R}_1\mathcal{R}_2}\,(1-\cos\theta)} \\ &\approx \frac{\mathcal{T}_1\mathcal{T}_2}{\left(\frac{\mathcal{T}_1+\mathcal{T}_2}{2}\right)^2 + 2\,(1-\cos\theta)},\end{aligned} \tag{2.203}$$

where we assume $\mathcal{T}_1 \approx \mathcal{T}_2 \ll 1$.

This shows that resonance occurs when the phase shift $\theta(\epsilon)$ is close to a multiple of 2π. Suppose that such resonance appears close to ϵ_i, then we can expand the cosine function close to this resonance in a Taylor series

$$1-\cos\theta(\epsilon) \approx \frac{1}{2}\left(\frac{d\theta(\epsilon)}{d\epsilon}\right)^2 (\epsilon-\epsilon_i)^2 \tag{2.204}$$

to get the transmission probability

$$\begin{aligned}\mathcal{T}(\epsilon) &\approx \sum_i \frac{\Gamma_1\Gamma_2}{\left(\frac{\Gamma_1+\Gamma_2}{2}\right)^2 + (\epsilon-\epsilon_i)^2} \\ &= \left(\frac{\Gamma_1\Gamma_2}{\Gamma_1+\Gamma_2}\right)\sum_i \frac{\Gamma}{(\epsilon-\epsilon_i)^2 + (\Gamma/2)^2},\end{aligned} \tag{2.205}$$

where we define

$$\Gamma_{1,2} = \mathcal{T}_{1,2}\left[\frac{d\theta(\epsilon)}{d\epsilon}\right]^{-1} \tag{2.206}$$

and $\Gamma = \Gamma_1 + \Gamma_2$. This shows the Lorentzian form with an effective broadening given by Γ, which is a weighted broadening due to the coupling of the resonant state ϵ_i with the continuum of states in the electrodes. Thus, Γ shows the escaping rate for a particle in the state ϵ_i to scatter out of that state into left and right electrodes. We can write the transmission probability as

$$\mathcal{T}(\epsilon) = \frac{\Gamma_1\Gamma_2}{\Gamma_1 + \Gamma_2} \sum_i A(\epsilon - \epsilon_i) \tag{2.207}$$

where $A(\epsilon)$ is called the spectral function and has the Lorentzian form

$$A(\epsilon) = \frac{\Gamma}{\epsilon^2 + (\Gamma/2)^2}. \tag{2.208}$$

The Lorentzian form of the current is a characteristic nature of the resonant tunneling.

The current is obtained to integrate the transmission probability within chemical potentials of electrodes

$$\begin{aligned} I &= \frac{2e}{h} \int_{\mu_L}^{\mu_R} \mathcal{T}(\epsilon) d\epsilon \\ &= \frac{2e}{h} \frac{\Gamma_1\Gamma_2}{\Gamma_1 + \Gamma_2} \sum_i \int_{\mu_L}^{\mu_R} A(\epsilon - \epsilon_i) d\epsilon. \end{aligned} \tag{2.209}$$

The conductance shows the periodic oscillations having the Lorentzian form as a function of energy ϵ

$$G(\epsilon) = \frac{2e^2}{h} \frac{\Gamma_1\Gamma_2}{\Gamma_1 + \Gamma_2} \sum_i A(\epsilon - \epsilon_i). \tag{2.210}$$

This formula represents coherent electron transport corresponding to the wave nature of electrons as shown in Fig. 1.6 (right). This situation is realized when the phase coherence length is much larger than the system size as $L \ll \ell_{\mathrm{mfp}} \ll L_\phi$ in the ballistic regime.

2.2.5 *Incoherent Transport to Ohm's Law*

As an opposite situation, we will show how the Landauer formula we described in the previous sessions describes the Ohm's law in the classical regime as $L_\phi \ll \ell_{\mathrm{mfp}} \ll L$. We know that Ohm's law is due to the particle nature of electrons in the classical diffusive regime, which suggests that electrons lose their memory of the phase due to incoherent scatterings. In order to realize this situation, we assume that the dephasing processes happen

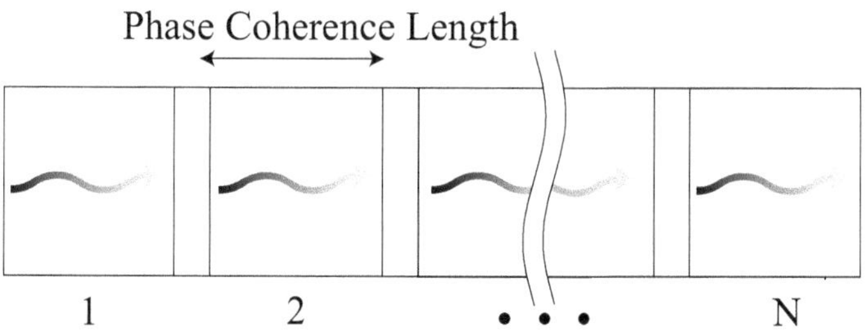

Figure 2.6 Schematic representation of junction system. Due to dephasing over the phase coherence length, the section is divided into small segments. Electron transport follows a classical particle nature, leading to Ohm's law.

inside the region between electrodes, which occur due to electron–phonon scatterings, electron–electron scatterings, and others. Thus, the scatterings are divided into small segments, in which the phase coherence of an electron is preserved with the coherence length L_ϕ as shown in Fig. 2.6.[a]

So we can average the transmission probability to obtain

$$\begin{aligned}\mathcal{T} &= \left\langle \frac{\mathcal{T}_1\mathcal{T}_2}{1+\mathcal{R}_1\mathcal{R}_2 - 2\sqrt{\mathcal{R}_1\mathcal{R}_2}\cos\theta} \right\rangle \\ &= \int \frac{\mathcal{T}_1\mathcal{T}_2}{1+\mathcal{R}_1\mathcal{R}_2 - 2\sqrt{\mathcal{R}_1\mathcal{R}_2}\cos\theta}\frac{d\theta}{2\pi} = \frac{\mathcal{T}_1\mathcal{T}_2}{1-\mathcal{R}_1\mathcal{R}_2}. \end{aligned} \tag{2.211}$$

This formula for the transmission probability is obtained more directly by summing the probabilities for transmissions with reflections as

$$\begin{aligned}\mathcal{T} &= \mathcal{T}_1\mathcal{T}_2 + \mathcal{T}_1\mathcal{T}_2\mathcal{R}_1\mathcal{R}_2 + \mathcal{T}_1\mathcal{T}_2\mathcal{R}_1^2\mathcal{R}_2^2 + \cdots \\ &= \frac{\mathcal{T}_1\mathcal{T}_2}{1-\mathcal{R}_1\mathcal{R}_2}. \end{aligned} \tag{2.212}$$

This result shows that an electron is transmitted between the segments via multiple scatterings without its memory of the phase

[a]The mechanism of the breaking of phase coherence due to coupling to an environment, especially at low temperature, is one of the fundamental problems in the quantum system. This is strongly related to the quantum information technology. The computational method using the quantum master equation for an electron transport with coupling to an environment is presented in Chapter 4.

coherence as if it behaves like a classical particle. We note that Eq. (2.212) is written as

$$\frac{1-\mathcal{T}}{\mathcal{T}} = \frac{1-\mathcal{T}_1}{\mathcal{T}_1} + \frac{1-\mathcal{T}_2}{\mathcal{T}_2}, \tag{2.213}$$

which shows that when we place the incoherent scatters, the quantity of $(1-\mathcal{T})/\mathcal{T}$ has an additive property.

We can generalize this result to N segments to couple incoherently in series as

$$\frac{1-\mathcal{T}}{\mathcal{T}} = \frac{1-\mathcal{T}_1}{\mathcal{T}_1} + \frac{1-\mathcal{T}_2}{\mathcal{T}_2} + \cdots + \frac{1-\mathcal{T}_N}{\mathcal{T}_N}. \tag{2.214}$$

This is simplified if we assume all the segments have the same transmission probability, say $\bar{\mathcal{T}}$, then we obtain

$$\mathcal{T}(N) = \frac{\bar{\mathcal{T}}}{N(1-\bar{\mathcal{T}})+\bar{\mathcal{T}}}. \tag{2.215}$$

Here $\bar{\mathcal{T}}$ is the transmission probability for each segment. The fact that $(1-\bar{\mathcal{T}})/\bar{\mathcal{T}}$ has an additive property means that the resistance is proportional to it.

$$R = \frac{h}{2e^2}\frac{1}{\mathcal{T}(N)} = \frac{h}{2e^2} + \frac{h}{2e^2}N\left(\frac{1-\bar{\mathcal{T}}}{\bar{\mathcal{T}}}\right). \tag{2.216}$$

This formula shows that the total resistance is expressed by the sum of two contributions, the sequence of its *intrinsic* resistance $(h/2e^2)(1-\bar{\mathcal{T}})/\bar{\mathcal{T}}$ in each segment and the *contact* resistance $h/2e^2$. Note that the resistance of the four-terminal system expressed by $R = 1/G = (h/2e^2)(1-\mathcal{T})/\mathcal{T}$ in Eq. (2.180) measures the *intrinsic* resistance of the scattering region without effects from the *contact* scattering between electrodes and reservoirs in the coherent $(N=1)$ regime.

When we introduce the density of scatterers $\nu \equiv N/L$, the resistance R is written as

$$R = \frac{h}{2e^2}\frac{L+L_0}{L_0}, \tag{2.217}$$

where the characteristic length L_0 is defined by $L_0 = \bar{\mathcal{T}}/\nu(1-\bar{\mathcal{T}})$. When we consider the weak scattering case $\bar{\mathcal{T}} \sim 1$, the characteristic length $L_0 \sim 1/\nu(1-\bar{\mathcal{T}})$ represents the distance where an electron can travel before it is scattered. Thus, L_0 is the size of the MFP ℓ_{mfp} as $L_0 \sim \ell_{\mathrm{mfp}}$ in the weak scattering case.

For the macroscopic materials, the size L is much larger than ℓ_{mfp} as $L \gg \ell_{\mathrm{mfp}}$, the first term in Eq. (2.217) dominates the resistance and the second term for the contact becomes negligible. The above formula in Eq. (2.217) is the single-channel case. Since the number of channels is proportional to the cross section S, we finally obtain the formula Eq. (2.6)

$$R = \frac{1}{\sigma}\frac{L}{S} \tag{2.218}$$

for the macroscopic Ohmic conductors in the classical regime. Here σ is the constant. This is the microscopic view of the macroscopic Ohmic conductors. The important points are that due to the loss of the phase of an electron, the resistance becomes proportional to the system size. We note that since the system size is much larger than the MFP, we can neglect the effects from the contact on transport properties. In the nanometer-scale systems, the contact effect becomes essential.

2.2.6 *Localization and Conductance Fluctuation*

Electron devices recently become smaller and smaller and it becomes possible to fabricate device structures with the size smaller than the phase coherence length. In such devices, electrons travel without inelastic scatterings with phonons or other electrons over L_ϕ and quantum interference effects manifest. Here we consider the intermediate regime $\ell_{\mathrm{mfp}} \ll L \ll L_\phi$, where electrons are elastically scattered with disorder and thus the system size is much larger than the MFP, but the memory of the phase coherence is preserved. This regime is the quantum diffusive regime and is often called the *mesoscopic* regime, bridging the gap between atomic scale and macroscopic scale. The transport properties are sensitive to the device geometry and configuration of impurities due to the phase coherence. We need to use the statistical approach for the ensemble average for transport properties.

We note that in the *classical* diffusive regime, the dimensionless resistance of $\rho = (1 - \mathcal{T})/\mathcal{T}$ has an additive property as

$$\rho_{12} = \rho_1 + \rho_2 \tag{2.219}$$

with $\rho(L) = L/L_0$, where we take the average of $\rho = (1 - \langle\mathcal{T}\rangle)/\langle\mathcal{T}\rangle$. In the *quantum* diffusive regime, we need to take the average for the dimensionless resistance itself

$$\begin{aligned}\rho_{12} &= \left\langle \frac{1-\mathcal{T}}{\mathcal{T}} \right\rangle \\ &= \int \frac{1+\mathcal{R}_1\mathcal{R}_2 - 2\sqrt{\mathcal{R}_1\mathcal{R}_2}\cos\theta - \mathcal{T}_1\mathcal{T}_2}{\mathcal{T}_1\mathcal{T}_2}\frac{d\theta}{2\pi} \\ &= \frac{1+\mathcal{R}_1\mathcal{R}_2 - \mathcal{T}_1\mathcal{T}_2}{\mathcal{T}_1\mathcal{T}_2}, \end{aligned} \tag{2.220}$$

which results in the relation

$$\rho_{12} = \rho_1 + \rho_2 + 2\rho_1\rho_2. \tag{2.221}$$

We see that ρ_{12} is not additive anymore. The last term $2\rho_1\rho_2$ appears due to the phase coherence, which produces the exponential divergence of ρ as a function of L.

Let us derive an expression for the resistance as a function of the system length L in this case. When we add a short segment with the length of ΔL to a segment L, we have $\rho_{12} = \rho(L + \Delta L)$, $\rho_1 = \rho(L)$, and $\rho_2 = \Delta L / L_0$. Then we have

$$\frac{\rho(L+\Delta L) - \rho(L)}{\Delta L} = \frac{d\rho(L)}{dL} = \frac{1+2\rho(L)}{L_0}, \tag{2.222}$$

which is solved as

$$\rho(L) = \frac{1}{2}\left[e^{2L/L_0} - 1\right] \to \infty \quad (L \to \infty). \tag{2.223}$$

This shows that the resistance becomes infinite and the material becomes an insulator in the limit of large system size L. This is because the wavefunction near the Fermi energy becomes localized due to the interference of the phase from scatterings with randomly distributed impurities. Then the transmission probability from one electrode to the other becomes zero. The length L_0 is characterized by the localization length ξ. This is called the Anderson localization [14].[a]

2.2.6.1 Strong localization and weak localization

When the localization length ξ is much smaller than the phase coherence length L_ϕ as $\xi \ll L_\phi$, then the strong localization effect appears and the resistance become exponentially divergent as

$$\rho(L) \sim e^{2L/\xi}. \tag{2.224}$$

---

[a]This is different from the ordinary insulator due to the band gap in the band structure. There is also another insulator called the Mott insulator, where electrons cannot move due to the strong Coulomb interaction.

The system is in the insulating regime with $\langle G \rangle \ll 2e^2/h$. This implies that the wavefunction becomes localized, in an average sense, like

$$|\Psi(\mathbf{r})| \sim \mathrm{e}^{-|\mathbf{r}-\mathbf{r}_0|/\xi} \tag{2.225}$$

with an exponential envelope. In this strong localization regime, the transport of electrons arises from the thermal hopping from one localized site to another with a thermal activation factor.

Let us estimate the conductivity for hopping regime. Since the overlap of wavefunctions are proportional to $\exp(-2r/\xi)$, the conductivity due to hopping should behave as

$$\sigma \sim \sigma_0 \, \mathrm{e}^{-2r/\xi - \Delta\epsilon/k_B T}, \tag{2.226}$$

which is schematically shown in Fig. 1.6 (left). Since the electronic states behaves randomly $\Delta\epsilon \sim 1/\nu(\epsilon)r^d$ as r changes in the disordered systems, the temperature dependence of conductivity for d dimensional system is estimated by

$$\sigma \sim \exp\left[-(T_0/T)^{1/(d+1)}\right]. \tag{2.227}$$

This formula represents the transport behaviors in Mott's variable-range hopping regime. Here we don't consider phonon effects.[a]

In the strong localization regime, we note that the coupling of electrons with phonons forming polaronic states becomes important. Carrier transport due to the phonon-assisted thermal hopping mechanism is observed in the heavily doped semiconductor and in the organic semiconductor materials.[b]

On the other hand, when the localization length ξ is much larger than the phase coherence length L_ϕ as $\xi \gg L_\phi$, then the system is called the weak localization regime. In this regime, we can write the resistance Eq. (2.223) as

$$\rho(L) = \frac{L}{\xi} + \left(\frac{L}{\xi}\right)^2 + \cdots. \tag{2.228}$$

[a] There is also the Coulomb gap problem in the insulating regime.

[b] Recent rapid progress of the fabrication technologies for the single crystal of organic semiconductors makes it possible to observe the mobility of transport larger than that of amorphous silicons. In such situations, there is a big problem as to the localization nature of wavefunctions whether it is localized for hopping conduction or extended for band-like conduction. These problems are treated in Chapter 5 briefly.

The first term is just the classical Ohm's resistance where the resistance is proportional to the system size. Since the system size L is much less than ξ, this system is in the metallic regime with $\langle G \rangle \gg 2e^2/h$. The second term represents the deviation from the Ohm's law, which appears due to the quantum interference and is called the quantum correction. Since the correction in the resistance is proportional to the square of the resistance, we have $\Delta(1/\rho) = -\Delta\rho/\rho^2 \sim -1$. This means that the quantum correction to the conductance is of the order of

$$\Delta G \sim -\frac{2e^2}{h}. \tag{2.229}$$

This shows that the average quantum conductance due to interference effect is always lower than the classical conductance in the weak localization regime.[a]

The quantum correction to the conductance in the weak localization regime is well understood when we consider the wavefunction $\Psi(m \to n)$ for the path from site m to site n. The contribution from the phase of various paths in the ensemble average due to interference effects cancels out usually for $|\Psi(m \to n)|^2$. However, there is a special path when the initial and final sites are the same. Consider a process of $m \to m_1 \cdots \to m_q \to m$ and its time-reversed process as $m \to m_q \cdots \to m_1 \to m$. These have the identical phases. When the phase coherence is preserved, this becomes $|\Psi(m \to m)|^2 = 4|A|^2$ where A is the amplitude of the wavefunction. On the other hand, when the phase coherence is lost, it becomes $|\Psi(m \to m)|^2 = 2|A|^2$. Thus, the coherence between the pairs of time-reversed paths gives the doubling of the probability to go back to the same site. This is known as the enhanced backscattering, implying a reduced probability to find an electron at the destination and therefore a decrease of the conductance. A magnetic field, which puts an extra phase $\exp[i(e/c\hbar)\int A(\mathbf{r})d\mathbf{r}]$ to the wavefunction, serves to destroy the coherence and the weak localization. This is observed as a negative magnetoresistance, that is, a magnetic field contributes to enhance the conductance in contrast to the classical expectation.

---

[a] The transition from metallic to insulating regimes due to disorder is described by the scaling theory [15].

2.2.6.2 Universal conductance fluctuations

In the weak localization regime, the magnitude of fluctuation of conductance has a striking property due to coherence effects known as the universal conductance fluctuation (UCF). As we see, for weak disorder in the metallic regime, the average conductance is $\langle G\rangle \gg 2e^2/h$ and the quantum correction becomes $\Delta G \sim -2e^2/h$. The fluctuation of conductance in this regime has a property as

$$\langle(\delta G)^2\rangle \equiv \langle G^2\rangle - \langle G\rangle^2 \sim \left(\frac{2e^2}{h}\right)^2. \tag{2.230}$$

The conductance fluctuates from sample to sample because of differences of the distribution of disorder impurities. The above relation implies that magnitude of the fluctuation $\sqrt{\langle(\delta G)^2\rangle}$ is of the order of $2e^2/h$ [16].

Let us derive this property phenomenologically. We assume the number of channels is $M(\gg 1)$ in the metallic regime. First we note that if these channels are independent, the fluctuation $\sqrt{\langle(\delta G)^2\rangle}/\langle G\rangle$ for M channel should be proportional to $\sim 1/\sqrt{M}$ and thus becomes zero for $M \gg 1$. Therefore, the property $\sqrt{\langle(\delta G)^2\rangle}/\langle G\rangle \sim 1$ in Eq. (2.230) indicates that there is any correlation between the channels. From the Landauer formula, since the dimensionless conductance $g = G/(2e^2/h)$ becomes $\langle g\rangle = M - \langle\mathcal{R}\rangle$ and thus $\langle g^2\rangle = M^2 - 2M\langle\mathcal{R}\rangle + \langle\mathcal{R}^2\rangle$, the fluctuation of conductance $\langle(\delta g)^2\rangle = \langle g^2\rangle - \langle g\rangle^2$ is expressed by

$$\langle(\delta g)^2\rangle = \langle\mathcal{R}^2\rangle - \langle\mathcal{R}\rangle^2 = M^2\langle\delta\mathcal{R}^2\rangle \tag{2.231}$$

where $\mathcal{R}$ is the reflection matrix with $M \times M$ dimensions. Here the average quantity of the resistance $\mathcal{R}$ is written as

$$\langle\mathcal{R}\rangle = \langle|\Psi(m \to n)|^2\rangle = \sum_{p,p'}\langle A_p A_{p'}^*\rangle = \sum_{p}\langle|A_p|^2\rangle \tag{2.232}$$

and

$$\begin{aligned}\langle\mathcal{R}^2\rangle &= \sum_{p,p',p'',p'''}\langle A_p A_{p'} A_{p''}^* A_{p'''}^*\rangle \\ &= \sum_{p,p',p'',p'''}\langle|A_p|^2|A_{p'}|^2\rangle\left[\delta_{p,p''}\delta_{p',p'''} + \delta_{p,p'''}\delta_{p',p''}\right] \\ &= 2\langle\mathcal{R}\rangle^2,\end{aligned} \tag{2.233}$$

where A_p is the amplitude of the wavefunction for the p-th path with multiple scattering processes. Therefore, we have

$$\langle \delta \mathcal{R}^2 \rangle = \langle \mathcal{R} \rangle^2. \tag{2.234}$$

Since the probability to choose one channel from M channels is $\langle \mathcal{R} \rangle \sim 1/M$, we finally have

$$\langle (\delta g)^2 \rangle = M^2 \langle \mathcal{R} \rangle^2 \approx 1. \tag{2.235}$$

This shows that the fluctuation of conductance in the weak localization regime is of the order of quantum conductance

$$\sqrt{\langle (\delta G)^2 \rangle} \sim \frac{2e^2}{h}, \tag{2.236}$$

which does not depend on the channel number M or any particular choice of material parameters. This fluctuation cannot be avoided when the phase coherence is preserved over the sample as $L \ll L_\phi$ and changes sample to sample with different impurities. Since this fluctuation is reproducible as a function of an external parameter such as magnetic field or Fermi energy, it represents a fingerprint of a realization of disorder.

We note that the fluctuation of conductance is smoothed out as the temperature increases. One reason is because the phase coherence length L_ϕ becomes smaller due to electron–phonon interaction at higher temperatures. Another reason is the energy-averaging. Since the current flows over an energy range of a few $k_B T$ around the Fermi energy, this energy range is regarded as $N(= k_B T/(\hbar/\tau_\phi))$ uncorrelated units coupled in series for $k_B T > \hbar/\tau_\phi$, where τ_ϕ is a lifetime of an electron with phase-coherent motion. For ensembles of a large number N of phase-coherent units, the fluctuation tends to be reduced and smoothed out. In chapter 6, we treat the fluctuation of conductance in view of the energy spectrum, where the spectral correlation manifests as a repulsion of the energy levels.

2.3 Problems

(1) Consider the quantum transport of an electron to transmit the potential barrier.

(a) First let us treat the potential step in one-dimension $V(x)$ with V_L for $x < 0$ and V_R for $x > 0$. For an energy $\epsilon > \max\{V_L, V_R\}$ where an electron is incident either from the left or from the right, write down the transfer matrix M from the Schrödinger equations, then calculate the transmission $\mathcal{T}(\epsilon)$ as a function of the energy ϵ.

(b) Next consider the square potential barrier $V(x)$ with V_0 for $|x| < a$ and 0 for $|x| > a$. For an energy $\epsilon < V_0$, write down the transfer matrix $M = M(L \leftarrow C)M(C \leftarrow R)$ from the Schrödinger equations, then calculate the transmission $\mathcal{T}(\epsilon)$ as a function of the energy ϵ.

(c) Construct the S-matrix for the square potential and relate it with the transfer matrix M.

(2) When there is a temperature gradient in an external electric field, the electric current accompanies the thermal current in metallic systems. Let us derive the relationship between the electric and thermal currents within the Boltzmann approximation.

(a) In an electric field E with a temperature gradient of $\partial T/\partial x$, write down the Boltzmann equation within a relaxation approximation for the electric current density j and the thermal current density j_q.

(b) Calculate the electrical conductivity from $\sigma = j/E$, the Seebeck coefficient from $S = \partial V/\partial T$, and the thermal conductivity from $\kappa = j_q/(-\partial T/\partial x)$.

(c) The heat transfer density Q in the system is given by

$$Q = jE + \frac{\partial j_q}{\partial x}.$$

Write down the heat transfer density Q using σ and κ. These become the basic formulas for the thermoelectric relations between electric and thermal currents.

(3) Let us derive the Kubo formula by evaluating energy absorption rate P due to electronic transitions induced by an applied field,

$$P = \langle \mathbf{E} \cdot \mathbf{J} \rangle = \lim_{\omega \to 0} \mathrm{Re}\,[\sigma(\omega)]\,|E|^2$$

where $\mathrm{Re}\,[\sigma(\omega)]$ determines the dissipative part of the current.

(a) Evaluate P within the lowest order perturbation theory for summing all possibilities of the transition rate $P_{\alpha\beta}$, which is expressed by the multiplication of an energy absorbed in the system for the field-induced potential $V = eEz$.

(b) Drive the formula for the conductivity Re $[\sigma(\omega)]$. Then show that the conductivity σ in the limit of $\omega \to 0$ is expressed by

$$\sigma = 2\pi\hbar\left(\frac{e}{m}\right)^2 \int \sum_{\alpha\beta} |\langle\alpha|p_z|\beta\rangle|^2 \times \left(-\frac{\partial f}{\partial \epsilon}\right)\delta\left(\epsilon - \epsilon_\alpha\right)\delta\left(\epsilon - \epsilon_\beta\right)d\epsilon,$$

which reveals the Kubo formula for the conductivity to the equilibrium current fluctuations.

(4) Let us consider a particle with a coordinate and a momentum of $(x,\ p)$ in the external potential $V(x)$ coupled to the harmonic oscillator bath $(x_i,\ p_i)$. The Hamiltonian is written as

$$H = \frac{p^2}{2m} + V(x) + \sum_i \left[\frac{p_i^2}{2m_i} + \frac{1}{2}m_i\omega_i^2\left(x_i - \frac{\gamma_i}{m_i\omega_i^2}x\right)^2\right]$$

where ω_i is the frequency of the i-th oscillator and γ_i measures the strength of linear coupling of a particle to the i-th oscillator.

(a) Construct the equations of motion for x and x_i.

(b) Obtain $x_i(t)$ with an initial condition of $x_i(0)$ and $p_i(0)$.

(c) Construct the generalized Langevin equation for a particle

$$m\frac{dv(t)}{dt} = -m\int_{-\infty}^{t} \gamma(t-t')v(t')dt' + \eta(t) + F$$

with external force $F = -\partial V(x)/\partial x$, friction function $\gamma(t)$

$$\gamma(t) = \Theta(t)\frac{1}{m}\sum_i \frac{\gamma_i^2}{m_i\omega_i^2}\cos(\omega_i t),$$

and fluctuating force $\eta(t)$ as

$$\eta(t) = \sum_i \gamma_i \left[\left(x_i(0) - \frac{\gamma_i}{m_i\omega_i^2}x(0)\right)\cos(\omega_i t) + \frac{p_i(0)}{m\omega_i}\sin(\omega_i t)\right].$$

(d) Show the fluctuation–dissipation relation

$$\langle \eta(t) \rangle = 0, \quad \langle \eta(t)\eta(t') \rangle = mk_B T \gamma(t - t'),$$

for the initial distribution of $f_{\rm eq} = \exp(-H_B^{(0)}/k_B T)$ where the initial Hamiltonian of the bath is given by

$$H_B^{(0)} = \sum_i \left[\frac{p_i(0)^2}{2m_i} + \frac{1}{2} m_i \omega_i^2 \left(x_i(0) - \frac{\gamma_i}{m_i \omega_i^2} x(0) \right)^2 \right].$$

(5) Let us consider the continuous version of master equation for the probability $P(v, t)$ as

$$\frac{\partial P(v, t)}{\partial t} = -\int W(v \to v') P(v, t) dv' + \int W(v' \to v) P(v', t) dv'.$$

(a) For the Kramers–Moyal expansion of

$$\frac{\partial P(v, t)}{\partial t} = \sum_{n=0}^{\infty} \frac{(-1)^n}{n!} \left(\frac{\partial}{\partial v} \right)^n C_n(v) P(v, t),$$

show that $C_0(v) = 0$ and $C_{n\geq 1}(v) = \int W(v \to v + r) r^n dr$. In case of the Langevin equation Eq. (2.37) without external force $\mathbf{F} = 0$, derive $C_1 = -\gamma v$, $C_2 = 2k_B T \gamma / m$, and $C_{n\geq 3} = 0$. Show that the obtained Fokker–Planck equation has the stationary solution $P_{eq}(v) = e^{-mv^2/2k_B T}$ of the Maxwell–Boltzmann distribution for $\partial P / \partial t = 0$.

(b) Consider the external force is present as $F = -\partial V / \partial x$. For large friction γ, the Langevin equation approaches $dv/dt \to 0$. Show that the Fokker–Planck equation is expressed by

$$\frac{\partial P(v, t)}{\partial t} = \frac{1}{m\gamma} \frac{\partial}{\partial x} \left[\frac{\partial V}{\partial x} + k_B T \frac{\partial}{\partial x} \right] P(v, t),$$

which is called the Smoluchowski equation.
Defining the probability current J by $\partial P / \partial t = -\partial J / \partial x$, show $P \sim e^{-V/k_B T}$ when the probability current $J = 0$. Derive the formula of J at the saddle point of potential V.

Chapter 3

Green's Function Techniques for Electron Transport

In the previous chapter, we introduced the basic formalisms of quantum transport calculations. The Kubo formula is derived from particle description and the Landauer formula is derived from wave description. These are expressed in mathematics using Green's functions in the framework of quantum field approach, which describes both particle and wave natures in the second quantization operators. Linear-response conductance is described by the equilibrium Green's functions, while we need nonequilibrium Green's function (NEGF) to describe the transport with a finite bias voltage.

In this chapter, we introduce Green's functions for the mathematical tools to describe the quantum transport. We derive the formula of linear-response conductance in the particle–hole propagation by the retarded Green's function and express the nonlinear conductance using the NEGF formulation. The temperature Green's function in imaginary-time formalism, which is related to the retarded Green's function by the analytical continuation, is briefly described in the problem. We note that there are a number of excellent textbooks for these Green's function techniques [17–28].

Quantum Transport Calculations for Nanosystems
Kenji Hirose and Nobuhiko Kobayashi

ISBN 978-981-4267-32-8 (Hardcover), 978-981-4267-59-5 (eBook)
www.panstanford.com

3.1 Equilibrium Green's Function

3.1.1 *Second Quantization for Many-Electron Systems*

Let us consider a system composed of N interacting electrons. First, the system satisfies the Fermi–Dirac statistics since electrons are fermions. The wavefunction is created from a fully anti-symmetric nature for an exchange of the coordinates of any two electrons, which is satisfied by introducing the Slater determinant of $N \times N$ matrix with single-particle states $\psi_{k_j}(\mathbf{r}_i)$ such as

$$\Psi_0^{\mathcal{K}}(\mathbf{r}_1, \mathbf{r}_2, \cdots, \mathbf{r}_N) = \frac{1}{\sqrt{N!}} \sum_P (-1)^P P \prod_{i=1}^{N} \psi_{k_i}(\mathbf{r}_i)$$

$$= \frac{1}{\sqrt{N!}} \begin{vmatrix} \psi_{k_1}(\mathbf{r}_1) & \psi_{k_1}(\mathbf{r}_2) & \dots & \psi_{k_1}(\mathbf{r}_N) \\ \psi_{k_2}(\mathbf{r}_1) & \psi_{k_2}(\mathbf{r}_2) & \dots & \psi_{k_2}(\mathbf{r}_2) \\ \vdots & \vdots & \ddots & \vdots \\ \psi_{k_N}(\mathbf{r}_1) & \psi_{k_N}(\mathbf{r}_2) & \dots & \psi_{k_N}(\mathbf{r}_N) \end{vmatrix}. \tag{3.1}$$

Next, we sum up all the sets of Slater determinants corresponding to all the configurations of electronic states $\mathcal{K}$

$$\Psi(\mathbf{r}_1, \mathbf{r}_2, \cdots, \mathbf{r}_N) = \sum_{\text{all config.}\mathcal{K}_i} C_{\mathcal{K}_i} \Psi_0^{\mathcal{K}_i}(\mathbf{r}_1, \mathbf{r}_2, \cdots, \mathbf{r}_N). \tag{3.2}$$

This is the general many-body state of the system where N electrons interact with each other within an external field. We will show in Chapter 4 how to derive an effective single-particle equation to calculate electronic structures for the realistic materials and thereby to obtain transport properties from an atomistic point of views. Here we consider the formulation for the calculations using Green's function. This helps us understand the fundamental physics of quantum transport and treat many-body interaction effects beyond the mean field approximation systematically.

Instead of treating the Slater determinant explicitly, we can represent many-body state by the number of particles in a specific single-particle state $\psi_k(\mathbf{r})$, since electrons are indistinguishable. When we denote the occupation number n_k as a number of particles in $\psi_k(\mathbf{r})$, the many-body state is represented by

$$|n_{k_1}, n_{k_2}, \cdots, n_{k_N}\rangle, \qquad n_{k_i} = 0, 1 \tag{3.3}$$

with the total sum of $\sum_{k_i} n_{k_i} = N$.

Now we introduce the creation and annihilation operators $\hat{c}_k^\dagger$ and $\hat{c}_k$, which are defined by

$$\hat{c}_{k_i}|n_{k_1}, \cdots, n_{k_i}, \cdots\rangle = (-1)^{s_i}\sqrt{n_{k_i}}|n_{k_1}, \cdots, n_{k_i} - 1, \cdots\rangle,$$
$$\hat{c}_{k_i}^\dagger|n_{k_1}, \cdots, n_{k_i}, \cdots\rangle = (-1)^{s_i}\sqrt{1 - n_{k_i}}|n_{k_1}, \cdots, n_{k_i} + 1, \cdots\rangle$$

with $s_i = \sum_{j=1}^{i-1} n_{k_j}$. Then many-body state is constructed from

$$|n_{k_1}, n_{k_2}, \cdots, n_{k_N}\rangle = (\hat{c}_{k_1}^\dagger)^{n_{k_1}}(\hat{c}_{k_2}^\dagger)^{n_{k_2}} \cdots (\hat{c}_{k_N}^\dagger)^{n_{k_N}}|0\rangle, \tag{3.4}$$

where $|0\rangle$ is the vacuum state with no electrons. The successive applications of creation and annihilation operators on the same state exchange the electrons and become

$$\hat{c}_{k_i}^\dagger\hat{c}_{k_i}|n_{k_1}, \cdots, n_{k_i}, \cdots\rangle = n_{k_i}|n_{k_1}, \cdots, n_{k_i}, \cdots\rangle$$
$$\hat{c}_{k_i}\hat{c}_{k_i}^\dagger|n_{k_1}, \cdots, n_{k_i}, \cdots\rangle = (1 - n_{k_i})|n_{k_1}, \cdots, n_{k_i}, \cdots\rangle, \tag{3.5}$$

which shows that the occupation number operator $\hat{n}_k$ is written as

$$\hat{n}_{k_i} = \hat{c}_{k_i}^\dagger\hat{c}_{k_i} \tag{3.6}$$

and the total number operator is

$$\hat{N} = \sum_{k_i}\hat{n}_{k_i} = \sum_{k_i}\hat{c}_{k_i}^\dagger\hat{c}_{k_i}. \tag{3.7}$$

From Eq. (3.5), it is easy to show that creation and annihilation operators obey

$$\left[\hat{c}_k, \hat{c}_{k'}^\dagger\right] = \hat{c}_k\hat{c}_{k'}^\dagger + \hat{c}_{k'}^\dagger\hat{c}_k = \delta_{k,k'},$$
$$[\hat{c}_k, \hat{c}_{k'}] = 0, \quad \left[\hat{c}_k^\dagger, \hat{c}_{k'}^\dagger\right] = 0, \tag{3.8}$$

where $[\cdots]$ denotes the anti-commutator relation.

Here we introduce field operators in the position representation

$$\hat{\psi}(\mathbf{r}) = \sum_k \hat{c}_k\varphi_k(\mathbf{r}), \quad \hat{\psi}^\dagger(\mathbf{r}) = \sum_k \hat{c}_k^\dagger\varphi_k^*(\mathbf{r}), \tag{3.9}$$

where $\varphi_k(\mathbf{r})$ is single-particle wavefunctions to form a complete set of eigenfunctions with quantum labels k. The meaning of these operators are $\hat{\psi}^\dagger(\mathbf{r})$ creates a particle at position $\mathbf{r}$ and $\hat{\psi}(\mathbf{r})$

annihilates a particle at position $\mathbf{r}$. Then by using the commutator relations for $\hat{c}_k^\dagger$ and $\hat{c}_k$, the commutator relations for the field operators become

$$\left[\hat{\psi}(\mathbf{r}), \hat{\psi}^\dagger(\mathbf{r}')\right] = \sum_{k,k'} \left[\hat{c}_k, \hat{c}_{k'}^\dagger\right] \varphi_k(\mathbf{r})\varphi_{k'}^\dagger(\mathbf{r}') = \delta(\mathbf{r}-\mathbf{r}')$$

$$\left[\hat{\psi}(\mathbf{r}), \hat{\psi}(\mathbf{r}')\right] = 0, \quad \left[\hat{\psi}^\dagger(\mathbf{r}), \hat{\psi}^\dagger(\mathbf{r}')\right] = 0, \tag{3.10}$$

and single-particle number operator is written as

$$\hat{n}(\mathbf{r}) = \hat{\psi}^\dagger(\mathbf{r})\hat{\psi}(\mathbf{r}) = \sum_{k,k'} \varphi_k^\dagger(\mathbf{r})\varphi_{k'}(\mathbf{r})\hat{c}_{k'}^\dagger\hat{c}_k. \tag{3.11}$$

The Hamiltonians for the kinetic part $\hat{K}$, one-body operator part $\hat{V}$, and two-body interaction operator part $\hat{U}$ are expressed using these field operators by

$$\hat{K} = \int \hat{\psi}^\dagger(\mathbf{r})\left[-\frac{\hbar^2}{2m}\nabla^2\right]\hat{\psi}(\mathbf{r})d\mathbf{r}$$

$$\hat{V} = \int \hat{\psi}^\dagger(\mathbf{r})V(\mathbf{r})\hat{\psi}(\mathbf{r})d\mathbf{r} = \sum_{k,k'} V_{k,k'}\hat{c}_k^\dagger\hat{c}_{k'}$$

$$\hat{U} = \int\int \hat{\psi}^\dagger(\mathbf{r})\hat{\psi}^\dagger(\mathbf{r}')U(\mathbf{r},\mathbf{r}')\hat{\psi}(\mathbf{r}')\hat{\psi}(\mathbf{r})d\mathbf{r}d\mathbf{r}'$$

$$= \sum_{k,k',k'',k'''} U_{k,k':k'',k'''}\hat{c}_k^\dagger\hat{c}_{k'}^\dagger\hat{c}_{k''}\hat{c}_{k'''}, \tag{3.12}$$

where we define

$$V_{k,k'} = \langle k|V|k'\rangle = \int \varphi_k^\dagger(\mathbf{r})V(\mathbf{r})\varphi_{k'}(\mathbf{r})d\mathbf{r}$$

$$U_{k,k',k'',k'''} = \langle k,k'|U|k'',k'''\rangle$$

$$= \int\int \varphi_k^\dagger(\mathbf{r})\varphi_{k'}^\dagger(\mathbf{r}')U(\mathbf{r},\mathbf{r}')\varphi_{k''}(\mathbf{r}')\varphi_{k'''}(\mathbf{r})d\mathbf{r}d\mathbf{r}'.$$

The Coulomb energy is two-body interaction $U(\mathbf{r},\mathbf{r}') = e^2/|\mathbf{r}-\mathbf{r}'|$.

For the boson such as phonons, the state is expressed as

$$|n_{k_1}, n_{k_2}, \cdots, n_{k_N}\rangle, \qquad n_{k_i} = 0, 1, 2, 3, \cdots \tag{3.13}$$

and the creation and annihilation operators $\hat{a}_k^\dagger$, $\hat{a}_k$ are defined by

$$\hat{a}_{k_i}|n_{k_1}, \cdots, n_{k_i}, \cdots\rangle = \sqrt{n_{k_i}}|n_{k_1}, \cdots, n_{k_i}-1, \cdots\rangle,$$

$$\hat{a}_{k_i}^\dagger|n_{k_1}, \cdots, n_{k_i}, \cdots\rangle = \sqrt{n_{k_i}+1}|n_{k_1}, \cdots, n_{k_i}+1, \cdots\rangle$$

with the relations of

$$\left[\hat{a}_k, \hat{a}_{k'}^\dagger\right]_- = \hat{a}_k\hat{a}_{k'}^\dagger - \hat{a}_{k'}^\dagger\hat{a}_k = \delta_{k,k'},$$

$$\left[\hat{a}_k, \hat{a}_{k'}\right]_- = 0, \quad \left[\hat{a}_k^\dagger, \hat{a}_{k'}^\dagger\right]_- = 0, \tag{3.14}$$

where $[\cdots]_-$ denotes the commutator relation.

3.1.2 *Green's Function*

Here we define the single-particle time-ordered Green's function for electrons

$$G(\mathbf{r}, t : \mathbf{r}', t') = -\frac{i}{\hbar}\langle T \left\{\hat{\psi}(\mathbf{r}, t)\hat{\psi}^{\dagger}(\mathbf{r}', t')\right\}\rangle, \tag{3.15}$$

where field operators are written in the Heisenberg representation

$$\begin{aligned}\hat{\psi}(\mathbf{r}, t) &= e^{i\hat{H}t/\hbar}\hat{\psi}(\mathbf{r})e^{-i\hat{H}t/\hbar}\\ \hat{\psi}^{\dagger}(\mathbf{r}, t) &= e^{-i\hat{H}t/\hbar}\hat{\psi}^{\dagger}(\mathbf{r})e^{i\hat{H}t/\hbar}\end{aligned} \tag{3.16}$$

and the time-ordered operator T is defined by

$$\begin{aligned}T\left\{\hat{\psi}(\mathbf{r}, t)\hat{\psi}^{\dagger}(\mathbf{r}', t')\right\} &= \theta(t - t')\hat{\psi}(\mathbf{r}, t)\hat{\psi}^{\dagger}(\mathbf{r}', t')\\ &\quad - \theta(t' - t)\hat{\psi}^{\dagger}(\mathbf{r}', t')\hat{\psi}(\mathbf{r}, t).\end{aligned} \tag{3.17}$$

With this expression, the number of electrons is written as

$$\begin{aligned}n(\mathbf{r}) = \langle\hat{n}(\mathbf{r})\rangle &= -i\hbar G(\mathbf{r}, t : \mathbf{r}, t + \epsilon)\\ &= \sum_k f(\epsilon_k)|\varphi_k(\mathbf{r})|^2.\end{aligned} \tag{3.18}$$

Here $f(\epsilon_k)$ is the Fermi–Dirac distribution function

$$f(\epsilon_k) = \langle\hat{c_k}^{\dagger}\hat{c_k}\rangle = \frac{1}{e^{(\epsilon_k-\mu)/k_BT} + 1}, \tag{3.19}$$

which determines the energy-dependent occupation of electrons with respect to the chemical potential μ.

Let us derive the equation of motion for Green's function $G(\mathbf{r}, t : \mathbf{r}', t')$. Since $\partial\theta(t - t')/\partial t = \delta(t - t')$ and $\partial\theta(t' - t)/\partial t = -\delta(t - t')$, the derivative of Green's function becomes

$$\begin{aligned}i\hbar\frac{\partial G(\mathbf{r}, t : \mathbf{r}', t')}{\partial t} &= \delta(t - t')\langle\left[\hat{\psi}(\mathbf{r}, t), \hat{\psi}(\mathbf{r}', t)\right]\rangle\\ &\quad - \frac{i}{\hbar}\langle T\left\{i\hbar\frac{\partial\hat{\psi}(\mathbf{r}, t)}{\partial t}\hat{\psi}^{\dagger}(\mathbf{r}', t)\right\}\rangle.\end{aligned} \tag{3.20}$$

When we consider a single-particle electron case, the Hamiltonian becomes

$$\hat{H} = \int\hat{\psi}^{\dagger}(\mathbf{r}, t)\left(-\frac{\hbar^2}{2m}\nabla^2\right)\hat{\psi}(\mathbf{r}, t)d\mathbf{r} + \int\hat{\psi}^{\dagger}(\mathbf{r}, t)V(\mathbf{r})\hat{\psi}(\mathbf{r}, t)d\mathbf{r} \tag{3.21}$$

and thus the equation of motion is

$$i\hbar\frac{\partial\hat{\psi}(\mathbf{r},t)}{\partial t} = \left[\hat{\psi}(\mathbf{r},t), \hat{H}\right] = \left(-\frac{\hbar^2\nabla^2}{2m} + V(\mathbf{r})\right)\hat{\psi}(\mathbf{r},t). \quad (3.22)$$

We put this into Eq. (3.20) and obtain the equation for Green's function to satisfy

$$\left(i\hbar\frac{\partial}{\partial t} + \frac{\hbar^2\nabla^2}{2m} - V(\mathbf{r})\right)G(\mathbf{r},t:\mathbf{r}',t') = \delta(t-t')\delta(\mathbf{r}-\mathbf{r}'). \quad (3.23)$$

If the system is in equilibrium and not time dependent, Green's function depends only on the time difference $t - t'$ and we can use the Fourier space to represent

$$G(\mathbf{r},t:\mathbf{r}',t') = \frac{1}{2\pi\hbar}\int G(\mathbf{r},\mathbf{r}')e^{-i\epsilon(t-t')/\hbar}d\epsilon. \quad (3.24)$$

Since the delta function is expressed by

$$\delta(t-t') = \frac{1}{2\pi\hbar}\int e^{-i\epsilon(t-t')/\hbar}d\epsilon, \quad (3.25)$$

we can show that

$$\left(\epsilon + \frac{\hbar^2\nabla^2}{2m} - V(\mathbf{r})\right)G(\mathbf{r},\mathbf{r}',\epsilon) = \delta(\mathbf{r}-\mathbf{r}'), \quad (3.26)$$

which reduces to

$$(\epsilon - H)\,G(\mathbf{r},\mathbf{r}',\epsilon) = \delta(\mathbf{r}-\mathbf{r}'). \quad (3.27)$$

Next we define the different Green's functions, the retarded Green's function G^r, and the advanced Green's function G^a by

$$G^r(\mathbf{r},t:\mathbf{r}',t') = -\frac{i}{\hbar}\theta(t-t')\langle\left[\hat{\psi}(\mathbf{r},t), \hat{\psi}^\dagger(\mathbf{r}',t')\right]\rangle$$

$$G^a(\mathbf{r},t:\mathbf{r}',t') = \frac{i}{\hbar}\theta(t'-t)\langle\left[\hat{\psi}(\mathbf{r},t), \hat{\psi}^\dagger(\mathbf{r}',t')\right]\rangle. \quad (3.28)$$

It should be noted that the following relation is satisfied:

$$G^r(\mathbf{r},t:\mathbf{r}',t') = \left[G^a(\mathbf{r}',t':\mathbf{r},t)\right]^*. \quad (3.29)$$

We can show that $G^{r,a}(\mathbf{r},t : \mathbf{r}',t')$ have analytic structure in one half-plane and are well suited for calculating a physical response. Information about spectral properties, densities of states, and scattering rates are contained in $G^{r,a}(\mathbf{r},t:\mathbf{r}',t')$.

The equations of motions for these Green's functions become

$$i\hbar\frac{\partial G^{r,a}(\mathbf{r},t:\mathbf{r}',t')}{\partial t} = \delta(t-t')\langle[\hat{\psi}(\mathbf{r},t),\hat{\psi}(\mathbf{r}',t)]\rangle$$
$$\mp\frac{i}{\hbar}\theta\left[\pm(t-t')\right]\langle\left[i\hbar\frac{\partial\hat{\psi}(\mathbf{r},t)}{\partial t},\hat{\psi}^\dagger(\mathbf{r}',t)\right]\rangle, \tag{3.30}$$

which reduce to

$$\left(i\hbar\frac{\partial}{\partial t}+\frac{\hbar^2\nabla^2}{2m}-V(\mathbf{r})\pm i0^+\right)G^{r,a}(\mathbf{r},t:\mathbf{r}',t') = \delta(t-t')\delta(\mathbf{r}-\mathbf{r}'). \tag{3.31}$$

If the system is in equilibrium and not time dependent, we can use the Fourier space to represent

$$\left(\epsilon - H \pm i0^+\right)G^{r,a}(\mathbf{r},\mathbf{r}',\epsilon) = \delta(\mathbf{r}-\mathbf{r}'), \tag{3.32}$$

and therefore the retarded and the advanced Green's functions are obtained from

$$G^{r,a}(\mathbf{r},\mathbf{r}',\epsilon) = \sum_k \frac{\varphi_k^*(\mathbf{r}')\varphi_k(\mathbf{r})}{\epsilon-\epsilon_k\pm i0^+}. \tag{3.33}$$

With these formulas, the local density of states (LDOS) defined by

$$\nu(\mathbf{r},\epsilon) = \sum_k |\varphi_k(\mathbf{r})|^2\delta(\epsilon-\epsilon_k) \tag{3.34}$$

is expressed as

$$\nu(\mathbf{r},\epsilon) = -\frac{1}{\pi}\mathrm{Im}\sum_k\frac{|\varphi_k(\mathbf{r})|^2}{\epsilon-\epsilon_k+i0^+}$$
$$= -\frac{1}{\pi}\mathrm{Im}G^r(\mathbf{r},\mathbf{r},\epsilon). \tag{3.35}$$

Here we use the relation of

$$\frac{1}{\epsilon-\epsilon_k\pm i0^+} = P\left(\frac{1}{\epsilon-\epsilon_k}\right)\mp i\pi\delta(\epsilon-\epsilon_k). \tag{3.36}$$

Then the electron density becomes

$$n(\mathbf{r}) = \sum_k f(\epsilon_k)|\varphi_k(\mathbf{r})|^2 = \int f(\epsilon)\sum_k|\varphi_k(\mathbf{r})|^2\delta(\epsilon-\epsilon_k)d\epsilon$$
$$= \int f(\epsilon)\nu(\mathbf{r},\epsilon)d\epsilon = -\frac{1}{\pi}\mathrm{Im}\int f(\epsilon)G^r(\mathbf{r},\mathbf{r},\epsilon)d\epsilon. \tag{3.37}$$

So we calculate the density of states (DOS) $\nu(\epsilon)$ from

$$\nu(\epsilon) = \int \nu(\mathbf{r},\epsilon)d\mathbf{r} = -\frac{1}{\pi}\mathrm{Im}\int G^r(\mathbf{r},\mathbf{r},\epsilon)d\mathbf{r} \tag{3.38}$$

and the electron density $n(\mathbf{r})$ from the retarded Green's function.

3.1.3 *Dyson Equation*

To obtain Green's functions for the full Hamiltonian $\hat{H}$, formally it is possible to construct it from the partial Hamiltonian first and then the remaining interaction effects are included introducing the self-energy $\hat{\Sigma}(\epsilon)$ in the interaction representation.

We assume that the Hamiltonian $\hat{H}$ can be separated into an unperturbed part $\hat{H}_0$ and a perturbation $\hat{V}$ such as $\hat{H} = \hat{H}_0 + \hat{V}$. Usually $\hat{H}_0$ is taken so that we obtain the eigenvalues and eigenfunctions. Following Eq. (3.27) or Eq. (3.32), Green's function for $\hat{H}$ is defined by $\left[\epsilon\hat{I} - \hat{H}\right]\hat{G}(\epsilon) = \hat{I}$ and is constructed from the series expansion of $\hat{V}$ as

$$\begin{aligned}\hat{G}(\epsilon) &= \left(\epsilon\hat{I} - \hat{H}_0 - \hat{V}\right)^{-1} \\ &= \left\{\left(\epsilon\hat{I} - \hat{H}_0\right)\left[\hat{I} - (\epsilon\hat{I} - \hat{H}_0)^{-1}\hat{V}\right]\right\}^{-1} \\ &= \left[\hat{I} - \hat{G}_0\hat{V}\right]^{-1}\hat{G}_0(\epsilon) \\ &= \hat{G}_0(\epsilon) + \hat{G}_0(\epsilon)\hat{V}\hat{G}_0(\epsilon) + \hat{G}_0(\epsilon)\hat{V}\hat{G}_0(\epsilon)\hat{V}\hat{G}_0(\epsilon) + \cdots,\end{aligned} \tag{3.39}$$

where the operator $\left[\hat{I} - \hat{G}_0\hat{V}\right]^{-1}$ is expanded in the power series and we define $\hat{G}_0$ as $\left[\epsilon\hat{I} - \hat{H}_0\right]\hat{G}_0(\epsilon) = \hat{I}$. When we introduce the matrix $\hat{T}$ (called T-matrix) corresponding to the unperturbed Hamiltonian $\hat{H}_0$ as

$$\hat{T}(\epsilon) = \hat{V} + \hat{V}\hat{G}_0(\epsilon)\hat{V} + \hat{V}\hat{G}_0(\epsilon)\hat{V}\hat{G}_0(\epsilon)\hat{V} + \cdots, \tag{3.40}$$

Green's function for the full Hamiltonian $\hat{H}$ is given by

$$\hat{G}(\epsilon) = \hat{G}_0(\epsilon) + \hat{G}_0(\epsilon)\hat{T}(\epsilon)\hat{G}_0(\epsilon). \tag{3.41}$$

Here $\hat{T}(\epsilon)$ includes all the multiple scatterings from interaction $\hat{V}$. We note that the lowest term of $\hat{T}$ produces the Fermi golden rule.

Next we define the (irreducible) self-energy $\hat{\Sigma}(\epsilon)$ from $\hat{T}(\epsilon)$, whose parts include only the interaction that cannot be separated into two parts by the cut of a single line. Then, since all the multiple scattering effects are represented in the series of the partial irreducible term as

$$\hat{T}(\epsilon) = \hat{\Sigma}(\epsilon) + \hat{\Sigma}(\epsilon)\hat{G}_0(\epsilon)\hat{\Sigma}(\epsilon) + \hat{\Sigma}(\epsilon)\hat{G}_0(\epsilon)\hat{\Sigma}(\epsilon)\hat{G}_0(\epsilon)\hat{\Sigma}(\epsilon) + \cdots, \tag{3.42}$$

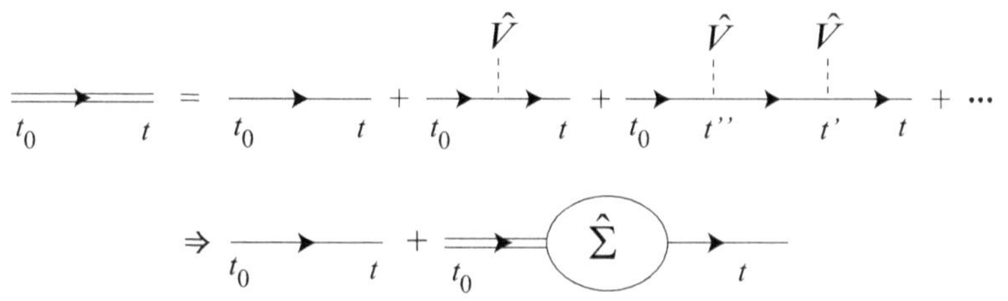

Figure 3.1 Schematic diagram of the perturbation expansion for Green's function and the construction of the Dyson equation. The double line represents the full Green's function $\hat{G}(\epsilon)$ for $\hat{H}$ and the single line represents $\hat{G}_0(\epsilon)$ for $\hat{H}_0$. Here the self-energy $\hat{\Sigma}$ is constructed only from the irreducible parts that cannot be separated into two parts by the cut of a single line.

we can rewrite Eq. (3.41) as

$$\begin{aligned}\hat{G}(\epsilon) &= \hat{G}_0(\epsilon) + \hat{G}_0(\epsilon)\left\{\hat{\Sigma}(\epsilon) + \hat{\Sigma}(\epsilon)\hat{G}_0(\epsilon)\hat{\Sigma}(\epsilon) + \cdots\right\}\hat{G}_0(\epsilon)\\ &= \hat{G}_0(\epsilon) + \hat{G}_0(\epsilon)\Sigma(\epsilon)\left\{\hat{G}_0(\epsilon) + \hat{G}_0(\epsilon)\hat{\Sigma}(\epsilon)\hat{G}_0(\epsilon) + \cdots\right\}\\ &= \hat{G}_0(\epsilon) + \hat{G}_0(\epsilon)\hat{\Sigma}(\epsilon)\hat{G}(\epsilon). \qquad (3.43)\end{aligned}$$

This is called the Dyson equation (Fig. 3.1) and Green's function is obtained from

$$\hat{G}(\epsilon) = \left[\hat{G}_0(\epsilon)^{-1} - \hat{\Sigma}(\epsilon)\right]^{-1}. \qquad (3.44)$$

It is important to note that introducing the *irreducible* matrix $\hat{\Sigma}(\epsilon)$, the double countings of the various interaction terms in Eq. (3.39) are avoided.

The Dyson equation is a general form and all the interaction effects are included in the self-energy term $\hat{\Sigma}(\epsilon)$. So to obtain Green's function for the full Hamiltonian $\hat{H}$ is equivalent to obtain the self-energy $\hat{\Sigma}(\epsilon)$. We give two examples for the transport problem.

One example is the isolated nanosystems, which couple to the left and right semi-infinite electrodes for the transport. We can include the electrodes effects as self-energy term $\hat{\Sigma}^{L/R}(\epsilon)$. Then the nanosystems involving electrodes are treated by solving the Dyson equation. This method, extended to nonequilibrium situations using the NEGF, is more precisely described in Chapter 4. Another example is when we consider two-body Coulomb interaction U described in Eq. (3.12) beyond the single-particle Hamiltonian Eq. (3.21). We can show that the equation motion for the single-particle Green's function Eq. (3.15) produces the two-particle Green's function and is not solved analytically. In the Dyson equation, it is formally included in $\hat{\Sigma}(\epsilon)$. There are various methods to treat the Coulomb interaction approximately, such as Hartree and Hartree–Fock approximations, which are also described in Chapter 4. As for the topic beyond the single-particle approximations, we describe the Kondo effect for the transport introducing the two-particle Green's function in Chapter 6.

To find Green's function $\hat{G}(t, t')$ or equivalently the self-energy $\hat{\Sigma}(t, t')$, it is possible to construct it in the series expansion of the S matrix with the perturbation technique in the time domain. Using the interaction representation for $\hat{H} = \hat{H}_0 + \hat{V}(t)$, we can analyze series expansion generally in the field operator, which evolves as

$$i\hbar \frac{\partial \hat{S}(t, t')}{\partial t} = \hat{V}(t)\hat{S}(t, t') \tag{3.45}$$

with the boundary condition $\hat{S}(t_0, t_0) = 1$. $\hat{S}(t, t')$ is integrated such as

$$\hat{S}(t, t') = T \exp\left[-\frac{i}{\hbar}\int_{t'}^{t} \hat{V}(t_1)dt_1\right] \tag{3.46}$$

where T is the time-ordered operator and the field operator is obtained using $\hat{S}(t, t')$ such as

$$\hat{\psi}(t) = \hat{S}(t, t')\hat{\psi}(t'). \tag{3.47}$$

Then the time-ordered Green's function is constructed from

$$G(\mathbf{r}, t : \mathbf{r}', t') = -\frac{i}{\hbar}\frac{{}_0\langle T\left\{\hat{S}(\infty, -\infty)\hat{\psi}(\mathbf{r}, t)\hat{\psi}^{\dagger}(\mathbf{r}', t')\right\}\rangle_0}{{}_0\langle \hat{S}(\infty, -\infty)\rangle_0}, \tag{3.48}$$

where ${}_0\langle\cdots\rangle_0$ implies to use wavefunctions for $\hat{H}_0$.

An important assumption here is that in equilibrium we use

$${}_0\langle \hat{S}(\infty, -\infty)\rangle_0 = e^{i\phi}, \tag{3.49}$$

which means that we turn on the interaction adiabatically at $t = -\infty$ and turn off it slowly at $t = \infty$, resulting that the system only changes its phase factor ϕ for the large times. In nonequilibrium, this assumption is not valid and we need to change the time loop of the S-matrix. This will be described in the following section.

Here time-ordered S matrix is calculated by the expansion as

$$\hat{S}(\infty,-\infty) = T\exp\left[-\frac{i}{\hbar}\int_{-\infty}^{\infty}\hat{V}(t_1)dt_1\right]$$
$$= \sum_{n=0}^{\infty}\left(\frac{-i}{\hbar}\right)^n\frac{1}{n!}\int_{-\infty}^{\infty}\cdots\int_{-\infty}^{\infty}T\left\{\hat{V}(t_1)\cdots\hat{V}(t_n)\right\}dt_1\cdots dt_n. \tag{3.50}$$

We can show that the denominator in Eq. (3.48) is canceled out for the terms that are disconnected in the expansion of the numerator (vacuum polarization) and we only need to take the terms connected to the lines of wavefunctions $\hat{\psi}(\mathbf{r},t)\hat{\psi}^\dagger(\mathbf{r}',t')$ in the evaluation of expansions.

Thus, Green's function is written as

$$G(\mathbf{r},t:\mathbf{r}',t') = -\frac{i}{\hbar}\langle T\left\{\hat{\psi}(\mathbf{r},t)\hat{\psi}^\dagger(\mathbf{r}',t')\right\}\rangle$$
$$= -\frac{i}{\hbar}\sum_{n=0}^{\infty}\left(\frac{-i}{\hbar}\right)^n\int_{-\infty}^{\infty}\cdots\int_{-\infty}^{\infty}$$
$${}_0\langle T\left\{\hat{\psi}(\mathbf{r},t)\hat{V}(t_1)\cdots\hat{V}(t_n)\hat{\psi}^\dagger(\mathbf{r}',t')\right\}\rangle_0^{\text{connect}}dt_1\cdots dt_n. \tag{3.51}$$

Here we use the property that there are $n!$ terms that produce the same contributions to Green's function and are thus omitted from the summation in Eq. (3.51). So the summation is only for the connected terms with different topological structures, which are evaluated by Wick's theorem and are schematically described in the Feynman diagrams. This corresponds to taking the irreducible terms for the self-energy $\hat{\Sigma}$ in the Dyson equation.

3.1.4 *Spectral Representation*

Here we show the spectral properties from the retarded and the advanced Green's functions $G^{r,a}(\epsilon)$. First we introduce the complete eigenstates of H as $H|n\rangle = \epsilon_n|n\rangle$. Note that this is a formal description since usually we cannot obtain the exact form of $|n\rangle$ explicitly. Then we have

$$\begin{aligned} G^r(\epsilon) &= -\frac{i}{\hbar}\int_{-\infty}^{\infty}\theta(t)\langle\left[\hat{\psi}(t), \hat{\psi}^\dagger(0)\right]\rangle e^{i\epsilon t/\hbar}dt \\ &= -\frac{i}{\hbar}\frac{1}{Z}\int_0^{\infty}\sum_{n,m}|\langle n|\hat{\psi}(0)|m\rangle|^2 e^{i(\epsilon+\epsilon_n-\epsilon_m+i0^+)t/\hbar} \\ &\qquad \times\left(e^{-\beta(\epsilon_n-\mu N_n)}+e^{-\beta(\epsilon_m-\mu N_m)}\right)dt \\ &= \frac{1}{Z}\sum_{n,m}|\langle n|\hat{\psi}(0)|m\rangle|^2\frac{e^{-\beta(\epsilon_n-\mu N_n)}+e^{-\beta(\epsilon_m-\mu N_m)}}{\epsilon+\epsilon_n-\epsilon_m+i0^+}. \end{aligned} \tag{3.52}$$

Here we take the thermodynamic average $\langle\cdots\rangle$ by $e^{-\beta(H-\mu N)}/Z$ with the partition function of $Z = \mathrm{Tr}(e^{-\beta(H-\mu N)})$ and $\beta = 1/k_B T$. Also we use the following relation:

$$\langle n|\hat{\psi}(t)|m\rangle = \langle n|e^{iHt/\hbar}\hat{\psi}(0)e^{-iHt/\hbar}|m\rangle = e^{i(\epsilon_n-\epsilon_m)t/\hbar}\langle n|\hat{\psi}(0)|m\rangle \tag{3.53}$$

and exchange the variables m and n in the summation of the second term of the anti-commutator $[\cdots]$.

The spectral function $A(\epsilon)$ is defined by

$$A(\epsilon) = i\left[G^r(\epsilon) - G^a(\epsilon)\right] = -2\mathrm{Im}\left[G^r(\epsilon)\right]. \tag{3.54}$$

Using the relation of Eq. (3.36), we have

$$\begin{aligned} A(\epsilon) &= \frac{2\pi}{Z}\sum_{n,m}|\langle n|\hat{\psi}(0)|m\rangle|^2\left(e^{-\beta(\epsilon_n-\mu N_n)}+e^{-\beta(\epsilon_m-\mu N_m)}\right) \\ &\quad \times\delta\left(\epsilon+\epsilon_n-\epsilon_m\right) \end{aligned} \tag{3.55}$$

and the retarded Green's function $G^r(\epsilon)$ is expressed by

$$G^r(\epsilon) = \frac{1}{2\pi}\int_{-\infty}^{\infty}\frac{A(\epsilon')}{\epsilon-\epsilon'+i0^+}d\epsilon'. \tag{3.56}$$

This formula, called the Lehmann representation, shows that $A(\epsilon)$ is the weight how electronic states are changed due to interaction effects.

There are important properties for $A(\epsilon)$. The first one is that

$$A(\epsilon) \geq 0 \tag{3.57}$$

for all the energy ϵ. This is apparent from Eq. (3.55) and indicates that $A(\epsilon)$ is interpreted as the probability function. The second one is the sum rule such as

$$\int_{-\infty}^{\infty}\frac{1}{2\pi}A(\epsilon)d\epsilon = 1. \tag{3.58}$$

This is proved directly from

$$\begin{aligned}\int_{-\infty}^{\infty}\frac{1}{2\pi}A(\epsilon)d\epsilon &= \frac{1}{Z}\sum_{n,m}|\langle n|\hat{\psi}(0)|m\rangle|^2 \\ &\quad\times\left(e^{-\beta(\epsilon_n-\mu N_n)}+e^{-\beta(\epsilon_m-\mu N_m)}\right) \\ &= \frac{1}{Z}\sum_{n,m}\Big\{e^{-\beta(\epsilon_n-\mu N_n)}\langle n|\psi\hat{(0)}|m\rangle\langle m|\hat{\psi}^{\dagger}(0)|n\rangle \\ &\qquad +e^{-\beta(\epsilon_m-\mu N_m)}\langle n|\psi\hat{(0)}|m\rangle\langle m|\hat{\psi}^{\dagger}(0)|n\rangle\Big\} \\ &= \frac{1}{Z}\sum_{n}\left\{e^{-\beta(\epsilon_n-\mu N_n)}\langle n|\left[\hat{\psi}(0),\hat{\psi}^{\dagger}(0)\right]|n\rangle\right\} \\ &= 1, \end{aligned} \tag{3.59}$$

where we exchange n and m in the second term of the second equation. From Eq. (3.37), electron density is expressed by

$$n(\mathbf{r}) = \int f(\epsilon)\frac{1}{2\pi}A(\mathbf{r},\epsilon)d\epsilon \tag{3.60}$$

and the DOS becomes from Eq. (3.38)

$$\nu(\epsilon) = \int\frac{1}{2\pi}A(\mathbf{r},\epsilon)d\mathbf{r} = \sum_k\frac{1}{2\pi}A(k,\epsilon). \tag{3.61}$$

Now we express the spectral function $A(k,\epsilon)$ by the self-energy $\Sigma(k,\epsilon)$. From Eq. (3.44), we have

$$[G^a(k,\epsilon)]^{-1} - [G^r(k,\epsilon)]^{-1} = \Sigma^r(k,\epsilon) - \Sigma^a(k,\epsilon), \tag{3.62}$$

then the spectral representation of $A(k,\epsilon)$ is expressed from the retarded and the advanced Green's functions as

$$A(k,\epsilon) = i(G^r(k,\epsilon) - G^a(k,\epsilon)) = G^r(k,\epsilon)\Gamma(k,\epsilon)G^a(k,\epsilon), \tag{3.63}$$

where we define the coupling coefficient $\Gamma(k,\epsilon)$ by

$$\Gamma(k,\epsilon) = i\left[\Sigma^r(k,\epsilon) - \Sigma^a(k,\epsilon)\right]. \tag{3.64}$$

The shape of the spectral function is expressed directly from the Dyson equation,

$$G^{r,a}(k,\epsilon) = G_0^{r,a}(k,\epsilon) + G_0^{r,a}(k,\epsilon)\Sigma^{r,a}(k,\epsilon)G^{r,a}(k,\epsilon). \tag{3.65}$$

We thus have

$$A(k,\epsilon) = -2\mathrm{Im}G^r(k,\epsilon) = \frac{\Gamma(k,\epsilon)}{(\epsilon-\epsilon_k-\mathrm{Re}\Sigma^r(k,\epsilon))^2+(\Gamma(k,\epsilon)/2)^2}, \tag{3.66}$$

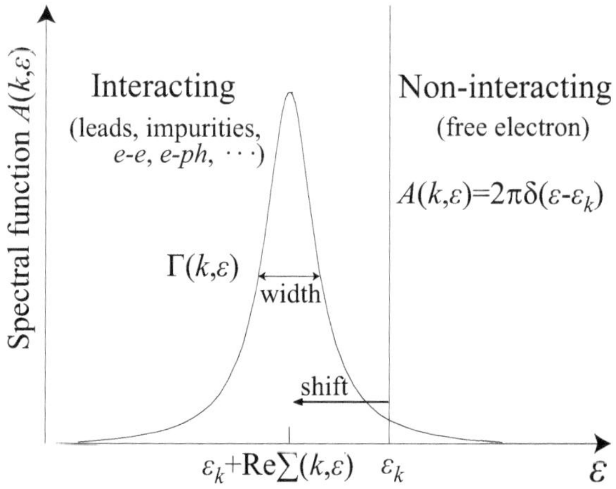

Figure 3.2 Schematic picture of the spectral function $A(k, \epsilon)$. In the absence of interaction, the spectral function takes a δ-function at the resonance of ϵ_k. Due to interaction effects with leads, impurities, other electrons (e-e), phonons (e-ph) $\cdots$, this energy shifts by $\mathrm{Re}\,\Sigma(k, \epsilon_k)$ and broadens into a Lorentzian form with a width of coupling coefficient $\Gamma(k, \epsilon_k)$. This determines the lifetime of the quasi-particle.

where we use $G_0^{r,a}(k, \epsilon) = 1/(\epsilon - \epsilon_k \pm i0^+)$. Since $\Gamma(k, \epsilon) > 0$, $A(k, \epsilon) > 0$. This shows that the spectral function $A(k, \epsilon)$ shifts due to interaction effects by an amount given by the real part of self-energy. The shifted level broadens into a Lorentzian form with the width of the imaginary part of the self-energy, which determines the lifetime of electrons (Fig. 3.2).

Here we show the behavior of electron as a quasi-particle near the Fermi level. When we expand the self-energy $\Sigma(k, \epsilon)$ close to the chemical potential μ,

$$\Sigma(k, \epsilon) = \Sigma(k, \mu) + \left.\frac{\partial \Sigma(k, \epsilon)}{\partial \epsilon}\right|_{\epsilon=\mu} (\epsilon - \mu) + \cdots \tag{3.67}$$

and define the renormalization factor by

$$Z_k = \left[1 - \partial \Sigma(k, \epsilon)/\partial \epsilon|_{\epsilon=\mu}\right]^{-1}, \tag{3.68}$$

Green's function becomes

$$G(k, \epsilon) = \frac{1}{\epsilon - \epsilon_k - \Sigma(k, \epsilon)} \simeq \frac{Z_k}{\epsilon - \mu - \tilde{\epsilon}_k}, \tag{3.69}$$

where the energy is renormalized by $\tilde{\epsilon_k} = Z_k(\epsilon_k - \mu + \Sigma(k, \mu))$ with the weight of the renormalization factor $Z_k \leq 1$. This relation shows the quasi-particle behavior close to the Fermi level. The electron density $n(\mathbf{r})$ is renormalized by Z_k compared to the noninteracting case and effective mass m^*/m due to interaction effects is also expressed by the self-energy $\Sigma(k, \epsilon)$.

3.1.4.1 Application to free electron

Using the obtained techniques for Green's functions, we give several simple applications here. The first is the simplest one: electronic states for free electrons. The Hamiltonian becomes

$$H_0 = \sum_k \epsilon_k \hat{c}_k^\dagger \hat{c}_k. \tag{3.70}$$

The retarded and advanced Green's functions $G^{r,a}$ is obtained as

$$G_0^{r,a}(k, \epsilon) = \frac{1}{\epsilon - \epsilon_k \pm i0^+}. \tag{3.71}$$

From these expressions, the spectral function becomes

$$A(k, \epsilon) = i(G^r(k, \epsilon) - G^a(k, \epsilon)) = 2\pi\delta(\epsilon - \epsilon_k), \tag{3.72}$$

with the LDOS of

$$\nu(\epsilon) = \sum_k \frac{1}{2\pi} A(k, \epsilon) = \sum_k \delta(\epsilon - \epsilon_k). \tag{3.73}$$

To obtain the time-ordered Green's function G_0, we construct the equation of motion for $\hat{c}_k$ and $\hat{c}_k^\dagger$,

$$i\hbar \frac{d\hat{c}_k}{dt} = [\hat{c}_k, H_0] = \sum_{k'} \left[\hat{c}_k, \epsilon_{k'} \hat{c}_{k'}^\dagger \hat{c}_{k'}\right] = \epsilon_k \hat{c}_k$$

$$i\hbar \frac{d\hat{c}_k^\dagger}{dt} = \left[\hat{c}_k^\dagger, H_0\right] = \sum_{k'} \left[\hat{c}_k^\dagger, \epsilon_{k'} \hat{c}_{k'}^\dagger \hat{c}_{k'}\right] = -\epsilon_k \hat{c}_k^\dagger, \tag{3.74}$$

which can be integrated to produce

$$\hat{c}_k(t) = e^{-i\epsilon_k t/\hbar} \hat{c}_k, \quad \hat{c}_k^\dagger(t) = e^{i\epsilon_k t/\hbar} \hat{c}_k^\dagger. \tag{3.75}$$

Then the time-ordered Green's function is constructed as

$$\begin{aligned} G_0(k, t) &= -(i/\hbar)\langle \theta(t - t')\hat{c}_k(t)\hat{c}_k^\dagger(t') - \theta(t' - t)\hat{c}_k^\dagger(t')\hat{c}_k(t)\rangle \\ &= -(i/\hbar)e^{-i\epsilon_k(t-t')/\hbar}\left[\theta(t - t')\langle \hat{c}_k\hat{c}_k^\dagger\rangle - \theta(t' - t)\langle \hat{c}_k^\dagger \hat{c}_k\rangle\right], \end{aligned} \tag{3.76}$$

which becomes, in a Fourier space,

$$\begin{aligned} G_0(k, \epsilon) &= -\frac{i}{\hbar}\int_0^{\infty} e^{i(\epsilon-\epsilon_k+i0^+)t/\hbar}(1-f_k)dt \\ &\quad +\frac{i}{\hbar}\int_{-\infty}^{0} e^{i(\epsilon-\epsilon_k-i0^+)t/\hbar} f_k dt \\ &= \frac{1-f_k}{\epsilon-\epsilon_k+i0^+}+\frac{f_k}{\epsilon-\epsilon_k-i0^+}. \end{aligned} \tag{3.77}$$

This relation becomes the basic formula for the perturbation expansion in the next sections.

3.1.4.2 Application to resonant tunneling

The following example is the resonant tunneling problem through a resonant energy level ϵ_0. The Hamiltonian becomes

$$H = \sum_k \epsilon_k \hat{c}_k^{\dagger}\hat{c}_k + \epsilon_0 \hat{d}^{\dagger}\hat{d} + \sum_k \left[V_k \hat{c}_k^{\dagger}\hat{d} + V_k^* \hat{d}^{\dagger}\hat{c}_k\right], \tag{3.78}$$

where $\hat{d}^{\dagger}$ and $\hat{d}$ are creation and annihilation operators of the resonant site and V_k is the transfer energy of electrons.

We define two Green's functions for the operator $\hat{d}$ as

$$\begin{aligned} G^r(t) &= -\frac{i}{\hbar}\theta(t)\langle[\hat{d}(t), \hat{d}^{\dagger}(0)]\rangle \\ K^r(t) &= -\frac{i}{\hbar}\theta(t)\langle[\hat{c}_k(t), \hat{d}(0)]\rangle. \end{aligned} \tag{3.79}$$

The equations of motions for the operator $\hat{d}$ and $\hat{c}$ become

$$\begin{aligned} i\hbar\frac{d\hat{d}(t)}{dt} &= \epsilon_0\hat{d}(t) + \sum_k V_k^* \hat{c}_k(t) \\ i\hbar\frac{d\hat{c}_k(t)}{dt} &= \epsilon_k\hat{c}_k(t) + V_k\hat{d}(t), \end{aligned} \tag{3.80}$$

and using these relations, the equations of motions for two Green's functions are obtained as

$$\begin{aligned} i\hbar\frac{\partial G^r(t)}{\partial t} &= \delta(t) + \epsilon_0 G^r(t) + \sum_k V_k^* K^r(t) \\ i\hbar\frac{\partial K^r(t)}{\partial t} &= \epsilon_k K^r(t) + V_k G^r(t). \end{aligned} \tag{3.81}$$

These formulas are represented in a Fourier space as

$$\left(\epsilon - \epsilon_0 + i0^+\right) G^r(\epsilon) = 1 + \sum_k V_k^* K^r(\epsilon)$$

$$\left(\epsilon - \epsilon_k + i0^+\right) K^r(\epsilon) = V_k G^r(\epsilon). \tag{3.82}$$

Therefore, Green's function for $\hat{d}$ is obtained from

$$G^r(\epsilon) = G_0^r(\epsilon) + G_0^r(\epsilon)\left(\sum_k |V_k|^2 g_0^r(\epsilon)\right) G^r(\epsilon), \tag{3.83}$$

where we use Green's functions for free electrons in the previous section

$$G_0^r(\epsilon) = \frac{1}{\epsilon - \epsilon_0 + i0^+}, \qquad g_0^r(k, \epsilon) = \frac{1}{\epsilon - \epsilon_k + i0^+}. \tag{3.84}$$

From these relations, we can define the self-energy as

$$\Sigma^r(\epsilon) = \sum_k |V_k|^2 g_0^r(\epsilon) = \sum_k \frac{|V_k|^2}{\epsilon - \epsilon_k + i0^+} \tag{3.85}$$

and correspondingly the coupling constant $\Gamma(\epsilon)$ becomes

$$\Gamma(\epsilon) = i\left[\Sigma^r(\epsilon) - \Sigma^a(\epsilon)\right] = 2\pi \sum_k |V_k|^2 \delta(\epsilon - \epsilon_k). \tag{3.86}$$

It should be noted that the retarded Green's function $G^r(\epsilon)$ satisfies the Dyson equation

$$\begin{aligned} G^r(\epsilon) &= G_0^r(\epsilon) + G_0^r(\epsilon)\Sigma^r(\epsilon)G^r(\epsilon) \\ &= G_0^r(\epsilon) + G_0^r(\epsilon)\Sigma^r(\epsilon)G_0^r(\epsilon) \\ &\quad + G_0^r(\epsilon)\Sigma^r(\epsilon)G_0^r(\epsilon)\Sigma^r(\epsilon)G_0^r(\epsilon) + \cdots \end{aligned} \tag{3.87}$$

and is obtained directly from

$$G^r(\epsilon) = \frac{1}{\left[G_0^r(\epsilon)\right]^{-1} - \Sigma^r(\epsilon)} = \frac{1}{\epsilon - \epsilon_0 - \Sigma^r(\epsilon)}. \tag{3.88}$$

The DOS of the resonant level ϵ_0 becomes Lorentzian form due to coupling effects with conducting electrons ϵ_k

$$\begin{aligned} \nu(\epsilon) &= \frac{1}{2\pi} A(\epsilon) = -\frac{1}{\pi}\mathrm{Im}G^r(\epsilon) \\ &= \frac{1}{2\pi} \frac{\Gamma(\epsilon)}{\left[\epsilon - \epsilon_0 - \mathrm{Re}\left\{\Sigma^r(\epsilon)\right\}\right]^2 + \left[\Gamma(\epsilon)/2\right]^2}. \end{aligned} \tag{3.89}$$

The coupling constant $\Gamma(\epsilon)$ works for the "broadening" of DOS $\nu(\epsilon)$. The corresponding states are "quasi-particle" states with the renormalized energies of $\epsilon_0 \to \epsilon_0 + \mathrm{Re}\left\{\Sigma^r(\epsilon_0)\right\}$, and the coupling coefficient $\Gamma(\epsilon)$ is related to the escape rate or the lifetime of quasi-particle $\Sigma^r = -i\hbar/2\tau$ as

$$\frac{\hbar}{\tau} = \frac{1}{\Gamma(\epsilon_0)}. \tag{3.90}$$

3.1.5 *Kubo–Greenwood Formula*

Here we recall the Kubo formula for the quantum transport and rewrite it using Green's function. Then we will show how the Kubo formula is utilized to obtain various transport properties in the next sections.

We remind that the conductivity σ is expressed in the Kubo formula as

$$\sigma = 2\pi\hbar\left(\frac{e}{m}\right)^2 \int \sum_{\alpha,\beta} |\langle\alpha|p_z|\beta\rangle|^2 \left(-\frac{df}{d\epsilon}\right)\delta\left(\epsilon-\epsilon_\alpha\right)\delta\left(\epsilon-\epsilon_\beta\right)d\epsilon. \tag{3.91}$$

Introducing the retarded and the advanced Green's functions,

$$G^{r,a}(k,k',\epsilon) = \left\langle k\right|\frac{1}{\epsilon-H\pm i0^+}\left|k'\right\rangle = \sum_\alpha \frac{\langle k|\alpha\rangle\langle\alpha|k'\rangle}{\epsilon-\epsilon_\alpha\pm i\delta}, \tag{3.92}$$

which have the property of

$$G^a(k,k',\epsilon) - G^r(k,k',\epsilon) = 2\pi i\sum_\alpha \langle k|\alpha\rangle\langle\alpha|k'\rangle\delta(\epsilon-\epsilon_\alpha). \tag{3.93}$$

Here we use the relation

$$\delta(x) = \frac{1}{2\pi i}\left(\frac{1}{x-i\delta}-\frac{1}{x+i\delta}\right). \tag{3.94}$$

We can rewrite the conductivity formula as

$$\begin{aligned}\sigma &= \frac{2\pi\hbar^3e^2}{m^2}\left(\frac{1}{2\pi i}\right)^2\int\sum_{k,k'}\left(-\frac{df}{d\epsilon}\right)k_zk'_z\left(G^a(k,k',\epsilon)-G^r(k,k',\epsilon)\right)\\ &\qquad\times\left(G^a(k',k,\epsilon)-G^r(k',k,\epsilon)\right)d\epsilon\\ &= \frac{\hbar e^2}{2\pi}\int\sum_{k,k'}\left(-\frac{df}{d\epsilon}\right)\left(\frac{\hbar}{m}\right)^2 k_zk'_zG^r(k,k',\epsilon)G^a(k',k,\epsilon)d\epsilon.\end{aligned} \tag{3.95}$$

This is the Kubo formula in terms of Green's functions (Kubo–Greenwood formula). Here we assume that the terms from G^rG^r and G^aG^a are negligible.

Usually, the observable quantity is the average of parameters, such as impurities. So we use the statistical average[a]

$$\sigma = \frac{\hbar e^2}{2\pi}\int\sum_{k,k'}\left(-\frac{df}{d\epsilon}\right)\left(\frac{\hbar}{m}\right)^2\left\langle k_zk'_zG^r(k,k',\epsilon)G^a(k',k,\epsilon)\right\rangle d\epsilon. \tag{3.96}$$

[a]We note that $\langle G^r(k,k',\epsilon)G^a(k',k,\epsilon)\rangle$ represent electron–hole propagators.

When we take

$$\langle G^r(k, k', \epsilon) G^a(k', k, \epsilon)\rangle = \langle G^r(k, k', \epsilon)\rangle \langle G^a(k', k, \epsilon)\rangle, \tag{3.97}$$

this corresponds to the situation where interference of the phase is neglected and we have the classical formula for the transport relation. We will show that this approximation produces Drude relation in the impurity scattering case later. Also the higher order expansion for Green's function produces phase interference effect, which leads to the localization of electrons.

3.1.5.1 Relevance with Landauer formula

We see how the Kubo formula is related to the Landauer formula here. In order to analyze the conductance of a finite sample, we change the Kubo formula using the current operator. Since $\langle\alpha|j(\mathbf{r})|\beta\rangle = (e/m)\langle\alpha|p_z|\beta\rangle$, we obtain

$$\begin{aligned}\sigma(\mathbf{r}, \mathbf{r}', 0) = 2\pi\hbar \int \sum_{\alpha\beta} \langle\alpha|j(\mathbf{r})|\beta\rangle\langle\beta|j(\mathbf{r}')|\alpha\rangle \\ \times \left(-\frac{\partial f}{\partial \epsilon}\right) \delta(\epsilon - \epsilon_\alpha)\, \delta\left(\epsilon - \epsilon_\beta\right) d\epsilon.\end{aligned} \tag{3.98}$$

The total current is given by

$$\begin{aligned} I &= \int_S \mathbf{j}(\mathbf{r}) dS = \int_S dS \int \sigma(\mathbf{r}, \mathbf{r}', 0) E_z(\mathbf{r}') d\mathbf{r}' \\ &= \int_S dS \int dz' \int_{S'} dS' \sigma(\mathbf{r}, \mathbf{r}', 0) E_z(\mathbf{r}'). \end{aligned} \tag{3.99}$$

Here we consider the current conversion rule expressed by

$$\int_S \langle\alpha|\hat{j}_z(\mathbf{r})|\beta\rangle dS = \int_{S'} \langle\alpha|\hat{j}_z(\mathbf{r}')|\beta\rangle dS', \tag{3.100}$$

which means that we can interchange the integration over z'. Therefore, we get

$$I = \int_S dS \int_{S'} dS' \sigma(\mathbf{r}, \mathbf{r}', 0) \int E_z(\mathbf{r}') dz'. \tag{3.101}$$

The term coming from an electric field gives the voltage along z' direction where the bias is applied. So from the relation $I = GV$, we have

$$G = \int_S dS \int_{S'} dS' \sigma(\mathbf{r}, \mathbf{r}', 0). \tag{3.102}$$

This is independent of positions of the cross section along the z direction. We can take thesc surfaces deep inside electrodes far away from nanosystems where the voltage drop happens. Deep inside electrodes, we assume that electronic states become plane waves along the z direction.

When we expand Green's functions as

$$G^{r,a}(\mathbf{r}, \mathbf{r}', \epsilon) = \sum_{n,m} \chi_m(\mathbf{r}_{||})\chi_n^*(\mathbf{r}'_{||})G^{r,a}_{n,m}(z, z', \epsilon), \tag{3.103}$$

where the indexes n and m refer to modes on the left and right electrodes, respectively. $\mathbf{r}_{||}$ indicates the parallel positions to the surfaces. Deep inside the left and right electrodes, we can take

$$G^{r,a}_{n,m}(z, z', \epsilon) \sim e^{\pm ik_n z}e^{\mp ik_m z'}. \tag{3.104}$$

So by evaluating the surface integrations at regions deep inside electrodes, we obtain the following formula using Green's functions:

$$G = \int_S dS \int_{S'} dS' \frac{\hbar e^2}{2\pi} \int \sum_{k,k'} \left(-\frac{df}{d\epsilon}\right)\left(\frac{\hbar}{m}\right)^2 \times k_z k'_z G^r(k, k', \epsilon)G^a(k', k, \epsilon)d\epsilon. \tag{3.105}$$

This is evaluated as

$$G = \frac{\hbar e^2}{2\pi} \int \sum_{n,m} \left(-\frac{df}{d\epsilon}\right) v_n v_m G^r_{n,m}(z, z', \epsilon)G^a_{m,n}(z, z', \epsilon)d\epsilon$$
$$= \frac{2e^2}{h} \int \left(-\frac{df}{d\epsilon}\right) \mathcal{T}(\epsilon)d\epsilon, \tag{3.106}$$

where

$$\mathcal{T}(\epsilon) = \hbar^2 \sum_{nm} v_n v_m G^r_{nm}(z, z', \epsilon)G^a_{nm}(z, z', \epsilon). \tag{3.107}$$

Here we put 2 for spin degeneracy. This is the equivalent form with the Landauer formula.

3.1.6 *Conductivity with Impurity Scatterings*

Here we show how the Kubo formula recovers the classical result of Drude conductivity. As for the collisions, we consider impurity scatterings for the diffusive transport. Here we take $T \to 0$ limit, for simplicity, where the conductivity is represented by electron–hole propagators as

$$\sigma = \frac{\hbar e^2}{2\pi} \sum_{k,k'} \left(\frac{\hbar}{m}\right)^2 \left\langle k_z k'_z G^r(k, k', \epsilon) G^a(k', k, \epsilon) \right\rangle_{\text{imp}}. \tag{3.108}$$

The impurity potential is expressed by

$$V_{\text{imp}}(\mathbf{r}) = \sum_i v_{\text{imp}}(\mathbf{r} - \mathbf{R}_i), \tag{3.109}$$

where the sites of impurities $\mathbf{R}_i$ are randomly situated and we take the average. Taking plane waves for the unperturbed wavefunctions $|k\rangle = e^{i\mathbf{k}\cdot\mathbf{r}}$, the unperturbed Green's function G_0 is

$$G_0(k, k') = G_0(k)\delta_{k,k'} = \frac{1}{\epsilon - \epsilon_k}\delta_{k,k'} \tag{3.110}$$

and

$$\langle k'|V_{\text{imp}}(\mathbf{r})|k\rangle = \left[\int v_{\text{imp}}(\mathbf{r}) e^{i(\mathbf{k}-\mathbf{k}')\cdot\mathbf{r}} d\mathbf{r}\right] \sum_i e^{i(\mathbf{k}-\mathbf{k}')\cdot\mathbf{R}_i}. \tag{3.111}$$

We will consider Green's function $G(k, \epsilon)$ in the series expansion involving interaction effects from impurity scatterings (Fig. 3.3). Since the average of impurity sites has the property

$$\left\langle \sum_{i,j} e^{i(\mathbf{k}-\mathbf{k}')\cdot(\mathbf{R}_i - \mathbf{R}_j)} \right\rangle_{\text{imp}} = n_i \delta_{i,j} \tag{3.112}$$

where n_i is the density of impurities per volume, we can neglect the first-order term from a single scattering with an impurity. The second-order term from the double scattering process with a single impurity contribute to Green's function and have to be included in the summation (Fig. 3.3 (left)). Thus, the self-energy is obtained from the diagram of two scatterings as (Fig. 3.3 (right))[a]

$$\Sigma(k, \epsilon) = n_i \sum_{k'} |v_{\text{imp}}(k - k')|^2 G_0(k'). \tag{3.113}$$

If we assume the short-range potentials for impurities and ignore the k dependence, then

$$\Sigma^{r,a}(k, \epsilon) = n_i |v_{\text{imp}}|^2 \sum_k G_0^{r,a}(k) = \mp i\pi\gamma\nu(\epsilon), \tag{3.114}$$

[a]Contribution from the other forms of double scattering process to Green's function is shown to be small or is included in the diagram with a change of $G_0(k')$ to $G(k')$ in the self-energy $\Sigma(k, \epsilon)$ calculation in Eq. (3.113).

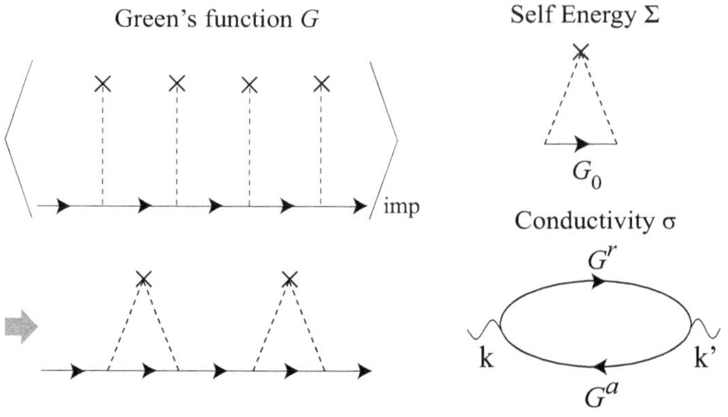

Figure 3.3 (Left) Diagram for impurity scatterings. A cross and dotted lines represent an impurity and its scattering. Thin solid lines represent the bare Green's function G_0 for free electrons. Due to an ensemble average, the double scattering process with a single impurity contributes to Green's function $G^{r,a}$. (Right) Self-energy Σ for impurity scatterings and the diagram of the Kubo formula representing electron–hole propagators for the conductivity. Here thick solid lines represent the (dressed) Green's function $G^{r,a}$, including the impurity scatterings.

where the DOS $\nu(\epsilon)$ is

$$\nu(\epsilon) = \sum_k \delta(\epsilon - \epsilon_k) = -\frac{1}{\pi} \sum_k \mathrm{Im} G_0^r(k) \tag{3.115}$$

and $\gamma = n_i |v_{\mathrm{imp}}|^2$.

Since the imaginary part of self-energy $\Sigma^{r,a}(k, \epsilon)$ gives

$$\mathrm{Im}\Sigma^{r,a}(k, \epsilon) = \mp \frac{\hbar}{2\tau}, \tag{3.116}$$

where τ is the scattering time of

$$\tau = \frac{\hbar}{2\pi \nu \gamma}, \tag{3.117}$$

we have Green's function

$$G^{r,a}(k, \epsilon) = \frac{1}{G_0^{r,a}(k, \epsilon)^{-1} - \Sigma^{r,a}(k, \epsilon)} = \frac{1}{\epsilon - \epsilon_k \pm i\hbar/2\tau}. \tag{3.118}$$

The conductivity σ in the lowest order is

$$\sigma = \frac{\hbar e^2}{2\pi}\sum_k \left(\frac{\hbar}{m}\right)^2 \langle k_z^2\rangle \langle G^r(k,\epsilon)\rangle\langle G^a(k,\epsilon)\rangle$$
$$= \frac{\hbar e^2}{2\pi}\sum_k \left(\frac{\hbar}{m}\right)^2 \langle k_z^2\rangle \frac{\tau}{i\hbar}\left(\frac{1}{\epsilon-\epsilon_k - i\hbar/2\tau} - \frac{1}{\epsilon-\epsilon_k + i\hbar/2\tau}\right)$$
$$= e^2\tau\sum_k \left(\frac{\hbar}{m}\right)^2 \langle k_z^2\rangle \delta(\epsilon-\epsilon_k) = e^2\nu D, \tag{3.119}$$

where the diffusion constant is $D = \langle v^2\rangle\tau$. Thus, we can derive the Drude conductivity for impurity scatterings from the lowest-order expansion of the Kubo formula. We note that we assume isotropic scattering of impurities in the present case.

3.1.6.1 Quantum correction in higher order

Since the evaluation of the lowest-order correlation function $\langle G^r(k,\epsilon)\rangle\langle G^a(k,\epsilon)\rangle$ provides the classical Drude conductivity, we expect that the calculation of electron motion in higher-order expansion of $\langle G^r(k,\epsilon)\rangle\langle G^a(k,\epsilon)\rangle_{\rm imp}$ includes the quantum correction to electron transport,[a] interference effect due to disorder [29]. The contribution in the higher-order diagram is written as

$$\Delta\sigma = \frac{\hbar e^2}{2\pi}\sum_{k,k'} \left(\frac{\hbar}{m}\right)^2 \langle k_z k_z'\rangle \langle G^r(k,\epsilon)\rangle\langle G^a(k,\epsilon)\rangle$$
$$\times\Gamma(k,k',\epsilon)\langle G^r(k',\epsilon)\rangle\langle G^a(k',\epsilon)\rangle, \tag{3.120}$$

where $\Gamma(k,k',\epsilon)$ represents various multiple-scattering effects in electron–hole propagator. Here we consider two diagrams, the ladder diagram and the maximally crossed diagram.

The first is the ladder diagram called diffuson shown in the top of Fig. 3.4. When we assume that impurity scattering has no k dependence, the contribution from one impurity scattering is $\Gamma_d^{(0)}(k,k',\epsilon) = n_i|v_{\rm imp}|^2$ and then $\Delta\sigma$ becomes zero due to the

[a] Since we assume that scattering is weak as $\lambda_F \ll \ell_{\rm mfp}$ or $k_F\ell_{\rm mfp} \gg 1$, we treat the metallic weak localization regime in the perturbative approach.

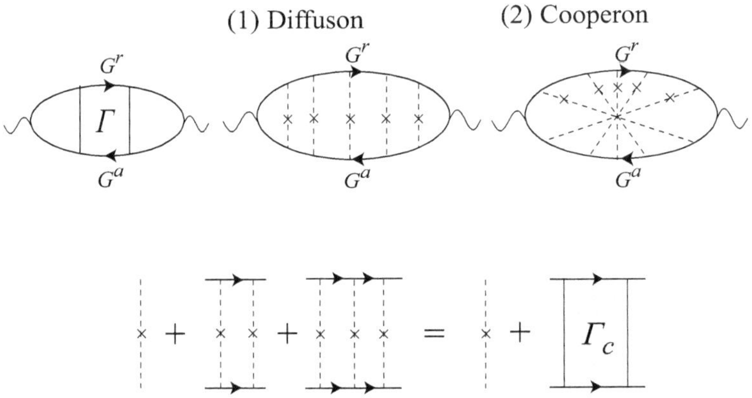

Figure 3.4 (Top) (left) Diagrammatic summation of the higher-order contributions. Here Γ represents vertex correction by impurity scatterings, changing the momentum in the higher-order contributions. (Center) Ladder diagram called diffuson, and (right) maximally crossed diagram called cooperon. The dotted lines represent scatterings by impurities between two propagators. (Bottom) Dyson equation for particle–hole propagators. Here Γ_c represents the vertex correction in maximally crossed diagram in the higher-order contributions, changing the momentum due to back-scattering with the hole-line turned around.

summation of k, k' for $\langle k_z k'_z \rangle$. This observation holds for the higher-order of the summation of $\Gamma_d(k, k', \epsilon)$ and $\Delta\sigma_d = 0$ for the ladder diagram. We note that $\Delta\sigma_d = 0$ is due to no k dependence.[a]

The quantum correction in terms of interference effect emerges in the maximally crossed diagram in the right panel of Fig 3.4. When we consider the range of $k + k' = q \approx 0$, the energy in Green's function is approximated by

$$\epsilon_{k'} = \frac{\hbar^2 k'^2}{2m} = \frac{\hbar^2}{2m}(-k+q)^2 \approx \epsilon - \hbar v q, \tag{3.121}$$

[a]When we consider the k dependence, the result becomes to replace scattering time τ by τ_{tr}, which is called the transport lifetime of conduction electron. Physically, this corresponds to the non-isotropic scattering effects that contribute to the conductivity by $1 - \cos\theta_{k,k'}$. When there is no k dependence as $\theta_{k,k'} = 0$, the contribution disappears. Therefore, the effects on conductivity from the ladder diagram is regarded as classical.

where $v = \hbar k/m$ and thus for small q we have

$$\begin{aligned}
&\sum_k \langle G^r(k)\rangle\langle G^a(k)\rangle\langle G^r(-k+q)\rangle\langle G^a(-k+q)\rangle \\
&= 2\pi i\nu \frac{1}{i\hbar/\tau}\frac{1}{\hbar v q}\frac{-2\hbar v q}{(i\hbar/\tau)^2-(\hbar v q)^2} \\
&\approx 4\pi\nu\left(\frac{\tau}{\hbar}\right)^3.
\end{aligned} \tag{3.122}$$

Using $\langle k_z k_z'\rangle \simeq -(m/\hbar)^2\langle v^2\rangle$, the conductivity is given by

$$\Delta\sigma_c = -\frac{e^2}{\pi\hbar}\frac{D\tau}{\gamma}\sum_q \Gamma_c(q) \tag{3.123}$$

with $D = \langle v^2\rangle\tau$ and scattering strength $\gamma = n_i|v_{\mathrm{imp}}|^2 = \hbar/2\pi\nu\tau$.

We note that $\Gamma_c(q)$ for the maximally crossed diagram is expressed by a two-particle propagator with the hole-line turned around by $k' \approx -k$ called cooperon shown in the bottom of Fig. 3.4. Thus, $\Gamma_c(q)$ is the product of Green's functions and the matrix elements between electron–hole propagators to become a Dyson equation

$$\Gamma_c(q) = \gamma + \gamma\Pi\gamma + \gamma\Pi\gamma\Pi\gamma + \cdots = \frac{\gamma}{1-\gamma\Pi}. \tag{3.124}$$

The propagator Π is evaluated for $q \approx 0$ by

$$\begin{aligned}
\Pi &= \sum_q \langle G^r(k)\rangle\langle G^a(-k+q)\rangle \\
&= \nu\int\left(\frac{1}{\epsilon-\epsilon_k-i\hbar/2\tau}\right)\cdot\left(\frac{1}{\epsilon-\epsilon_k-\hbar v q+i\hbar/2\tau}\right)d\epsilon \\
&= \frac{2\pi\nu}{\hbar}\tau\left\langle\frac{1}{1+iqv\tau}\right\rangle \approx \frac{1}{\gamma}\left(1-Dq^2\tau\right),
\end{aligned} \tag{3.125}$$

and thus we obtain the vertex correction

$$\Gamma_c(q) = \frac{\gamma}{Dq^2}\frac{1}{\tau}, \tag{3.126}$$

which diverges as $q \to 0$. This shows a diffusion pole.[a]

Then we have the quantum correction of conductivity as

$$\Delta\sigma_c = -\frac{e^2}{\pi\hbar}\sum_q \frac{1}{q^2}. \tag{3.127}$$

[a]In the time-dependent case, it represents the diffusion pole of $1/(-i\omega + Dq^2)$.

Here the sum of diffusion pole $\sum_q (1/q^2)$ depends on the dimensionality. For example, when we consider thin film with a thickness d and a mean free path $\ell_{\text{mfp}} (\gg d)$, we observe a two-dimensional system. Since the sum of q is restricted by the cutoff of L_0^{-1} for the lower limit where L_0 is set by the sample size L, the diffusion length, or the phase coherence length $L_\phi = \sqrt{D\tau_\phi}$, we have

$$\Delta\sigma_c^{(d)} = -\frac{e^2}{\pi\hbar}\frac{2}{(2\pi)^d}\int_{L_0^{-1}}^{\infty}\frac{1}{q^2}d\mathbf{q} = -\frac{e^2}{\pi^d\hbar}\times\begin{cases} -1/L_0 & \text{for } d = 3 \\ \ln\left(L_0/\ell_{\text{mfp}}\right) & \text{for } d = 2 \\ L_0 & \text{for } d = 1. \end{cases} \tag{3.128}$$

We note that the quantum correction diverges for $d = 2$ for the upper limit of ∞; thus we put ℓ_{mfp}^{-1} corresponding to the diffusion during a collision time. Also it diverges for $d \leq 2$ for $L_0 \to \infty$ and we include electron phase breaking processes occurring as a result of inelastic scatterings. Since the phase coherence time τ_ϕ increases with decreasing temperature, we have the temperature dependence of quantum correction of conductance[a] by setting $1/\tau_\phi \propto T^p$. For example, the temperature dependence of the negative quantum correction becomes $\Delta\sigma_c^{(2)} \propto -(1/2)\ln(\tau_\phi/\tau) \propto (p/2) \times \ln T$.

Since the quantum correction is due to the back-scattering of electron by $k' \approx -k$ leading to the weak localization of wavefunction by interference,[b] it can be estimated using the return probability of electron $P(t) = (Dt)^{-d/2}$ and cross-section λ_F^2 by

$$\frac{\Delta\sigma}{\sigma} \sim -\int_\tau^{\tau_\phi} v_F\lambda_F^2 P(t)dt = -\frac{2}{2-d}\frac{\lambda_F^2}{\ell_{\text{mfp}}}(L_\phi^{2-d} - \ell_{\text{mfp}}^{2-d}). \tag{3.129}$$

---

[a] At finite temperature, electron–electron interaction due to Coulomb interaction contributes to the quantum correction.

[b] Quantum phase interference is modified by the magnetic field B, resulting in the negative magnetoresistance. It is especially important for 2D system as $\Delta\sigma_c^{(2)} = -(e^2/2\pi^2\hbar)\left\{\psi(1/2 + (L_B/\ell_{\text{mfp}})^2) - \psi(1/2 + (L_B/L_\phi)^2)\right\}$ with $L_B = \sqrt{\hbar c/4eB}$. Here $\psi(z)$ is the digamma function $\psi(z) = -\gamma - \sum_{n=0}^{\infty}\left[(n+z)^{-1} - (n+1)^{-1}\right]$ with an asymptotic limit of $\psi(z) \sim \ln(z)$ for $z \to \infty$. In a strong magnetic field $L_B \ll L_\phi$, $\Delta\sigma_c^{(2)} = -(e^2/\pi^2\hbar)\ln(L_B/\ell_{\text{mfp}})$.

3.2 Nonequilibrium Green's Function

3.2.1 *Contour Ordering*

When we consider nonequilibrium systems, we need to develop more complicated formulas. Since the nonequilibrium system is not restored to the original system for asymptotically large times, it is not sufficient to describe the nonequilibrium system only by the time-ordered Green's function as in the previous section. As for the time integral of S-matrix for equilibrium systems, we take

$$ {}_0\langle \hat{S}(\infty, -\infty)\rangle_0 = e^{i\phi}, \tag{3.130}$$

which assumes that the system only changes its phase factor ϕ for large times. However, this assumption is no more applicable for the nonequilibrium system, since the system does not return to its ground state as $t \to +\infty$ and irreversible effects occur in the process of $t = -\infty$ and $t = +\infty$.

One way to circumvent this problem in the nonequilibrium situations is that we take the time integral of S matrix such that the system evolves from $t = -\infty$ then continues the time evolution up to t_0 and returns back to $t = -\infty$. This contour order is expressed as

$$\hat{S}(-\infty, -\infty) = T_C \exp\left[-\frac{i}{\hbar}\int_C \hat{V}(t)dt\right], \tag{3.131}$$

where the contour-ordering operator T_C orders the operator following the contour C, which is taken to come back after a long time to the original point at $t = -\infty$, as shown in Fig 3.5. Then we define the contour-ordering Green's function by

$$G(\mathbf{r}, t : \mathbf{r}', t') = -\frac{i}{\hbar}\langle T_C \left\{\hat{\psi}(\mathbf{r}, t)\hat{\psi}^\dagger(\mathbf{r}', t')\right\}\rangle. \tag{3.132}$$

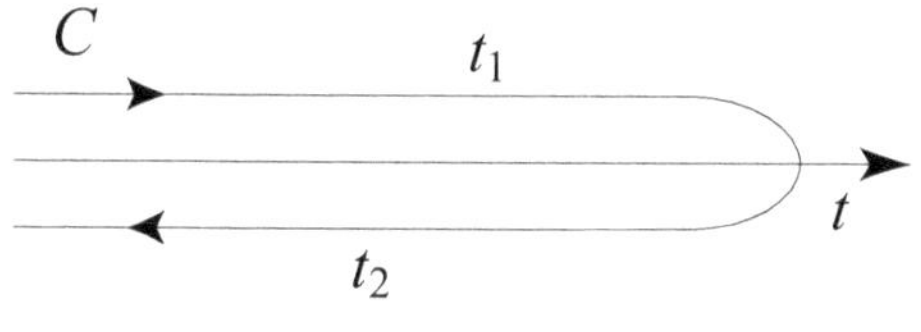

Figure 3.5 Time contour C to construct the NEGF. The system evolves from $t = -\infty$ and then continues the evolution up to t_0. Then it returns back to $t = -\infty$.

The important point of the formal theory of the NEGFs is to show that the perturbation expansion in the interaction picture using the contour-ordered S-matrix has the same structure as the equilibrium expansion. Thus, the self-energy operators are specified, the contour-ordering Green's function obeys the same Dyson equation as in Eq. (3.43).

However, the contour-ordered Green's function is not convenient for practical calculations. The solutions for this inconvenience is to introduce the 2×2 Green's function matrices, which requires at least four Green's functions. This corresponds to treat two time branches for the NEGF. Here we introduce six Green's functions, which are related to each other by the two equations. Thus, effectively four independent functions are present to describe the nonequilibrium systems [30]. In addition to the time-ordered Green's function $G^t(\mathbf{r}, t : \mathbf{r}', t')$ in Eq. (3.15), we define the lesser and the greater Green's functions as

$$\begin{aligned} G^{<}(\mathbf{r}, t : \mathbf{r}', t') &= \frac{i}{\hbar}\langle \hat{\psi}^{\dagger}(\mathbf{r}', t')\hat{\psi}(\mathbf{r}, t)\rangle \\ G^{>}(\mathbf{r}, t : \mathbf{r}', t') &= -\frac{i}{\hbar}\langle \hat{\psi}(\mathbf{r}, t)\hat{\psi}^{\dagger}(\mathbf{r}', t')\rangle \end{aligned} \tag{3.133}$$

and the anti-time-ordered Green's function $G^{\tilde{t}}(\mathbf{r}, t : \mathbf{r}', t')$. These constitute four Green's functions for the Keldysh formalism (Fig. 3.6) [31]. The other two are the retarded and the advanced Green's functions as described before

$$\begin{aligned} G^{r}(\mathbf{r}, t : \mathbf{r}', t') &= -\frac{i}{\hbar}\theta(t - t')\langle\left[\hat{\psi}(\mathbf{r}, t), \hat{\psi}^{\dagger}(\mathbf{r}', t')\right]\rangle \\ G^{a}(\mathbf{r}, t : \mathbf{r}', t') &= \frac{i}{\hbar}\theta(t' - t)\langle\left[\hat{\psi}(\mathbf{r}, t), \hat{\psi}^{\dagger}(\mathbf{r}', t')\right]\rangle. \end{aligned} \tag{3.134}$$

These Green's functions are expressed using the lesser and the greater Green's functions as

$$\begin{aligned} G^{t}(\mathbf{r}, t : \mathbf{r}', t') &= \theta(t - t')G^{>}(\mathbf{r}, t : \mathbf{r}', t') + \theta(t' - t)G^{<}(\mathbf{r}, t : \mathbf{r}', t')) \\ G^{\tilde{t}}(\mathbf{r}, t : \mathbf{r}', t') &= \theta(t' - t)G^{>}(\mathbf{r}, t : \mathbf{r}', t') + \theta(t - t')G^{<}(\mathbf{r}, t : \mathbf{r}', t')) \\ G^{r}(\mathbf{r}, t : \mathbf{r}', t') &= \theta(t - t')\left[G^{>}(\mathbf{r}, t : \mathbf{r}', t') - G^{<}(\mathbf{r}, t : \mathbf{r}', t')\right] \\ G^{a}(\mathbf{r}, t : \mathbf{r}', t') &= -\theta(t' - t)\left[G^{>}(\mathbf{r}, t : \mathbf{r}', t') - G^{<}(\mathbf{r}, t : \mathbf{r}', t')\right]. \end{aligned} \tag{3.135}$$

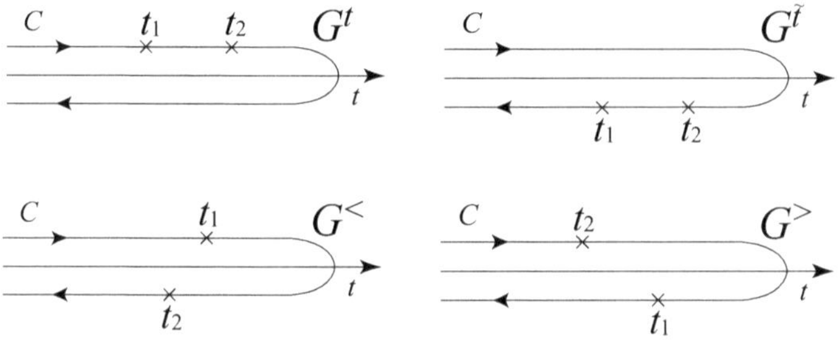

Figure 3.6 Time contour C for four Green's functions G^t, $G^{\bar{t}}$, $G^<$, and $G^>$ to constitute the Keldysh formalism. The system evolves from $t = -\infty$ and then continues the evolution up to t_0. Then it returns back to $t = -\infty$.

Note that since $\theta(t - t') + \theta(t' - t) = 1$, the following two relations connect six Green's functions:

$$\begin{aligned} G^r - G^a &= G^> - G^< \\ G^t + G^{\bar{t}} &= G^> + G^<. \end{aligned} \tag{3.136}$$

Thus, we have effectively four independent Green's functions.

The observables are expressed by these NEGFs. The electron density is written as

$$\hat{n}(\mathbf{r}, t) = -i\hbar G^<(\mathbf{r}, t : \mathbf{r}, t) \tag{3.137}$$

and the current density is expressed by

$$\begin{aligned} \hat{\mathbf{j}}(\mathbf{r}, t) &= \left\langle \frac{e\hbar}{2mi} \left[\hat{\psi}(\mathbf{r})^* \frac{\partial \hat{\psi}(\mathbf{r})}{\partial \mathbf{r}} - \hat{\psi}(\mathbf{r}) \frac{\partial \hat{\psi}(\mathbf{r})^*}{\partial \mathbf{r}} \right] \right\rangle \\ &= \left\langle \frac{e\hbar}{2mi} \lim_{\mathbf{r} \to \mathbf{r}'} \left(\nabla_{\mathbf{r}} - \nabla_{\mathbf{r}'} \right) \hat{\psi}^*(\mathbf{r}') \hat{\psi}(\mathbf{r}) \right\rangle \\ &= \frac{e\hbar^2}{2m} \lim_{\mathbf{r}' \to \mathbf{r}} \left(\nabla_{\mathbf{r}'} - \nabla_{\mathbf{r}} \right) G^<(\mathbf{r}, t : \mathbf{r}', t). \end{aligned} \tag{3.138}$$

The spectral function is expressed by the lesser and greater Green's function as

$$\begin{aligned} A(k, \epsilon) &= i \left[G^r(k, \epsilon) - G^a(k, \epsilon) \right] \\ &= i \left[G^>(k, \epsilon) - G^<(k, \epsilon) \right]. \end{aligned} \tag{3.139}$$

We can connect the spectral function $A(\epsilon)$ to the particle propagator $G^<(\epsilon)$ in the equilibrium system as follows. Using the complete set of eigenstates $|n\rangle$, we have

$$\begin{aligned} G^<(\epsilon) &= \frac{i}{\hbar}\int_{-\infty}^{\infty}\langle\hat{\psi}^\dagger(t)\hat{\psi}(0)\rangle e^{i\epsilon t/\hbar}dt \\ &= \frac{i}{\hbar}\frac{1}{Z}\int_{-\infty}^{\infty}\sum_{n,m}|\langle n|\hat{\psi}(0)|m\rangle|^2 e^{i(\epsilon+\epsilon_n-\epsilon_m)t/\hbar}e^{-\beta(\epsilon_n-\mu N_n)} \\ &= \frac{2\pi i}{Z}\sum_{n,m}|\langle n|\hat{\psi}(0)|m\rangle|^2 e^{-\beta(\epsilon_m-\mu N_m)}\delta(\epsilon+\epsilon_n-\epsilon_m). \end{aligned} \tag{3.140}$$

Here we assume the thermodynamic average defined by the chemical potential μ. This equation is satisfied for $N_m - N_n = 1$. In the same way,

$$\begin{aligned} G^>(\epsilon) &= -\frac{i}{\hbar}\int_{-\infty}^{\infty}\langle\hat{\psi}(t)\hat{\psi}^\dagger(0)\rangle e^{i\epsilon t/\hbar}dt \\ &= -\frac{2\pi i}{Z}\sum_{n,m}|\langle m|\hat{\psi}(0)|n\rangle|^2 e^{-\beta(\epsilon_m-\mu N_m)}\delta(\epsilon+\epsilon_m-\epsilon_n). \end{aligned} \tag{3.141}$$

Comparing these two equations, we realize that the difference is only the thermal factors. Substituting $\epsilon_m - \epsilon_n = \epsilon$ and $N_m - N_n = 1$, we obtain

$$G^>(\epsilon) = -e^{\beta(\epsilon-\mu)}G^<(\epsilon). \tag{3.142}$$

Thus, from the definition of $A(\epsilon)$ in Eq. (3.139), we have

$$G^<(\epsilon) = if(\epsilon)A(\epsilon) \tag{3.143}$$

for the electron propagation and

$$G^>(\epsilon) = -i\,(1 - f(\epsilon))\,A(\epsilon) \tag{3.144}$$

for the hole propagation. Here $f(\epsilon) = 1/(e^{\beta(\epsilon-\mu)} + 1)$ is the Fermi–Dirac distribution function. These relations show that $G^<$ and $G^>$ contain information about the occupation of the state in the presence of the distribution $f(\epsilon)$. On the other hand, the spectral function A is related to the decay of the system, since it contains information about how a particle in a state scatters out of that state. Therefore, these relations of Eqs. (3.143) and (3.144) are the representations for the *fluctuation–dissipation* theorem.

3.2.2 *Keldysh Formalism*

To calculate the lesser and the greater Green's functions, it is convenient to introduce 2 × 2 matrix [31] as

$$\hat{G} = \begin{bmatrix} G^t & -G^< \\ G^> & -G^{\tilde{t}} \end{bmatrix}, \qquad \hat{\Sigma} = \begin{bmatrix} \Sigma^t & -\Sigma^< \\ \Sigma^> & -\Sigma^{\tilde{t}} \end{bmatrix}, \tag{3.145}$$

where we express the self-energy also in 2 × 2 matrix form.

The Dyson equation for these 2 × 2 matrix becomes

$$\hat{G} = \hat{G}_0 + \hat{G}_0 \hat{\Sigma} \hat{G}. \tag{3.146}$$

To obtain explicit forms for $G^<$ and $G^>$, it is convenient if we introduce the following unitary transformation U:

$$\hat{U} = \frac{1}{\sqrt{2}} \begin{bmatrix} 1 & 1 \\ -1 & 1 \end{bmatrix}, \quad \hat{U}^{-1} = \hat{U}^{\dagger} = \frac{1}{\sqrt{2}} \begin{bmatrix} 1 & -1 \\ 1 & 1 \end{bmatrix}, \tag{3.147}$$

Then we obtain

$$\hat{U} \hat{G} \hat{U}^{\dagger} = \begin{bmatrix} G^r & -(G^< + G^>) \\ 0 & G^a \end{bmatrix}. \tag{3.148}$$

From the diagonal elements, we have

$$G^{r,a} = G_0^{r,a} + G_0^{r,a} \Sigma^{r,a} G^{r,a} \tag{3.149}$$

and from the off-diagonal elements, we have

$$\begin{aligned} G^> + G^< &= G^r (\Sigma^> + \Sigma^<) G^a \\ &\quad + [1 + G^r \Sigma^r] \left(G_0^> + G_0^<\right) [1 + \Sigma^a G^a]. \end{aligned} \tag{3.150}$$

Using the relation $G^> - G^< = G^r - G^a$, we obtain

$$\begin{aligned} G^< &= G^r \Sigma^< G^a + [1 + G^r \Sigma^r] G_0^< [1 + \Sigma^a G^a] \\ G^> &= G^r \Sigma^> G^a + [1 + G^r \Sigma^r] G_0^> [1 + \Sigma^a G^a]. \end{aligned} \tag{3.151}$$

Since the following expression is satisfied:

$$G_0^<(\epsilon) = \frac{i}{\hbar} \int_{-\infty}^{\infty} \langle \hat{c}^{\dagger} \hat{c} \rangle e^{i(\epsilon - \epsilon_0)t/\hbar} dt = 2\pi i f(\epsilon_0) \delta(\epsilon - \epsilon_0), \tag{3.152}$$

the second term does not contribute to electron density $\hat{n}(\mathbf{r})$ in usual cases, and thus we do not taken into account. So we obtain the explicit forms

$$\begin{aligned} G^<(\epsilon) &= G^r(\epsilon) \Sigma^<(\epsilon) G^a(\epsilon) \\ G^>(\epsilon) &= G^r(\epsilon) \Sigma^>(\epsilon) G^a(\epsilon), \end{aligned} \tag{3.153}$$

where electron propagator $G^<(\epsilon)$ gives electron density and hole propagator $G^>(\epsilon)$ gives hole density.

By evaluating the contour integration along the time loop in Fig. 3.5, we construct multiplication of two variables for the NEGF $A(t, t') = \int_C B(t, \tau)C(\tau, t')d\tau$. First let us consider $A^<(t, t')$ where t is on the upper half C_1 and t' is on the lower half C_2 by definition. Then we have

$$\begin{aligned}
A^<(t, t') &= \int_{C_1} B(t, \tau)C^<(\tau, t')d\tau + \int_{C_2} B^<(t, \tau)C(\tau, t')d\tau \\
&= \int_{-\infty}^{t} B^>(t, t_1)C^<(t_1, t')dt_1 + \int_{t}^{-\infty} B^<(t, t_1)C^<(t_1, t')dt_1 \\
&\quad + \int_{-\infty}^{t} B^<(t, t_1)C^<(t_1, t')dt_1 + \int_{t}^{-\infty} B^<(t, t_1)C^>(t_1, t')dt_1 \\
&= \int_{-\infty}^{\infty} \left[B^r(t, t_1)C^<(t_1, t') + B^<(t, t_1)C^a(t_1, t')\right] dt_1.
\end{aligned}$$

In the same way, for $A^>(t, t')$ where t' is on the upper half C_1 and t is on the lower half C_2 as in Fig. 3.6, we have the relation

$$A^>(t, t') = \int_{-\infty}^{\infty} \left[B^r(t, t_1)C^>(t_1, t') + B^>(t, t_1)C^a(t_1, t')\right] dt_1.$$

Next we consider the retarded Green's function $A^r(t, t')$. With use of the relations for $A^<(t, t')$ and $A^>(t, t')$, we have

$$\begin{aligned}
A^r(t, t') &= \theta(t - t')\left[A^>(t, t') - A^<(t, t')\right] \\
&= \theta(t - t') \int_{-\infty}^{\infty} \Big[B^r(t, t_1)\left(C^>(t_1, t') - C^<(t_1, t')\right) \\
&\qquad\qquad + \left(B^>(t, t_1) - B^<(t, t_1)\right) C^a(t_1, t')\Big] dt_1 \\
&= \theta(t - t')\left[\int_{-\infty}^{t} (B^> - B^<)(C^> - C^<)\, dt_1 \right. \\
&\qquad\qquad \left. + \int_{-\infty}^{t'} (B^> - B^<)(C^> - C^<)\, dt_1\right] \\
&= \int_{t'}^{t} B^r(t, t_1)C^r(t_1, t')dt_1.
\end{aligned}$$

These formulas are abbreviated as $A^< = \int(B^rC^< + B^<C^a)$, $A^r = \int B^rC^r$, $A^< = B^<C^<$, $A^r = B^<C^r + B^rC^< + B^rC^r$ and construct some of the Langreth theorem for the NEGFs [32].

3.2.3 *NEGF Formulation for Electron Transport*

Here we derive the transport formula from the following Hamiltonian, which is connected to the left (L) and right (R) electrodes with different chemical potentials $\mu_{L,R}$ [33] as

$$\begin{aligned}\hat{H} &= \hat{H}_L + \hat{H}_R + \hat{H}_C + \hat{V}_{LC} + \hat{V}_{LC}^{\dagger} + \hat{V}_{RC} + \hat{V}_{RC}^{\dagger} \\ &= \sum_k \epsilon_{kL}\hat{c}_{kL}^{\dagger}\hat{c}_{kL} + \sum_k \epsilon_{kR}\hat{c}_{kR}^{\dagger}\hat{c}_{kR} + \sum_n \epsilon_n \hat{d}_n^{\dagger}\hat{d}_n \\ &\quad + \sum_{k,n}\left[V_{kL,n}\hat{c}_{kL}^{\dagger}\hat{d}_n + V_{kL,n}^{*}\hat{d}_n^{\dagger}\hat{c}_{kL}\right] \\ &\quad + \sum_{k,n}\left[V_{kR,n}\hat{c}_{kR}^{\dagger}\hat{d}_n + V_{kR,n}^{*}\hat{d}_n^{\dagger}\hat{c}_{kR}\right]. \end{aligned} \tag{3.154}$$

Here $\hat{c}_k^{\dagger}\hat{c}_k$ represents electrons in the electrodes with continuous states ϵ_k labeled by k, while $\hat{d}_n^{\dagger}\hat{d}_n$ represents electrons in the nanosystem with discretized states ϵ_n labeled by n.

3.2.3.1 Expression for electric current

Electrical current is calculated from the time-dependent relation of the number of electrons $\hat{N}_L = \sum_k \hat{c}_{kL}^{\dagger}\hat{c}_{kL}$ by

$$I_L = -e\langle\frac{d\hat{N}_L}{dt}\rangle = -\frac{ie}{\hbar}\langle[\hat{H}, \hat{N}_L]\rangle = -\frac{ie}{\hbar}\langle\left[\hat{V}_{LC} + \hat{V}_{LC}^{\dagger}, \hat{N}_L\right]\rangle.$$

In the case of the Hamiltonian above, the current formula becomes

$$I_L = \frac{ie}{\hbar}\sum_{k,n}\left[V_{kL,n}\langle\hat{c}_{kL}^{\dagger}\hat{d}_n\rangle - V_{kL,n}^{*}\langle\hat{d}_n^{\dagger}\hat{c}_{kL}\rangle\right]. \tag{3.155}$$

Defining the NEGFs as

$$G_{n,kL}^{<}(t,t') = \frac{i}{\hbar}\langle\hat{c}_{kL}^{\dagger}(t')\hat{d}_n(t)\rangle, \quad G_{kL,n}^{<}(t,t') = \frac{i}{\hbar}\langle\hat{d}_n^{\dagger}(t')\hat{c}_{kL}(t)\rangle,$$

which satisfies the properties $G_{kL,n}^{<}(t=t') = -G_{n,kL}^{<}(t=t')$, we express the current for a steady state by

$$I_L = \frac{2e}{\hbar}\int\frac{d\epsilon}{2\pi}\mathrm{Re}\left(\sum_{k,n}V_{kL,n}G_{n,kL}^{<}(\epsilon)\right). \tag{3.156}$$

From the equation of motion, $G_{n,kL}^{<}(\epsilon)$ is expressed by the multiples of Green's function for electrodes and those for nanosystems

$$G_{n,kL}^{<}(\epsilon) = \sum_m V_{kL,m}^{*}\left[G_{nm}^{r}(\epsilon)g_{kL}^{<}(\epsilon) + G_{nm}^{<}(\epsilon)g_{kL}^{a}(\epsilon)\right], \tag{3.157}$$

where we use the Langreth theorem with the NEGFs of nanosystems defined by

$$G^r_{n,m}(t,t') = -\frac{i}{\hbar}\theta(t-t')\langle\left[\hat{d}_n(t), \hat{d}^\dagger_m(t')\right]\rangle$$

$$G^<_{n,m}(t,t') = \frac{i}{\hbar}\langle\hat{d}^\dagger_n(t')\hat{d}_n(t)\rangle \tag{3.158}$$

and those of electrodes by

$$g^<_{kL}(t,t') = \frac{i}{\hbar}\langle\hat{c}^\dagger_{kL}(t')\hat{c}_{kL}(t)\rangle$$

$$g^a_{kL}(t,t') = \frac{i}{\hbar}\theta(t'-t)\langle\left[\hat{c}_{kL}(t), \hat{c}^\dagger_{kL}(t')\right]\rangle. \tag{3.159}$$

Then the current formula for I_L becomes

$$I_L = \frac{2e}{\hbar}\int\frac{d\epsilon}{2\pi}\mathrm{Re}\left(\sum_{k,n,m} V_{kL,n}V^*_{m,kL}\left[G^r_{n,m}(\epsilon)g^<_{kL}(\epsilon) + G^<_{n,m}(\epsilon)g^a_{kL}(\epsilon)\right]\right).$$

Here we assume that electrons are noninteracting inside the electrodes keeping a steady state and thus Green's functions of electrodes remain in a Fourier space such as

$$g^<_{kL}(\epsilon) = 2\pi i f_L(\epsilon)\delta(\epsilon-\epsilon_{kL}),\quad g^a_{kL}(\epsilon) = i\pi\delta(\epsilon-\epsilon_{kL}).$$

Defining the coupling to electrodes with multiple modes by

$$\Gamma^L_{n,m}(\epsilon) = 2\pi\sum_k \nu^L(\epsilon)V_{kL,n}(\epsilon)V^*_{m,kL}(\epsilon), \tag{3.160}$$

we have the formula for the matrix equation of

$$I_L = \frac{ie}{\hbar}\int\frac{d\epsilon}{2\pi}\mathrm{Tr}\left\{\Gamma^L(\epsilon)\left[G^<(\epsilon) + f_L(\epsilon)\left(G^r(\epsilon) - G^a(\epsilon)\right)\right]\right\}.$$

Since in a steady state $I_R = -I_L$, the total current is obtained from $I = (I_L - I_R)/2$, which gives

$$I = \frac{ie}{2h}\int\mathrm{Tr}\left\{\left[\Gamma^L(\epsilon) - \Gamma^R(\epsilon)\right]G^<(\epsilon) + \left[f_L(\epsilon)\Gamma^L(\epsilon) - f_R(\epsilon)\Gamma^R(\epsilon)\right]\left[G^r(\epsilon) - G^a(\epsilon)\right]\right\}d\epsilon. \tag{3.161}$$

This is the expression for the current flowing between electrodes. The assumption is that electrons are noninteracting inside electrodes where steady states are kept even if electron loses its energy via various inelastic processes in the electrodes. Note that I is expressed in terms of the distribution function $f_{L/R}$ in the electrodes, the DOS $G^r - G^a$, and occupation $-i\hbar G^<$ in the nanosystems. In equilibrium, since $f_L(\epsilon) = f_R(\epsilon) = f(\epsilon)$, by using the fluctuation–dissipation relation $G^<(\epsilon) = if(\epsilon)A(\epsilon) = -f(\epsilon)\left(G^r(\epsilon) - G^a(\epsilon)\right)$, the current is vanishing with $I = 0$.

3.2.4 *Applications to Various Systems*

Here we give several transport formulas using Green's functions for various nanosystems. These include transports through resonant tunneling, under on-site Coulomb interaction, with electron–phonon interaction, and time-dependent dynamics. We note that the retarded Green's function $G^r(\epsilon)$ is obtained from the temperature Green's function $\mathcal{G}(\epsilon)$, which enables us to obtain electron transport within the linear-response regime.[a]

3.2.4.1 Transport through resonance levels

First, we consider that electrons in the nanosystem are noninteracting and the current flows through the resonance levels. We can use the former results for the resonance tunneling model and extend it to the nonequilibrium situation. To obtain the current, we need $G^<(\epsilon)$ and $\Sigma^<(\epsilon)$, which are given by

$$\begin{aligned}\Sigma^<(\epsilon) &= \sum_k |V_{kL}|^2 g^<_{kL}(\epsilon) + \sum_k |V_{kR}|^2 g^<_{kR}(\epsilon)\\ &= i\left(f_L(\epsilon)\Gamma^L(\epsilon) + f_R(\epsilon)\Gamma^R(\epsilon)\right)\\ G^<(\epsilon) &= G^r(\epsilon)\Sigma^<(\epsilon)G^a(\epsilon)\\ &= G^r(\epsilon)\left(if_L(\epsilon)\Gamma^L(\epsilon) + if_R(\epsilon)\Gamma^R(\epsilon)\right)G^a(\epsilon). \qquad (3.162)\end{aligned}$$

Putting the relation

$$G^r(\epsilon) - G^a(\epsilon) = -iG^r(\epsilon)\left(\Gamma^L(\epsilon) + \Gamma^R(\epsilon)\right)G^a(\epsilon), \qquad (3.163)$$

we have (including 2 for the spin)

$$I = \frac{2e}{h}\int \left(f_L(\epsilon) - f_R(\epsilon)\right)\mathrm{Tr}\left[G^r(\epsilon)\Gamma^L(\epsilon)G^a(\epsilon)\Gamma^R(\epsilon)\right]d\epsilon. \qquad (3.164)$$

This corresponds to the Landauer formula with the transmission coefficient $\mathcal{T}(\epsilon)$ obtained from

$$\mathcal{T}(\epsilon) = \mathrm{Tr}\left[G^r(\epsilon)\Gamma^L(\epsilon)G^a(\epsilon)\Gamma^R(\epsilon)\right]. \qquad (3.165)$$

This is called the Fisher–Lee relation [34].

[a] Numerical methods by the Quantum Monte Carlo (QMC) technique are effective for obtaining $\mathcal{G}(\epsilon)$ in the interacting case. The temperature Green's function $\mathcal{G}(\epsilon)$ and QMC are treated briefly in the problems.

This formula can also be obtained from the direct calculations of the current density (here we include 2 for the spin)

$$\begin{aligned}\hat{\mathbf{j}}(\mathbf{r}) &= 2\frac{e\hbar^2}{2m}\lim_{\mathbf{r}'\to\mathbf{r}}\left(\nabla_{\mathbf{r}'}-\nabla_{\mathbf{r}}\right)G^<(\mathbf{r},t:\mathbf{r}',t)\\ &= 2\frac{e\hbar}{2m}\int\frac{d\epsilon}{2\pi}\lim_{\mathbf{r}'\to\mathbf{r}}\left(\nabla_{\mathbf{r}'}-\nabla_{\mathbf{r}}\right)G^<(\mathbf{r},\mathbf{r}',\epsilon)\end{aligned} \tag{3.166}$$

such as

$$\begin{aligned}I &= \int_S \hat{\mathbf{j}}(\mathbf{r})d\mathbf{S} = \int_\Omega \nabla\cdot\hat{\mathbf{j}}(\mathbf{r})d\mathbf{r}\\ &= \int_\Omega\left[2\frac{e\hbar}{2m}\int\frac{d\epsilon}{2\pi}\lim_{\mathbf{r}'\to\mathbf{r}}\left(\nabla^2_{\mathbf{r}'}-\nabla^2_{\mathbf{r}}\right)G^<(\mathbf{r},\mathbf{r}',\epsilon)\right]d\mathbf{r}\\ &= \frac{2e}{h}\int \mathrm{Tr}\left[H(\epsilon)G^<(\epsilon)-G^<(\epsilon)H(\epsilon)\right]d\epsilon\\ &= \frac{2e}{h}\int \mathrm{Tr}\left[\Sigma^<(\epsilon)G^>(\epsilon)-G^>(\epsilon)\Sigma^<(\epsilon)\right]d\epsilon\\ &= \frac{2e}{h}\int \left(f_L(\epsilon)-f_R(\epsilon)\right)\mathrm{Tr}\left[G^r(\epsilon)\Gamma^L(\epsilon)G^a(\epsilon)\Gamma^R(\epsilon)\right]d\epsilon.\end{aligned}$$

3.2.4.2 Transport with Coulomb interactions

The Coulomb interaction is in general described as

$$U(\mathbf{r},\mathbf{r}') = \frac{e^2}{|\mathbf{r}-\mathbf{r}'|}. \tag{3.167}$$

Here we consider that Coulomb interaction is effective only in the nanosystem, which is expressed by

$$\hat{U}_{\text{Coulomb}} = \sum_{\sigma\sigma'}\int\int \hat{n}_{\sigma'}(\mathbf{r}')U(\mathbf{r},\mathbf{r}')\hat{n}_\sigma(\mathbf{r})d\mathbf{r}d\mathbf{r}' \tag{3.168}$$

with $\hat{n}_\sigma(\mathbf{r}) = \psi^\dagger_\sigma(\mathbf{r})\psi_\sigma(\mathbf{r})$ and in the other parts such as in the electrodes and in the coupling between electrodes and nanosystem, it is treated as renormalized $\langle U_{\text{Coulomb}}\rangle$ in the single-particle energy.

Let us consider a simple case so that the Coulomb interaction is effective only at one site as $\hat{U}_{\text{Coulomb}} = U\hat{n}_\uparrow\hat{n}_\downarrow$. In the presence of the Coulomb interaction, we need to include the spin explicitly (Fig. 3.7). This is well seen when we change this in the mean-field (Hartree–Fock) approximation[a] as

$$U\hat{n}_\uparrow\hat{n}_\downarrow \approx U\sum_\sigma\langle\hat{n}_{\bar{\sigma}}\rangle\hat{n}_\sigma - U\langle\hat{n}_\uparrow\rangle\langle\hat{n}_\downarrow\rangle, \tag{3.169}$$

[a]We neglect the fluctuation term by $U(\hat{n}_\uparrow - \langle\hat{n}_\uparrow\rangle)(\hat{n}_\downarrow - \langle\hat{n}_\downarrow\rangle)$, which is important for the higher-order approximation resulting in the Kondo transport.

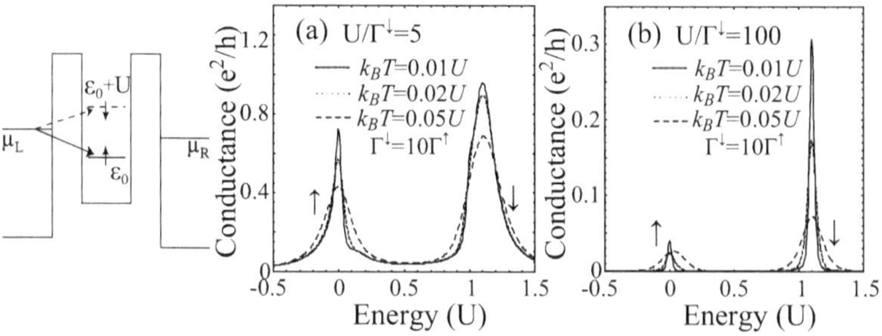

Figure 3.7 (Left) Schematic view of the band diagram of junction system with strong Coulomb interaction. When one electron occupies a quantum state ϵ_0, an electron coming from the left electrode feels higher energy by an amount of U due to the Coulomb interaction. (Right) Conductance as a function of chemical potential energy in units of U at three temperatures for (a) $U/\Gamma^{\downarrow} = 5$ and (b) $U/\Gamma^{\downarrow} = 100$ with $\Gamma^{\downarrow} = 10\Gamma^{\uparrow}$. Here we split the levels by $\epsilon_{\downarrow} - \epsilon_{\uparrow} = \Delta_0 = 0.1U$. Two peaks of conductance are seen at $\epsilon_{\uparrow}$ and $\epsilon_{\downarrow} + U$ with the splitting by $U + \Delta_0 = 1.1U$. The situation in (b) corresponds to the Coulomb blockade regime.

which shows that electrons feel the effective potential of $U\langle n_{\bar{\sigma}}\rangle$ from the opposite spin state (problem 2).

Thus, we define Green's functions, including the spin σ, as

$$G^r_\sigma(t,t') = -\frac{i}{\hbar}\theta(t-t')\langle\left[\hat{d}_\sigma(t), \hat{d}^\dagger_\sigma(t')\right]\rangle. \qquad (3.170)$$

When the Coulomb interaction in the nanosystem is very strong compared with the hopping energy $U \gg V$, we expect that the nature of Green's functions in the nanosystem $G_\sigma(\epsilon)$ determines the behavior of the current. However, in the present form, it looks very difficult to see it from the formula. In order to see the relevance with non-interacting cases, we change the formula with an assumption that the coupling to electrodes differs only by a constant factor as

$$\Gamma^L_\sigma(\epsilon) = \lambda\Gamma^R_\sigma(\epsilon). \qquad (3.171)$$

In this case, the current expression is greatly simplified as

$$\begin{aligned} I &= \frac{ie}{h}\sum_\sigma \int [f_L(\epsilon) - f_R(\epsilon)]\,\mathrm{Tr}\left\{\tilde{\Gamma}_\sigma(\epsilon)\left(G^r_\sigma(\epsilon) - G^a_\sigma(\epsilon)\right)\right\} d\epsilon \\ &= \frac{e}{\hbar}\sum_\sigma \int [f_L(\epsilon) - f_R(\epsilon)]\,\mathrm{Tr}\left\{\tilde{\Gamma}_\sigma(\epsilon)\left[-\frac{1}{\pi}\mathrm{Im}G^r_\sigma(\epsilon)\right]\right\} d\epsilon \end{aligned}$$

where the proportional coupling constant $\tilde{\Gamma}_\sigma(\epsilon)$ is

$$\tilde{\Gamma}_\sigma(\epsilon) = \frac{\Gamma^L_\sigma(\epsilon)\Gamma^R_\sigma(\epsilon)}{\Gamma^L_\sigma(\epsilon) + \Gamma^R_\sigma(\epsilon)}. \tag{3.172}$$

This means that LDOS $-(1/\pi)\mathrm{Im}G^r_\sigma(\epsilon)$ determines the behavior of the current [33]. We will see later that $G^r_\sigma(\epsilon)$ has resonance peaks at ϵ_n and $\epsilon_n + U$.

Figure 3.7 shows the conductance as a functional of chemical potential energy for (a) $U/\Gamma^\downarrow = 5$ and (b) $U/\Gamma^\downarrow = 100$. We note that the situation in (b) with $U \gg \Gamma$ corresponds to the Coulomb blockade regime. The linear-response conductance is computed after we obtain the self-consistent solution for $\langle \hat{n}_\sigma \rangle = \int (-1/\pi)\mathrm{Im}G^r_\sigma(\epsilon) f(\epsilon) d\epsilon$. We see two peaks of conductance at $\epsilon_\uparrow$ and $\epsilon_\downarrow + U$ with the splitting by $U + \Delta_0 = 1.1U$. The temperature dependence of the peaks shows $\propto 1/T$ for height and $\propto T$ for width. The conductance behaviors in the Coulomb blockade regime for quantum dots are treated more precisely in Chapter 6.

3.2.4.3 Transport with electron–phonon interactions

For treating the scatterings from phonons in the weak coupling case, we limit the small displacement of atomic positions where the Hamiltonian for electron–phonon coupling systems is described by

$$\hat{H} = \hat{H}_0 + \sum_j \int \hat{\psi}^\dagger(\mathbf{r}) \left. \frac{\partial V_{\mathrm{ion}}(\mathbf{r} - R_j)}{\partial R_j} \right|_{\delta R_j = 0} \delta \hat{R}_j \hat{\psi}(\mathbf{r}) d\mathbf{r}. \tag{3.173}$$

Here $\delta \hat{R}_j = \hat{R} - \hat{R}_j$ is displacement variables around equilibrium positions of atoms $\hat{R}_j$ where the variable j runs all nuclei. Usually the normal coordinates $\hat{u}$ with $\delta \hat{R}_j = \sum_q e^{iq\cdot R_j} \hat{u}_q / \sqrt{M_j}$ are calculated from the classical Newton's equation with the frequency ω_q as $\partial^2 \hat{u}_q / \partial t^2 = -\omega_q^2 \hat{u}_q$ and are expressed quantum-mechanically by

$$\delta \hat{R}_j = \sum_q \sqrt{\frac{\hbar}{2 M_j \omega_q}} \left(\hat{a}_q e^{iq\cdot R_j} + \hat{a}^\dagger_q e^{-iq\cdot R_j} \right). \tag{3.174}$$

Here $\hat{a}^\dagger$ and $\hat{a}$ are the creation and annihilation operators of phonons, which obey the bosonic commutator relations as

$$\left[\hat{a}_q, \hat{a}^\dagger_{q'}\right]_- = \hat{a}_q \hat{a}^\dagger_{q'} - \hat{a}^\dagger_{q'} \hat{a}_q = \delta_{q,q'}, \quad \left[\hat{a}_q, \hat{a}_{q'}\right]_- = 0, \quad \left[\hat{a}^\dagger_q, \hat{a}^\dagger_{q'}\right]_- = 0. \tag{3.175}$$

We explicitly write the commutator relation for a phonon as

$$\left[\hat{A},\, \hat{B}\right]_{-} = \hat{A}\hat{B} - \hat{B}\hat{A} \tag{3.176}$$

to distinguish from that of an electron $\left[\hat{A},\, \hat{B}\right] = \hat{A}\hat{B} + \hat{B}\hat{A}$. Then the occupation number is obtained from

$$N_q = \langle \hat{a}_q^\dagger \hat{a}_q \rangle = \frac{1}{e^{\hbar\omega_q/k_B T} - 1}. \tag{3.177}$$

The Hamiltonian with electron–phonon interaction is written for plane-wave $\hat{\psi}(\mathbf{r}) = \sum_k \hat{c}_k e^{ik\cdot r}$ in the weak coupling regime[a] as

$$\hat{V}_{\text{e-ph}} = \sum_{k,q} \gamma_q \hat{c}_{k+q}^\dagger \hat{c}_k \left(\hat{a}_q + \hat{a}_{-q}^\dagger\right) \tag{3.178}$$

where the electron–phonon coupling γ_q is given by

$$\gamma_q = \sum_j \sqrt{\frac{\hbar}{2M_j\omega_q}} \left(\frac{\partial V_{\text{ion}}(q)}{\partial R_j}\right). \tag{3.179}$$

First we consider a resonant tunneling model with a single resonant level ϵ_0 coupled to electron–phonon interactions,

$$\begin{aligned}\hat{H} = &\sum_{k\alpha\in LR} \epsilon_{k\alpha} \hat{c}_{k\alpha}^\dagger \hat{c}_{k\alpha} + \epsilon_0 \hat{d}^\dagger \hat{d} + \sum_{k\alpha\in LR} \left[V_{k\alpha,n} \hat{c}_{k\alpha}^\dagger \hat{d} + h.c.\right] \\ &+ \sum_q \gamma_q \hat{d}^\dagger \hat{d} \left(\hat{a}_q + \hat{a}_{-q}^\dagger\right).\end{aligned} \tag{3.180}$$

Within the proportional coupling approximation in Eq. (3.172) with energy-independent Γ, the elastic current is obtained from

$$I = \frac{e}{h}\frac{\Gamma^L\Gamma^R}{\Gamma^L + \Gamma^R} \int \left[f_L(\epsilon) - f_R(\epsilon)\right] d\epsilon \int_{-\infty}^{\infty} A(t) e^{i\epsilon t/\hbar} dt, \tag{3.181}$$

where the spectral function is $A(t) = i\left(G^r(t) - G^a(t)\right)$ and we include the spin degeneracy. The analytical solution to $G^r(t)$ is possible for this Hamiltonian if we neglect the Fermi sea as

$$G^r(t) = -\frac{i}{\hbar}\theta(t)\exp\left[-\frac{i}{\hbar}(\epsilon_0 - \Delta)t - \Phi(t) - \frac{1}{\hbar}\frac{\Gamma}{2}t\right], \tag{3.182}$$

---

[a] It is expressed as $\hat{V}_{\text{e-ph}} = \sum_{n,m,q} \gamma_{n,m,q} \hat{c}_n^\dagger \hat{c}_m \left(\hat{a}_q + \hat{a}_{-q}^\dagger\right)$ with $\gamma_{n,m,q} = \sum_j \sqrt{\hbar/2M_j\omega_q}\langle n|\partial V_{\text{ion}}/\partial R_j|m\rangle$, where we take $\hat{\psi} = \sum_m \hat{c}_m|m\rangle$ and consider the transitions between states due to electron–phonon interactions. This expression is used in the quantum master equation in Chapter 4.

with

$$\Phi(t) = \sum_q \frac{|\gamma_q|^2}{(\hbar\omega_q)^2} \left[N_q\left(1 - e^{i\omega_q t}\right) + (N_q + 1)\left(1 - e^{-i\omega_q t}\right)\right].$$

$$\Delta = \sum_q \frac{|\gamma_q|^2}{\hbar\omega_q}. \tag{3.183}$$

Let us consider a model [35] with an optical phonon ($\hbar\omega_0$) couples with electrons at $T = 0$ ($N_q = 0$), where we have

$$G^r(t) = -\frac{i}{\hbar}\theta(t)e^{-g}\sum_n \frac{g^n}{n!}\exp\left[\frac{i}{\hbar}\left(-\epsilon_0 + \Delta - n\hbar\omega_0 + i\frac{\Gamma}{2}\right)t\right]. \tag{3.184}$$

Here $g = \sum_q |\gamma_q|^2/(\hbar\omega_q)^2$ is the coupling constant with $\Delta = g\hbar\omega_0$ and n denotes the number of phonons. Figure 3.8 shows the transmission curves with and without the electron–phonon interaction. We see that the resonance peak shifts from ϵ_0 to $\epsilon = \epsilon_0 - g\hbar\omega_0$ due to phonon emissions. Since the sum rule for the spectral function by $\int_{-\infty}^{\infty} A(\epsilon)d\epsilon/2\pi = e^{-g}\sum_n g^n/n! = 1$ is satisfied, the resonance peak shift accompanies the sideband peaks at $\epsilon + n\hbar\omega_0$, which are visible up to the second sideband peaks.

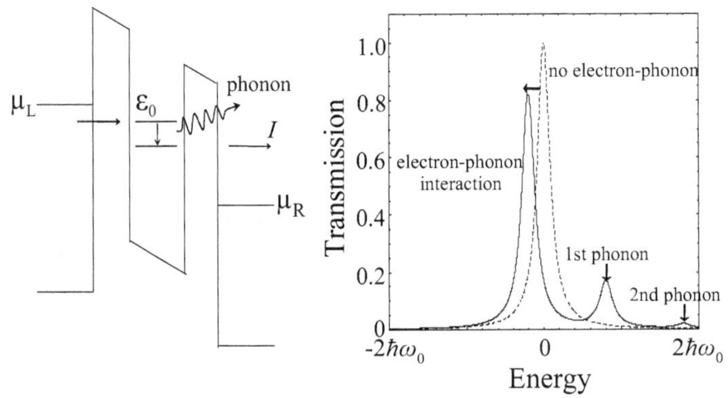

Figure 3.8 (Left) Schematic view of the band diagram of double-barrier resonant-tunneling system with electron–phonon interaction. (Right) Total transmission as a function of energy with (solid line) and without (dashed line) electron–phonon interaction for the coupling constant of $g = 0.2$ with $\epsilon_0 = 0$ and $\Gamma = 0.2\hbar\omega_0$. The sidebands due to phonon emissions are observed for $n = 1, 2$.

Next we consider general formulas for currents by putting the effects of electron–phonon scatterings into Green's functions through the self-energies

$$G^<(\epsilon) = G^r(\epsilon)\left(i f_L(\epsilon)\Gamma^L(\epsilon) + i f_R(\epsilon)\Gamma^R(\epsilon) + \Sigma^<_{\text{e-ph}}(\epsilon)\right) G^a(\epsilon)$$

$$G^r(\epsilon) - G^a(\epsilon) = -iG^r(\epsilon)\left(\Gamma^L(\epsilon) + \Gamma^R(\epsilon) + \Gamma_{\text{e-ph}}(\epsilon)\right) G^a(\epsilon)$$

where

$$\Gamma_{\text{e-ph}}(\epsilon) = i\left(\Sigma^r_{\text{e-ph}}(\epsilon) - \Sigma^a_{\text{e-ph}}(\epsilon)\right). \tag{3.185}$$

We have the following two contributions for the current [27]:

$$\begin{aligned} I_{\text{elastic}} &= \frac{2e}{h}\int (f_L(\epsilon) - f_R(\epsilon))\,\mathrm{Tr}\left[G^r(\epsilon)\Gamma^L(\epsilon)G^a(\epsilon)\Gamma^R(\epsilon)\right] d\epsilon \\ I_{\text{inelastic}} &= \frac{e}{h}\int \mathrm{Tr}\left\{i\left[\Gamma^L(\epsilon) - \Gamma^R(\epsilon)\right] G^r(\epsilon)\Sigma^<_{\text{e-ph}}(\epsilon)G^a(\epsilon)\right. \\ &\quad \left. + \left[f_L(\epsilon)\Gamma^L(\epsilon) - f_R(\epsilon)\Gamma^R(\epsilon)\right] G^r(\epsilon)\Gamma_{\text{e-ph}}(\epsilon)G^a(\epsilon)\right\} d\epsilon. \end{aligned} \tag{3.186}$$

The first equation has the same form for a non-interacting case and thus describes an elastic current. The second equation is new and describes an inelastic current causing energy relaxation. Here Green's function $G^r(\epsilon)$, including electron–phonon scatterings, is obtained from the Dyson equation

$$G^r(\epsilon) = G^r_0(\epsilon) + G^r_0(\epsilon)\Sigma^r_{\text{e-ph}}(\epsilon)G^r(\epsilon), \tag{3.187}$$

where $G^r_0(\epsilon)$ is Green's function without phonon scatterings. It should be noted that even if the present formula for the elastic current in Eq. (3.186) has the same form with the noninteracting case, it includes the effects from phonon scatterings through Green's function $G^r(\epsilon)$ in Eq. (3.187), and that this acts as a source of resistance. The self-energies of $\Sigma^{r,a}_{\text{e-ph}}(\epsilon)$ and $\Sigma^<_{\text{e-ph}}(\epsilon)$ are evaluated from the well-known bubble diagrams.[a] In the simple BA, we evaluate it as

$$\Sigma_{\text{e-ph}}(t, t') = i\sum_q |\gamma_q|^2 G_0(t, t')D_q(t, t'), \tag{3.188}$$

[a] Migdal's theorem states that the higher-order contributions for the vertex corrections are at least of the order $\gamma\sqrt{m/M}$ where m is the mass of electrons and M is the mass of atoms.

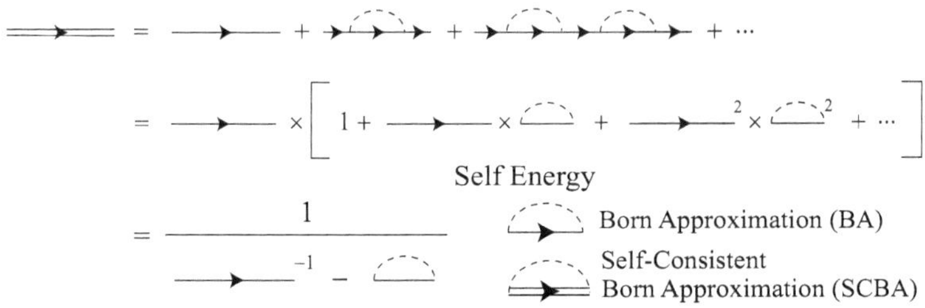

Figure 3.9 Diagrammatic summation of one-phonon diagram. One-phonon self-energy is constructed from the Born approximation (BA). When electron free propagator is replaced by the full propagator, it corresponds to the self-consistent Born approximation (SCBA).

where $D_q(t,t') = -(i/\hbar)\langle T_C\{\hat{A}_q(t)\hat{A}_{-q}(t')\}\rangle$ is the free-phonon Green's function with the normalized normal coordinates of $\hat{A}_q(t) = \hat{a}_q e^{-i\omega_q t} + \hat{a}^\dagger_{-q}e^{i\omega_q t}$. Then the Fourier transforms of $D_q(t,t')$ are expressed in the same way as Eqs. (3.133) and (3.134) by

$$D_q^<(\omega) = -2\pi i\left[(N_q+1)\delta(\omega+\omega_q) + N_q\delta(\omega-\omega_q)\right]$$
$$D_q^r(\omega) = \frac{1}{\hbar\omega - \hbar\omega_q + i0^+} - \frac{1}{\hbar\omega + \hbar\omega_q + i0^+}, \qquad (3.189)$$

with $\langle\hat{a}_q\hat{a}_q^\dagger\rangle = N_q + 1$ and $\langle\hat{a}_q^\dagger\hat{a}_q\rangle = N_q$. The Fourier transforms of $\Sigma_{\text{e-ph}}$ are obtained using the Keldysh contour techniques as [24]

$$\Sigma^<_{\text{e-ph}}(\epsilon) = \frac{i}{2\pi}\sum_q\int|\gamma_q|^2 G_0^<(\epsilon-\epsilon')D_q^<(\epsilon')d\epsilon'$$
$$\Sigma^r_{\text{e-ph}}(\epsilon) = \frac{i}{2\pi}\sum_q\int|\gamma_q|^2\left[G_0^<(\epsilon-\epsilon')D_q^r(\epsilon')\right.$$
$$\left. + G_0^r(\epsilon-\epsilon')D_q^<(\epsilon') + G_0^r(\epsilon-\epsilon')D_q^r(\epsilon')\right]d\epsilon'. \qquad (3.190)$$

The procedure is that we use the bare Green's functions $G_0^{r,<}(\epsilon)$ without electron–phonon scatterings and obtain the self-energies for phonon $\Sigma^{r,<}_{\text{e-ph}}(\epsilon)$ first. Then these are used to obtain the dressed Green's functions $G^{r,<}(\epsilon)$ and the currents are calculated from these Green's functions. In the SCBA, we use the dressed Green's functions for the self-energy $\Sigma_{\text{e-ph}}(t,t')$ instead of Eq. (3.188) by $\Sigma_{\text{e-ph}}(t,t') = i\sum_q|\gamma_q|^2 G(t,t')D_q(t,t')$, and then the equations are solved self-consistently (Fig. 3.9).

3.2.5 *Electron Dynamics in NEGF Formalism*

Here we treat the calculation methods for time-dependent transport. Electron dynamics in time-dependent phenomena are very important and central issues for many device applications, including *ac* response current, pumping, response to the femtosecond laser, quantum information devices, and others to control electron dynamics between ground and excited energy states. We note that the present formalism for electron dynamics is also important for the TDDFT approach to transport problems in Chapter 4.

We extend the NEGF formalism to treat time-dependent dynamical transport for the time-dependent Hamiltonian [36],

$$\hat{H}(t) = \sum_k \epsilon_{kL}(t)\hat{c}^\dagger_{kL}\hat{c}_{kL} + \sum_k \epsilon_{kR}(t)\hat{c}^\dagger_{kR}\hat{c}_{kR} + \sum_n \epsilon_n(t)\hat{d}^\dagger_n\hat{d}_n$$
$$+ \sum_{k,n}\left[V_{kL,n}(t)\hat{c}^\dagger_{kL}\hat{d}_n + \text{h.c.}\right] + \sum_{k,n}\left[V_{kR,n}(t)\hat{c}^\dagger_{kR}\hat{d}_n + \text{h.c.}\right].$$

The time-dependent current $I_L(t)$ given by

$$I_L(t) = -e\langle\frac{d\hat{N}_L}{dt}\rangle = 2e\,\text{Re}\left\{\sum_{k,n} V_{kL,n}(t)G^<_{n,kL}(t,t)\right\} \qquad (3.191)$$

with $\hat{N}_L = \sum_k \hat{c}^\dagger_{kL}\hat{c}_{kL}$ and $G^<_{n,kL}(t,t') \equiv (i/\hbar)\langle\hat{c}^\dagger_{kL}(t')\hat{d}_n(t)\rangle$. Here using the Langreth theorem for contour integration, we obtain

$$G^<_{n,kL}(t,t') = \sum_m \int V^*_{kL,m}(t_1)\left[G^r_{n,m}(t,t_1)g^<_{kL}(t_1,t')\right.$$
$$\left.+G^<_{n,m}(t,t_1)g^a_{kL}(t_1,t')\right]dt_1 \quad (3.192)$$

with $G^r_{n,m}(t,t') = -(i/\hbar)\theta(t-t')\langle\left[\hat{d}_n(t), \hat{d}^\dagger_m(t')\right]\rangle$ and $G^<_{n,m}(t,t') = (i/\hbar)\langle\hat{d}^\dagger_n(t')\hat{d}_n(t)\rangle$ and Green's functions of electrodes are

$$g^<_{kL}(t,t') = \frac{i}{\hbar}f(\epsilon_{kL})\exp\left[-\frac{i}{\hbar}\int_{t'}^t \epsilon_{kL}(t_1)dt_1\right]$$
$$g^a_{kL}(t,t') = \frac{i}{\hbar}\theta(t'-t)\exp\left[-\frac{i}{\hbar}\int_{t'}^t \epsilon_{kL}(t_1)dt_1\right]. \qquad (3.193)$$

Defining the time-dependent coupling constant $\left[\Gamma^L(\epsilon,t',t)\right]_{n,m}$ by

$$\left[\Gamma^L(\epsilon,t',t)\right]_{n,m} = 2\pi\sum_k \nu^L_k(\epsilon)V_{kL,n}(t)V^*_{kL,m}(t')$$
$$\times\exp\left[\frac{i}{\hbar}\int_{t'}^t \epsilon_{kL}(t_1)dt_1\right], \qquad (3.194)$$

where $\nu_k^L(\epsilon)$ is the DOS for the L electrode, the formula for the time-dependent current $I_L(t)$ becomes

$$I_L(t) = -\frac{2e}{\hbar}\int_{-\infty}^{t} dt' \int \frac{d\epsilon}{2\pi} \mathrm{ImTr}\left\{e^{i\epsilon(t-t')/\hbar}\Gamma^L(\epsilon, t', t)\right.$$
$$\left.\times\left[G^<(t, t') + f_L(\epsilon)G^r(t, t')\right]\right\}. \qquad (3.195)$$

Here the first term indicates the out-tunneling rate and the second term indicates the in-tunneling rate.[a]

In the non-interacting case, the retarded Green's function is obtained from the Dyson equation as

$$G^r(t, t') = g^r(t, t') + \int\int g^r(t, t_1)\Sigma^r(t_1, t_2)G^r(t_2, t')dt_1dt_2, \quad (3.196)$$

where the retarded self-energy $\Sigma^r(t_1, t_2)$ is given by

$$\Sigma^r_{n,n'}(t_1, t_2) = \sum_{k\alpha\in L,R} V^*_{k\alpha,n}(t_1)g^r_{k\alpha}(t_1, t_2)V_{k\alpha,n'}(t_2), \qquad (3.197)$$

and the lesser Green's function is obtained from

$$G^<(t, t') = \int\int G^r(t, t_1)\Sigma^<(t_1, t_2)G^a(t_2, t')dt_1dt_2, \qquad (3.198)$$

with the lesser self-energy $\Sigma^<(t_1, t_2)$ given by

$$\Sigma^<(t_1, t_2) = \frac{i}{\hbar}\sum_{L,R}\int \frac{d\epsilon}{2\pi} e^{i\epsilon(t_2-t_1)/\hbar} f_{L/R}(\epsilon)\Gamma^{L/R}(\epsilon, t_1, t_2).$$

From these Green's functions, $I_L(t)$ is obtained.

---

[a] For a steady state, from the Fourier transform with $\Gamma^L(\epsilon, t', t) \to \Gamma^L(\epsilon)$,

$$\int_{-\infty}^{t} dt' \int \frac{d\epsilon}{2\pi}\mathrm{ImTr}\left\{e^{i\epsilon(t-t')/\hbar}\Gamma^L(\epsilon)G^<(t-t')\right\}$$
$$= \frac{1}{2i}\int \frac{d\epsilon}{2\pi}\mathrm{Tr}\left\{\Gamma^L(\epsilon)G^<(\epsilon)\right\},$$

$$\int_{-\infty}^{t} dt' \int \frac{d\epsilon}{2\pi}\mathrm{ImTr}\left\{e^{i\epsilon(t-t')/\hbar}\Gamma^L(\epsilon)f_L(\epsilon)G^r(t-t')\right\}$$
$$= \frac{1}{2i}\int \frac{d\epsilon}{2\pi}\mathrm{Tr}\left\{\Gamma^L(\epsilon)f^L(\epsilon)\left[G^r(\epsilon) - G^a(\epsilon)\right]\right\},$$

and correspondingly the current (3.195) becomes

$$I_L = \frac{ie}{\hbar}\int \frac{d\epsilon}{2\pi}\mathrm{Tr}\left\{\Gamma^L(\epsilon)(G^<(\epsilon) + f_L(\epsilon)\left[G^r(\epsilon) - G^a(\epsilon)\right])\right\},$$

which reduces to the time-independent formula in the previous section.

Let us consider the double-barrier structure with a single state with the resonant-level of $\epsilon_0(t) = \epsilon_0 + \Delta(t)$. We assume that the coupling to electrodes are energy-independent (called the wide-band limit) such as

$$\Gamma_{L/R}(t', t) = \Gamma_{L/R}\exp\left[\frac{i}{\hbar}\int_{t'}^{t}\Delta_{L/R}(t_1)dt_1\right]. \qquad (3.199)$$

Within this approximation, application of a bias voltage corresponds to a shift of energies in the left electrode with respect to the right one and correspondingly a shift of the resonant level. Also we assume that the energy shift due to coupling is negligible as

$$\Sigma^r(\epsilon) = \sum_{k\alpha\in L,R}\frac{|V_{k\alpha}|^2}{\epsilon - \epsilon_{k\alpha} + i0^+} \approx -\frac{i}{2}(\Gamma^L + \Gamma^R). \qquad (3.200)$$

Thus, $\Sigma^r(t_1, t_2) = -i(\hbar/2)(\Gamma^L + \Gamma^R)\delta(t_1 - t_2)$. From the Dyson equation (3.196), we have

$$G^r(t, t') = g^r(t, t')e^{-(\Gamma^L+\Gamma^R)(t-t')/2} \qquad (3.201)$$

with $g^r(t, t') = -(i/\hbar)\theta(t - t')\exp\left[-(i/\hbar)\int_{t'}^{t}\epsilon_0(t_1)dt_1\right]$.

This is used to obtain $G^<(t, t')$ from Eq. (3.198) and then we have the occupation of the level $n(t) = -i\hbar G^<(t, t)$ and the current formula $I_{L/R}(t)$ from Eq. (3.196)[a] as

$$n(t) = \int\frac{d\epsilon}{2\pi}\left(f_L(\epsilon)\Gamma_L|A_L(\epsilon, t)|^2 + f_R(\epsilon)\Gamma_R|A_R(\epsilon, t)|^2\right),$$

$$I_{L/R}(t) = -\frac{e}{\hbar}\Gamma_{L/R}\left[n(t) + \frac{1}{\pi}\int f_{L/R}(\epsilon)\mathrm{Im}\left\{A_{L/R}(\epsilon, t)\right\}d\epsilon\right],$$

where

$$A_{L/R}(\epsilon, t) = \int_{-\infty}^{t}G^r(t, t')\exp\left[\frac{i}{\hbar}\epsilon(t - t') + \frac{i}{\hbar}\int_{t'}^{t}\Delta_{L/R}(t_1)dt_1\right]dt'.$$

Here we see two examples of the time-dependent current $I(t)$ for the response of an electron based on the NEGF formalism [36]. The first is the response to a rectangular bias pulse. In Fig. 3.10(a), we set μ_L, μ_R, ϵ_0 to be all zero initially. A bias pulse (dashed line) suddenly increases energies by $\Delta_L = 10$ and by $\Delta = 5$ at $t = 0$ and ends off at

[a] In the time-independent case as $t \to \infty$ with $\Delta(t) = \Delta_{L/R}(t) = 0$, $A(\epsilon, t)$ is the Fourier transform of $G^r(t, t')$ and we have the usual steady current formula $I = (e/h)\int d\epsilon\,(f_L(\epsilon) - f_R(\epsilon))\,\Gamma_L\Gamma_R/\left\{(\epsilon - \epsilon_0)^2 + (\Gamma/2)^2\right\}$ for resonance of ϵ_0.

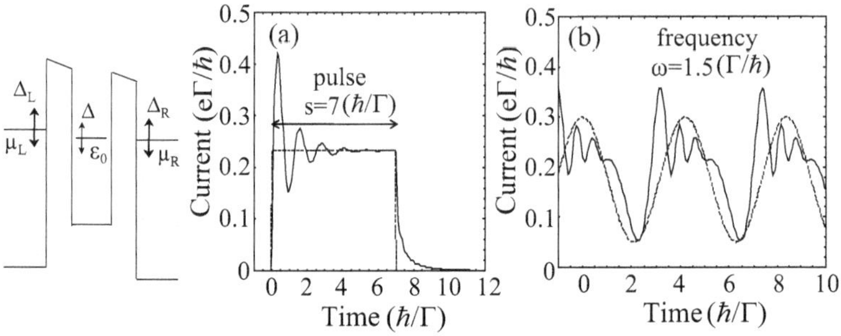

Figure 3.10 (Left) Schematic view of energy diagram for the time-dependent dynamical transport through nanosystems with double-barrier structures. (Right) (a) Time-dependent $I(t)$ current in response to a rectangular bias pulse. At $t = 0$, a bias pulse (dashed line) switches on by $\Delta_L = 10$, $\Delta = 5$, $\Delta_R = 0$ and ends off at $t = 7$, where the current decays to zero. (b) $I(t)$ for an *ac* bias with frequency of $\omega = 1.5$ (dashed line) with $\Delta_{L/R}(t) = 10\cos(\omega t)$, $\Delta(t) = 5\cos(\omega t)$. Here $\mu_L = 10$, $\epsilon_0 = 5$, $\mu_R = 0$. All the energies are in units of Γ, the currents in units of $e\Gamma/\hbar$, and the time in units of $\hbar/\Gamma$. The temperature is $k_B T = 0.1\Gamma$.

$t = 7$; thus the duration of the pulse is $s = 7$. In this case, $A_{L/R}(\epsilon, t)$ is given by

$$A_{L/R} = \begin{cases} 1/\alpha(\epsilon) & \text{for } t < 0 \\ \left\{\alpha(\epsilon) - (\Delta - \Delta_{L/R})e^{i\beta(\epsilon)t}\right\}/\alpha(\epsilon)\beta(\epsilon) & \text{for } 0 < t < s \\ \left\{\alpha(\epsilon) - (\Delta - \Delta_{L/R})\left[1 - e^{i\alpha(\epsilon)(t-s)}\left(1 - e^{i\beta(\epsilon)s}\right)\right]\right\} & \\ \qquad /\alpha(\epsilon)\beta(\epsilon) & \text{for } t > s \end{cases}$$

with $\alpha(\epsilon) = \epsilon - \epsilon_0 + i\Gamma/2$ and $\beta(\epsilon) = \epsilon - \epsilon_0 - \Delta + \Delta_{L/R} + i\Gamma/2$. We see that the current has settled to a new steady value during the pulse and then decays back to zero.

The second is the response to an *ac* bias voltage. In Fig. 3.10(b), we apply an *ac* bias with frequency ω by $\Delta_{L/R}(t) = \Delta_{L/R}\cos(\omega t)$ and $\Delta(t) = \Delta\cos(\omega t)$ with $\Delta_{L/R} = 10$ and $\Delta = 5$. Using the relation of $\exp[i\alpha\sin(\omega t)] = \sum_k e^{ik\omega t} J_k(\alpha)$, the time-dependent $A_{L/R}(\epsilon, t)$ is given by the sum of the Bessel function J_k. We see that the current has the same period $2\pi/\omega$ for an *ac* bias with additional complex time dependence inside each period, similar to the pulse case in (a). These are due to the photon-assisted resonant tunneling peaks at $\epsilon_0 + k\hbar\omega = \mu_{L/R}$, which are observed at low temperature $k_B T < \hbar\omega$.

3.3 Problems

(1) For the Schrödinger representation, the Heisenberg representation, and the interaction representation,

(a) write down the equation of motion;

(b) write down the expectation value of the operator $\langle \hat{A} \rangle$ for each representation.

(2) Consider a system with a magnetic impurity embedded in a non-magnetic metal.

(a) Within the Hartree–Fock approximation as

$$U\hat{n}_\uparrow \hat{n}_\downarrow \approx U \sum_\sigma \langle \hat{n}_{-\sigma} \rangle \hat{n}_\sigma - U\langle \hat{n}_\uparrow \rangle \langle \hat{n}_\downarrow \rangle,$$

write down the occupation number $\langle \hat{n}_\uparrow \rangle$ and $\langle \hat{n}_\downarrow \rangle$.

(b) Derive the condition that the magnetic moment appears $\langle \hat{n}_\uparrow \rangle \neq \langle \hat{n}_\downarrow \rangle$ when the impurity site energy ϵ_d is situated at the symmetric case of $\epsilon_F = \epsilon_d + U/2$.

(3) Consider a system with electrons plus phonons in a metal.

(a) Show that the coupling constant of the electron–phonon interaction $\hat{V}_{\text{e-ph}} = \sum_{k,q} \gamma_q \hat{c}^\dagger_{k+q} \hat{c}_k (\hat{a}_q + \hat{a}^\dagger_{-q})$ due to a displacement of the ionic potential

$$V_{\text{ion}}(\mathbf{r} - R_j) = -\frac{Ze^2}{\epsilon |\mathbf{r} - R_j|}$$

becomes

$$\gamma_q = iA\frac{1}{q} = i\left(\frac{4\pi Ze^2}{q\epsilon}\right)\sqrt{\frac{\hbar}{2M\omega_0}}$$

with one mode of optical phonons to approximate $\omega_q \simeq \omega_o$.

(b) Derive the single-particle energy ϵ_k within the second-order perturbation. Show that the effective mass of this state m_{pol} is enhanced with the coupling constant g as

$$m_{\text{pol}} = m^*\left(1 + \frac{g}{6}\right) \quad \text{with} \quad g = \frac{A^2}{2\pi\hbar^2\omega_o}\sqrt{\frac{m^*}{2\hbar\omega_o}}$$

This is called the Fröhlich polaron [37].

(c) Derive the effective potential for paring of electrons in the second-order perturbation with electron–phonon interaction. There are energy regimes for the attractive interaction, which induces the instability to form the Cooper pair.

(4) Let us consider Green's function defined by

$$\mathcal{G}(\tau, \tau') = -\langle T \{\hat{c}(\tau)\hat{c}^\dagger(\tau')\}\rangle.$$

Here T is the operator for time-ordering as

$$\mathcal{G}(\tau, \tau') = \begin{cases} -\langle \hat{c}(\tau)\hat{c}^\dagger(\tau')\rangle & \text{for } \tau > \tau' \\ \langle \hat{c}^\dagger(\tau')\hat{c}(\tau)\rangle & \text{for } \tau < \tau', \end{cases}$$

where $\hat{c}(\tau)$ and $\hat{c}^\dagger(\tau)$ are the Heisenberg representation of the operators $\hat{c}$ and $\hat{c}^\dagger$

$$\hat{c}(\tau) = e^{\tau\hat{H}}\hat{c}e^{-\tau\hat{H}}, \quad \hat{c}^\dagger(\tau) = e^{\tau\hat{H}}\hat{c}^\dagger e^{-\tau\hat{H}},$$

and $\langle\cdots\rangle$ represents the statistical average by the canonical distribution at temperature $\beta = 1/k_B T$ as

$$\langle\cdots\rangle = \mathrm{Tr}(\cdots e^{-\beta\hat{H}})/\mathrm{Tr}(e^{-\beta\hat{H}}).$$

This is called the temperature (Matsubara) Green's function [38] with an imaginary time τ.

(a) Show that $\mathcal{G}(\tau, \tau')$ is a function of $\tau - \tau'$.

(b) Show that $\mathcal{G}(\tau + \beta) = -\mathcal{G}(\tau)$, which enables us to represent

$$\mathcal{G}(\tau) = \frac{1}{\beta}\sum_{\ell} \mathcal{G}(i\omega_\ell)e^{-i\omega_\ell\tau}$$

$$\mathcal{G}(i\omega_\ell) = \int_0^\beta \mathcal{G}(\tau)e^{i\omega_\ell\tau}d\tau$$

with $\omega_\ell = (2\ell + 1)\pi/\beta, \quad \ell = 0, \pm1, \pm2, \cdots$.

(c) Show that the equation of motion for $\mathcal{G}$ satisfies

$$\left[\frac{\partial}{\partial\tau} + \epsilon_0\right]\mathcal{G}(\tau, \tau') = -\delta(\tau - \tau')$$

for a free electron $\hat{H} = \epsilon_0\hat{c}^\dagger\hat{c}$.

(d) For the retarded Green's function G^r of electrons

$$G^r(t, t') = -\frac{i}{\hbar}\theta(t - t')\langle\hat{c}(t)\hat{c}^\dagger(t') + \hat{c}^\dagger(t')\hat{c}(t)\rangle,$$

show that the retarded Green's function is obtained by the analytical continuation of the temperature Green's function by

$$G^r(\epsilon) = \mathcal{G}(\epsilon + i0^+).$$

Since the linear-response transport property is expressed by the retarded Green's function as $\sigma \propto \int \langle G^r(\epsilon) G^a(\epsilon) \rangle d\epsilon$, this shows that the temperature Green's function $\mathcal{G}$ is useful for the transport at finite temperature.

(5) Consider the correlated system with the Hamiltonian given by

$$\hat{H} = \sum_{k\sigma} \epsilon_{k\sigma} \hat{c}^\dagger_{k\sigma} \hat{c}_{k\sigma} + \sum_{k\sigma} \left(V_k \hat{c}^\dagger_{k\sigma} \hat{d}_\sigma + V_k^* \hat{d}^\dagger_\sigma \hat{c}_{k\sigma} \right) + U \hat{n}_{d\uparrow} \hat{n}_{d\downarrow}$$

where $\hat{n}_{d\sigma} = \hat{d}^\dagger_\sigma \hat{d}_\sigma$ is the number operator with spin σ and U is the repulsive Coulomb interaction between electrons. Let us consider to calculate numerically the temperature Green's function $\mathcal{G}_{dd} = \langle \hat{d}_{i\sigma} \hat{d}^\dagger_{j\sigma} \rangle$ where we assume the canonical ensemble for the thermal average as $\langle \cdots \rangle = \mathrm{Tr}(\cdots e^{-\beta \hat{H}}) / \mathrm{Tr} e^{-\beta \hat{H}}$.

(a) For the non-interacting Hamiltonian,

$$\hat{H}_0 = \sum_{i\sigma} \epsilon_{i\sigma} \hat{c}^\dagger_{i\sigma} \hat{c}_{i\sigma} + \sum_{k\sigma} V_k \left(\hat{c}^\dagger_{k\sigma} \hat{d}_\sigma + V_k^* \hat{d}^\dagger_\sigma \hat{c}_{k\sigma} \right),$$

give the temperature Green's function $\mathcal{G}^0_{dd}$.

(b) The bilinear formula of $e^{-\beta \hat{H}_1}$ for the interacting term $\hat{H}_1 = U \hat{n}_\uparrow \hat{n}_\downarrow$ is expressed by introducing auxiliary Ising variables σ in the Hubbard–Stratonovich transformation

$$e^{-\Delta\tau U \hat{n}_\uparrow \hat{n}_\downarrow} = \frac{1}{2} \sum_{\sigma=\pm1} e^{\lambda\sigma(\hat{n}_\uparrow - \hat{n}_\downarrow) - \Delta\tau U(\hat{n}_\uparrow + \hat{n}_\downarrow)/2}$$

with an imaginary-time step $\Delta\tau = \beta/L$ and the parameter of $\cosh(\lambda) = e^{\Delta\tau U/2}$. Using this formula, the partition function taking the imaginary-time path-integral form as

$$Z = \mathrm{Tr}\, e^{-\beta \hat{H}} \simeq \mathrm{Tr} \prod_{l=1}^{L} e^{-\Delta\tau \hat{H}_0} e^{-\Delta\tau \hat{H}_1}$$

is constructed as a product of the terms of exponents. Then the calculation of the temperature Green's function $\mathcal{G}_{dd}$ is written as the sum of all auxiliary variables, which increases exponentially as N becomes large. Use the Metropolis algorithm of the Monte Carlo procedure for sampling of the calculations of $\mathcal{G}_{dd}$.

This is called the quantum Monte Carlo (QMC) method for the temperature Green's function within the imaginary-time formalism [39]. We note that recently the QMC method is extended to study the dynamical properties directly within the real-time path-integral formalism. This enables us to treat nonequilibrium transport problems by the strongly correlated model Hamiltonian [40].

Chapter 4

Numerical Methods Based on Density Functional Theory

4.1 Overview of Density Functional Theory

In the previous chapter, we introduce the Green's function method for treating many-electron systems. To obtain the transport properties of nanometer-scale materials, we need to obtain electronic states, taking the realistic atomic crystal structures into account. The complete solutions to the Schrödinger equation for many-electron systems are impossible to obtain, even if we use the fastest supercomputers for the calculations. Therefore, a number of practical computation methods have been proposed with various approximations to solve the Schrödinger equation for electronic structures of many-electron systems.

The density functional theory (DFT) has been one of the most effective computational methods. This method treats electronic states of an interacting many-electron system by solving the equations for the *fictitious* system of non-interacting electrons in an effective potential, whose density is precisely the same as that of the real system. The DFT has been shown to be very useful to obtain various physical properties of the realistic materials. Recently it has

Quantum Transport Calculations for Nanosystems
Kenji Hirose and Nobuhiko Kobayashi

ISBN 978-981-4267-32-8 (Hardcover), 978-981-4267-59-5 (eBook)
www.panstanford.com

also been applied to the transport problems for nanometer-scale structures. Here wc review the computational methods for many-electron systems briefly and introduce the concepts of the DFT.[a]

4.1.1 *Many-Electron System from Wavefunction*

4.1.1.1 Many-body wavefunction

The main purpose here is to study the properties of many-electron systems from the wavefunction approach. For that purpose, we need to solve the Schrödinger equation with the Hamiltonian;

$$\hat{H} = \sum_{i=1}^{N} \left[-\frac{\hbar^2}{2m}\nabla_i^2 + V_{\text{ext}}(\mathbf{r}_i) \right] + \sum_{i<j}^{N} \frac{e^2}{|\mathbf{r}_i - \mathbf{r}_j|}, \tag{4.1}$$

which describes a system composed of N electrons in an external potential $V_{\text{ext}}(\mathbf{r}_i)$, interacting with other electrons through the two-body Coulomb interaction.

When we treat only the first term, which contains the one-body interaction and is referred to as a non-interacting electron system

$$\hat{H}_0 = \sum_{i=1}^{N} \left[-\frac{\hbar^2}{2m}\nabla_i^2 + V_{\text{ext}}(\mathbf{r}_i) \right], \tag{4.2}$$

the eigenfunctions of $\hat{H}_0$ can be written as a product of single-particle wavefunctions of

$$\left[-\frac{\hbar^2}{2m}\nabla^2 + V_{\text{ext}}(\mathbf{r}) \right] \psi_{k_i}(\mathbf{r}) = \epsilon_{k_i} \psi_{k_i}(\mathbf{r}). \tag{4.3}$$

Here the infinite number of single-particle states $\psi_{k_i}(\mathbf{r})$, which are characterized by the quantum index number $\mathcal{K} = (k_1, k_2, \cdots)$, satisfies the orthonormalized condition $\int \psi_{k_i}^*(\mathbf{r})\psi_{k_j}(\mathbf{r})d\mathbf{r} = \delta_{i,j}$. Then the wavefunction $\Psi_0^{\mathcal{K}}$ and the energy E_0 for $\hat{H}_0\Psi_0^{\mathcal{K}} = E_0\Psi_0^{\mathcal{K}}$ become in general

$$\Psi_0^{\mathcal{K}}(\mathbf{r}_1, \mathbf{r}_2, \cdots, \mathbf{r}_N) = \prod_{i=1}^{N} \psi_{k_i}(\mathbf{r}_i) \quad \text{with} \quad E_0 = \sum_{i=1}^{N} \epsilon_{k_i}. \tag{4.4}$$

However, since electrons are fermions, we need to create a fully anti-symmetric wavefunction under an exchange of coordinates of two

[a]There are a number of excellent books for the DFT [41–46]. The readers should refer these books for more details.

particles. The anti-symmetrized wavefunction becomes using the permutation P such as

$$\Psi_0^{\mathcal{K}}(\mathbf{r}_1, \mathbf{r}_2, \cdots, \mathbf{r}_N) = \frac{1}{\sqrt{N!}} \sum_P (-1)^P P \prod_{i=1}^{N} \psi_{k_i}(\mathbf{r}_i)$$

$$= \frac{1}{\sqrt{N!}} \begin{vmatrix} \psi_{k_1}(\mathbf{r}_1) & \psi_{k_1}(\mathbf{r}_2) & \ldots & \psi_{k_1}(\mathbf{r}_N) \\ \psi_{k_2}(\mathbf{r}_1) & \psi_{k_2}(\mathbf{r}_2) & \ldots & \psi_{k_2}(\mathbf{r}_2) \\ \vdots & \vdots & \ddots & \vdots \\ \psi_{k_N}(\mathbf{r}_1) & \psi_{k_N}(\mathbf{r}_2) & \ldots & \psi_{k_N}(\mathbf{r}_N) \end{vmatrix}. \quad (4.5)$$

In the non-interacting case, the ground state for a system with N electrons are obtained by occupying N *lowest-energy* single-particle states ψ_{k_i} characterized by the sets of quantum numbers $\mathcal{K}_0 = (k_1, k_2, \cdots, k_N)$ and the highest-energy occupied state defines the Fermi level and its energy is the Fermi energy.

When we include the two-body Coulomb interaction, the wavefunction of the ground state for $\hat{H}$ cannot be described only one Slater determinant with lowest-energy single-particle states characterized by $\mathcal{K}_0$. Instead we need to sum up a number of Slater determinants with various configuration of the states like $\mathcal{K}_1 = (k_1, k_2, \cdots, k_{N-1}, k_{N+1}), \mathcal{K}_2 = (k_1, k_2, \cdots, k_{N-1}, k_{N+2}) \cdots$ such that

$$\Psi(\mathbf{r}_1, \mathbf{r}_2, \cdots, \mathbf{r}_N) = \sum_{\text{all config.}\mathcal{K}_i} C_{\mathcal{K}_i} \Psi_0^{\mathcal{K}_i}(\mathbf{r}_1, \mathbf{r}_2, \cdots, \mathbf{r}_N). \quad (4.6)$$

Constructing the Hamiltonian matrix and diagonalizing it, we obtain all the ground and excited energies. However, since the number of configurations become exponentially large as we increase the number of electrons N, the solution to Schrödinger equation becomes very difficult for increasing N. The calculation method taking the full configuration interactions are frequently called an exact diagonalization method, which is usually limited to treat a small number of electrons. Instead, various approximate calculation methods have been developed to reduce the computation.

For obtaining the ground-state energy, it is important to note that the variational principle is applied such that

$$\langle \Psi_{\text{trial}} | \hat{H} | \Psi_{\text{trial}} \rangle = E_{\text{trial}} \geq E_0 = \langle \Psi_0 | \hat{H} | \Psi_0 \rangle, \quad (4.7)$$

which means that the expectation value of the Hamiltonian $\hat{H}$ from any guessed $\Psi_{\rm trial}$ will be an upper bound to the true energy of the ground state. The equality holds if and only if $\Psi_{\rm trial}$ is identical to Ψ_0. Thus, we need to minimize the functional $E[\Psi]$ by searching through all N-electron wavefunctions. The function that gives the lowest energy will be Ψ_0 and the energy will be the true ground-state energy E_0. However, since a search over all functions is obviously not possible, we apply the variational principle to some subsets. The result will be the best approximation to the exact wavefunction that can be obtained from this particular subset.

4.1.1.2 Hartree approximation and Hartree–Fock approximation

Here we will show the typical examples to obtain the ground-state wavefunction and the energy of the many-electron systems, that is, the Hartree approximation and the Hartree–Fock approximation.

In the Hartree approximation, we minimize the total energy functional $E_{\rm H}[\Psi]$ using the wavefunctions from Eq. (4.4),

$$E_{\rm H} = \sum_i \int \psi_i^*(\mathbf{r}) \left[-\frac{\hbar^2}{2m}\nabla^2 + V_{\rm ext}(\mathbf{r}) \right] \psi_i(\mathbf{r}) d\mathbf{r}$$

$$+ \frac{1}{2}\sum_{i,j} \int\int \psi_i^*(\mathbf{r})\psi_j^*(\mathbf{r}') \frac{e^2}{|\mathbf{r}-\mathbf{r}'|} \psi_i(\mathbf{r})\psi_j(\mathbf{r}') d\mathbf{r} d\mathbf{r}'$$

with the constraint of $\int |\psi_i(\mathbf{r})|^2 d\mathbf{r} = 1$ such as

$$\delta\left(E_{\rm H} - \sum_i \epsilon_i \int |\psi_i(\mathbf{r})|^2 d\mathbf{r} \right) / \delta\psi_i^*(\mathbf{r}) = 0. \tag{4.8}$$

Here ϵ_i is the Lagrange multiplier and we write $k_i \to i$ for simplicity. This variational equation produces

$$\left[-\frac{\hbar^2}{2m}\nabla^2 + V_{\rm ext}(\mathbf{r}) + e^2 \int \frac{n(\mathbf{r}')}{|\mathbf{r}-\mathbf{r}'|} d\mathbf{r}' \right] \psi_i(\mathbf{r}) = \epsilon_i \psi_i(\mathbf{r})$$

$$n(\mathbf{r}) = \sum_{i=1}^{N} |\psi_i(\mathbf{r})|^2, \tag{4.9}$$

which is called the Hartree equation. These two equations are solved self-consistently. From the variational principle, we should take the lowest-energy N single-particle states $\psi_i(\mathbf{r})$ from Eq. (4.9) to construct electron density $n(\mathbf{r})$. Using Eq. (4.9) and taking the lowest energy ϵ_i, the ground-state energy becomes

$$E_{\rm H} = \sum_{i=1}^{N} \epsilon_i - \frac{e^2}{2} \int\int \frac{n(\mathbf{r})n(\mathbf{r}')}{|\mathbf{r}-\mathbf{r}'|} d\mathbf{r}d\mathbf{r}'. \tag{4.10}$$

It is apparent from the trial wavefunction in Eq. (4.4) that the Hartree approximation is lacking the Pauli principle for fermions. The Hartree equation Eq. (4.9) shows that the effective potential for electron–electron interaction is approximated by the classical Coulomb potential $V_H(\mathbf{r}) = e^2 \int \frac{n(\mathbf{r}')}{|\mathbf{r}-\mathbf{r}'|} d\mathbf{r}'$, which is obtained also from the Poisson equation

$$\nabla^2 V_H(\mathbf{r}) = -4\pi e^2 n(\mathbf{r}). \tag{4.11}$$

Here $V_H(\mathbf{r})$ is called the Hartree potential.

In the Hartree–Fock (HF) method, using Eq. (4.5) as a trial wavefunction, we take Pauli principle into account and derive the HF equation from minimizing the total energy functional of $E_{\rm HF}[\Psi]$;

$$E_{\rm HF} = \sum_i \int \psi_i^*(\mathbf{r}) \left[-\frac{\hbar^2}{2m}\nabla^2 + V_{\rm ext}(\mathbf{r}) \right] \psi_i(\mathbf{r}) d\mathbf{r}$$

$$+\frac{1}{2}\sum_{i,j} \int\int \psi_i^*(\mathbf{r})\psi_j^*(\mathbf{r}') \frac{e^2}{|\mathbf{r}-\mathbf{r}'|} \left[\psi_i(\mathbf{r})\psi_j(\mathbf{r}') - \psi_i(\mathbf{r}')\psi_j(\mathbf{r}) \right] d\mathbf{r}d\mathbf{r}'$$

with the constraint of $\int |\psi_i(\mathbf{r})|^2 d\mathbf{r} = 1$ and Lagrange multiplier ϵ_i

$$\delta \left(E_{\rm HF} - \sum_i \epsilon_i \int |\psi_i(\mathbf{r})|^2 d\mathbf{r} \right) / \delta\psi_i^*(\mathbf{r}) = 0. \tag{4.12}$$

This variational equation produces

$$\left[-\frac{\hbar^2}{2m}\nabla^2 + V_{\rm ext}(\mathbf{r}) + e^2 \int \frac{n(\mathbf{r}')}{|\mathbf{r}-\mathbf{r}'|} d\mathbf{r}' \right] \psi_i(\mathbf{r})$$

$$-e^2 \sum_j \int \frac{\psi_j^*(\mathbf{r}')\psi_i(\mathbf{r}')}{|\mathbf{r}-\mathbf{r}'|} d\mathbf{r}' \times \psi_j(\mathbf{r}) = \epsilon_i \psi_i(\mathbf{r})$$

$$n(\mathbf{r}) = \sum_{i=1}^{N} |\psi_i(\mathbf{r})|^2, \tag{4.13}$$

which is called the Hartree–Fock equation. These two equations are solved self-consistently. It should be noted that the term $i = j$ is allowed since the Coulomb integral in $E_{\rm HF}$ cancels to each other, which means that the self-interaction is eliminated in the HF equation. The additional term, called a Fock term or an exchange term, has no classical analog. This term leads to an exchange of the variables in the two orbitals ψ_i, ψ_j.

In operating $e^2 \int \frac{\psi_j^*(\mathbf{r}')\psi_i(\mathbf{r}')}{|\mathbf{r}-\mathbf{r}'|} d\mathbf{r}'$ on $\psi_j(\mathbf{r})$, we need the values of ψ_j on all points in space since ψ_j is now related to $\mathbf{r}'$ and therefore the exchange term is called non-local term. It is important to note that the appearance of the exchange term is entirely due to anti-symmetry of the Slater determinant applying to the fermions. The resulting ground-state energy becomes using Eq. (4.13) as

$$E_{\rm HF} = \sum_{i=1}^{N} \epsilon_i - \frac{e^2}{2} \int\int \frac{n(\mathbf{r})n(\mathbf{r}')}{|\mathbf{r}-\mathbf{r}'|} d\mathbf{r}d\mathbf{r}' + \frac{e^2}{2} \sum_{i,j} \int\int \frac{\psi_i^*(\mathbf{r})\psi_j(\mathbf{r})\psi_j^*(\mathbf{r}')\psi_i(\mathbf{r}')}{|\mathbf{r}-\mathbf{r}'|} d\mathbf{r}d\mathbf{r}'. \tag{4.14}$$

The true wavefunction of an interacting many-electron system is not described by a single Slater determinant. The difference between the true ground-state energy E_0 and $E_{\rm HF}$ is called the correlation energy $E_C = E_0 - E_{\rm HF}$, which is mainly caused by the instantaneous repulsion of electrons.

It should be noted that there are various calculation methods to include the correlation energy effectively for better ground-state and also for excited-state energies. Examples include configuration interaction (CI) method, GW approximation, variational Monte Carlo (VMC), diffusion Monte Carlo (DMC), and transcorrelated method. Such calculation methods are very important for the accurate electronic states of various molecules and solids with strongly correlated systems, but are usually very demanding and time-consuming. The DFT described in the next section treats many-electron systems to include correlation effects in the framework of (static) mean-field theory.[a]

[a]The extension for applications of DFT to treat strongly correlated electron systems is desired and is making much progress recently. These include, for example, LDA+U and LDA+DMFT (dynamical mean-field theory) [47].

4.1.2 *Many-Electron System from Charge Density: Kohn–Sham Equation*

4.1.2.1 Hohenberg–Kohn theorem

The charge density $n(\mathbf{r})$ of electrons is defined from the many-body wavefunction that is anti-symmetric in an exchange of orbitals of two electrons as follows:

$$n(\mathbf{r}) = N \int \cdots \int |\Psi(\mathbf{r}, \mathbf{r}_2, \cdots, \mathbf{r}_N)|^2 d\mathbf{r}_2 \cdots d\mathbf{r}_N. \qquad (4.15)$$

Here the multiple integrals are performed over all but one of the spatial variables. $n(\mathbf{r})$ is a non-negative function $n(\mathbf{r}) \geq 0$ of at most only three spatial variables and satisfies the condition

$$\int n(\mathbf{r}) d\mathbf{r} = N. \qquad (4.16)$$

Unlike the wavefunction, electron density $n(\mathbf{r})$ is observable and can be measured experimentally, e.g., by X-ray diffraction.

The central concept of the DFT for an interacting N-electron system is that the ground-state energy in a general external potential $V_{\text{ext}}(\mathbf{r})$ is determined from electron density $n(\mathbf{r})$ uniquely and thus $n(\mathbf{r})$ is the basic variable. This is based on the Hohenberg–Kohn theorem, which states

(1) Two external potentials, which differ by more than a constant, cannot give the same ground-state density.
(2) The ground-state energy takes its minimum value with respect to variation of electron density subject to the normalization condition when electron density has its correct value.

The proof of the theorem proceeds by *reductio ad absurdum* [48], which is described in various textbooks.

It is important to note that the minimum value of the ground-state energy is obtained in the Hartree and Hartree–Fock equations with respect to variation of the wavefunction, which is a complex function of N variables, while it is obtained in the DFT with respect to variation of electron density, which is a real function of at most three variables. It is surprisingly remarkable that it is sufficient to vary, in principle, electron density to obtain the ground-state energy of an interacting many-electron system.

4.1.2.2 Single-particle Kohn–Sham equation

From these theorems, we construct an effective equation for obtaining the ground-state energy and electron density. The total energy of the system in its ground state is a functional of electron density $n(\mathbf{r})$ in the form of

$$\begin{aligned} E_{\mathrm{DFT}}[n] &= F[n] + \int V_{\mathrm{ext}}(\mathbf{r})n(\mathbf{r})d\mathbf{r} \\ &= T[n] + \frac{e^2}{2}\int\int \frac{n(\mathbf{r})n(\mathbf{r}')}{|\mathbf{r}-\mathbf{r}'|}d\mathbf{r}d\mathbf{r}' + E_{xc}[n] \\ &\quad + \int V_{\mathrm{ext}}(\mathbf{r})n(\mathbf{r})d\mathbf{r}, \end{aligned} \tag{4.17}$$

where $T[n]$ is the kinetic energy of a non-interacting system with the same ground-state density $n(\mathbf{r})$ as the interacting one. Kohn and Sham introduced a set of N orthonormal single-particle functions [49], called Kohn–Sham orbitals, $\psi_i(\mathbf{r})$, $i = 1, 2, \cdots, N$ as auxiliary variables to satisfy the condition

$$\sum_{i=1}^{N} |\psi_i(\mathbf{r})|^2 = n(\mathbf{r}), \tag{4.18}$$

which is intended to carry out the variation of electron density by varying the orbital function $\psi_i(\mathbf{r})$. Then the kinetic part $T[n]$ is written approximately by the form of a single-particle kinetic energy $T_s[n]$ as

$$\begin{aligned} T[n] \approx T_s[n] &= \frac{\hbar^2}{2m}\sum_{i=1}^{N}\int \nabla\psi_i^*(\mathbf{r})\nabla\psi_i(\mathbf{r})d\mathbf{r} \\ &= \frac{\hbar^2}{2m}\sum_{i=1}^{N}\int \psi_i^*(\mathbf{r})(-\nabla^2)\psi_i(\mathbf{r})d\mathbf{r}. \end{aligned} \tag{4.19}$$

Note that since we are not sure that a non-interacting system having the same ground-state density $n(\mathbf{r})$ as the interacting one always exists, this is an approximate form based on the single-particle approximation. On the other hand, $E_{xc}[n]$ is called the exchange-correlation energy functional, which includes information on the exchange energy and all the other correlation effects of a many-electron system. We are also not sure the exact form for it, nor does the exact form surely exist. The variation is carried out $\delta E_{\rm DFT}[n]/\delta n(\mathbf{r}) = 0$ to obtain

$$\left[-\frac{\hbar^2}{2m}\nabla^2 + V_{\rm ext}(\mathbf{r}) + e^2\int\frac{n(\mathbf{r}')}{|\mathbf{r}-\mathbf{r}'|}d\mathbf{r}' + \frac{\delta E_{xc}[n]}{\delta n(\mathbf{r})}\right]\psi_i(\mathbf{r}) = \epsilon_i\psi_i(\mathbf{r})$$

$$n(\mathbf{r}) = \sum_i^N |\psi(\mathbf{r})|^2. \tag{4.20}$$

This effective single-particle equation is the Kohn–Sham equation [49].

These two equations are solved self-consistently for an effective potential $V_{\rm eff}(\mathbf{r}) = V_{\rm ext}(\mathbf{r}) + V_H(\mathbf{r}) + \delta E_{xc}[n]/\delta n$ and the electron density $n(\mathbf{r})$. Then the ground-state energy is given by

$$E_{\rm DFT} = \sum_{i=1}^N \epsilon_i - \frac{e^2}{2}\int\int\frac{n(\mathbf{r})n(\mathbf{r}')}{|\mathbf{r}-\mathbf{r}'|}d\mathbf{r}d\mathbf{r}' + E_{xc}[n] - \int\frac{\delta E_{xc}[n]}{\delta n(\mathbf{r})}n(\mathbf{r})d\mathbf{r}. \tag{4.21}$$

The Hohenberg–Kohn theorem and the Kohn–Sham equation can be generalized to various systems as

(1) Thermal DFT: electrons in grand-canonical equilibrium case where the ensemble-averaged density $n(\mathbf{r}) = \langle n(\mathbf{r})\rangle$.
(2) Spin DFT: electrons with spin polarization case where $n(\mathbf{r}, \zeta)$, ζ is the relative spin polarization.
(3) Current DFT: electrons in magnetic field case where $n(\mathbf{r}, \mathbf{j}_p(\mathbf{r}))$, $\mathbf{j}_p$ is the orbital current due to the magnetic field.
(4) Time-dependent DFT: electrons in excited states case where $n(\mathbf{r}, t)$ is the time-dependent electron density obtained from the time-dependent Kohn–Sham equation,

$$i\hbar\frac{\partial}{\partial t}\psi_i(\mathbf{r}, t) = \left[-\frac{\hbar^2}{2m}\nabla^2 + V_{\rm ext}(\mathbf{r}, t) + V_H(\mathbf{r}, t) + \frac{\delta E_{xc}[n(\mathbf{r}, t)]}{\delta n(\mathbf{r}, t)}\right]\psi_i(\mathbf{r}, t).$$

4.1.3 *Exchange-Correlation Potential*

In order to solve the Kohn–Sham equation, we need the exchange-correlation energy functional $E_{xc}[n]$, which is unknown and only approximate forms are possible. The simplest one is the local-density approximation (LDA)

$$E_{xc}[n] = \int n(\mathbf{r})\epsilon_{xc}(n(\mathbf{r}))d\mathbf{r}, \tag{4.22}$$

where $\epsilon_{xc}(n(\mathbf{r}))$ is the exchange-correlation energy for the homogeneous electron liquid with local electron density of n. The local electron density n and corresponding ϵ_{xc} is expressed by the parameter r_s, which is the interatomic distance defined by $r_s = (3/4\pi n)^{1/3}$ for the 3D case. Then the exchange-correlation potential in the LDA case is written as

$$\begin{aligned} V_{xc}[n(\mathbf{r})] &= \frac{\delta E_{xc}[n]}{\delta n(\mathbf{r})} = \epsilon_{xc}(\mathbf{r}) + \epsilon_{xc}(\mathbf{r})\frac{\partial \epsilon_{xc}(\mathbf{r})}{\partial n(\mathbf{r})} \\ &= \left(1 - \frac{r_s}{3}\frac{d}{dr_s}\right)\epsilon_{xc}(r_s). \end{aligned} \tag{4.23}$$

Approximate formulas for the exchange-correlation energy $\epsilon_{xc}(r_s) = \epsilon_x(r_s) + \epsilon_c(r_s)$ have been proposed[a] based on the interpolation of numerical diffusion Monte Carlo results [50]:

$$\begin{aligned} \epsilon_x(r_s) &= -\frac{3}{4\pi}\left(\frac{9\pi}{4}\right)^{1/3}\frac{1}{r_s} \\ \epsilon_c(r_s) &= \begin{cases} \dfrac{\gamma}{1 + \beta_1\sqrt{r_s} + \beta_2 r_s} & r_s \geq 1 \\ A\ln r_s + B + C r_s \ln r_s + D r_s & r_s < 1 \end{cases}, \end{aligned}$$

where γ, $\beta_{1,2}$, and A, B, C, D are parameters [51].

When we treat electronic states with spin-polarized cases in the spin DFT, we introduce the relative spin polarization

$$\zeta = \frac{n_\uparrow - n_\downarrow}{n_\uparrow + n_\downarrow}, \tag{4.24}$$

which ranges from 0 for an unpolarized case to ± 1 for a fully spin-polarized case. The exchange energy becomes

$$\epsilon_x(r_s, \zeta) = -\frac{3}{8\pi}\left(\frac{9\pi}{4}\right)^{1/3}\frac{1}{r_s}\left[(1+\zeta)^{4/3} + (1-\zeta)^{4/3}\right] \tag{4.25}$$

---

[a] We give the analytical treatments of $\epsilon_x(r_s)$ and $\epsilon_c(r_s)$ for an interacting many-electron gas system in problem 1.

and the correlation energy is interpolated as

$$\epsilon_c(r_s, \zeta) = \epsilon_c(r_s, 0) + \frac{(1+\zeta)^{4/3} + (1-\zeta)^{4/3} - 2}{(2^{4/3} - 2)} \left(\epsilon_c(r_s, 1) - \epsilon_c(r_s, 1)\right), \tag{4.26}$$

where $\epsilon_c(r_s, 0)$ is the correlation energy for an unpolarized case and $\epsilon_c(r_s, 1)$ is the correlation energy for a fully spin-polarized case. In the 2D case, we will give the exchange-correlation energy in Chapter 6.

The use of the gradient of electron density $\nabla n(\mathbf{r})$, in addition to electron density at a particular point $n(\mathbf{r})$ for the exchange-correlation energy, is considered to be the next step beyond the local-density approximation. When we restrict the correct sum rule for an electron hole, it is known that a better form for the exchange-correlation energy functional is constructed. Such approximation is called the generalized gradient approximation (GGA) [52],

$$E_{xc}^{GGA}[n, \zeta] = \int f(n_\uparrow(\mathbf{r}), n_\downarrow(\mathbf{r}), \nabla n_\uparrow(\mathbf{r}), \nabla n_\downarrow(\mathbf{r}))d\mathbf{r}. \tag{4.27}$$

The GGA and its hybrid functionals produce better ground-state energy, lattice constant, bulk modulus, cohesive energy, and so on compared with the local-density approximation (LDA).

4.1.4 *Pseudopotential for Atom*

In the atomic systems, the main contribution to an external potential $V_{\text{ext}}(\mathbf{r})$ comes from the potential of each atom, which behaves such as $V_{\text{ion}}(\mathbf{r}) = -Ze^2/r$ with the atomic number Z. Here we divide electrons in two groups, valence electrons and core electrons. The core electrons are strongly bound in the inner shells and do not contribute to the chemical bonding and transport properties. Chemical bonding and electron transport are in usual cases due to the valence electrons, especially in metal and semiconductor cases. So it is favorable to construct an effective potential, pseudopotential, which acts on the valence electrons only.

We denote wavefunctions of the Schrödinger equation as $|\psi_c\rangle$ for the core electrons and $|\psi_v\rangle$ for the valence electrons, that is, $\hat{H}|\psi_c\rangle = \epsilon_c|\psi_c\rangle$ and $\hat{H}|\psi_v\rangle = \epsilon_v|\psi_v\rangle$. Here the wavefunction for the valence electrons $|\psi_v\rangle$ behaves smoothly in space, while $|\psi_c\rangle$ has an oscillating behavior close to the core, which is due to the orthogonalization of the valence wavefunction to the core wavefunction. Introducing the pseudo-wavefunction $|\varphi_v\rangle$ as

$$|\psi_v\rangle = |\varphi_v\rangle - \sum_c \langle\psi_c|\varphi_v\rangle|\psi_c\rangle, \tag{4.28}$$

which satisfies the orthogonalization relation $\langle\psi_c|\psi_c\rangle = 0$, the Schrödinger equation for $|\varphi_v\rangle$ becomes

$$\begin{aligned}\hat{H}\,|\varphi_v\rangle &= \hat{H}\,|\psi_v\rangle + \sum_c \hat{H}\,|\psi_c\rangle\langle\psi_c|\varphi_v\rangle \\ &= \epsilon_v|\varphi_v\rangle + \sum_c (\epsilon_c - \epsilon_v)|\psi_c\rangle\langle\psi_c|\varphi_v\rangle.\end{aligned} \tag{4.29}$$

This equation shows that the pseudo-wavefunction $|\varphi_v\rangle$ satisfies the relation $\hat{H}^{ps}|\varphi_v\rangle = \epsilon_v|\varphi_v\rangle$ with

$$\hat{H}^{ps} = \hat{H} - \sum_c (\epsilon_c - \epsilon_v)|\psi_c\rangle\langle\psi_c| \tag{4.30}$$

and thus we can construct a pseudopotential by

$$V^{ps}(E) = V_{\text{ion}} - \sum_c (\epsilon_c - \epsilon)|\psi_c\rangle\langle\psi_c| \tag{4.31}$$

as an effective potential that the valence electrons feel.

Since $|\psi_c\rangle$ has an amplitude only close to the core region, the pseudopotential V^{ps} becomes V_{atom} at a certain distance from the core due to the decay of $|\psi_c\rangle$. In the region close to the core, due to the orthogonalization constraint of the valence wavefunction to the oscillating core wavefunction, the valence electrons feel a short-ranged repulsive potential from the second term, which makes the pseudopotential much weaker than the bare atomic potential in the core region. From this cancellation of the potential, the pseudo-wavefunction becomes smooth and does not show oscillating behavior. This is a desired form for the pseudopotential to reduce the computational cost.

The pseudopotential is spatially nonlocal, which depends on $\mathbf{r}$ and $\mathbf{r}'$ since the second term is written

$$\sum_c (\epsilon_c - \epsilon)|\psi_c(\mathbf{r})\rangle\langle\psi_c(\mathbf{r}')| \tag{4.32}$$

and this expression is reserved for the angular momentum l. Therefore, the pseudopotential is written in the form as

$$V^{ps}(\mathbf{r},\mathbf{r}') = V_s + V_p + V_d + \cdots = \sum_l \sum_{m=-l}^{l} Y_{lm}(\hat{\Omega})^* V_l^{ps}(r,r') Y_{lm}(\hat{\Omega}'). \tag{4.33}$$

Since the form of the pseudopotential in (4.31) depends on the energy of an electronic state that is solved, other empirical models have been proposed.

In the empirical pseudopotentials, an analytical form is constructed with the parameters fitted to experimental data. The simplest one, which reflects the cancellation completely, is the empty-core pseudopotential [53],

$$V_{\mathrm{loc}}^{ps}(r) = \begin{cases} -\dfrac{Z_v e^2}{r} & r > r_c \\ 0 & r \leq r_c \end{cases}, \tag{4.34}$$

where Z_v is the number of valence electrons and includes one adjustable parameter r_c to fit atomic data. For the band calculations of semiconductor materials, the Fourier transform is more appropriate. Since the Fourier transform of empty-core potential takes

$$V_{\mathrm{loc}}^{ps}(q) \sim -\frac{4\pi Z_v e^2}{q^2} \cos q r_c, \tag{4.35}$$

the following form has been proposed and is proved effective for the local pseudopotential [54]:

$$V_{\mathrm{loc}}^{ps}(q) = -\frac{4\pi Z_v e^2}{q^2} \left[\cos(a_2 q) + a_3\right] e^{a_4 q^4}, \tag{4.36}$$

which fits the band structure of semiconductors very well. These are the local pseudopotentials.

To obtain electronic states more accurately, we need the nonlocal pseudopotential, which is described by the angular momentum l. Separating the long range component as local potential from the short range potential as nonlocal potential

$$\delta V_l^{ps}(r) = V_l^{ps}(r) - V_{\mathrm{loc}}^{ps}(r), \tag{4.37}$$

Kleinman and Bylander showed that the nonlocal part $V_l^{ps}(r,r')$ in Eq. (4.33) is written as a fully separable pseudopotential [55] such as

$$V^{ps}(\mathbf{r},\mathbf{r}') = V_{\mathrm{loc}}^{ps}(\mathbf{r})\delta(\mathbf{r}-\mathbf{r}') + \sum_l \sum_{m=-l}^{l} \frac{|\phi_{lm}\delta V_l^{ps}\rangle\langle\delta V_l^{ps}\phi_{lm}|}{\langle\phi_{lm}|\delta V_l^{ps}|\phi_{lm}\rangle}, \tag{4.38}$$

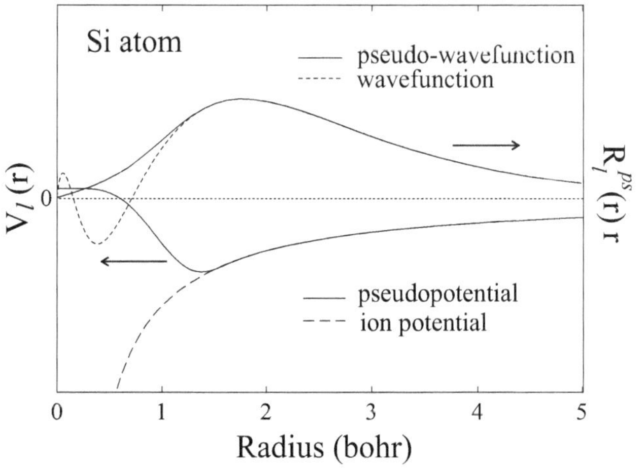

Figure 4.1 (Left) Pseudopotential for silicon atom. Dashed line corresponds to $\sim -Z_v/r$ and the solid line corresponds to $V_{l=0}^{ps}(r)$. (Right) Pseudo-wavefunction for silicon with an angular momentum of $l = 0$. Dashed curve corresponds to the wavefunction for the full-core atom and the solid curve corresponds to the pseudopotential atom.

where $\phi_{lm} = R_l^{ps}(r)Y_{lm}(\hat{\Omega})$ denotes the pseudo-wavefunction. We note that short range $\delta V_l^{ps}(r)$ have finite values only close to the atomic core region. The l-dependent pseudopotentials $V_l^{ps}(r)$ are constructed from first-principles for all atoms, called ab initio pseudopotentials. There are several forms. The famous ones are by Bachelet, Hamman, and Schlüter [56] and by Troullier and Martins [57]. The effectiveness and the transferability to various environmental situations of the pseudopotentials have been studied by electronic structure calculations and from the comparison to experiments.

4.1.5 *Application to Transport Problem*

The DFT is originally developed for a theory of electronic structures of atoms, molecules, and solids in their ground states and has been shown to be very effective for electronic state calculations. Recently, as the sizes of electron devices become small to the nanometer scale and even the devices using single molecules become realized, much interest has been attracted for the calculations of electron transport through the atomic-scale nanostructures between electrodes applying finite bias voltages. Since DFT has successfully been applied to electronic state calculations of the atomic-scale nanostructures, there has been a growing interest to apply DFT to the transport problem. Before we present several calculation methods to apply DFT to the transport problems of nanostructures and show the calculation results, we describe here some features for electronic-state calculations for the transport problem and also their fundamental problems.

4.1.5.1 Modeling

In order to treat the quantum transport problems, as we see in the previous chapters, there are two types of the methods. One is based on the Landauer-type approach, which describes electron transport as the scattering problems. The other is based on the Kubo-type approach, which treats electron transport as the response to electric fields. Both methods have successfully described the transport problems for various systems and they should produce the same results when appropriate boundary conditions are imposed.

Usually, when the system is large enough to have many scatterings in the diffusive regime, the transport problems have been analyzed by the Kubo-type or semi-classical Boltzmann-type approach. On the other hand, as the system becomes small in the ballistic regime, the Landauer-type approach has been used for the transport calculations. In this chapter, we will present five computational methods. The first three methods (RTM, Lippmann–Schwinger, NEGF) are categorized to the scattering approach between electrodes and the last two methods (time-dependent wave-packet diffusion [TD-WPD] and quantum master equation [QME]) are categorized to the response approach. We also describe the formalisms for thermal transport and for the time-dependent dynamical transport aiming at an extension to time-dependent DFT for transport.

Here we describe the characteristics for the modeling of a system when we apply DFT to the transport problem in a scattering approach. The traditional way to obtain electronic states of materials using DFT is to construct the Kohn–Sham Hamiltonian matrix using some basis functions under appropriate boundary conditions. For example, in the solid case, with the periodic boundary conditions imposed for all the directions, the wavefunctions are expanded to satisfy the Bloch condition. Plane-wave expansion of $\psi(\mathbf{r}) = \sum_j c_j e^{i(\mathbf{k}+\mathbf{G}_j)\cdot\mathbf{r}}$ together with the fast Fourier transform (FFT) are frequently used to construct the Kohn–Sham equation, which is solved by matrix diagonalization, by molecular dynamics approach [58], by the direct energy minimization technique using the conjugate-gradient method [59], etc. A number of efficient computation methods have so far been developed.

For the transport problem as a scattering approach, the main difference from the above approach is that we have to treat open systems with boundary conditions to connect to left (L) and right (R) semi-infinite electrodes with different Fermi energies $\epsilon_F^{L/R}$. Correspondingly, the Kohn–Sham eigenvalue ϵ becomes continuous and the Kohn–Sham orbital is solved as a scattering state for each continuous energy ϵ, giving the boundary conditions deep inside the left and right electrodes. This is a quite different approach from the conventional matrix diagonalization problem.

Electric fields are determined from the Poisson equation

$$\nabla^2 V_H(\mathbf{r}) = -4\pi e^2 n(\mathbf{r}) \tag{4.39}$$

with the boundary conditions imposed deep inside left and right electrodes

$$\epsilon_F^L - \epsilon_F^R = eV, \tag{4.40}$$

where V is an applied bias voltage. The effective potential $V_{\text{eff}}(\mathbf{r})$ is constructed by adding the pseudopotentials and exchange-correlation potential to $V_H(\mathbf{r})$.

4.1.5.2 Contour integral for electron density

Electron density $n(\mathbf{r})$ is constructed to occupy electronic states up to ϵ_F^L for the wavefunctions $\psi_\epsilon^L(\mathbf{r})$ coming from the left electrode and to occupy electronic states up to ϵ_F^R for the wavefunctions $\psi_\epsilon^R(\mathbf{r})$ coming from the right electrode (we assume $\epsilon_F^R < \epsilon_F^L$). Since $\psi_\epsilon^{L/R}(\mathbf{r})$ are continuous for ϵ, the summation becomes integrals as

$$n(\mathbf{r}) = \int_{-\infty}^{\epsilon_F^L} |\psi_\epsilon^L(\mathbf{r})|^2 d\epsilon + \int_{-\infty}^{\epsilon_F^R} |\psi_\epsilon^R(\mathbf{r})|^2 d\epsilon. \tag{4.41}$$

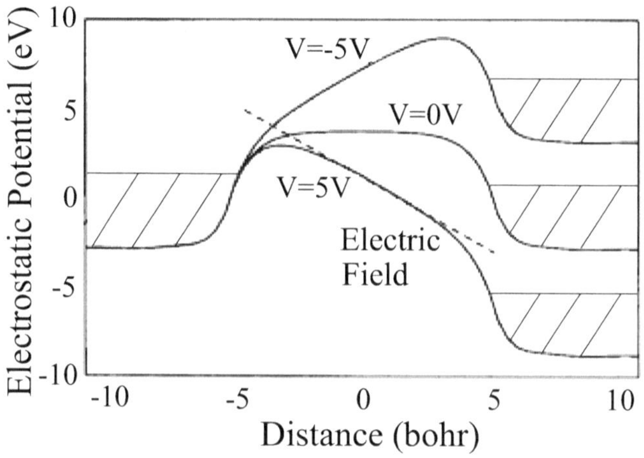

Figure 4.2 Schematic illustration of effective potentials in nanosystems. Different Fermi energies $\epsilon_F^{L/R}$ are assigned to the left and right electrodes to match to applied bias voltage. The Kohn–Sham equation is solved as a scattering problem with continuous energy ϵ for these potentials. The slopes of the potential correspond to an electric field, which is determined after we obtain the effective potential.

Since the electron number $N = \int n(\mathbf{r})d\mathbf{r}$ is not conserved in an open system problem, this integration must be done carefully. The abundance or deficiency of electron density prohibits to perform the self-consistent calculations. When an effective potential is smooth and there are no localized states, that is no steep peaks in energies, this integral can be performed by taking a precise mesh for ϵ.

A better approach for this is to use the analytical continuation of the retarded Green's function $G^r(\mathbf{r}, \mathbf{r}, \epsilon)$. Since the retarded Green's function $G^r(\mathbf{r}, \mathbf{r}, \epsilon)$ is constructed using the wavefunctions $\psi_\epsilon(\mathbf{r})$ or by solving directly the second-order differential equation,

$$\left[\epsilon + \frac{\hbar^2\nabla^2}{2m} - V_{\text{eff}}[n(\mathbf{r})] + i0^+\right] G^r(\mathbf{r}, \mathbf{r}', \epsilon) = \delta(\mathbf{r} - \mathbf{r}'), \qquad (4.42)$$

where the effective potential is the sum of three terms

$$V_{\text{eff}}[n(\mathbf{r})] = V_{\text{ps}}(\mathbf{r}) + V_H(\mathbf{r}) + \frac{\delta E_{xc}[n(\mathbf{r})]}{\delta n(\mathbf{r})}, \qquad (4.43)$$

electron density is obtained from

$$n(\mathbf{r}) = -\frac{1}{\pi}\text{Im}\int_Z G^r(\mathbf{r}, \mathbf{r}, z)dz + \int_{\epsilon_F^R}^{\epsilon_F^L} |\psi_\epsilon^L(\mathbf{r})|^2 d\epsilon, \qquad (4.44)$$

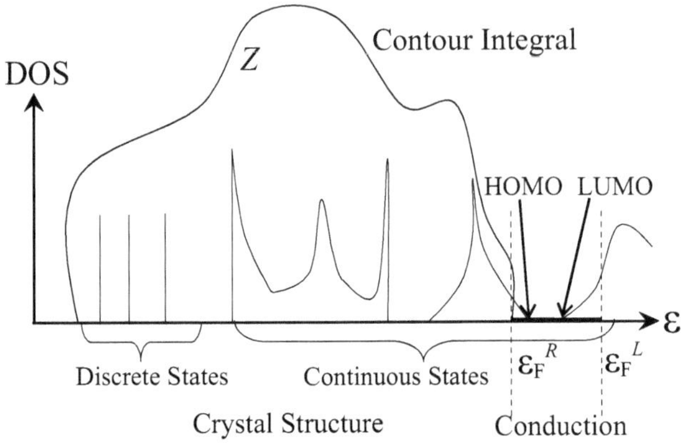

Figure 4.3 Schematic illustration of the contour integration to obtain electron density $n(\mathbf{r})$. For the equilibrium energy mesh points for $-\infty < \epsilon < \epsilon_F^R$, the integration is performed in the complex plane Z using the retarded Green's function technique. For nonequilibrium energy mesh points for $\epsilon_F^R < \epsilon < \epsilon_F^L$, the integration should be done in the real axis.

where the integral calculation for the first term of Eq. (4.44) is performed across the upper complex energy plane (Z) for $-\infty \leq z \leq \epsilon_F^R$. With this technique, the energy mesh points to construct electron density are greatly reduced and $n(\mathbf{r})$ is accurately obtained, which stabilize the self-consistent calculations of the DFT procedures.

It should be noted that in the nonequilibrium Green's function (NEGF) method, the second term of Eq. (4.44) can be treated by the integration of $(1/2\pi i)G^{<}(\mathbf{r}, \mathbf{r}, \epsilon)d\epsilon$. However, the integral must be performed in the real energy mesh points, since $G^{<}$ cannot be mapped to complex plane.

4.1.5.3 Basis functions for Kohn–Sham orbitals

To construct the Kohn–Sham Hamiltonian, we need to expand the Kohn–Sham orbitals $\psi(\mathbf{r})$. Since there is no periodicity in the direction where the current flows, the Fourier transform cannot be used in that direction.

The most accurate way for an expansion is to use the Laue representation, where the wavefunction is expanded by the plane-wave in the lateral direction and use the real-grid mesh in the current direction as

$$\psi_i(\mathbf{r}) = \sum_j \psi_i(\mathbf{G}_{||}^j, z)e^{i(\mathbf{k}_{||}+\mathbf{G}_{||}^j)\cdot\mathbf{r}_{||}}, \qquad (4.45)$$

where the z-dependent coefficients $\psi_i(\mathbf{G}_{||}^j, z)$ are determined. In this expansion, we can systematically increase the accuracy by increasing the number j of plane-waves and by reducing the real mesh spacing for z. The merit of this expansion is that since it does not depend on the atomic positions, it can be used for various atomic configurations. For example, a good contact condition where the atomic bonds to electrode materials are well formed and a bad contact condition where atomic bonds are no more realized thus in the tunneling regime. On the other hand, the demerit of this expansion is that the computation becomes too much time-consuming.

In the recursion-transfer-matrix (RTM) method, the Kohn–Sham equation is discretized into three terms recursive equation using the higher-order Numerov expansion, then the transfer matrix defined as a ratio matrix between sequential mesh points is calculated recursively. This method enables us to suppress always the notorious exponentially diverging scattering waves and to perform numerically stable scattering matrix computations.

In the Lippmann–Schwinger (LS) equation method, the second-order differential Kohn–Sham equation is transformed into the integral equation to obtain scattering waves. We solve the linear algebraic equation of the huge matrix when we use the plane-wave Laue representation.

An alternative way for the expansion of Kohn–Sham orbitals is to use the atomically localized basis sets $\phi(\mathbf{r})$, where the orbitals are expanded as a linear combination of atomic orbitals

$$\psi_i(\mathbf{r}) = \sum_j C_{ij}\phi_j(\mathbf{r}) \qquad (4.46)$$

with the coefficients C_{ij}. There are various forms for the basis sets of $\phi(\mathbf{r})$, such as Gaussian basis sets, molecular orbitals, and numerical atomic basis sets. In the NEGF method, which treats the effects of semi-infinite electrodes with use of self-energies, we show the numerical atomic basis sets later.

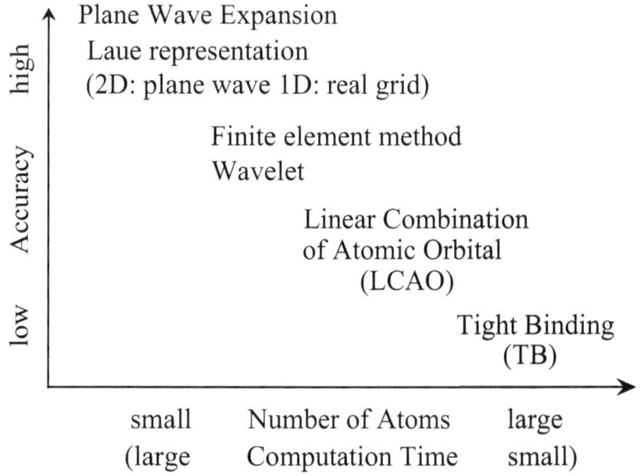

Figure 4.4 Comparison of the basis functions for the Kohn–Sham orbitals. Plane-wave expansion using the Laue representation is the most accurate, but most time-consuming. Expansion by the linear combination of atomic orbital can treat larger systems with decreasing the matrix dimension compared with the plane-wave expansion.

To treat larger systems, we can use the simple tight-binding Hamiltonian where the on-site energies and the nearest-neighbor transfer energies are treated in the simplest case. The TD-WPD method, which is an effective method of order-N (O(N)) computation for the transport of huge systems, is applicable to the tight-binding DFT or when we use the localized basis sets. We will present examples of the calculations using the tight-binding approximation.

Finally, the QME method using the density matrix, we use the periodic boundary conditions for the systems and accelerate electrons by the time-dependent vector potential. Accelerated electrons are damped due to the dissipative, no energy-conserving scatterings with phonons. In this method, due to the ordinary periodic boundary conditions, we can use any basis sets, including the usual plane-wave basis sets, localized basis set, and tight-binding approach. The difficult problem is that we have to treat the whole regions, which includes not only the nanosystem between electrodes but also the region deep inside the electrodes.

4.1.5.4 Problems for the DFT applying to transport

Here we consider the fundamental problems of DFT when we apply it to the transport phenomena. As we show, DFT is derived as a ground-state theory generally based on the variational principles. However, the electric current is not derived from a variational quantity. This means that to be accurate, a set of trial wavefunctions that produces a reasonable ground-state energy might not correspond to the true wavefunctions for obtaining the current.

In the high bias case, it is well known that the single-particle energy spectra of resonant states obtained using the conventional ground-state DFT take smaller values than the correct excitation energy levels. To describe such excitation energy levels is indispensable to obtain more accurate electric current and thus I–V curves. Presently, the time-dependent density functional theory (TDDFT), which treats the dynamics of electrons in the time domain, is the most promising way for that purpose.

We will show the way to extend the NEGF method and the QME method to the TDDFT formalism in the last section. Also, we will treat the TDDFT for excitation energies of quantum dots in Chapter 6.

4.2 Recursion-Transfer-Matrix Method

Here we present a method for the transport calculations using the RTM method [60]. We consider a system in which several layers of periodic atomic structures are attached smoothly to the left and right semi-infinite bulk electrodes, which have different Fermi energies ϵ_F^L, ϵ_F^R and satisfy the condition of $\epsilon_F^L - \epsilon_F^R = eV$, where V is an applied bias voltage.

4.2.1 *Recursive Equation for Transfer Matrix*

First, we consider the case in which an effective potential is expressed by the local potential $V_{\text{eff}}(\mathbf{r})$ only. We will show the method for the nonlocal pseudopotential term $V^{ps}(\mathbf{r}, \mathbf{r}')$ later.

The Kohn–Sham equation becomes

$$\left[-\frac{h^2}{2m}\nabla^2 + V_{\text{eff}}(\mathbf{r})\right]\psi(\mathbf{r}) = \epsilon\psi(\mathbf{r}), \tag{4.47}$$

where the total effective potential V_{eff} is composed of the Hartree V_{H}, the ion-core local pseudopotential V_{loc}^{ps} with atom index α, and the exchange-correlation potential V_{xc} by

$$V_{\text{eff}}(\mathbf{r}) = \sum_{\alpha} V_{\text{loc}}^{ps}(\mathbf{r}-\mathbf{r}_\alpha) + V_{\text{H}}(\mathbf{r}) + V_{\text{xc}}[n(\mathbf{r})]. \tag{4.48}$$

Taking the z-axis in which the current is flowing and assuming the periodic atomic structures in the x, y directions, we can expand the effective potential as

$$V_{\text{eff}}(\mathbf{r}) = \sum_{i=0}^{\infty} V_{\text{eff}}(\mathbf{G}_{||}^{i}, z)e^{i\,\mathbf{G}_{||}^{i}\cdot\mathbf{r}_{||}}, \tag{4.49}$$

where $\mathbf{G}_{||}^{i}$ is a set of 2D reciprocal lattice vectors in the x, y directions. The wavefunction $\psi(\mathbf{r})$, which obeys the Bloch theorem, is expressed by the Laue representation as

$$\psi(\mathbf{r}) = e^{i\,\mathbf{k}_{||}\cdot\mathbf{r}_{||}} \sum_{j=0}^{\infty} \psi(\mathbf{G}_{||}^{j}, z)\, e^{i\,\mathbf{G}_{||}^{j}\cdot\mathbf{r}_{||}}. \tag{4.50}$$

Then the Kohn–Sham equation becomes a set of coupled differential equation[a] for $u(\mathbf{G}_{||}^{i}, z)$,

$$\left[-\frac{1}{2}\frac{d^2}{dz^2} + \frac{1}{2}|\mathbf{k}_{||} + \mathbf{G}_{||}^{i}|^2 + V_{\text{eff}}(\mathbf{G}_{||}^{0}, z) - \epsilon\right]\psi(\mathbf{G}_{||}^{i}, z)$$
$$+\sum_{j}^{\infty} V_{\text{eff}}(\mathbf{G}_{||}^{i} - \mathbf{G}_{||}^{j}, z)\psi(\mathbf{G}_{||}^{j}, z) = 0. \tag{4.51}$$

This coupled equation is solved for a given value of the energy ϵ.

Numerically, instead of treating an infinite number of $\psi(\mathbf{G}_{||}^{j}, z)$, we take a finite number n for $\psi(\mathbf{G}_{||}^{j}, z)(0 \le j \le n)$ for a convergent representation of $\psi(\mathbf{r})$. We use the cutoff energy E_{cut} defined by $|\mathbf{k}_{||} + \mathbf{G}_{||}^{n}|^2 \le E_{\text{cut}}$ for the number n. In this case, there are $2n$ different states for an energy ϵ, which are classified by an incident channel number i $(0 \le i \le n)$ such that

$$\psi_i(\mathbf{r}) = e^{i\,\mathbf{k}_{||}\cdot\mathbf{r}_{||}} \sum_{j=0}^{n} \psi_i(\mathbf{G}_{||}^{j}, z)\, e^{i\,\mathbf{G}_{||}^{j}\cdot\mathbf{r}_{||}}. \tag{4.52}$$

---

[a] We use the atomic units with $|e| = \hbar = m = 1$.

All information of states for ϵ are included in the matrix $\hat{U}(z)$,

$$\hat{U}(z) = \left[\ldots \mathbf{u}_i(z) \ldots\right], \qquad (4.53)$$

where the vector $\mathbf{u}_i(z)$,

$$\mathbf{u}_i(z) = \begin{pmatrix} \psi_i(\mathbf{G}_{||}^0, z) \\ \psi_i(\mathbf{G}_{||}^1, z) \\ \vdots \\ \psi_i(\mathbf{G}_{||}^n, z) \end{pmatrix} \qquad (4.54)$$

denotes the state in channel i with ϵ. The row of $\hat{U}(z)$ represents the Fourier component j of the wavefunction and the column of $\hat{U}(z)$ represents the incident channel number i. Using $\hat{U}(z)$, we can rewrite the second-order coupled differential equation in the matrix form of

$$\frac{1}{2}\frac{d^2}{dz^2}\hat{U}(z) = \hat{V}(z, \epsilon)\hat{U}(z). \qquad (4.55)$$

Here the matrix $\hat{V}(z, \epsilon)$ are expressed by

$$\hat{V}(z, \epsilon) =$$

$$\begin{pmatrix} V_{\text{eff}}(\mathbf{G}_{||}^0, z) + \frac{1}{2}|\mathbf{k}_{||} + \mathbf{G}_{||}^0|^2 - \epsilon & V_{\text{eff}}(\mathbf{G}_{||}^0 - \mathbf{G}_{||}^1, z) & \cdots & V_{\text{eff}}(\mathbf{G}_{||}^0 - \mathbf{G}_{||}^n, z) \\ V_{\text{eff}}(\mathbf{G}_{||}^1 - \mathbf{G}_{||}^0, z) & V_{\text{eff}}(\mathbf{G}_{||}^0, z) + \frac{1}{2}|\mathbf{k}_{||} + \mathbf{G}_{||}^1|^2 - \epsilon & \cdots & V_{\text{eff}}(\mathbf{G}_{||}^1 - \mathbf{G}_{||}^n, z) \\ \cdots & \cdots & \cdots & \cdots \\ V_{\text{eff}}(\mathbf{G}_{||}^n - \mathbf{G}_{||}^0, z) & V_{\text{eff}}(\mathbf{G}_{||}^n - \mathbf{G}_{||}^1, z) & \cdots & V_{\text{eff}}(\mathbf{G}_{||}^0, z) + \frac{1}{2}|\mathbf{k}_{||} + \mathbf{G}_{||}^n|^2 - \epsilon \end{pmatrix}. \qquad (4.56)$$

Let us show the method to solve the coupled differential equation. First, we divide the z-axis into fine meshes $\{z_p;\, p = 0 \cdots l+1\}$ with the width Δz, where z_0, z_{l+1} are taken as $z_0 = z_L$ and $z_{l+1} = z_R$. The mesh size Δz is defined by

$$\Delta z = z_{p+1} - z_p = \frac{z_{l+1} - z_0}{l+1}. \qquad (\text{for } \forall p) \qquad (4.57)$$

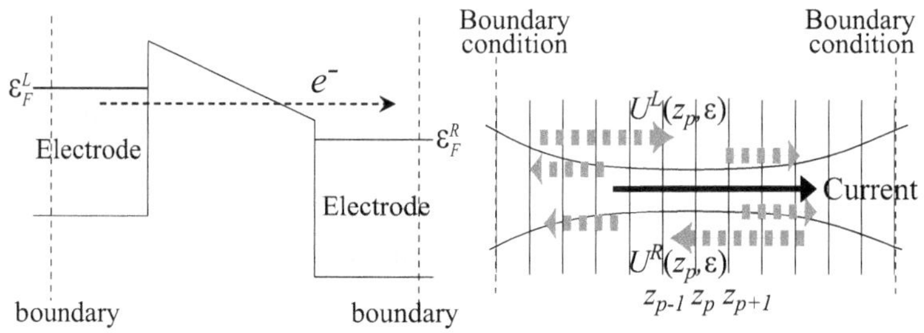

Figure 4.5 (Left) Schematic view of an effective potential for open, nonequilibrium junction system. (Right) Discretization of real mesh in the direction z where the current is flowing.

Then the second-order coupled differential matrix equation Eq. (4.55) is transformed into the following three-term difference-matrix equations:

$$\hat{a}(z_p, \epsilon)\hat{U}(z_{p+1}) - \hat{b}(z_p, \epsilon)\hat{U}(z_p) + \hat{c}(z_p, \epsilon)\hat{U}(z_{p-1}) = 0, \quad (4.58)$$

where

$$\begin{cases} \hat{a}(z_p, \epsilon) = \hat{I} - \frac{1}{6}\hat{V}(z_{p+1}, \epsilon)\Delta z^2 \\ \hat{b}(z_p, \epsilon) = 2\hat{I} + \frac{5}{3}\hat{V}(z_p, \epsilon)\Delta z^2 \\ \hat{c}(z_p, \epsilon) = \hat{I} - \frac{1}{6}\hat{V}(z_{p-1}, \epsilon)\Delta z^2. \end{cases} \quad (4.59)$$

Here $\hat{I}$ is the $n \times n$ unit matrix.

This discretization, called as Numerov expansion, is performed as follows. Expanding the wavefunction $\hat{U}(z)$ at the z_p,

$$\hat{U}(z_{p\pm1}) = \hat{U}(z_p) \pm \Delta z \left[\frac{d}{dz}\hat{U}(z)\right]_{z_p} + \frac{\Delta z^2}{2}\left[\frac{d^2}{dz^2}\hat{U}(z)\right]_{z_p} \pm \frac{\Delta z^3}{6}\left[\frac{d^3}{dz^3}\hat{U}(z)\right]_{z_p} + \frac{\Delta z^4}{24}\left[\frac{d^4}{dz^4}\hat{U}(z)\right]_{z_p} + O(\Delta z^5), \quad (4.60)$$

we obtain

$$\hat{U}(z_{p+1}) - 2\hat{U}(z_p) + \hat{U}(z_{p-1}) = \Delta z^2\left[\frac{d^2}{dz^2}\hat{U}(z)\right]_{z_p} + \frac{\Delta z^4}{12}\left[\frac{d^4}{dz^4}\hat{U}(z)\right]_{z_p} + O(\Delta z^6). \quad (4.61)$$

On the other hand, from the Kohn–Sham equation, the following equations are derived:

$$\left[\frac{d^2}{dz^2}U(z)\right]_{z_p} = 2V(z_p, \epsilon)U(z_p)$$
$$\left[\frac{d^4}{dz^4}U(z)\right]_{z_p} = 2\left[\frac{d^2}{dz^2}V(z, \epsilon)U(z)\right]_{z_p}$$
$$= 2\frac{V(z_{p+1}, \epsilon)U(z_{p+1}) - 2V(z_p, \epsilon)U(z_p) + V(z_{p-1}, \epsilon)U(z_{p-1})}{\Delta z^2}. \tag{4.62}$$

Inserting these formulas into Eq. (4.61), we obtain Eq. (4.58). Note that the error of this discretization is $O(\Delta z^6)$, which shows that the RTM method is very accurate for the real-mesh discretization.

Next, we introduce the transfer matrix defined by

$$\hat{S}(z_p, \epsilon) = \hat{U}(z_{p+1})\hat{U}(z_p)^{-1}, \tag{4.63}$$

then the coupled difference-matrix equation Eq. (4.58) satisfies the following recursive relation as

$$\hat{S}(z_{p-1}, \epsilon) = \left[\hat{b}(z_p, \epsilon) - \hat{a}(z_p, \epsilon)\hat{S}(z_p, \epsilon)\right]^{-1}\hat{c}(z_p, \epsilon). \tag{4.64}$$

Here note that the matrices $\hat{a}(z_p, \epsilon)$, $\hat{b}(z_p, \epsilon)$ and $\hat{c}(z_p, \epsilon)$ are all determined once the matrix $\hat{V}(z_p, E)$ is given.

Using this recursion-matrix equation, we obtain the wavefunctions $\hat{U}(z_p)$. The scattering states are classified into two categories of the linearly independent solutions, one whose incident waves come from the left electrodes $\hat{U}^L(z_p)$ and the other whose incident waves come from the right $\hat{U}^R(z_p)$. These states are assumed to be obtained independently through an effective potential $\hat{V}(z_p)$.

Let us describe the calculation method for $\hat{U}^L(z_p)$. Note that $\hat{U}^R(z_p)$ is obtained in the same way by changing the index $p \to l + 1 - p$. When we find the transfer matrix $\hat{S}^L(z_{l+1}, \epsilon)$ at the boundary z_{l+1}, $\hat{S}^L(z_p, \epsilon)$ is obtained at an arbitrary point in the scattering region. Since we have the wavefunctions $\hat{U}^L(z_0)$ at the boundary z_0 from the usual band calculations for bulk materials, $\hat{U}^L(z_p, E^L)$ at an arbitrary point in the scattering region is calculated from

$$\hat{U}^L(z_{p+1}) = \hat{S}^L(z_p, \epsilon)\hat{U}^L(z_p). \tag{4.65}$$

In the above procedure of solving the Kohn–Sham equation for scattering states, the ratios of wavefunctions $\hat{S}^L(z_p, \epsilon)$ are first determined using the boundary conditions deep in electrodes instead of treating the wavefunctions directly. This means that we effectively treat only a derivative at an arbitrary point p. On the condition that the amplitude of an incident wave is unity, the normalized wavefunction is obtained. In this calculation, we take plane-waves or decaying evanescent waves in the asymptotic electrode regions. Therefore, growing evanescent waves, which are often the cause of instability for the numerical calculation of transfer matrix, are excluded in the present procedure. Namely in both asymptotic electrode regions, we can avoid the appearance of growing evanescent waves. This is an important merit of this method compared with the other numerical methods of transfer matrix, in which the wavefunctions are obtained on the condition that amplitude and its derivative are matched with the asymptotic solutions at a single boundary point.

When we consider the jellium model for electrodes, the boundary conditions become very simple. Since we can take a plane wave $e^{ik^{L(R)}z}$ in the electrodes, we obtain the boundary conditions as $\left[S^L\right]_{ij} = e^{ik_i^R \Delta z}\delta_{ij}$ and $\left[S^R\right]_{ij} = e^{-ik_i^L \Delta z}\delta_{ij}$ where the wavevector for an incident channel number i is given by $k_i^{L(R)} = \sqrt{2m(\epsilon - V_{L(R)})/\hbar^2 - |k_z + G_{||}^i|^2}$ and $V_{L(R)}$ are the constant potentials inside the electrodes $L(R)$. The normalized wavefunctions at boundaries are $\left[U^L(z_0)\right]_{ij} = e^{ik_i^L z_0}\delta_{ij}$ and $\left[U^R(z_{l+1})\right]_{ij} = e^{-ik_i^R z_{l+1}}\delta_{ij}$.

To show how accurate the RTM method is, let us consider a simple one-dimensional potential barrier $V_{\rm eff}(z) = 1/\cosh^2(z)$ and calculate the transmission coefficient t. For this potential, we have exact solutions and approximate solutions from WKB approximation are also possible. Since the wavefunctions are written in the asymptotic regions as

$$\psi^L(z) = \begin{cases} e^{ik_z^L z} + re^{-ik_z^L z} & (z \le z_L) \\ te^{ik_z^R z} & (z \ge z_R), \end{cases} \tag{4.66}$$

$$\psi^R(z) = \begin{cases} t'e^{-ik_z^L z} & (z \le z_L) \\ e^{-ik_z^R z} + r'e^{ik_z^R z} & (z \ge z_R), \end{cases} \tag{4.67}$$

we can use the RTM method to obtain the transmission t with the boundary conditions at $z_{L,R}$. Figure 4.6 shows the transmission as

Transmission Coefficient for $V_{eff}(z) = \dfrac{1}{\cosh^2 z}$

Energy	Exact	$\Delta z = 1.0$	$\Delta z = 0.5$	$\Delta z = 0.2$	WKB
0.2	0.012413383	0.012472732	0.012417504	0.012413383	0.007358269
0.4	0.062986591	0.063653976	0.063035725	0.062987932	0.038162036
0.6	0.192860287	0.194673260	0.193006772	0.192864218	0.134945913
0.8	0.409617734	0.411489872	0.409833778	0.409657631	0.391372587
1.0	0.639483981	0.639971053	0.639558977	0.639485784	1.000000000

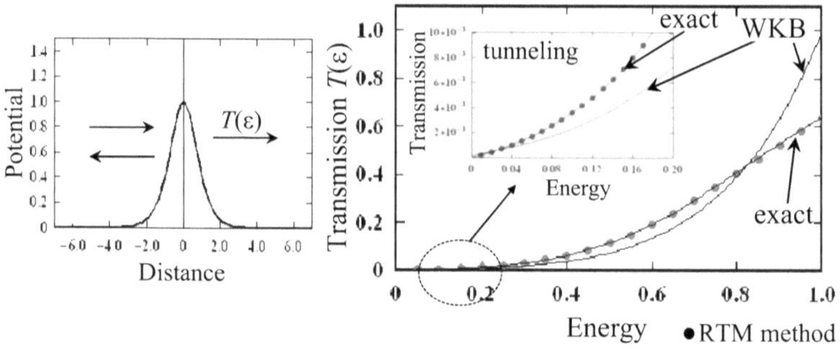

Figure 4.6 (Top) Transmission coefficients for the repulsive potential of $V_{\text{eff}}(z) = 1/\cosh^2(z)$, changing the mesh spacings Δz and the incident energies. Compared with the Wentzel–Kramers–Brillouin (WKB) approximation, the RTM method provide much accurate solutions as the mesh spacings become small. (Bottom) Schematic view of the potential and the result of transmission as a function of energy. For all the energy regime, including the tunneling region, the RTM method gives accurate results.

a function of incident energies. We can see that the RTM method provides very accurate transmission coefficient results for all the energy regimes.

4.2.1.1 Boundary condition connecting to bulk electrode

Now let us consider how to obtain the starting transfer matrix $\hat{S}^L(z_{l+1}, \epsilon)$ and the wavefunction $\hat{U}^L(z_0)$ at boundary points z_0 and z_{l+1} for $\hat{U}^L(z_p)$. These are determined from the calculations of bulk electrode on the left and right electrodes, whose electronic states are constructed from the Bloch states. Thus, we need a method to determine the conditions to connect to the Bloch states of bulk materials. From those calculations, we obtain the starting transfer matrix $\hat{S}^R(z_0, \epsilon)$ and the wavefunction $\hat{U}^R(z_0)$ for $\hat{U}^R(z_p)$. Since the unit cell for bulk electrode is not so large, the calculations are not so time-consuming. Once we have those matrices for the energy mesh points ϵ, we fix them for the self-consistent procedures in the RTM method to determine electron density $n(\mathbf{r})$ and the effective potential $V_{\text{eff}}(\mathbf{r})$.

In order to find the boundary conditions for the wavefunctions to connect to the Bloch states of bulk electrodes, we first assume that the bulk electrodes are periodic $\hat{V}_{\text{eff}}^{L(R)}(\mathbf{G}_{||}^i, z+a) = \hat{V}_{\text{eff}}^{L(R)}(\mathbf{G}_{||}^i, z)$ where a are the periods along the z-axis. Second, we solve the scattering states for this periodic potential. This is done as follows.

We assume the constant potentials for $z < z_0$ and $z > z_{l+1}$;

$$\begin{cases} z \leq z_0; & \hat{V}_{\text{eff}}(\mathbf{r}) = V_L = \text{const} \\ z \geq z_{l+1}; & \hat{V}_{\text{eff}}(\mathbf{r}) = V_R = \text{const.} \end{cases}$$

Thus, the Fourier components of $\hat{V}_{\text{eff}}(\mathbf{r})$ take the following values:

$$\begin{cases} \hat{V}_{\text{eff}}(\mathbf{G}_{||}^i - \mathbf{G}_{||}^j, z \leq z_0) = V_L \delta_{i,j} \\ \hat{V}_{\text{eff}}(\mathbf{G}_{||}^i - \mathbf{G}_{||}^j, z \geq z_{l+1}) = V_R \delta_{i,j}. \end{cases} \tag{4.68}$$

The matrix $\hat{V}(z, \epsilon)$ becomes diagonal in these regions and the difference equations are decoupled for each $\psi_i(\mathbf{G}_{||}^j, z)$ component as

$$\left[1 - \frac{1}{6}\beta_{L(R)}^i\right] \psi_i(\mathbf{G}_{||}^i, z_{p+1}) - \left[2 + \frac{5}{3}\beta_{L(R)}^i\right] \psi_i(\mathbf{G}_{||}^i, z_p) + \left[1 - \frac{1}{6}\beta_{L(R)}^i\right] \psi_i(\mathbf{G}_{||}^i, z_{p-1}) = 0, \tag{4.69}$$

where

$$\beta_{L(R)}^i = \Delta z^2 \left[V_{L(R)} + \frac{1}{2}|\mathbf{k}_{||} + \mathbf{G}_{||}^i|^2 - \epsilon\right]. \tag{4.70}$$

These equations are solved as

$$\psi_i(\mathbf{G}_{||}^i, z_p) \sim \left[g_{L(R)}^{(\pm)}(\mathbf{G}_{||}^i)\right]^p \tag{4.71}$$

with

$$g^{(\pm)}_{L(R)}(\mathbf{G}^i_{||}) = \frac{1+\frac{5}{6}\beta^i_{L(R)}}{1-\frac{1}{6}\beta^i_{L(R)}} \pm i\sqrt{1-\left(\frac{1+\frac{5}{6}\beta^i_{L(R)}}{1-\frac{1}{6}\beta^i_{L(R)}}\right)^2}$$
$$\left(\text{for} \quad \epsilon - \frac{1}{2}|\mathbf{k}_{||}+\mathbf{G}^i_{||}|^2 > V_{L(R)}\right) \tag{4.72}$$

$$g^{(\pm)}_{L(R)}(\mathbf{G}^i_{||}) = \frac{1+\frac{5}{6}\beta^i_{L(R)}}{1-\frac{1}{6}\beta^i_{L(R)}} \mp \sqrt{\left(\frac{1+\frac{5}{6}\beta^i_{L(R)}}{1-\frac{1}{6}\beta^i_{L(R)}}\right)^2 - 1}$$
$$\left(\text{for} \quad \epsilon - \frac{1}{2}|\mathbf{k}_{||}+\mathbf{G}^i_{||}|^2 < V_{L(R)}\right). \tag{4.73}$$

These states correspond to right-going (+) or left-going (–) plane-waves Eq. (4.72) or to growing or decaying evanescent waves Eq. (4.73). Note that in the case of plane waves Eq. (4.72), $|g^{(\pm)}_{L(R)}| = 1$. Using these relations, $\hat{U}(z_p)$ in the asymptotic electrode region is expressed as a linear combination of $\left[\hat{K}^{(\pm)}_{L(R)}\right]^p \hat{\Lambda}$, where $\hat{K}^{(\pm)}_{L(R)}$ is

$$\hat{K}^{(\pm)}_{L(R)} = \begin{pmatrix} g^{(\pm)}_{L(R)}(\mathbf{G}^0_{||}) & 0 & \ldots & 0 \\ 0 & g^{(\pm)}_{L(R)}(\mathbf{G}^1_{||}) & \ldots & 0 \\ \ldots & \ldots & \ldots & \ldots \\ 0 & 0 & \ldots & g^{(\pm)}_{L(R)}(\mathbf{G}^n_{||}) \end{pmatrix} \tag{4.74}$$

and $\hat{\Lambda}$ is an arbitrary matrix whose components are constant complex values. Note the relation between $\hat{K}^{(\pm)}_{L(R)}$ by

$$\hat{K}^{(+)}_{L(R)} \cdot \hat{K}^{(-)}_{L(R)} = \hat{I}. \tag{4.75}$$

$\psi_i^L(\mathbf{r})$ is composed of an incident wave (channel number i) and the reflected waves in the range of $z \le z_0$, while in the range of $z \ge z_{l+1}$ it is composed of only the transmitted waves

$$\psi_i^L(\mathbf{r}) = \begin{cases} e^{i\,k^i_z z} e^{i\,(\mathbf{k}_{||}+\mathbf{G}^i_{||})\cdot\mathbf{r}_{||}} + \sum_j r_{i,j} \begin{Bmatrix} e^{-i\,k^j_z z} \\ e^{\kappa^j_z z} \end{Bmatrix} e^{i\,(\mathbf{k}_{||}+\mathbf{G}^j_{||})\cdot\mathbf{r}_{||}} & (z \le z_0) \\ \sum_j t_{i,j} \begin{Bmatrix} e^{i\,k^j_z z} \\ e^{-\kappa^j_z z} \end{Bmatrix} e^{i\,(\mathbf{k}_{||}+\mathbf{G}^j_{||})\cdot\mathbf{r}_{||}} & (z \ge z_{l+1}). \end{cases} \tag{4.76}$$

Combining all the incident channel states, we express the wavefunctions as a matrix form such that

$$\hat{U}^{L}(z_p) = \begin{cases} \left[\hat{K}_L^{(+)}\right]^p \hat{\Lambda}_0 + \left[\hat{K}_L^{(-)}\right]^p \hat{R} & (\text{for} \quad z_p \le z_0) \\ \left[\hat{K}_R^{(+)}\right]^p \hat{T} & (\text{for} \quad z_p \ge z_{l+1}), \end{cases} \tag{4.77}$$

where $\hat{\Lambda}_0$ denotes the diagonal initial phase matrix and $\hat{R}$, $\hat{T}$ are unknown reflection and transmission matrices.

Similarly, $\psi_j^R(\mathbf{r})$ is composed of the transmitted waves in the range of $z \le z_0$, while in the range of $z \ge z_{l+1}$ it is composed of an incident wave (channel number i) and the reflected waves,

$$\psi_i^R(\mathbf{r}) = \begin{cases} \sum_j t'_{i,j} \begin{Bmatrix} e^{-i\,k_z^j z} \\ e^{\kappa_z^j z} \end{Bmatrix} e^{i\,(\mathbf{k}_{\|}+\mathbf{G}_{\|}^j)\cdot\mathbf{r}_{\|}} & (z \le z_0) \\ e^{-i\,k_z^i z} e^{i\,(\mathbf{k}_{\|}+\mathbf{G}_{\|}^i)\cdot\mathbf{r}_{\|}} + \sum_j r'_{i,j} \begin{Bmatrix} e^{i\,k_z^j z} \\ e^{-\kappa_z^j z} \end{Bmatrix} e^{i\,(\mathbf{k}_{\|}+\mathbf{G}_{\|}^j)\cdot\mathbf{r}_{\|}} & \\ & (z \ge z_{l+1}). \end{cases} \tag{4.78}$$

Combining all the incident channel states, we express the wavefunctions as a matrix form such that

$$\hat{U}^{R}(z_p, E^R) = \begin{cases} \left[\hat{K}_L^{(-)}\right]^p \hat{T}' & (\text{for} \quad z_p \le z_0) \\ \left[\hat{K}_R^{(-)}\right]^p \hat{\Lambda}_{l+1} + \left[\hat{K}_R^{(+)}\right]^p \hat{R}' & (\text{for} \quad z_p \ge z_{l+1}). \end{cases} \tag{4.79}$$

These are the asymptotic forms of wavefunctions for $z \le z_0$, $z \ge z_{l+1}$.

Next, consider the region $z_0 < z_p < z_{l+1}$, where the Fourier components $V_{\text{eff}}(\mathbf{G}_{\|}^i - \mathbf{G}_{\|}^j, z_0 < z < z_{l+1})$ remain finite for $i \ne j$ and thus the coupled difference-matrix equations are not decoupled as those in the asymptotic regions. By the matrix form at z_{l+1}, it follows that

$$\begin{aligned} \hat{S}(z_{l+1}, \epsilon) &= \hat{U}^{L}(z_{l+2})\hat{U}^{L}(z_{l+1})^{-1} \\ &= ([\hat{K}_R^{(+)}]^{l+2}\,\hat{T})([\hat{K}_R^{(+)}]^{l+1}\,\hat{T})^{-1} \\ &= \hat{K}_R^{(+)}. \end{aligned} \tag{4.80}$$

Starting the calculation with this matrix and furthermore using the matrix form at z_{-1}, it follows that

$$\begin{aligned} \hat{S}(z_{-1}, \epsilon) &= \hat{U}^{L}(z_0)\hat{U}^{L}(z_{-1})^{-1} \\ &= (\hat{\Lambda}_0 + \hat{R})([\hat{K}_L^{(+)}]^{-1}\hat{\Lambda}_0 + [\hat{K}_L^{(-)}]^{-1}\,\hat{R})^{-1}. \end{aligned} \tag{4.81}$$

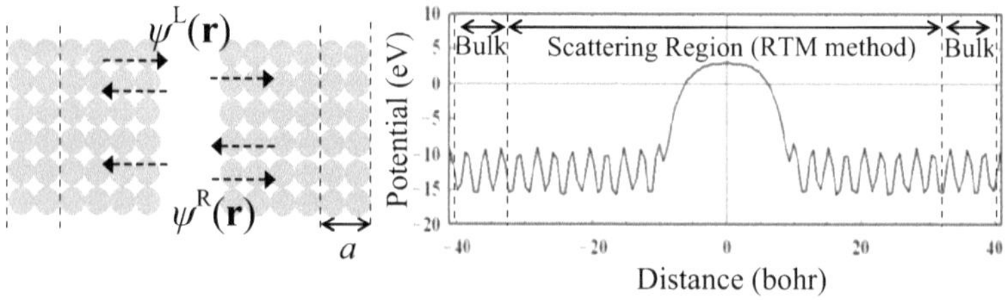

Figure 4.7 (Left) Schematic representation of junction system with a vacuum region between the electrodes. Scattering waves coming from left and from right are solved in the scattering region to construct electron density. At the boundaries, these states are connected to the bulk Bloch wavefunctions with the period of a. (Right) An example of the effective potential. At boundaries, the effective potential smoothly connects to the bulk potential. In the vacuum region between electrodes, the effective potential forms the potential barrier for electrons to tunnel.

Thus, by using the relation $\hat{K}_L^{(+)} \cdot \hat{K}_L^{(-)} = \hat{I}$, we obtain

$$\begin{aligned}\hat{U}^L(z_0) &= \hat{\Lambda}_0 + \hat{R} \\ &= \hat{S}(z_{-1})[\hat{K}_L^{(+)}\hat{S}(z_{-1}, \epsilon) - I]^{-1}[\hat{K}_L^{(+)} - \hat{K}_L^{(-)}]\hat{\Lambda}_0.\end{aligned} \tag{4.82}$$

It is apparent that the diagonal initial phase matrix $\hat{\Lambda}_0$ does not affect electron density or current density; thus we can omit it.

To obtain the Bloch states from these solutions [61], we introduce a $2N \times 2N$ matrix defined by

$$\hat{W}(z) = \begin{pmatrix} \hat{U}^L(z) & \hat{U}^R(z) \\ \dfrac{d}{dz}\hat{U}^L(z) & \dfrac{d}{dz}\hat{U}^R(z) \end{pmatrix}. \tag{4.83}$$

This implies that solutions to a second-order differential equation are determined by specifying a value and a first derivative. In order that the solutions constitute a complete set of $2N$-dimensional space, the columnar vectors of $\hat{W}(z)$ should be linearly independent. So when we use the translation operator T, the columnar vectors at $z + a$ (period of a) can be expressed by the linear combinations of vectors as

$$\hat{T}\hat{W}(z) = \hat{W}(z + a) = \hat{W}(z)\hat{C}, \tag{4.84}$$

where $\hat{C}$ is a representation matrix of the translation operator $\hat{T}$.

Then we can obtain the Bloch states by diagonalizing the representation matrix $\hat{C}$ as

$$\hat{C}\hat{D} = \hat{D}\lambda. \tag{4.85}$$

Here λ is a diagonal matrix for eigenvalues of $\hat{C}$ and $\hat{D}$ is a transformation matrix. The Bloch states are obtained from columnar vectors of $\hat{W}(z)\hat{D}$, which are eigenvectors of the translation operator.

We note that since the process for the representation matrix $\hat{C}$ in Eq. (4.84) becomes singular as the matrix dimension $2N$ of basis functions increases, in practice it is favorable to solve a generalized eigenvalue problem of

$$\hat{W}(z+a)\hat{D} = \hat{W}(z)\hat{D}\lambda \tag{4.86}$$

instead of solving Eq. (4.85), which is numerically more stable.

Next, we need to classify eigenstates into two classes, the waves coming from the left and the waves coming from the right. This can be done to see the absolute values of eigenvalues. When $|\lambda_n| = 1$, they represent the propagating waves in bulk crystals and the eigenvalues are written as $\lambda_n = e^{ik_n a}$, where k_n is a real number. When $\lambda_n \neq 1$, then they represent the evanescent waves. We can classify N right-going eigenstates from the condition $J_z^n > 0$, where J_z^n is the z component of electric currents defined by

$$J_z^n = \int d\mathbf{r}_{||} \mathrm{Im}\left[\psi_n^*(\mathbf{r})\frac{\partial \psi_n(\mathbf{r})}{\partial z}\right]. \tag{4.87}$$

Similarly, N left-going states are defined by $J_z^n < 0$. Then, by rearranging columnar vectors of the matrix $\hat{D}$, we can write the eigenstates by the $2N \times 2N$ matrix

$$\hat{W}_B(z) = \begin{pmatrix} \hat{U}_B^L(z) & \hat{U}_B^R(z) \\ \frac{d}{dz}\hat{U}_B^L(z) & \frac{d}{dz}\hat{U}_B^R(z) \end{pmatrix}, \tag{4.88}$$

where $\hat{U}_B^L(z)$ and $\hat{U}_B^R(z)$ are the right- and left-going Bloch states.

Thus, initial conditions for the RTM method is obtained, for example for the right-going states such as

$$\hat{S}(z_{l+1}) = \hat{U}_B^R(z_{l+2})\left[\hat{U}_B^R(z_{l+1})\right]^{-1}, \tag{4.89}$$

where $\hat{U}_B^R(z_{l+2})$ is obtained from $\hat{U}_B^R(z_{l+1})$ and $d\hat{U}_B^R(z)/dz$ at z_{l+1}, and the value $\hat{U}_B^L(z_0)$ at z_0.

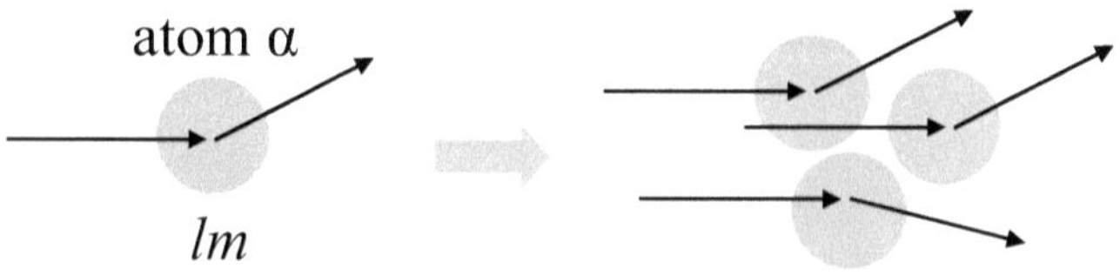

Figure 4.8 Schematic views to construct the Kohn–Sham orbital, including the nonlocal pseudopotentials. First, we solve the particular solution to a nonlocal pseudopotential term for one atom α with l and m. Then we collect each contribution to obtain the total orbital.

4.2.1.2 Inclusion of nonlocal pseudopotential

Here we consider how to include the nonlocal pseudopotentials in the RTM formulation [62]. When we include a nonlocal term $V_{nl}(\mathbf{r}, \mathbf{r}')$, the Kohn–Sham equation becomes

$$\left[-\frac{\hbar^2}{2m}\nabla^2 + V_{\text{eff}}(\mathbf{r})\right]\psi(\mathbf{r}) + \int V_{nl}^{ps}(\mathbf{r}, \mathbf{r}')\psi(\mathbf{r}')d\mathbf{r}' = \epsilon\psi(\mathbf{r}). \quad (4.90)$$

For the nonlocal pseudopotentials, we use the fully separable norm-conserving form of

$$V_{nl}^{ps}(\mathbf{r}, \mathbf{r}') = \frac{|\phi_{lm}(\mathbf{r})\delta V_l^{ps}(\mathbf{r})\rangle\langle\delta V_l^{ps}(\mathbf{r}')\phi_{lm}(\mathbf{r}')|}{\langle\phi_{lm}|\delta V_l^{ps}|\phi_{lm}\rangle}, \quad (4.91)$$

where δV_l^{ps} is the difference in the l-th component of the nonlocal pseudopotential from $V_{loc}^{ps}(\mathbf{r})$, and ϕ_{lm} is the pseudo-wavefunction of a pseudoatom.

For these pseudopotentials, we can derive the following inhomogeneous Kohn–Sham equation:

$$\left[-\frac{\hbar^2}{2m}\nabla^2 + V_{\text{eff}}(\mathbf{r})\right]\psi(\mathbf{r}) + \sum_{\alpha,lm} Z_{lm} f_{lm}(\mathbf{r} - \mathbf{r}_\alpha) \times \left[\int f_{lm}(\mathbf{r}' - \mathbf{r}_\alpha)^*\psi(\mathbf{r}')d\mathbf{r}'\right] = \epsilon\psi(\mathbf{r}), \quad (4.92)$$

where

$$f_{lm}(\mathbf{r}) = \frac{|\delta V_l^{ps}(\mathbf{r})\phi_{lm}(\mathbf{r})\rangle}{\sqrt{\langle\phi_{lm}|\delta V_l^{ps}|\phi_{lm}\rangle}}, \quad (4.93)$$

Z_{lm} is the sign of $\langle\phi_{lm}|\delta V_l|\phi_{lm}\rangle$ and the local effective potential becomes

$$V_{\text{eff}}(\mathbf{r}) = \sum_\alpha V_{\text{loc}}^{ps}(\mathbf{r} - \mathbf{r}_\alpha) + V_{\text{H}}(\mathbf{r}) + V_{\text{xc}}[n(\mathbf{r})]. \quad (4.94)$$

We note that since $\delta V_l^{ps}(\mathbf{r})$ is only present around the region close to each atom α, $f_{lm}(\mathbf{r}-\mathbf{r}_\alpha)$ is only present close to the atomic position $\mathbf{r}_\alpha$.

This inhomogeneous differential equation is solved as follows. First, we solve the homogeneous differential equation

$$\left[-\frac{\hbar^2}{2m}\nabla^2 + V_{\mathrm{eff}}(\mathbf{r})\right]\psi_0^{L(R)}(\mathbf{r}) = \epsilon\psi_0^{L(R)}(\mathbf{r}) \tag{4.95}$$

by the RTM method as we explained in the previous section.

Next, we solve the equation with one nonlocal term αlm as

$$\left[-\frac{\hbar^2}{2m}\nabla^2 + V_{\mathrm{eff}}(\mathbf{r})\right]\varphi^{\alpha lm}(\mathbf{r}) + Z_{lm}f_{lm}(\mathbf{r}-\mathbf{r}') = \epsilon\varphi^{\alpha lm}(\mathbf{r}), \tag{4.96}$$

where $\varphi^{\alpha lm}(\mathbf{r})$ is a particular solution for each atom αlm. This is transformed into the integral form as

$$\varphi^{\alpha lm}(\mathbf{r}) = \int G(\mathbf{r},\mathbf{r}',\epsilon)Z_{lm}f_{lm}(\mathbf{r}'-\mathbf{r}_\alpha)d\mathbf{r}', \tag{4.97}$$

where the definition of Green's function $G(\mathbf{r},\mathbf{r}',\epsilon)$ is

$$\left[\epsilon + \frac{\hbar^2}{2m}\nabla^2 - V_{\mathrm{eff}}(\mathbf{r})\right]G(\mathbf{r},\mathbf{r}',\epsilon) = \delta(\mathbf{r}-\mathbf{r}') \tag{4.98}$$

and is constructed from the solutions to local potential $V_{\mathrm{eff}}(\mathbf{r})$ by the RTM method, $\psi_0^{L(R)}(\mathbf{r})$ coming from the left and right electrodes with appropriate boundary conditions.

Finally, the general solutions for the inhomogeneous differential equation Eq. (4.91) are expressed by the sum of solutions to the homogeneous equation Eq. (4.95) and solutions to the inhomogeneous equation Eq. (4.96) as

$$\psi^{L(R)}(\mathbf{r}) = \psi_0^{L(R)}(\mathbf{r}) + \sum_{\alpha lm} C_{\epsilon,i,\alpha lm}^{L(R)}\varphi^{\alpha lm}(\mathbf{r}), \tag{4.99}$$

where the coefficients $C_{\epsilon,i,\alpha lm}^{L,R}$ are determined from the condition

$$\begin{aligned} C_{\epsilon,i,\alpha lm}^{L(R)} &= \int f_{lm}(\mathbf{r}-\mathbf{r}_\alpha)^*\psi^{L(R)}(\mathbf{r})d\mathbf{r} \\ &= \int f_{lm}(\mathbf{r}-\mathbf{r}_\alpha)^*\psi_0^{L(R)}(\mathbf{r})d\mathbf{r} \\ &\quad + \sum_{\alpha'l'm'} C_{\epsilon,i,\alpha'l'm'}^{L(R)} \int f_{l'm'}(\mathbf{r}-\mathbf{r}_\alpha)^*\varphi^{\alpha'l'm'}(\mathbf{r})d\mathbf{r}. \end{aligned} \tag{4.100}$$

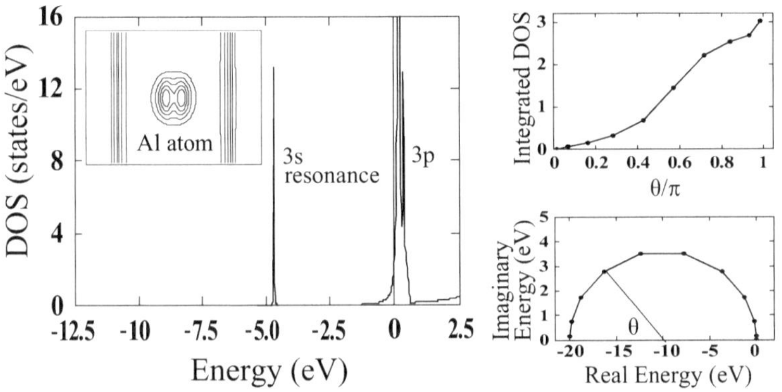

Figure 4.9 (Left) Density of states of a single Al atom located at the center of jellium electrodes. The $3s$ resonant state shows a very spiky structure, since it is much below the Fermi energy $\epsilon_F = 0$, whereas the $3p$ states are located around the Fermi energy, indicating partially occupied. The $3p$ states are seen to split due to breaking of the symmetry. (Right) Integrated density of states as a function of θ for the contour in the upper complex energy plane shown below. Due to the integration in the complex plane, a spiky resonance of the $3s$ is accurately treated.

When we take s and p orbitals as the nonlocal terms for example, there are 4 terms to each atoms. Then this linear algebraic equation becomes $4N_\alpha \times 4N_\alpha$ matrix, which are not so demanding to solve up to the atomic number N_α of several hundreds. We should note that since the nonlocal term $f_{lm}(\mathbf{r} - \mathbf{r}_\alpha)$ is only present at small regions around $\mathbf{r}_\alpha$, computations of the integrals are also not so time-consuming. These are the procedure that we obtain the solutions $\psi^{L,R}(\mathbf{r})$ for the inhomogeneous coupled-channel Kohn–Sham equation with the nonlocal pseudopotentials.

Electron density $n(\mathbf{r})$ is obtained from

$$n(\mathbf{r}) = \frac{2}{(2\pi)^2} \sum_{L(R)} \int d\mathbf{k} |\psi^{L(R)}(\mathbf{r})|^2, \qquad (4.101)$$

or equivalently by constructing Green's function from $\psi_Z^{L(R)}(\mathbf{r})$ to satisfy the boundary conditions and to calculate

$$n(\mathbf{r}) = -\frac{1}{\pi} \mathrm{Im} \int_Z G(\mathbf{r}, \mathbf{r}, z) dz, \qquad (4.102)$$

where we take the contour Z in the upper complex energy plane to reduce points for the integral to converge. Below we show a

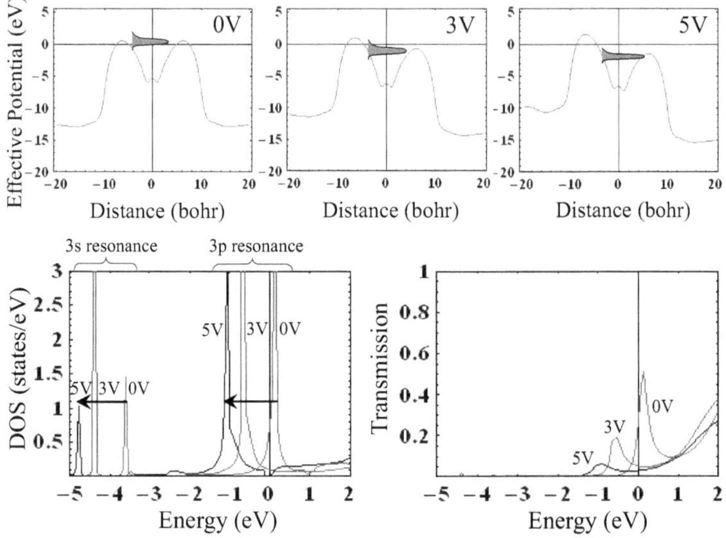

Figure 4.10 (Top) Effective potentials for a system where one Al atom is located at the center of jellium electrodes with applied bias voltages of $V = 0$ V, 3 V, and 5 V. The $3p$ resonant states around the Fermi energies go down as the applied voltages become large. (Bottom) Density of states and transmission coefficient as a function of energy for several applied bias voltages.

simple example for obtaining electron density $n(\mathbf{r})$ of a single Al atom located in the middle of the jellium electrodes.

4.2.2 *Poisson Equation in Laue Representation*

In this section, we explain how to solve the Poisson equation [60][a]

$$\nabla^2 V_{\mathrm{H}}(\mathbf{r}) = -4\pi n(\mathbf{r}). \tag{4.103}$$

In the 2D Fourier space, it becomes

$$\left[\frac{d^2}{dz^2} - |\mathbf{G}^i_{||}|^2\right] V_{\mathrm{H}}(\mathbf{G}^i_{||}, z) = -4\pi n(\mathbf{G}^i_{||}, z). \tag{4.104}$$

The numerical procedure depends on whether $|\mathbf{G}^i_{||}|$ is zero or finite.

[a]We note the atomic units with $|e| = 1$

For $|\mathbf{G}^i_{||}| \neq 0$, introducing Green's function $G(|\mathbf{G}^i_{||}|, z, z')$ as

$$\left[\frac{d^2}{dz^2} - |\mathbf{G}^i_{||}|^2\right] G(|\mathbf{G}^i_{||}|, z, z') = -\delta(z - z') \tag{4.105}$$

and with use of Green's theorem, we obtain

$$V_{\mathrm{H}}(\mathbf{G}^i_{||}, z) = 4\pi \int_{z_0}^{z_{l+1}} n(\mathbf{G}^i_{||}, z')G(|\mathbf{G}^i_{||}|, z, z')dz'$$
$$+ \left[G(|\mathbf{G}^i_{||}|, z, z')\frac{\partial}{\partial z'} V_{\mathrm{H}}(\mathbf{G}^i_{||}, z') - V_{\mathrm{H}}(\mathbf{G}^i_{||}, z')\frac{\partial}{\partial z'} G(|\mathbf{G}^i_{||}|, z, z')\right]_{z_0}^{z_{l+1}}.$$

Imposing the boundary conditions for $G(|\mathbf{G}^i_{||}|, z, z')$ such that

$$\begin{cases} G(|\mathbf{G}^i_{||}|, z_0, z') = 0 \\ G(|\mathbf{G}^i_{||}|, z_{l+1}, z') = 0, \end{cases} \tag{4.106}$$

we obtain Green's function as

$$G(|\mathbf{G}^i_{||}|, z, z') = \frac{e^{-|\mathbf{G}^i_{||}||z-z'|} + e^{-|\mathbf{G}^i_{||}|(2(z_{l+1}-z_0)-|z-z'|)}}{2|\mathbf{G}^i_{||}|(1 - e^{-2|\mathbf{G}^i_{||}|(z_{l+1}-z_0)})}$$
$$- \frac{e^{-|\mathbf{G}^i_{||}|(2z_{l+1}-z-z')} - e^{-|\mathbf{G}^i_{||}|(z+z'-2z_0)}}{2|\mathbf{G}^i_{||}|(1 - e^{-2|\mathbf{G}^i_{||}|(z_{l+1}-z_0)})} \tag{4.107}$$

and the final solution for $V_H(z, \mathbf{G}^i_{||})$ becomes

$$V_{\mathrm{H}}(\mathbf{G}^i_{||}, z) = \int_{z_0}^{z_{l+1}} 4\pi n(\mathbf{G}^i_{||}, z')G(|\mathbf{G}^i_{||}|, z, z')dz'$$
$$+ \left[V_{\mathrm{H}}(\mathbf{G}^i_{||}, z_0)\left(e^{-|\mathbf{G}^i_{||}|(z-z_0)} - e^{-|\mathbf{G}^i_{||}|(2z_{l+1}-z_0-z)}\right)\right.$$
$$\left.+ V_{\mathrm{H}}(\mathbf{G}^i_{||}, z_{l+1})\left(e^{-|\mathbf{G}^i_{||}|(z_{l+1}-z)} - e^{-K(z-2z_0+z_{l+1})}\right)\right]$$
$$/\left(1 - e^{-2|\mathbf{G}^i_{||}|(z_{l+1}-z_0)}\right). \tag{4.108}$$

In case of $|\mathbf{G}^0_{||}| = 0$, the Poisson equation becomes

$$\frac{d^2}{dz^2} V_{\mathrm{H}}(\mathbf{G}^0_{||}, z) = -4\pi n(\mathbf{G}^0_{||}, z). \tag{4.109}$$

The integration of this equation is performed as follows. First, we change the above equation into the Helmholtz-type equation,

$$\left[\frac{d^2}{dz^2} - K^2\right] V_{\mathrm{H}}(\mathbf{G}^0_{||}, z) = -4\pi n(\mathbf{G}^0_{||}, z) - K^2 V_{\mathrm{H}}(\mathbf{G}^0_{||}, z), \tag{4.110}$$

where K is an arbitrary constant. Introducing Green's function

$$\left[\frac{d^2}{dz^2} - K^2\right] G(K, z, z') = -\delta(z - z') \tag{4.111}$$

with the boundary conditions for $G(K, z, z')$ such that

$$\begin{cases} G(K, z_0, z') = 0 \\ G(K, z_{l+1}, z') = 0, \end{cases} \tag{4.112}$$

we obtain the same Green's function for $G(K, z, z')$. The solution is obtained by the self-consistent procedure for $V_{\mathrm{H}}(\mathbf{G}^0_{||}, z)$ as

$$\begin{aligned} V_{\mathrm{H}}(\mathbf{G}^0_{||}, z) = & \int_{z_0}^{z_{l+1}} \left(4\pi n(\mathbf{G}^o_{||}, z') + K^2 V_{\mathrm{H}}(\mathbf{G}^0_{||}, z')\right) G(K, z, z') dz' \\ & + \left[V_{\mathrm{H}}(\mathbf{G}^0_{||}, z_0)\left(e^{-K(z-z_0)} - e^{-K(2z_{l+1}-z_0-z)}\right)\right. \\ & + \left. V_{\mathrm{H}}(\mathbf{G}^0_{||}, z_{l+1})\left(e^{-K(z_{l+1}-z)} - e^{-K(z-2z_0+z_{l+1})}\right)\right] \\ & / \left(1 - e^{-2K(z_{l+1}-z_0)}\right). \end{aligned} \tag{4.113}$$

This self-consistent equation is remarkably stable and several iteration steps are sufficient for a good convergence of $V_{\mathrm{H}}(\mathbf{G}^0_{||}, z)$.

4.2.3 *Electric Current and Conservation Law*

After we obtain the self-consistent solutions for an effective potential $V_{\mathrm{eff}}(\mathbf{r})$ and electron density $n(\mathbf{r})$, we can calculate the electric current from the wavefunctions.

Let us show how to obtain the electric current and confirm the current conservation law. In general, the current density distribution $\mathbf{j}(\mathbf{r})$ is obtained from

$$\begin{aligned} \mathbf{j}(\mathbf{r}) &= \frac{2}{(2\pi)^3} \int d\mathbf{k} \left\{ f_L(\epsilon)\left(1 - f_R(\epsilon)\right) - f_R(\epsilon)\left(1 - f_L(\epsilon)\right) \right\} \\ &\qquad \times \langle \psi(\mathbf{r}) | e\mathbf{v} | \psi(\mathbf{r}) \rangle \\ &= \frac{2}{(2\pi)^3} \sum_i \int dk_z^i \int d\mathbf{k}_{||} \left(f_L(\epsilon) - f_R(\epsilon)\right) S_i(\mathbf{r}), \end{aligned} \tag{4.114}$$

where the index 2 comes from the spin degeneracy, $f(E)$ is the Fermi distribution function and

$$S_i(\mathbf{r}) = \frac{e\hbar}{2mi}\left[\psi_i^*(\mathbf{r})\nabla\psi_i(\mathbf{r}) - \psi_i(\mathbf{r})\nabla\psi_i^*(\mathbf{r})\right]. \tag{4.115}$$

The electric current is obtained from an integral of $\mathbf{j}(\mathbf{r})$ over any dividing surfaces as

$$I = \int \mathbf{j}(\mathbf{r}) \cdot d\mathbf{S}. \tag{4.116}$$

At the boundary $z = z_{l+1}$, the scattering state with an incident channel number i from Eq. (4.52) is changed under a unitary transformation into $\psi_i^L(\mathbf{r})$ as

$$\psi_i^L(\mathbf{r}) = \sum_j t_{i,j} \left\{ \begin{array}{c} e^{i\,k_z^j z} \\ e^{-\kappa_z^j z} \end{array} \right\} e^{i\,(\mathbf{k}_{||}+\mathbf{G}_{||}^j)\cdot \mathbf{r}_{||}} \quad (z = z_{l+1}). \tag{4.117}$$

Then we obtain for any incident channel number i,

$$\int S_i(\mathbf{r}) d\mathbf{r}_{||} = \frac{\hbar}{m} \sum_j k_z^j |t_{i,j}|^2. \tag{4.118}$$

Note that decaying evanescent waves $e^{-\kappa_z^j z}$ do not contribute to the current density. Therefore, the electric current through a surface area S is expressed as

$$\begin{aligned} I &= \frac{2e}{(2\pi)^3} \sum_{i,j} \int dk_z^i d\mathbf{k}_{||} \left(f_L(\epsilon) - f_R(\epsilon) \right) \frac{\hbar}{m} k_z^j |t_{i,j}|^2 S \\ &= \sum_{i,j} \int \frac{d\mathbf{k}_{||}}{(2\pi)^2} S \int_{\mu_R}^{\mu_L} \left(f_L(\epsilon) - f_R(\epsilon) \right) \frac{2e}{h} |\mathcal{T}_{i,j}(\epsilon)|^2 d\epsilon \\ &= \frac{2e}{h} \sum_{i,j} \int_{\mu_R}^{\mu_L} \left(f_L(\epsilon) - f_R(\epsilon) \right) |\mathcal{T}_{i,j}|^2 d\epsilon, \end{aligned} \tag{4.119}$$

where $\mathcal{T}_{i,j} = \sqrt{k_z^j / k_z^i}\, t_{i,j}$ is the matrix element of transmission $\hat{\mathcal{T}}$ and the linear-response conductance G becomes

$$G = -\left[\frac{dI}{dV} \right]_{V \to 0} = \frac{2e^2}{h} \sum_{i,j} |\mathcal{T}_{i,j}|^2, \tag{4.120}$$

which reduces to an analogous form of the Landauer formula, since we use a single-particle approximation of DFT.

On the other hand at the boundary $z = z_0$, the wavefunction with an incident channel number i is expressed under the same unitary transformation as

$$\psi_i^L(\mathbf{r}) = e^{i\,k_z^i z} e^{i\,(\mathbf{k}_{||}+\mathbf{G}_{||}^i)\cdot \mathbf{r}_{||}} + \sum_j r_{i,j} \left\{ \begin{array}{c} e^{-i\,k_z^j z} \\ e^{\kappa_z^j z} \end{array} \right\} e^{i\,(\mathbf{k}_{||}+\mathbf{G}_{||}^j)\cdot \mathbf{r}_{||}}$$

$$(z = z_0). \tag{4.121}$$

In this case, we can show that for any incident channel number i,

$$\int S_i(\mathbf{r})d\mathbf{r}_{||} = \frac{\hbar}{m}\left(k_z^i - \sum_j k_z^j |r_{i,j}|^2\right). \tag{4.122}$$

Since the electric current should be equal at an arbitrary z point, the current conservation law is expressed, for an incident channel number i, as

$$\sum_j |\mathcal{T}_{i,j}|^2 + \sum_j |\mathcal{R}_{i,j}|^2 = 1, \tag{4.123}$$

where $\mathcal{R}_{i,j} = \sqrt{k_z^j / k_z^i} r_{i,j}$ denotes the matrix elements of reflection $\hat{\mathcal{R}}$. This corresponds to the direct calculations of the S-matrix theory from the wavefunctions. For numerical calculations, we need to confirm that as a mesh size Δz decreases, the relation of $\sum_j |\mathcal{T}_{i,j}|^2 + \sum_j |\mathcal{R}_{i,j}|^2 = 1$ is satisfied for all the energies ϵ. This becomes a check of truncation errors due to discretization of the Kohn–Sham equation for fine mesh points.

As we see previously in the S-matrix theory, we can generalize this for the multichannel theory. For transmission and reflection matrices $\hat{\mathcal{T}}$, $\hat{\mathcal{R}}$ whose components are $\mathcal{T}_{i,j}$, $\mathcal{R}_{i,j}$, we can show that these matrices satisfy the unitarity relation of

$$\begin{aligned} &\hat{\mathcal{T}}^\dagger\hat{\mathcal{T}} + \hat{\mathcal{R}}^\dagger\hat{\mathcal{R}} = \hat{1} \\ &\hat{\mathcal{T}} = \hat{\mathcal{T}}^T \\ &\hat{\mathcal{R}} = \hat{\mathcal{R}}^T, \end{aligned} \tag{4.124}$$

where $\hat{\mathcal{T}}^T$, $\hat{\mathcal{R}}^T$ are transposed matrices and $\hat{\mathcal{T}}^\dagger$, $\hat{\mathcal{R}}^\dagger$ are complex conjugate matrices of $\hat{\mathcal{T}}$ and $\hat{\mathcal{R}}$.

4.2.3.1 Eigenchannel decomposition

The Hermitian $\hat{\mathcal{T}}^\dagger\hat{\mathcal{T}}$ is diagonalized using a unitary matrix $\hat{U}$ to transform original channels into eigenchannels. Then the conductance is expressed as a sum of the individual eigenchannel transmissions $\mathcal{T}_i$,

$$\mathrm{Tr}(\hat{U}^\dagger\hat{\mathcal{T}}^\dagger\hat{\mathcal{T}}\hat{U}) = \sum_i \mathcal{T}_i. \tag{4.125}$$

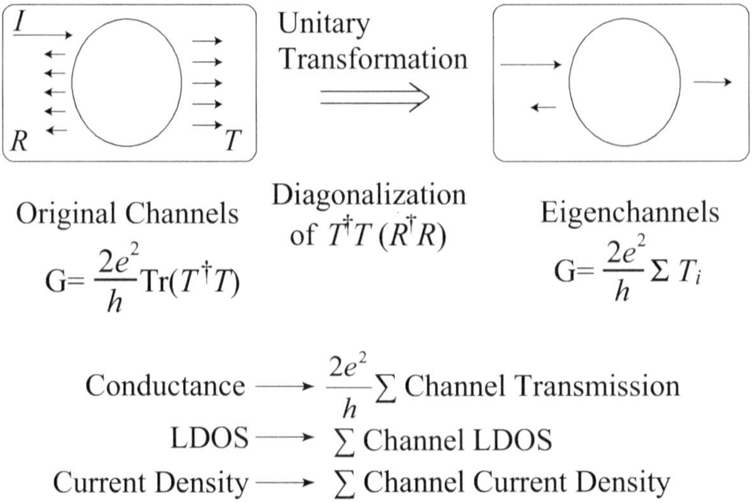

Figure 4.11 Conceptual illustration for the eigenchannel decomposition. The original channels are transformed by a unitary transformation into diagonalization of the matrix $\hat{T}^{\dagger}\hat{T}$. Conductance, LDOS, and current density are decomposed by the sum of the channels.

The local density of states (LDOS) and current density are also expressed as a sum of the individual eigenchannel components. This is the essence of the eigenchannel decomposition.

The Landauer formula is often employed for an investigation of the ballistic electron transport. In the multichannel systems, the conductance is calculated as

$$G = \frac{2e^2}{h}\mathrm{Tr}(\hat{T}^{\dagger}\hat{T}) = \frac{2e^2}{h}\sum_i T_i\,, \tag{4.126}$$

where the channel transmission T_i is an eigenvalue obtained by the diagonalization of the Hermitian matrix $\hat{T}^{\dagger}\hat{T}$ using a unitary matrix. Here $\hat{T}$ is the transmission matrix. The channels corresponding to the eigenvectors are called eigenchannels. Electrons propagating through the eigenchannels are not scattered into different channels in the ballistic limit. The properties of each eigenchannel give us a clear physical picture for the conductance and other relevant quantities. The eigenchannel analysis for transport through the constrictions have been performed for the channel transmissions and corresponding wavefunctions. LDOS and current density resolved into the eigenchannels give us a further clear viewpoint for quantum electron transport. These enable us to know a relation between LDOS and current through the channel concept. Moreover, the channel current density of local current distribution shows us paths for an electron flow in each channel.

The transmission and reflection matrices of the system $\hat{\mathcal{T}}$ and $\hat{\mathcal{R}}$, are obtained using coefficients $r_{i,j}$ and $t_{i,j}$ by

$$\hat{\mathcal{T}} = (k_z^{R\,1/2})^T \hat{t}\, k_z^{L\,-1/2}$$
$$\hat{\mathcal{R}} = (k_z^{R\,1/2})^T \hat{r}\, k_z^{L\,-1/2}. \qquad (4.127)$$

Here $k_z^{L(R)}$ is an $N \times M$ rectangular matrix whose i, j elements are given by $k_z^{L(R)}\delta_{i,j}$ and M is the number of open channels deep in the left and right electrodes. The unitary relation of scattering matrices

$$\hat{\mathcal{T}}^\dagger \hat{\mathcal{T}} + \hat{\mathcal{R}}^\dagger \hat{\mathcal{R}} = \hat{I} \qquad (4.128)$$

shows the conservation of the current in the system.

The LDOS is calculated from

$$\nu(\mathbf{r}, \epsilon) = \frac{m}{\pi h^2} \mathrm{Tr}\left(\int d\mathbf{k}_{||} (\psi(\mathbf{r}) k_z^{L\,-1/2})^\dagger (\psi(\mathbf{r}) k_z^{L\,-1/2}) \right). \qquad (4.129)$$

The current density is expressed as the probability current density by

$$\mathbf{j}(\mathbf{r}, \epsilon) = \frac{2e}{h} \mathrm{Tr}\left[\int \frac{d\mathbf{k}_{||}}{(2\pi)^2} \mathrm{Im}\left((\psi(\mathbf{r}) k_z^{L\,-1/2})^\dagger \nabla (\psi(\mathbf{r}) k_z^{L\,-1/2}) \right) \right]. \qquad (4.130)$$

The conductance per unit cell is obtained from

$$G(\epsilon) = \frac{2e^2}{h} \int \frac{d\mathbf{k}_{||}}{(2\pi)^2} S\mathrm{Tr}(\hat{\mathcal{T}}^\dagger \hat{\mathcal{T}}), \qquad (4.131)$$

where S is an area of supercell in the x, y directions. In case of large supercell compared with constriction, we treat only Γ point, $\mathbf{k}_{||} = (0, 0)$ for transmission properties. In such a case, the conductance is expressed similar to the Landauer formula as

$$G(\epsilon) = \frac{2e^2}{h} \mathrm{Tr}(\hat{\mathcal{T}}^\dagger \hat{\mathcal{T}}). \qquad (4.132)$$

The Hermitian matrix $\hat{\mathcal{T}}^\dagger \hat{\mathcal{T}}$ is diagonalized using a unitary matrix $\hat{U}$ to obtain eigenchannels as

$$\mathrm{Tr}\left[\hat{U}^\dagger \hat{\mathcal{T}}^\dagger \hat{\mathcal{T}} \hat{U} \right] = \sum_i \mathcal{T}_i. \qquad (4.133)$$

Correspondingly, LDOS and current density are decomposed into the eigenchannels after the transformation with $\hat{U}$. Namely, the channel LDOS and channel current density are written as

$$\nu_i(\mathbf{r}, \epsilon) = \frac{4\pi m}{h^2}\frac{1}{S}\left((\psi(\mathbf{r})k_z^{L^{-1/2}}\hat{U})^\dagger(\psi(\mathbf{r})k_z^{L^{-1/2}}\hat{U})\right)_{ii} \quad (4.134)$$

and

$$\mathbf{j}_i(\mathbf{r}, \epsilon) = \frac{2e^2}{h}\frac{1}{S}\mathrm{Im}\left((\psi(\mathbf{r})k_z^{L^{-1/2}}\hat{U})^\dagger\nabla(\psi(\mathbf{r})k_z^{L^{-1/2}}\hat{U})\right)_{ii}. \quad (4.135)$$

The sum of channel LDOS and channel current density yields the total LDOS and total current density. Since they are the diagonal elements, the sum corresponds to take the trace of the matrix [63].

4.2.3.2 Force acting on atoms

In the nonequilibrium system, since the Fermi levels in the left and right electrodes are not equal, electron density is polarized to satisfy the charge neutrality condition. This is due to an applied bias voltage, imposing different chemical potentials as the boundary conditions deep inside the electrodes. The total electron density is determined self-consistently with an effective potential where an applied bias voltage is included in the boundary conditions. Here we assume the system keeps its equilibrium situation deep inside the electrodes. Therefore, the forces acting to atoms are mainly due to the polarized electron density. We should note that this is a very subtle problem, since we cannot define the total energy in the open and nonequilibrium systems.[a]

Since the Laue representation constitutes a complete set of basis functions, we can calculate the force acting on an atom μ from the

[a] There is a controversy about the force from the electric current, which acts on the atoms working as a force. Here we will pass this problem and consider that all the forces acting on the atoms are reduced to redistribution of electron density due to the electric field from an applied bias voltage and that the local total energy $E = \langle\psi|H|\psi\rangle$ can be defined in the scattering regime.

local total energy E with use of the Hellman–Feynman theorem,

$$\begin{aligned}\mathbf{F}_\mu &= -\frac{\partial E}{\partial \mathbf{R}_\mu} \\ &= -\left[\langle \frac{\partial \psi}{\partial \mathbf{R}_\mu} | H | \psi \rangle + \langle \psi | \frac{\partial H}{\partial \mathbf{R}_\mu} | \psi \rangle + \langle \psi | H | \frac{\partial \psi}{\partial \mathbf{R}_\mu} \rangle \right] \\ &= -\langle \psi | \frac{\partial H}{\partial \mathbf{R}_\mu} | \psi \rangle. \end{aligned} \tag{4.136}$$

Here we show two terms related to the coordinate of atom μ. One is the interaction of the local pseudopotential $V_{\rm loc}(\mathbf{r})$ of atom μ with electron density, and the other is the interaction of the ion-core of atom μ with other atoms. Contribution from the interaction of the nonlocal pseudopotential with electron density will be left for an exercise of the readers.

The first term is expressed as

$$\begin{aligned}\mathbf{F}_\mu &= -\frac{\partial}{\partial \mathbf{R}_\mu} \int d\mathbf{r} n(\mathbf{r}) V_{\rm loc}(\mathbf{r} - \mathbf{R}_\mu) \\ &= -\frac{\partial}{\partial \mathbf{R}_\mu} \sum_{\mathbf{G}_{||}} e^{i\, \mathbf{G}_{||} \cdot \tau_\mu} \int_{-\infty}^{\infty} dz n(\mathbf{G}_{||}, z) V_{\rm loc}(-\mathbf{G}_{||}, z - z_\mu). \end{aligned} \tag{4.137}$$

Let us consider the simple empty-core local pseudopotential Eq. (4.35) for $V_{\rm loc}(r)$. It becomes

$$\begin{aligned}\mathbf{F}_\mu^{(x,y)} &= 2\pi Z_\mu e^2 \sum_{\mathbf{G}_{||}(\neq \mathbf{0})} i \frac{\mathbf{G}_{||}}{|\mathbf{G}_{||}|} e^{i\, \mathbf{G}_{||} \cdot \tau_\mu} \left[\int_{-\infty}^{\infty} dz e^{-|\mathbf{G}_{||}||z - z_\mu|} n(\mathbf{G}_{||}, z) \right. \\ &\quad \left. - |\mathbf{G}_{||}| \int_{z_\mu - r_c}^{z_\mu + r_c} dz \int_0^{\sqrt{r_c^2 - (z - z_\mu)^2}} dr \frac{r J_0(|\mathbf{G}_{||}| r)}{\sqrt{r^2 + (z - z_\mu)^2}} n(\mathbf{G}_{||}, z) \right], \\ \mathbf{F}_\mu^{(z)} &= 2\pi Z_\mu e^2 \sum_{\mathbf{G}_{||}(\neq \mathbf{0})} e^{i\, \mathbf{G}_{||} \cdot \tau_\mu} \left[-\int_{-\infty}^{z_\mu} dz e^{-|\mathbf{G}_{||}|(z_\mu - z)} n(\mathbf{G}_{||}, z) \right. \\ &\quad + \int_{z_\mu - r_c}^{z_\mu + r_c} dz (z_\mu - z) n(\mathbf{G}_{||}, z) \left(\frac{1}{r_c} J_0(|\mathbf{G}_{||}| \sqrt{r_c^2 - (z_\mu - z)^2}) \right. \\ &\quad \left. + \int_0^{\sqrt{r_c^2 - (z_\mu - z)^2}} \frac{r J_0(|\mathbf{G}_{||}| r) dr}{\left(r^2 + (z_\mu - z)^2\right)^{3/2}} \right) \\ &\quad \left. + \int_{z_\mu}^{\infty} dz e^{-|\mathbf{G}_{||}|(z - z_\mu)} n(\mathbf{G}_{||}, z) \right]. \end{aligned} \tag{4.138}$$

Derivation of the formula for other local pseudopotentials is straightforward and is left for an exercise.

The second term due to the interaction between ion-cores is

$$\begin{aligned}\mathbf{F}_\mu &= -\frac{\partial}{\partial \mathbf{R}_\mu}\sum_{\nu,\mu}\frac{1}{2}\frac{Z_\nu Z_\mu e^2}{|\mathbf{R}_\nu - \mathbf{R}_\mu|}\\ &= -\frac{\partial}{\partial \mathbf{R}_\mu}\sum_{\nu,\mathbf{R}_{||}}\frac{Z_\nu Z_\mu e^2}{\sqrt{|\mathbf{R}_{||}+\tau_\nu-\tau_\mu|^2+(z_\nu-z_\mu)^2}}\\ &= -\frac{\partial}{\partial \mathbf{R}_\mu}\sum_{\nu,\mathbf{R}_{||}}\frac{2Z_\nu Z_\mu e^2}{\sqrt{\pi}}\int_0^\infty d\rho e^{-\left(|\mathbf{R}_{||}+\tau_\nu-\tau_\mu|^2+(z_\nu-z_\mu)^2\right)\rho^2}.\end{aligned} \tag{4.139}$$

The use of the following relation accelerates the convergence:

$$\sum_{\mathbf{R}_{||}} e^{-\left(|\mathbf{R}_{||}+\mathbf{r}_{||}|^2+z^2\right)\rho^2} = \frac{\pi}{S}\sum_{\mathbf{G}_{||}}\frac{e^{-z^2\rho^2-\frac{|\mathbf{G}_{||}|^2}{4\rho^2}+i\mathbf{G}_{||}\cdot\mathbf{r}_{||}}}{\rho^2}. \tag{4.140}$$

This relation is verified from a Fourier transform of

$$\begin{aligned}\sum_{\mathbf{R}_{||}} e^{-|\mathbf{R}_{||}+\mathbf{r}_{||}|^2\rho^2} &= \sum_{\mathbf{G}_{||}}\frac{1}{S}\int_{\text{all}} d\mathbf{r}_{||} e^{-|\mathbf{r}_{||}|^2\rho^2+i\mathbf{G}_{||}\cdot\mathbf{r}_{||}}\\ &= \frac{2\pi}{S}\int_0^\infty drrJ_0(|\mathbf{G}_{||}|r)e^{-\rho^2r^2+i\mathbf{G}_{||}\cdot\mathbf{r}_{||}} = \frac{\pi}{S}\frac{e^{-\frac{|\mathbf{G}_{||}|^2}{4\rho^2}}}{\rho^2}e^{+i\mathbf{G}_{||}\cdot\mathbf{r}_{||}}.\end{aligned} \tag{4.141}$$

Then $\mathbf{F}_\mu$ is modified as

$$\begin{aligned}\mathbf{F}_\mu = -\frac{\partial}{\partial \mathbf{R}_\mu}\sum_{\nu}&\left[\frac{2\sqrt{\pi}Z_\nu Z_\mu e^2}{S}\sum_{\mathbf{G}_{||}} e^{i;\mathbf{G}_{||}\cdot(\tau_\nu-\tau_\mu)}\right.\\ &\left.\times\int_0^K\frac{e^{-(z_\nu-z_\mu)^2\rho^2-\frac{|\mathbf{G}_{||}|^2}{4\rho^2}}}{\rho^2}d\rho + Z_\nu Z_\mu e^2\sum_{\mathbf{R}_{||}}\frac{\text{erfc}(KR)}{R}\right].\end{aligned} \tag{4.142}$$

Here $R = \sqrt{|\mathbf{R}_{||} + \tau_\nu - \tau_\mu|^2 + (z_\nu - z_\mu)^2}$ and K is an arbitrary constant. Using the following relations

$$\int_0^K e^{-a^2\rho^2 - \frac{b^2}{\rho^2}} d\rho = \frac{\sqrt{\pi}}{4a}\left[e^{2ab}\left\{\mathrm{erf}(aK + \frac{b}{K}) - 1\right\}\right.$$
$$\left. + e^{-2ab}\left\{\mathrm{erf}(aK - \frac{b}{K}) + 1\right\}\right]$$

$$\int_0^K \frac{e^{-a^2\rho^2 - \frac{b^2}{\rho^2}}}{\rho^2} d\rho = -\frac{\sqrt{\pi}}{4b}\left[e^{2ab}\left\{\mathrm{erf}(aK + \frac{b}{K}) - 1\right\}\right.$$
$$\left. - e^{-2ab}\left\{\mathrm{erf}(aK - \frac{b}{K}) + 1\right\}\right], \tag{4.143}$$

we obtain the expressions of the force from the interaction between ion-cores such as

$$\mathbf{F}_\mu^{(x,y)} = -\sum_\nu Z_\nu Z_\mu e^2 \frac{\pi}{S} \sum_{\mathbf{G}_{||}(\neq 0)} i\frac{\mathbf{G}_{||}}{|\mathbf{G}_{||}|} e^{i;\mathbf{G}_{||}\cdot(\tau_\nu - \tau_\mu)}$$
$$\times \left[e^{|\mathbf{G}_{||}||z_\nu - z_\mu|}\left\{\mathrm{erf}(|z_\nu - z_\mu|K + \frac{|\mathbf{G}_{||}|}{2K}) - 1\right\}\right.$$
$$\left. + e^{-|\mathbf{G}_{||}||z_\nu - z_\mu|}\left\{\mathrm{erf}(|z_\nu - z_\mu|K - \frac{|\mathbf{G}_{||}|}{2K}) + 1\right\}\right]$$
$$- \sum_\nu Z_\nu Z_\mu e^2 \sum_{\mathbf{R}_{||}} (\mathbf{R}_{||} + \tau_\nu - \tau_\mu)\left[\frac{\mathrm{erf}(KR)}{R^3} + \frac{2KRe^{-k^2R^2}}{\sqrt{\pi}R^3}\right]$$

$$\mathbf{F}_\mu^{(z)} = -\sum_\nu Z_\nu Z_\mu e^2 \frac{\pi}{S} \sum_{\mathbf{G}_{||}} \frac{(z_\nu - z_\mu)}{|z_\nu - z_\mu|} e^{i;\mathbf{G}_{||}\cdot(\tau_\nu - \tau_\mu)}$$
$$\times \left[e^{|\mathbf{G}_{||}||z_\nu - z_\mu|}\left\{\mathrm{erf}(|z_\nu - z_\mu|K + \frac{|\mathbf{G}_{||}|}{2K}) - 1\right\}\right.$$
$$\left. - e^{-|\mathbf{G}_{||}||z_\nu - z_\mu|}\left\{\mathrm{erf}(|z_\nu - z_\mu|K - \frac{|\mathbf{G}_{||}|}{2K}) + 1\right\}\right]$$
$$- \sum_\nu Z_\nu Z_\mu e^2 \sum_{\mathbf{R}_{||}} (z_\nu - z_\mu)\left[\frac{\mathrm{erf}(KR)}{R^3} + \frac{2KRe^{-k^2R^2}}{\sqrt{\pi}R^3}\right]. \tag{4.144}$$

4.2.4 *Examples of Simple Systems*

In this section, we show examples of the analyses of electron transport through the constrictions between semi-infinite electrodes by using the RTM method. Figure 4.12 shows the DOS and the transmission for the structures of dot, point contact, and wire [63]. We can see that as the contact region becomes larger, the DOS changes from spiky resonant states to a broader Lorentzian form. Correspondingly, the transmission becomes from spiky resonant structure to broad structure. When the contact region is large enough for the wire case, the transmission reaches the unitary limit for an energy of the first channel and keeps in the completely open state. This shows an open channel in the wire case, whereas the channel is only open in the narrow resonance in the dot case. In the

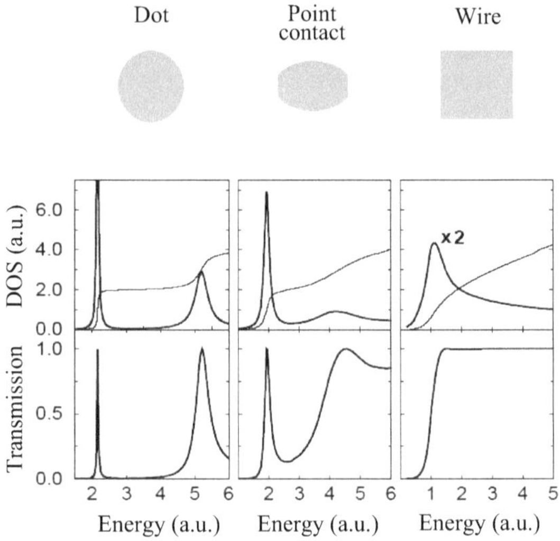

Figure 4.12 (Top) Schematic illustrations of the different structures between the electrodes; from left to right as dot, point contact, and wire structures. (Bottom) Density of states (DOS) and transmission for those structures. As the contact region to electrodes becomes large, resonances in the DOS become broad. The transmission for the wire shows that it keeps the unitary limit for all the energies once the channel is open. From N. Kobayashi, M. Brandbyge and M. Tsukada, *Jpn. J. Appl. Phys.* **38**, 336 (1999) [63].

point contact case, the behavior of the first channel is dot-like, while that of the second channel is more wire-like.

Transformation into the eigenchannels produces the independent channels with no inter-channels scattering. We note the present method is available for the system whose transmission channels are not obvious and thus the details of transport for the system such as atomic-scale systems can be analyzed from viewpoints of the eigenchannels. These will be given in the next chapter. We note that in the low-dimensional system such as in the dot and the point contact structures, the Coulomb interaction becomes important and electron transport character is strongly affected by the spin states duc to the exchange interaction. These problems will be treated in Chapter 6 more.

4.3 Lippmann–Schwinger Equation Method

4.3.1 *Integral Kohn–Sham Equation for Scattering*

Here we consider the calculation method based on the integral form of the Kohn–Sham equation. The Hamiltonian of the system composed of two semi-infinite jellium electrodes without any atomic structures becomes

$$\hat{H}^0 = -\frac{\hbar^2}{2m}\nabla^2 + V^0_{\text{eff}}(\mathbf{r}), \tag{4.145}$$

where we take the effective potential as a sum of the Hartree and the exchange-correlation potentials

$$V^0_{\text{eff}}(\mathbf{r}) = V^0_H(\mathbf{r}) + V_{xc}[n^0(\mathbf{r})]. \tag{4.146}$$

The Hamiltonian of the total system, including the atom structure, is composed of

$$\hat{H} = \hat{H}^0 + \Delta V(\mathbf{r}, \mathbf{r}'), \tag{4.147}$$

where the difference of potential is

$$\Delta V(\mathbf{r}, \mathbf{r}') = V_{\text{eff}}(\mathbf{r}, \mathbf{r}') - V^0_{\text{eff}}(\mathbf{r}) \tag{4.148}$$

with

$$V_{\text{eff}}(\mathbf{r}, \mathbf{r}') = \left\{ \sum_\alpha V^{ps}_{\text{loc}}(\mathbf{r} - \mathbf{r}_\alpha) + V_H(\mathbf{r}) + V_{xc}[n(\mathbf{r})] \right\} \delta(\mathbf{r} - \mathbf{r}') + V_{nl}(\mathbf{r}, \mathbf{r}'). \tag{4.149}$$

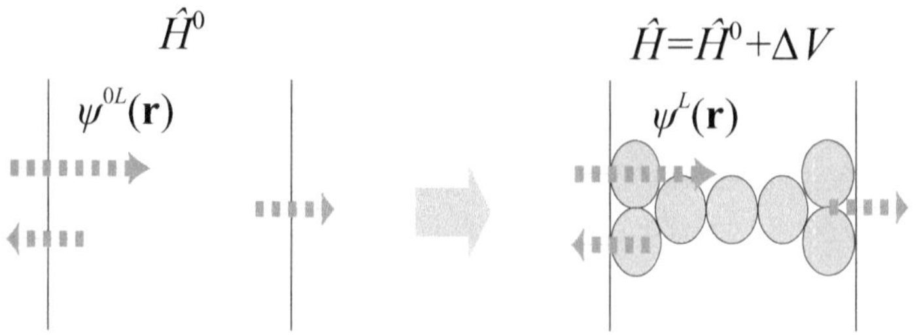

Figure 4.13 Schematic illustrations of the procedures of the LS method. First, we solve the Kohn–Sham equation for $\hat{H}^0$ without any atomic structures between electrodes. When the jellium electrodes are used, this is a simple one-dimensional problem and electronic states are obtained quickly. Then, we solve the LS equation for the total Hamiltonian $\hat{H} = \hat{H}^0 + \Delta V$, including the atomic structures between electrodes. The nonlocal pseudopotentials are included directly in the integration.

Here $V_{nl}(\mathbf{r}, \mathbf{r}')$ and $V_{loc}(\mathbf{r})$ are the nonlocal and the local parts of the pseudopotentials.

Scattering states of the system incident from the left and right electrodes are expressed as the plane-waves in the multichannel wavefunction $\psi^{L(R)}(\mathbf{r})$, which consist of a set of wavefunctions $\psi_i^{L(R)}(\mathbf{r})$ incident from the i-th channels of the left (right) electrode as

$$\psi^{L(R)}(\mathbf{r}) = \left[\psi_1^{L(R)}(\mathbf{r}), \psi_2^{L(R)}(\mathbf{r}), \cdots, \psi_N^{L(R)}(\mathbf{r})\right]. \tag{4.150}$$

For a system without atomic structures, there is no interchannel scattering for $i \neq j$. Thus, the wavefunctions are obtained from

$$\left(\epsilon\hat{I} - \hat{H}^0\right)\psi^{0L(R)}(\mathbf{r}) = 0. \tag{4.151}$$

The wavefunctions of the total system $\psi^{L(R)}(\mathbf{r})$ are obtained by solving the following LS equation [64]:

$$\psi^{L(R)}(\mathbf{r}) = \psi^{0L(R)}(\mathbf{r}) + \int\int d\mathbf{r}'d\mathbf{r}''G^0(\mathbf{r}, \mathbf{r}')\Delta V(\mathbf{r}', \mathbf{r}'')\psi^{L(R)}(\mathbf{r}''), \tag{4.152}$$

where $G^0(\mathbf{r}, \mathbf{r}')$ is Green's function for the system without atomic structures.

4.3.2 Green's Function in Laue Representation

Let us consider to solve the LS equation in the Laue representation [65]. We transform it into the matrix equation using the two-dimensional expansion as

$$\left[\hat{I} - (S\Delta z)^2 \hat{G}^0 \Delta V\right] \psi^{L(R)}(\mathbf{r}_{||}, z_p) = \psi^{L(R)0}(\mathbf{r}_{||}, z_p), \tag{4.153}$$

where S is an area of the unit cell in the x, y directions and

$$\Delta V(\mathbf{r}_{||}, z_p, \mathbf{r}'_{||}, z_{p'}) = \sum_{ij} e^{i(\mathbf{k}_{||}+\mathbf{G}^i_{||})\cdot\mathbf{r}_{||}} \Delta V_{ip,jp'} e^{-i(\mathbf{k}_{||}+\mathbf{G}^j_{||})\cdot\mathbf{r}'_{||}}. \tag{4.154}$$

We use a separable form of the nonlocal pseudopotentials Eq. (4.91) expanded in the two-dimensional plane-waves for the atomic ion-core potentials. Green's function $G(\mathbf{r}, \mathbf{r}')$ satisfying

$$(\epsilon\hat{I} - \hat{H}^0)G(\mathbf{r}, \mathbf{r}') = \delta(\mathbf{r} - \mathbf{r}') \tag{4.155}$$

is simply obtained by

$$G(\mathbf{r}, \mathbf{r}') = \frac{2m}{S\hbar^2} \sum_i e^{i(\mathbf{k}_{||}+\mathbf{G}^i_{||})\cdot(\mathbf{r}_{||}-\mathbf{r}'_{||})} \frac{\psi^{0L}_{ii}(z_<)\psi^{0R}_{ii}(z_>)}{W(\psi^{0L}_{ii}, \psi^{0R}_{ii})}, \tag{4.156}$$

where $z_> = \max(z_p, z_{p'})$, $z_< = \min(z_p, z_{p'})$, and $W(\psi^{0L}_{ii}, \psi^{0R}_{ii})$ is a Wronskian that is independent of z

$$W(\psi^{0L}_{ii}, \psi^{0R}_{ii}) = \psi^{0L}_{ii}(z)\frac{d}{dz}\psi^{0R}_{ii}(z) - \psi^{0R}_{ii}(z)\frac{d}{dz}\psi^{0L}_{ii}(z). \tag{4.157}$$

Therefore, the matrix form of Green's function $\hat{G}$ using the two-dimensional plane-wave basis set is expressed in the Laue representation as

$$G_{ip,jp'} = \delta_{ij}\frac{2m}{S\hbar^2}\frac{\psi^{0L}_{ii}(z_<)\psi^{0R}_{ii}(z_>)}{W(\psi^{0L}_{ii}, \psi^{0R}_{ii})}. \tag{4.158}$$

Electron density is calculated from

$$n(\mathbf{r}) = \frac{m}{\pi h^2}\sum_{L,R}\int d\epsilon\left(\int d\mathbf{k}_{||}[\psi^{L(R)}(\mathbf{r})k_z^{(-1/2)}]^\dagger[\psi^{L(R)}(\mathbf{r})k_z^{(-1/2)}]\right). \tag{4.159}$$

Here, k_z is the $N \times M$ rectangular matrix whose i, j elements are given by $(\mathrm{k}_z^{(-1/2)})_{ij} = \delta_{ij}/(k_z^{L(R)})^{1/2}$.

The matrix elements of transmission $\hat{\mathcal{T}}$ of states incident from the left electrode is calculated as

$$\mathcal{T}_{ij} = \sqrt{\frac{k_z^j}{k_z^i}}\psi_{ij}(z_R). \tag{4.160}$$

For a sufficiently large supercell compared with the size of wire, we can use only Γ point $\mathbf{k}_{\parallel} = (0, 0)$ to determine the transmission properties. In such a case, the electric current per unit cell is simply expressed by the Landauer-type formula as

$$I = \frac{2e}{h} \int_{\mu_R}^{\mu_L} (f_L(\epsilon) - f_R(\epsilon)) \, \mathrm{Tr}(\hat{\mathcal{T}}^{\dagger} \hat{\mathcal{T}}) d\epsilon. \tag{4.161}$$

In the same way as we discussed in the previous sections, we can construct the concept of the eigenchannel decompositions for the LS equation. The Hermitian matrix $\hat{\mathcal{T}}^{\dagger}\hat{\mathcal{T}}$ is diagonalized using a unitary matrix $\hat{U}$ to transform the original channels into the eigenchannels. The conductance is expressed as a sum of individual eigenchannel transmissions $\mathcal{T}_i$ as

$$\mathrm{Tr}(\hat{U}^{\dagger} \hat{\mathcal{T}}^{\dagger} \hat{\mathcal{T}} \hat{U}) = \sum_i \mathcal{T}_i. \tag{4.162}$$

The LDOS and current density are also expressed as a sum of individual eigenchannel components. The i-th channel LDOS and channel current density are respectively calculated from

$$\nu_i(\mathbf{r}, \epsilon) = \frac{4\pi m}{h^2} \frac{1}{S} ([\psi(\mathbf{r}) k_z^{-1/2} \hat{U}]^{\dagger} [\psi(\mathbf{r}) k_z^{-1/2} \hat{U}])_{ii}, \tag{4.163}$$

$$\mathbf{j}_i(\mathbf{r}, \epsilon) = \frac{2e^2}{h} \frac{1}{S} \mathrm{Im}([\psi(\mathbf{r}) k_z^{-1/2} \hat{U}]^{\dagger} \nabla [\psi(\mathbf{r}) k_z^{-1/2} \hat{U}])_{ii}. \tag{4.164}$$

4.4 Nonequilibrium Green's Function Method

In this section, we describe the calculation method for transport using the NEGF method. Fundamental derivations of the equations are precisely described in Chapter 3. Here we explain how the formulas are utilized for the nanometer-scale systems sandwiched between the electrodes. We also describe the numerical atomic orbital basis sets to construct an effective Hamiltonian.

4.4.1 *Modeling for Infinite System*

The system consists of the left (L) and right (R) semi-infinite electrodes and the center (C) region between the electrodes and

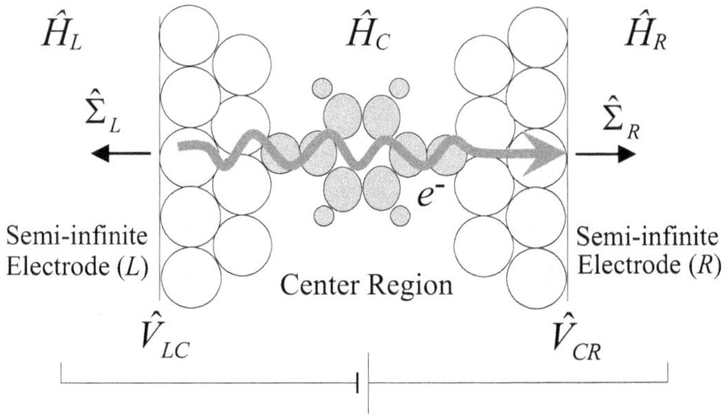

Figure 4.14 Schematic illustration of the junction systems in the NEGF approach. The total system is divided into three parts at boundaries. To construct an effective Hamiltonian in the nanosystem, the effects from semi-infinite electrodes are included in the Hamiltonian of the center region as self-energies $\Sigma_{L,R}$.

thus is divided into three regions. When we set the boundaries deep inside the electrodes, the Hamiltonian is partitioned as

$$\hat{H} = \begin{pmatrix} \hat{H}_L & \hat{V}_{LC} & 0 \\ \hat{V}_{LC}^{\dagger} & \hat{H}_C & \hat{V}_{CR}^{\dagger} \\ 0 & \hat{V}_{CR} & \hat{H}_R \end{pmatrix}. \tag{4.165}$$

Then the Kohn–Sham equation is written as a matrix equation

$$\begin{pmatrix} \hat{H}_L & \hat{V}_{LC} & 0 \\ \hat{V}_{LC}^{\dagger} & \hat{H}_C & \hat{V}_{CR}^{\dagger} \\ 0 & \hat{V}_{CR} & \hat{H}_R \end{pmatrix} \begin{pmatrix} |\psi_L\rangle \\ |\psi_C\rangle \\ |\psi_R\rangle \end{pmatrix} = \epsilon \hat{I} \begin{pmatrix} |\psi_L\rangle \\ |\psi_C\rangle \\ |\psi_R\rangle \end{pmatrix}. \tag{4.166}$$

This may be possible when we assume that the connecting potentials $\hat{V}_{LC}$ and $\hat{V}_{CR}$ are short range in the interface regions. We can expand the wavefunctions using the localized basis sets of $|\psi_{L,C,R}\rangle$. Then we have the coupled equations as

$$\begin{aligned} \left(\epsilon\hat{I} - \hat{H}_L\right)|\psi_L\rangle &= \hat{V}_{LC}|\psi_C\rangle \\ \left(\epsilon\hat{I} - \hat{H}_R\right)|\psi_R\rangle &= \hat{V}_{CR}|\psi_C\rangle \end{aligned} \tag{4.167}$$

for the electrode regions. When we define Green's functions $\hat{g}_{L,R}(\epsilon)$ for the electrode regions L, R as

$$\begin{aligned} \left(\epsilon\hat{I} - \hat{H}_L\right)\hat{g}_L(\epsilon) &= \hat{I} \\ \left(\epsilon\hat{I} - \hat{H}_R\right)\hat{g}_R(\epsilon) &= \hat{I}, \end{aligned} \tag{4.168}$$

the equation for the center becomes

$$
\begin{aligned}
\left(\epsilon\hat{I} - \hat{H}_C\right)|\psi_C\rangle &= \hat{V}_{LC}^{\dagger}|\psi_L\rangle + \hat{V}_{CR}^{\dagger}|\psi_R\rangle, \\
&= \hat{V}_{LC}^{\dagger}\hat{g}_L(\epsilon)\hat{V}_{LC}|\psi_C\rangle + \hat{V}_{CR}^{\dagger}\hat{g}_R(\epsilon)\hat{V}_{CR}|\psi_C\rangle \\
&= \hat{\Sigma}_L(\epsilon)|\psi_C\rangle + \hat{\Sigma}_R(\epsilon)|\psi_C\rangle,
\end{aligned}
\tag{4.169}
$$

where we define self-energies $\hat{\Sigma}_{L,R}(\epsilon)$ for the electrodes L and R by

$$
\begin{aligned}
\hat{\Sigma}_L(\epsilon) &= \hat{V}_{LC}^{\dagger}\hat{g}_L(\epsilon)\hat{V}_{LC} = \hat{V}_{LC}^{\dagger}\left(\epsilon\hat{I} - \hat{H}_L\right)^{-1}\hat{V}_{LC} \\
\hat{\Sigma}_R(\epsilon) &= \hat{V}_{CR}^{\dagger}\hat{g}_R(\epsilon)\hat{V}_{CR} = \hat{V}_{CR}^{\dagger}\left(\epsilon\hat{I} - \hat{H}_R\right)^{-1}\hat{V}_{CR}
\end{aligned}
\tag{4.170}
$$

and then we obtain

$$
\left[\epsilon\hat{I} - \hat{H}_C - \hat{\Sigma}_L(\epsilon) - \hat{\Sigma}_R(\epsilon)\right]|\psi_C\rangle = 0. \tag{4.171}
$$

This equation means that the effects from electrode regions are separated as self-energies $\hat{\Sigma}_{L(R)}(\epsilon)$ [66–68]. We should note that to construct self-energies $\hat{\Sigma}_{L(R)}(\epsilon)$, we need only the matrix elements of the electrode regions and those of boundary parts, not the matrix elements of the center region. Usually, the atomic configurations for the electrodes are taken as the bulk ones and thus we treat a small number of atoms to describe their electronic states. We calculate self-energies $\hat{\Sigma}_{L(R)}(\epsilon)$ only once at the beginning of the calculation procedure, since we assume that they are not affected deep inside the electrodes.

After we obtain self-energies for the electrodes $\hat{\Sigma}_{L,R}(\epsilon)$, we solve Green's function in the center region C, taking an inverse of the following matrix with an expansion of the localized basis sets of $|\psi\rangle$:

$$
\hat{G}_C^{r,a}(\epsilon) = \left[\epsilon\hat{I} - \hat{H}_C - \hat{\Sigma}_L(\epsilon) - \hat{\Sigma}_R(\epsilon) \pm i0^+\right]^{-1}. \tag{4.172}
$$

Note that the dimension of this matrix $\langle\psi_i|\hat{G}_C^{r,a}|\psi_j\rangle$ is finite. However, as the center region for the nanostructures become larger and larger, this matrix inversion procedure would become much more time-consuming, since the scaling of the matrix inversion increases as $O(N^3)$ with the number of basis sets to represent the Hamiltonian in the center region.

When we have the retarded Green's functions $\hat{G}_C^r(\epsilon)$ for the center region, we can construct electron density $n(\mathbf{r})$ from it. For the nonequilibrium situation, we need to obtain the lesser

Green's function $\hat{G}^{<}(\epsilon)$ for electron density where we have different chemical potentials μ_L and μ_R in the electrodes. Using the coupling constant $\hat{\Gamma}^{L(R)}(\epsilon)$ and the imaginary part of self-energy $\Sigma_{L(R)}(\epsilon)$ defined by

$$\hat{\Gamma}^{L(R)}(\epsilon) = i\left(\hat{\Sigma}_{L(R)}(\epsilon) - \hat{\Sigma}^{*}_{L(R)}(\epsilon)\right), \tag{4.173}$$

the lesser Green's function $\hat{G}^{<}(\epsilon)$ is obtained from

$$\hat{G}^{<}_{C}(\epsilon) = \hat{G}^{r}_{C}(\epsilon)\left(if_L(\epsilon)\hat{\Gamma}^{L}(\epsilon) + if_R(\epsilon)\hat{\Gamma}^{R}(\epsilon)\right)\hat{G}^{a}(\epsilon), \tag{4.174}$$

where we assign different $\mu_{L,R}$ for the Fermi distribution functions

$$f_{L/R}(\epsilon) = \frac{1}{\exp\left((\epsilon - \mu_{L/R})/k_B T\right) + 1}. \tag{4.175}$$

Then the total electron density in the center region is obtained from

$$n_C(\mathbf{r}) = \frac{1}{2\pi i}\int G^{<}_{C}(\mathbf{r},\mathbf{r},\epsilon)d\epsilon. \tag{4.176}$$

We note that $n_C(\mathbf{r})$ can be divided as

$$n_C(\mathbf{r}) = -\frac{1}{\pi}\mathrm{Im}\int_Z f(z)G^{r}(\mathbf{r},\mathbf{r},z)dz + \frac{1}{2\pi i}\int_{\mu_R}^{\mu_L} G^{<}(\mathbf{r},\mathbf{r},\epsilon)d\epsilon. \tag{4.177}$$

For obtaining the first term, we use the following relation to satisfy in the equilibrium region $f(\epsilon) = f_L(\epsilon) = f_R(\epsilon)$:

$$\begin{aligned}\hat{G}^{<}(\epsilon) &= if(\epsilon)\hat{G}^{r}(\epsilon)\left[\hat{\Gamma}^{L}(\epsilon) + \hat{\Gamma}^{R}(\epsilon)\right]\hat{G}^{a}(\epsilon)\\ &= f(\epsilon)\left[\hat{G}^{r}(\epsilon) - \hat{G}^{a}(\epsilon)\right] = -2if(\epsilon)\mathrm{Im}\hat{G}^{r}(\epsilon).\end{aligned} \tag{4.178}$$

Then the first term in Eq. (4.177) becomes electron density in the equilibrium situation where we can use the imaginary contour integral technique for the retarded Green's function $G^{r}(\epsilon)$ as usual. The second term in Eq. (4.177) is electron density in the nonequilibrium situation where we need a fine mesh for the direct energy integration.

Using the DFT, we can construct an effective potential for $\hat{H}_C$. These procedures are repeated as usual until we obtain the self-consistent solutions for $n_C(\mathbf{r})$ and the effective potential. After we obtain the self-consistent solutions, we can calculate the current flowing between the electrodes as

$$I = \frac{2e}{h}\int_{\mu_R}^{\mu_L}\left(f_L(\epsilon) - f_R(\epsilon)\right)\mathcal{T}(\epsilon)d\epsilon, \tag{4.179}$$

where

$$\mathcal{T}(\epsilon) = \mathrm{Tr}\left[\hat{G}_C^r(\epsilon)\hat{\Gamma}^L(\epsilon)\hat{G}_C^a(\epsilon)\hat{\Gamma}^R(\epsilon)\right] \tag{4.180}$$

and the conductance G in the zero bias limit is obtained from

$$\begin{aligned} G &= -\left[\frac{\partial I}{\partial V}\right]_{V\to 0} \\ &= \frac{2e^2}{h}\mathrm{Tr}\left[\hat{G}_C^r(\epsilon)\hat{\Gamma}^L(\epsilon)\hat{G}_C^a(\epsilon)\hat{\Gamma}^R(\epsilon)\right] \\ &= \frac{2e^2}{h}\mathrm{Tr}\left[\hat{\mathcal{T}}^\dagger\hat{\mathcal{T}}\right]. \end{aligned} \tag{4.181}$$

The Hermitian matrix $\hat{\mathcal{T}}^\dagger\hat{\mathcal{T}}$ formed from a product of the transmission matrices $\hat{\mathcal{T}}$ is diagonalized using a unitary matrix $\hat{U}$ to transform the original channels into the eigenchannels,

$$G = \frac{2e^2}{h}\mathrm{Tr}(\hat{U}^\dagger\hat{\mathcal{T}}^\dagger\hat{\mathcal{T}}\hat{U}) = \frac{2e^2}{h}\sum_i \mathcal{T}_i. \tag{4.182}$$

The total conductance is expressed as a sum of the eigenchannel transmissions $\mathcal{T}_i$ at the Fermi energy ϵ_F. In this basis, scattering state in each channel is independent of other states without interchannel scattering.

As we see in Chapter 3, it is possible to include the electron–phonon interaction effects into the NEGF and to treat scatterings with the ionic motions, which is one of the merits of using this method for the numerical calculations of transport properties. We also note that this method can be extended within the time-dependent DFT formalism to treat the electron dynamics in excited energies as the ab initio electron dynamics method. Formalism for such electron transport by the TDDFT formalism will be described later.

4.4.2 *Numerical Atomic Orbital Basis Sets*

Here we describe the numerical atomic orbital basis sets. As localized basis sets, analytical atomic orbitals such as Gaussian and Wannier functions are often used. To reduce the number of elements for the Hamiltonian matrix, the use of numerical pseudo-atomic orbitals (PAOs) for the pseudopotentials is effective as the localized basis sets $\phi(\mathbf{r})$, where its radial part is numerically constructed to confine in a finite range at the cutoff radius r_c^l [69] as

$$R_l^{ps}(r) = \begin{cases} \text{finite} & r \le r_c^l \\ 0 & r > r_c^l. \end{cases} \tag{4.183}$$

Here l is the angular momentum with $l = 0, 1, 2, \cdots$. Then Green's function is expanded using the localized radial part of pseudo-wavefunction $R_l^{ps}(r)$ with spherical harmonic functions $Y_{lm}(\hat{\Omega})$ $(m = -l, -l+1, \cdots, l-1, l)$ as

$$\begin{aligned} G_C^{r,a}(\mathbf{r}, \mathbf{r}', \epsilon) &= \sum_{ij} \phi_i(\mathbf{r}) G_{ij}(\epsilon) \phi_j^*(\mathbf{r}') \\ &= \sum_{ij} G_{ij}(\epsilon) \rho_{ij}(\mathbf{r}, \mathbf{r}') \end{aligned} \tag{4.184}$$

with

$$\phi_i(\mathbf{r}) = R_l^{ps}(r - r_\alpha) Y_{lm}(\hat{\Omega}_\alpha) \tag{4.185}$$

and the density matrix

$$\rho_{ij}(\mathbf{r}, \mathbf{r}') = |\phi_i\rangle\langle\phi_j| = \phi_i(\mathbf{r})\phi_j^*(\mathbf{r}'). \tag{4.186}$$

Here α denotes the atom and we use the composite index $\{\alpha l m\} \to i$. Then the matrix elements of Green's functions in the center region are obtained from

$$\sum_j \left[\epsilon S_{ij} - (H_C)_{ij} - \Sigma_L(\epsilon) - \Sigma_R(\epsilon) \pm i0^+\right] G_{jk}(\epsilon) = \delta_{ik}. \tag{4.187}$$

Here the overlap of $S_{ij} = \langle\phi_i(\mathbf{r})|\phi_j(\mathbf{r})\rangle$ is calculated by

$$\begin{aligned} S_{ij} &= 4\pi i^{l_i - l_j} \sum_{l=|l_i - l_j|}^{l_i + l_j} i^l c_l(l_i m_i, l_j m_j) Y_{l,-m_i+m_j}(\hat{\Omega}_{\alpha_i - \alpha_j}) \\ &\quad \times \int q^2 j_l(q|\mathbf{r}_{\alpha_i} - \mathbf{r}_{\alpha_j}|) R_i^{ps}(q) R_j^{ps}(q) dq, \end{aligned} \tag{4.188}$$

where $c_l(l_i m_i, l_j m_j)$ is the Gaunt coefficients from Clebsch–Gordan coefficients and $j_l(x)$ is the spherical Bessel functions.

The Hamiltonian in the center region H_C is divided by the kinetic part T and the potential parts V. The matrix elements are obtained

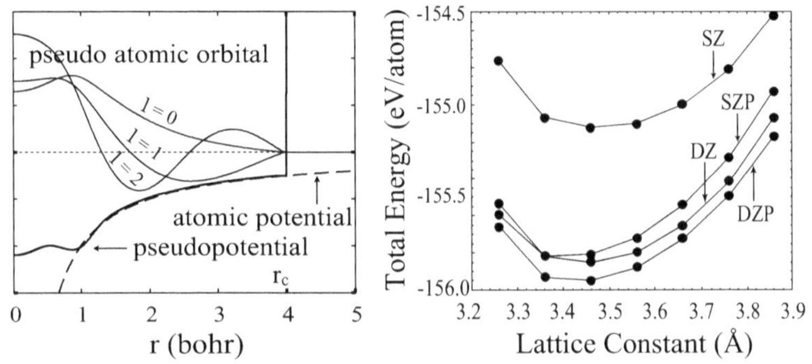

Figure 4.15 (Left) Radial pseudo atomic orbital for $l = 0, 1, 2$ of carbon atom, whose atomic potential and pseudopotential are also shown. (Right) Total energy of diamond as a function of lattice constant for SZ, SZP, DZ, and DZP orbitals with a cutoff radius of r_c as a basis set.

from

$$T_{ij} = 4\pi i^{l_i - l_j} \sum_{l=|l_i-l_j|}^{l_i+l_j} i^l c_l(l_i m_i, l_j m_j) Y_{l,-m_i+m_j}(\hat{\Omega}_{\alpha_i-\alpha_j}) \times \int (q^4/2) j_l(q|\mathbf{r}_{\alpha_i} - \mathbf{r}_{\alpha_j}|) R_i^{ps}(q) R_j^{ps}(q) dq$$

$$V_{ij} = \int \rho_{ij}(\mathbf{r}, \mathbf{r}) V_{\text{eff}}(\mathbf{r}) d\mathbf{r} + \sum_l \sum_{m=-l}^{l} \frac{\langle \phi_i | \phi_{lm} \delta V_l^{ps} \rangle \langle \delta V_l^{ps} \phi_{lm} | \phi_j \rangle}{\langle \phi_{lm} | \delta V_l^{ps} | \phi_{lm} \rangle} \tag{4.189}$$

with

$$V_{\text{eff}}(\mathbf{r}) = \sum_\alpha V_{\text{loc}}^{ps}(\mathbf{r} - \mathbf{r}_\alpha) + \int \frac{n_C(\mathbf{r}')}{|\mathbf{r} - \mathbf{r}'|} d\mathbf{r}' + V_{\text{xc}}[n_C(\mathbf{r})]. \tag{4.190}$$

The local potential part is calculated in the Fourier space using

$$\rho_{ij}(\mathbf{q}) = \sqrt{\frac{2}{\pi}} \sum_{l=|l_i-l_j|}^{l_i+l_j} (-i)^l c_l(l_i m_i, l_j m_j) Y_{l,-m_i+m_j}(\hat{\Omega}_{\mathbf{q}}) \times \int r^2 j_l(qr) R_i^{ps}(r) R_j^{ps}(r) dr, \tag{4.191}$$

which is effective for accurate treatments of the long-range Coulomb potentials. We note that in the zero bias case, the matrix elements of the Hamiltonian are constructed in the localized basis sets to perform the O(N) calculations using, for example, the divide-conquer method [70].

4.4.2.1 Surface Green's function

To obtain self-energies of the electrodes L/R, the use of the surface Green's function technique is effective [71]. The concept is that the surface is viewed as a semi-infinite stack of layers with the nearest-neighbor interactions. The matrix elements between layer orbitals are obtained from a set of equations for Green's function as

$$\begin{aligned}(\epsilon S_{00} - H_{00})G_{00} &= I + H_{01}G_{10} \\ (\epsilon S_{00} - H_{00})G_{10} &= H_{01}^{\dagger}G_{00} + H_{01}G_{20} \\ &\cdots \\ (\epsilon S_{00} - H_{00})G_{n0} &= H_{01}^{\dagger}G_{n-1,0} + H_{01}G_{n+1,0}.\end{aligned} \tag{4.192}$$

Here $S_{ij} = \langle\phi_i(\mathbf{r})|\phi_j(\mathbf{r})\rangle$, $H_{ij} = \langle\phi_i(\mathbf{r})|\hat{H}_{L/R}|\phi_j(\mathbf{r})\rangle$, and $G_{ij} = \langle\phi_i(\mathbf{r})|\hat{G}_{L/R}|\phi_j(\mathbf{r})\rangle$ are matrix elements of overlap, Hamiltonian, and Green's function between layer orbitals in the electrodes. We assume that in the bulk electrodes $S_{00} = S_{11} = \cdots$, $H_{00} = H_{11} = \cdots$, and $H_{01} = H_{12} = \cdots$. With the initial conditions of

$$\begin{aligned}t_0 &= (\epsilon S_{00} - H_{00})^{-1}H_{01}^{\dagger} \\ \tilde{t}_0 &= (\epsilon S_{00} - H_{00})^{-1}H_{01}\end{aligned} \tag{4.193}$$

and using an asymptotic relation

$$\begin{aligned}t_i &= (I - t_{i-1}\tilde{t}_{i-1} - \tilde{t}_{i-1}t_{i-1})^{-1}t_{i-1}^2 \\ \tilde{t}_i &= (I - t_{i-1}\tilde{t}_{i-1} - \tilde{t}_{i-1}t_{i-1})^{-1}\tilde{t}_{i-1}^2,\end{aligned} \tag{4.194}$$

we obtain self-energies from

$$\begin{aligned}\Sigma_L(\epsilon) &= H_{01}^{\dagger}(\tilde{t}_0 + t_0\tilde{t}_1 + t_0t_1\tilde{t}_2 + \cdots + t_0t_1t_2\cdots\tilde{t}_n) \\ \Sigma_R(\epsilon) &= H_{01}(t_0 + \tilde{t}_0t_1 + \tilde{t}_0\tilde{t}_1t_2 + \cdots + \tilde{t}_0\tilde{t}_1\tilde{t}_2\cdots t_n).\end{aligned} \tag{4.195}$$

The process is repeated until t_n, $\tilde{t}_n$ become less than an arbitrary small value δ. This is a quite fast method for convergence.

4.4.2.2 Poisson equation in localized basis sets

Here we describe the solution to the Poisson equation in the localized basis sets [68].

$$\nabla^2 V_H(\mathbf{r}) = -4\pi n_C(\mathbf{r}). \tag{4.196}$$

Electron density $n_C(\mathbf{r})$ is obtained from Eq. (4.177). Since the boundaries are taken deep inside the electrodes, screening effects are included on the metallic electrodes. Therefore, we assume that electron density is that of the bulk beyond the boundaries. When we assign the different chemical potentials to each electrode, the solution to the Poisson equation is given, including the linear term, by

$$V_H(\mathbf{r}) = \varphi(\mathbf{r}) + \mathbf{a} \cdot \mathbf{r} + b, \tag{4.197}$$

where $\varphi(\mathbf{r})$ is the usual solution to the Poisson equation, which may be obtained by the direct integration of the differential equation in real space. $\mathbf{a}$ and b are the parameters that must be determined from the boundary conditions to Poisson equation. If we use the periodic boundary conditions that fix the values of a_x and a_y, we need to determine the other two parameters a_z and b. Since the electronic potentials at the boundaries of the left and right electrodes are determined from the separate bulk calculations, we shift them relative to each other by a bias voltage of V. The Hartree potential in contacts are uniquely defined with these boundary conditions and could be computed from

$$V_H(z) = -4\pi \int_{z_0}^{z} dz'' \int_{z_0}^{z''} dz' n_C(z') + a_z(z - z_0) + b \tag{4.198}$$

with $V_H(z_0)(= b)$ and $V_H(z_{l+1}) = V_H(z_0) + eV$ at the boundaries.

4.4.3 *Thermoelectric Effects at Nanometer Scale*

4.4.3.1 Characteristics of nanoscale phonon transport

Let us treat phonon transport at the nanometer-scale systems using the NEGF method. For technological points of view, as electron devices become smaller, the number of transistors in a microchip becomes larger. If heating from each transistor is not reduced, the total amount of heating from one microchip becomes tremendous, which consume too much energies to operate. As a result, such electronic device will not work. This means that to construct the nanometer-scale electron devices, we cannot neglect heating problems anymore. Also, there is much interest in the development of electron devices that convert the thermal energy to the electric energy. Such electron devices are called thermoelectric devices, for which we need to search for material structure to enforce electronic performance and to suppress heating from phonons.

In metallic systems, thermal conductance is dominated by the electronic contribution κ_e, which is proportional to the electrical conductance via the Wiedemann–Franz law as we see in Chapter 2. On the other hand, in semiconductor or insulator systems, due to deficiency of the conducting electrons, the thermal conductance is dominated by the lattice contribution κ_ℓ. First, we present the formalism to treat phonon transport for thermal conductance κ_ℓ and treat the generation of phonon for local heating problem briefly. Then we derive the relations with the formula for efficiency of thermoelectric devices. An example is presented for phonon thermal transport of silicon nanowires (SiNWs).

First, we consider phonon transport as the lattice vibrations of nanosystem between the electrodes with different temperatures for the lattice thermal conductance κ_ℓ. We assume that inelastic scatterings of a phonon with other phonons inside the electrodes might be the main contribution to dissipate heat and thus lead to local equilibrium. This mechanism of heat transport is similar to ballistic electron transport expressed by the Landauer formula. Thus, here we consider phonon conduction as a ballistic transport, which means that generated heat is dissipated inside the electrodes.[a]

Then thermal current I_{ph} is derived as

$$I_{ph} = \sum_q \int_{\omega_{\min}}^{\omega_{\max}} \hbar\omega_q(k) v_q(k) \mathcal{T}_{ph}^q(k) \left(\eta_L - \eta_R\right) \frac{dk}{2\pi}, \tag{4.199}$$

where $\mathcal{T}_{ph}^q(k)$ is the transmission probability of phonon, $\omega_q(k)$ and $v_q(k)$ are the frequency and the velocity of the phonon mode q with wavevector k, and

$$\eta_{L/R}(\omega) = \frac{1}{e^{\hbar\omega/k_B T_{L/R}} - 1} \tag{4.200}$$

[a]This is based on the assumption that the size of nanosystem is less than the phonon mean free path determined by the phonon–phonon scatterings.

represents the thermal distributions of phonons in the electrodes with the temperatures $T_{L/R}$. Changing the variable from k to ω, we have

$$I_{ph} = \frac{1}{2\pi}\sum_q \int_{\omega_{\min}}^{\omega_{\max}} \hbar\omega \mathcal{T}_{ph}^q(\omega)\left(\eta_L(\omega) - \eta_R(\omega)\right) d\omega, \qquad (4.201)$$

since the phonon velocity $v_q(k) = \partial\omega_q/\partial k$ is canceled by the density of states $g(\omega_q) = \partial k/\partial\omega_q$. Then the thermal conductance $\kappa_\ell = I_{ph}/\Delta T$ becomes

$$\begin{aligned}\kappa_\ell &= \frac{1}{2\pi}\sum_q \int_{\omega_{\min}}^{\omega_{\max}} \hbar\omega \mathcal{T}_{ph}^q(\omega)\frac{\partial\eta(\omega)}{\partial T} d\omega \\ &= \frac{k_B}{2\pi}\sum_q \int_{\omega_{\min}}^{\omega_{\max}} \left(\frac{\hbar\omega}{k_B T}\right)^2 \frac{e^{\hbar\omega/k_B T}}{(e^{\hbar\omega/k_B T}-1)^2}\mathcal{T}_{ph}^q(\omega) d\omega,\end{aligned} \qquad (4.202)$$

where $T = (T_L + T_R)/2$.

When the transmission is perfect $\mathcal{T}_{ph}^q(\omega) = 1$ for all the phonon modes q as $\omega \to 0$ at low temperature, we have the thermal conductance for $T \to 0$ as

$$\kappa_\ell = \sum_q \frac{\pi^2 k_B^2 T}{3h} N_q. \qquad (4.203)$$

Here we use the integral evaluation of

$$\int_0^\infty \frac{x^2 e^x}{(e^x - 1)^2} dx = 2\sum_{n=1}^{\infty} \frac{1}{n^2} = \frac{\pi^2}{3}. \qquad (4.204)$$

This indicates that the thermal conductance is quantized with the fundamental quantum of $\kappa_0 = \pi^2 k_B^2 T/3h$ at low temperature $T \simeq 0$ with N_q [72, 73], representing the acoustic phonon branch number as $\omega \to 0$. This is called the universal thermal conductance.

As the temperature increases, the thermal phonons are excited and contribute to phonon transport, which induces the rapid increase of the thermal conductance.[a] Figure 4.16 shows the characteristics of the difference of electron and phonon distributions. For fermionic electron case obeying the Fermi–Dirac distribution (left), electrons with energies in the range of $|\epsilon - \mu| \sim k_B T$

[a]When the temperature is less than the Debye frequency, expanding Eq. (4.201) for $k_B T \ll \hbar\omega$, we have $I_{ph} \propto T^3$.

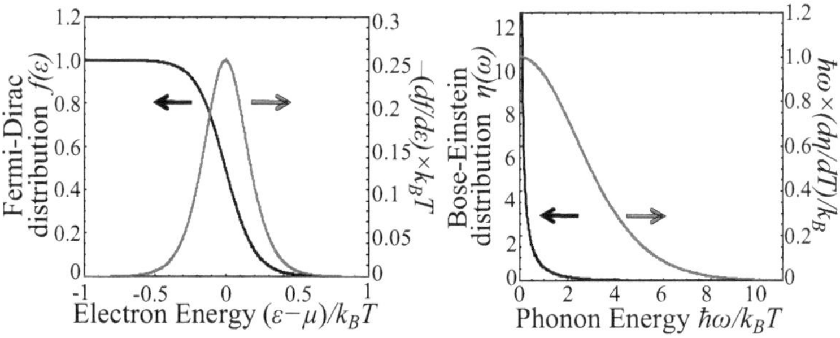

Figure 4.16 (Left) Fermi–Dirac distribution function $f(\epsilon)$ (solid line) and its differential form $-(df/d\epsilon)/k_BT$ (gray line) as a function of electron energy $(\epsilon-\mu)/k_BT$. (Right) Bose–Einstein distribution function $\eta(\omega)$ (solid line) and its differential form $\hbar\omega(d\eta/dT)/k_B$ (gray line) as a function of phonon energy $\hbar\omega/k_BT$.

contribute to electron conduction since $-(df/d\epsilon) \sim k_BT$. Thus, electronic states near μ is essential to electron transport. On the other hand, for bosonic phonon case obeying the Bose–Einstein distribution (right), phonons with energies in the range of $\hbar\omega \gg k_BT$ contribute to phonon conduction since $\hbar\omega(d\eta/dk_BT)$ has large values above k_BT. Thus, the phonon band structures with energies larger than k_BT are also important for phonon thermal transport. This shows the dependence on intrinsic properties of materials at room temperature and we need to obtain both acoustic and optical phonon modes for various materials.

In the realistic nanometer-scale materials, transmission of phonons have significant effects through reflections from contacts between electrodes and nanosystems [74]. Thus, we need to evaluate the transmission probability $\mathcal{T}^q_{ph}(\omega)$ for various phonon modes q to have the thermal conduction κ_ℓ. At high temperature, due to anharmonic effects the phonon–phonon scatterings accompanied by the Umklapp processes reduce the thermal conduction as $1/T$.

4.4.3.2 NEGF method for phonon transport

Thermal transport has been simulated with various methods, which include the studies based on the molecular dynamics or the Boltzmann equation in view of the particle picture as we described in Chapter 2. Here we introduce the NEGF method, one of the effective calculation methods for phonon transport problem [75, 76], which treats phonons as scattering waves for applications to nanosystems. The method is similar to the one we used in the previous section for electron transport.

We consider the Hamiltonian

$$\hat{H} = \sum_i \frac{1}{2}\dot{\hat{u}}_i^2 + \sum_{ij} \frac{1}{2!}k_{ij}\hat{u}_i\hat{u}_j + \sum_{ijk} \frac{1}{3!}t_{ijk}\hat{u}_i\hat{u}_i\hat{u}_k + \cdots, \qquad (4.205)$$

where $\hat{u}_i = \sqrt{M_i}\delta\hat{R}_i = \sum_q \sqrt{\hbar/2\omega_q}\left(\hat{a}_q e^{iq\cdot R_i} + \hat{a}_q^\dagger e^{-iq\cdot R_i}\right)$ is an operator for the normal coordinate of atomic displacement from equilibrium position of the i-th atom with mass M_i and k_{ij} is the spring constants between i-th and j-th atoms. We assume that the anharmonic term t_{ijk} and the higher order terms exist only in the center region.

When we divide the total system into the electrode parts $\hat{H}_{L/R}$, the center part $\hat{H}_C$, and the coupling parts $\hat{H}_{LC/RC}$ between electrode and center parts, and assign the different temperatures $T_{L/R}$ in $\hat{H}_{L/R}$, the thermal current flowing through the center part is obtained from

$$\begin{aligned} I_{ph} &= -\langle \frac{d}{dt}\hat{H}_L\rangle = -\frac{i}{\hbar}\langle\left[\hat{H}, \hat{H}_L\right]\rangle \\ &= i\hbar\sum_{ij}\frac{k_{ij}}{2}\left[\frac{d}{dt'}D_{ij}^<(t,t') + h.c.\right], \end{aligned} \qquad (4.206)$$

where the phonon Green's function is defined by

$$D_{ij}^<(t,t') = -\frac{i}{\hbar}\langle\hat{u}_j(t')\hat{u}_i(t)\rangle. \qquad (4.207)$$

Using the Fourier transform for steady states, I_{ph} are evaluated by

$$\begin{aligned} I_{ph} &= \frac{1}{2\pi}\int_{-\infty}^{\infty} \hbar\omega \mathrm{Tr}\left[D^r(\omega)\Sigma_L^<(\omega) + D^<(\omega)\Sigma_L^a(\omega)\right] d\omega, \\ &= \frac{i}{2}\frac{1}{2\pi}\int_0^{\infty} \hbar\omega\mathrm{Tr}\left\{\left[\Gamma^L(\omega) - \Gamma^R(\omega)\right] D^<(\epsilon)\right. \\ &\quad \left. + \left[\eta_L(\omega)\Gamma^L(\omega) - \eta_R(\omega)\Gamma^R(\omega)\right]\left[D^r(\omega) - D^a(\omega)\right]\right\} d\omega, \end{aligned} \qquad (4.208)$$

where self-energies $\Sigma_{L/R}(\omega)$ are due to the interaction with the electrodes and $\Gamma^{L/R}(\omega) = i\left[\Sigma_{L/R}(\omega) - \Sigma_{L/R}^*(\omega)\right]$. Here $D^<(\omega)$ satisfy the Keldysh form of

$$D^<(\omega) = D^r(\omega)(\Sigma_L^<(\omega) + \Sigma_R^<)D^a(\omega) \qquad (4.209)$$

and the retarded and the advanced phonon Green's functions $D^{r,a}(\omega)$ are obtained from the Dyson equation

$$\hat{D}^{r,a}(\omega) = \left[(\omega \pm i0)^2 M - K - (\Sigma_L^{r,a} + \Sigma_R^{r,a})\right]^{-1}. \tag{4.210}$$

M is a matrix with the mass of each atom as a diagonal element.

The dynamical matrix K is derived from the second derivative of the total energy E with respect to atomic displacements $\hat{u}$, whose elements k_{ij} are given by

$$k_{ij} = \frac{\partial^2 E}{\partial u_i \partial u_j} = \frac{-\left[F_i(+u_j) - F_i(-u_j)\right]}{2u_j}. \tag{4.211}$$

The total energy E can be obtained, in principle, from the ab initio calculations based on the DFT formalism, or alternatively from empirical parameters for the interactions between atoms, which are given later. Numerically the matrix elements of k_{ij} are calculated by finite difference of the force F_i, which is obtained from the derivative of the total energy E. Here $F_i(+u_j)$ indicates the force of the i-th atom generated by the j-th atom with a displacement of $+u_j$ from equilibrium positions. Note that there are three directions x, y, z; thus the dimension of K is $3N \times 3N$ where N is the number of atoms.

The total transmission probability $\mathcal{T}_{ph}(\omega)$ is obtained from

$$\mathcal{T}_{ph}(\omega) = \mathrm{Tr}\left[D^r(\omega)\Gamma^L(\omega)D^a(\omega)\Gamma^R(\omega)\right] \tag{4.212}$$

and then the thermal conductance is calculated from

$$\kappa_\ell = \frac{k_B}{2\pi}\int_0^\infty \left(\frac{\hbar\omega}{k_B T}\right)^2 \frac{e^{\hbar\omega/k_B T}}{(e^{\hbar\omega/k_B T} - 1)^2}\mathcal{T}_{ph}(\omega)d\omega. \tag{4.213}$$

For an ideal ballistic limit without any scattering from the anharmonic terms and thus for the perfect transmission case, $\mathcal{T}_{ph}(\omega)$ is equal to the number of phonon subbands N_α at the frequency ω. The thermal conductance is quantized at low temperature limit $T \to 0$ to give

$$\kappa_\ell = \sum_q \frac{\pi^2 k_B^2 T}{3h} N_q, \tag{4.214}$$

where q is the acoustic phonon modes for $\omega \to 0$. We will show later that electronic contribution to κ_e is given by $\kappa_e = \kappa_0 N_e - S^2 GT$ when the transmission for electrons $\mathcal{T}(\epsilon_F)$ is perfect. Here N_e is the channel number for electron transport. So the total thermal conductance becomes $\kappa = \kappa_e + \kappa_\ell = \kappa_0(N_e + N_\alpha/2) - S^2GT$ in the limit of $\mathcal{T}(\epsilon_F) = 1$ and $\mathcal{T}_{ph}(\omega \to 0) = 1$.

We note that this formalism is effective in the nanosystems where phonon–phonon scattering is negligible. The anharmonic effects from the higher term, for example $t_{ijk} = \partial^3 E/\partial u_i \partial u_j \partial u_k$, which produce the interaction of phonons with crystal structure, must be included in the phonon Green's function for large systems over the phonon mean free path at high temperature by the self-energy term Σ_{ph-ph}.[a]

4.4.3.3 Generation of phonon–local heating

Here we consider heating problem induced by phonon emission in nanosystems phenomenologically. When an electron interacts with phonons, it dissipates energy from inelastic electron–phonon scatterings. The transition rate by phonon scattering is written as

$$\left(\frac{\partial f}{\partial t}\right)_{col} = \begin{cases} \nu_{\text{ph}}(\epsilon_m - \epsilon_n)|\langle n|V_{e-ph}|m\rangle|^2 \left(N_{\epsilon_m-\epsilon_n} + 1\right) & \epsilon_n < \epsilon_m \\ \nu_{\text{ph}}(\epsilon_n - \epsilon_m)|\langle n|V_{e-ph}|m\rangle|^2 N_{\epsilon_n-\epsilon_m} & \epsilon_n > \epsilon_m, \end{cases} \tag{4.215}$$

where $\nu_{\text{ph}}(\epsilon)$ is the density of states for phonons and N_ϵ is the thermal mean occupation number of phonons with energy ϵ. The first term represents the phonon emission and the second term represents the phonon absorption, which satisfy the detailed balance of $\Gamma_{n,m}/\Gamma_{m,n} = \exp(-\hbar\omega_q/k_B T)$ with $(\partial f/\partial t)_{col} = \Gamma_{n,m}$.

We note that the electron–phonon matrix elements are approximately given by the matrix elements of derivative of the local part of effective potential in the DFT calculations. When electronic state changes from $|m\rangle$ to $|n\rangle$ through an exchange of a phonon whose energy is $\hbar\omega_q$, electron–phonon interaction is evaluated by

$$\left|\langle n|V_{e-ph}|m\rangle\right|^2 \approx \left|\int \psi_n^L(\mathbf{r})\nabla V_{loc}(\mathbf{r})\psi_m^R(\mathbf{r})d\mathbf{r}\right|^2, \tag{4.216}$$

[a] The inverse of phonon lifetime $1/\tau \propto \mathrm{Im}\,\Sigma_{\text{ph-ph}}$ due to phonon–phononn scatterings is evaluated from the 3rd-order t_{ijk} using the Fermi golden rule with momentum conservation $q_i + q_j + q_k = G$ where G is the reciprocal-lattice vector for Umklapp processes and with energy conservation $\delta(\omega_i - \omega_j - \omega_k)$ for phonon emission and $\delta(\omega_i + \omega_j - \omega_k)$ for phonon absorption. The phonon mean free path is estimated from τ.

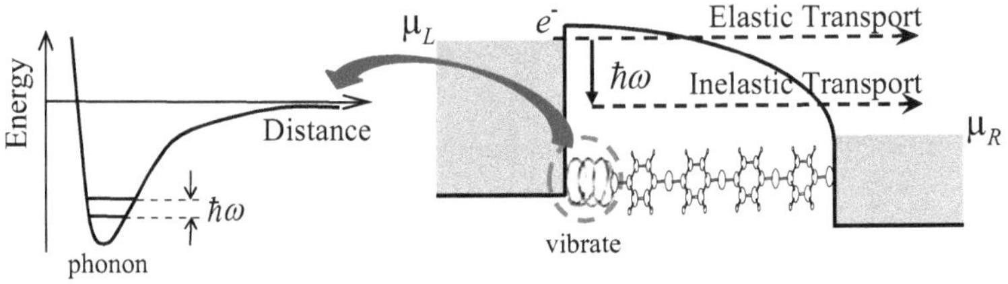

Figure 4.17 Schematic pictures of a molecule between electrodes with the left contact is not well formed, where the vibration with small frequency $\hbar\omega$ is induced within the gap region.

and the energy per unit time transferred from electrons to lattice atoms for that process is given by

$$\langle W_q \rangle = \hbar\omega_q \int \frac{2\pi}{\hbar} |\langle n|V_{e-ph}|m\rangle|^2 (1 + \langle N_q \rangle) \nu_{\rm ph}(\hbar\omega_q) \nu^L(\epsilon) \times \nu^R(\epsilon - \hbar\omega_q) d\epsilon.$$

In the bulk system where the crystal structure is almost perfect and the crystal momentum is preserved, phonon transfer is greatly suppressed since the inelastic phonon processes is estimated of order $\sim e^{-T_D/T_{ph}}$, where T_D is the Debye temperature of the bulk typically few hundred K and T_{ph} is the local ionic temperature at phonon, usually $T_{ph} \ll T_D$.

In the nanosystems, translational invariance of crystal is violated between electrodes, especially at contacts as in Fig. 4.17. Let us consider the junction nanosystem sandwiched between electrodes. The phonons generated in the nanosystem go away as heat flow. This is the heating process, including the spontaneous phonon emission. On the other hand, there is a process in which energy is transferred from lattice atoms to electrons; it is the cooling process. In the nanosystems, since there are two processes that electrons come from the left electrode and from the right electrode, totally there are four processes for an energy exchange. The total power energy per unit time for local heating is the sum of all phonon modes q in these four processes [27]. These energies generated in the nanosystem, which produces local heating, are transferred to electrodes and then the nanosystem is cooled down. The present treatment of phonon emission is based on the transition rate. We will describe the dissipation problem by phonon emission using the QME method later.

4.4.3.4 Thermoelectric effects

Since phonon transport at the nanometer scale might be different from that of the bulk, the knowledge of thermoelectric effects in nanosystems becomes important. Here we give the fundamental formulas.

First, we treat the Seebeck effect. Electron transport occurs when the different chemical potentials are assigned to electrodes. On the other hand, when different temperatures are assigned to the electrodes, an electric field is induced by the temperature gradient with the opposite direction. Such an electric field is the thermoelectric field known as the Seebeck effect. Figure 4.18 (left) shows a bimetallic system with different temperatures at two junctions [77]. Here we consider an open junction system with left and right electrodes keeping different temperatures $T_{L/R}$. Thermoelectric field is induced by the temperature gradient as $\mathcal{E} = E - \nabla\mu/e$. Due to this electric field, charge carriers diffuse from the hot region to the cold region and separate at the junction, creating the thermoelectric voltage V_T. The Seebeck coefficient or thermopower S is defined by the proportionality constant as

$$\Delta V_T = -\int_{\ell_R}^{\ell_L} \mathcal{E} \cdot d\ell = S(T_L - T_R) = S\Delta T. \tag{4.217}$$

We note that the thermopower S becomes both positive and negative depending on the carriers (here we take $e < 0$).

Let us derive the thermoelectric voltage from the Fermi–Dirac distribution functions $f_{L/R}(\epsilon)$ of the electrodes and obtain an expression for the thermopower S.[a] Electric current is calculated from the Landauer formula

$$\begin{aligned} I &= \frac{2e}{h}\int_{-\infty}^{\infty} (f_L(\epsilon) - f_R(\epsilon))\, \mathcal{T}(\epsilon) d\epsilon \\ &\simeq \frac{2e^2}{h}\mathcal{T}(\epsilon_F)\Delta V - \frac{2e}{h}\frac{\pi^2 k_B^2 T}{3}\left.\frac{\partial \mathcal{T}(\epsilon)}{\partial \epsilon}\right|_{\epsilon_F} \Delta T. \end{aligned} \tag{4.218}$$

[a] This represents the thermopower S for the metallic systems.

Here we expand the Fermi–Dirac distribution function

$$f_{L/R}(\epsilon) \simeq f(\epsilon) \mp \frac{\partial f(\epsilon)}{\partial \epsilon}\left[\frac{e\Delta V}{2} - (\epsilon - \epsilon_F)\frac{\Delta T}{2T}\right] \tag{4.219}$$

and use an expansion formula

$$\int_{-\infty}^{\infty}\left(-\frac{\partial f(\epsilon)}{\partial \epsilon}\right) g(\epsilon)d\epsilon = \int_{-\infty}^{\infty}\frac{\partial g(\epsilon)}{\partial \epsilon} f(\epsilon)d\epsilon$$

$$= g(\epsilon_f) + \frac{\pi^2}{6}(k_B T)^2 \left.\frac{\partial^2 g(\epsilon)}{\partial \epsilon^2}\right|_{\epsilon=\epsilon_f} + \cdots . \tag{4.220}$$

The first term gives the usual electron transport and the second term gives the thermal transport with the transmission coefficient $\mathcal{T}(\epsilon)$, which represents that the temperature gradient produces thermal transport. Since the Seebeck effect is measured on the condition of $I = 0$, the thermoelectric voltage is given by

$$\Delta V_T = \frac{\pi^2 k_B^2 T}{3e} \left.\frac{\partial \ln \mathcal{T}(\epsilon)}{\partial \epsilon}\right|_{\epsilon_F} \Delta T \tag{4.221}$$

and the Seebeck coefficient is obtained (Mott's formula) as

$$S = \frac{\pi^2 k_B^2 T}{3e} \left.\frac{\partial \ln \mathcal{T}(\epsilon)}{\partial \epsilon}\right|_{\epsilon_F} . \tag{4.222}$$

For free electrons, the thermopower S_0 becomes

$$S_0 = \frac{\pi^2 k_B}{2e}\left(\frac{k_B T}{\epsilon_F}\right). \tag{4.223}$$

We note that the total electric current of Eq. (4.218) is written as $I \simeq G\Delta V - SG\Delta T$ with electric conductance of $G = \left.\frac{2e^2}{h} T(\epsilon)\right|_{\epsilon_F}$. From this formula, we see that the Seebeck coefficient S depends on the scattering events included in $\mathcal{T}(\epsilon)$ and thus is sensitive to geometries of the systems, especially at contacts. On the other hand, when transmission coefficient does not change much, the Seebeck coefficient should be suppressed close to zero $S \simeq 0$. We note that in the atomic-scale contact systems, even tiny changes of the conductance are accompanied by large jumps of the Seebeck coefficient S, showing that S is sensitive to the geometries in nanosystems.

Next, we treat the Peltier effect. When two metals are connected with the electric current I driven at the temperature T, heat is

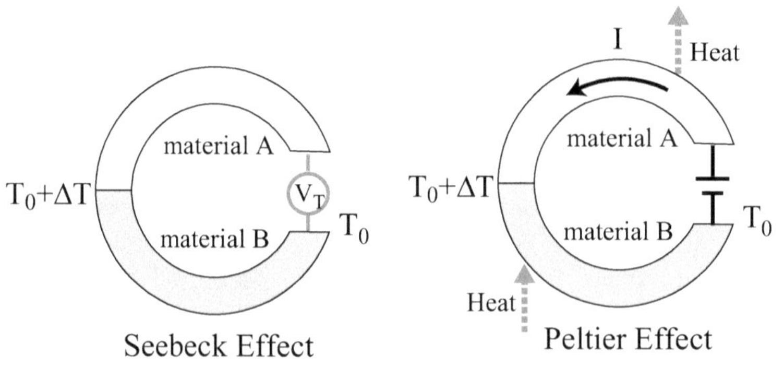

Figure 4.18 (Left) Schematic illustration of the Seebeck effect. When two metals are connected with one junction at temperature T_L and with other junction at $T_R(\neq T_L)$, there appears a thermoelectric voltage V_T at the junction. This shows an energy conversion from thermal to electric one. Proportionality constant $S = V_T/(T_L - T_R)$ is the thermopower or the Seebeck coefficient. (Right) Schematic illustration of the Peltier effect. When two metals are connected with the electric current I driven at uniform temperature T, the thermal current I_q is absorbed at one junction and is evolved at the other junction. This shows an energy conversion from electric to thermal one. Proportionality constant $\Pi = I_q/I$ is the Peltier coefficient. Peltier coefficient is related with the Seebeck coefficient by $\Pi = TS$.

evolved at one side and absorbed at the other side as is shown in the right panel of Fig. 4.18, the thermal current I_q is accompanied to maintain the temperature of the system. I_q is proportional to the electric current I as $I_q = \Pi I$ where proportionality constant Π is called the Peltier coefficient. We show that Π is related with the Seebeck coefficient or the thermopower by $\Pi = TS$.

Here we derive the thermal current I_q and the Peltier coefficient Π from the Fermi–Dirac distribution functions. We note that carrier of heat is an energy $\epsilon - \epsilon_F$ of an electron, instead of e in the electric current. Then we have

$$
\begin{aligned}
I_q &= \frac{2}{h}\int_{-\infty}^{\infty}(\epsilon - \epsilon_F)\left(f_L(\epsilon) - f_R(\epsilon)\right)\mathcal{T}(\epsilon)d\epsilon \\
&\simeq \frac{2e}{h}\frac{\pi^2 k_B^2 T^2}{3}\left.\frac{\partial \mathcal{T}(\epsilon)}{\partial \epsilon}\right|_{\epsilon_F}\Delta V - \frac{2}{h}\frac{\pi^2 k_B^2 T}{3}\mathcal{T}(\epsilon)\Big|_{\epsilon_F}\Delta T.
\end{aligned}
\tag{4.224}
$$

We note that this is expressed as $I_q = TSG\Delta V - \kappa_0 \Delta T$ where κ_0 is the electronic thermal conductance for zero electric field

$$\kappa_0 = \frac{2}{h}\frac{\pi^2 k_B^2 T}{3}\mathcal{T}(\epsilon)\bigg|_{\epsilon_F} . \tag{4.225}$$

It is important to note that the ratio of κ_0 and the electric conductance G takes a universal value of $\kappa_0/GT = (\pi^2/3)(k_B/e)^2$. This relation is well known as the Wiedemann–Franz law. Here the constant number $L = (\pi^2/3)(k_B/e)^2 = 2.44 \times 10^{-8}$ $(W\Omega/K^2)$ is called the Lorenz number [1].

Using $I \simeq G\Delta V - SG\Delta T$, the total thermal current is expressed as $I_q \simeq TSI - (\kappa_0 - S^2GT)\Delta T$. Since the Peltier effect is measured at a uniform temperature $\Delta T = 0$, the Peltier coefficient Π as proportionality constant is given by $\Pi = TS$. This shows that the Peltier coefficient Π is related to the thermopower S. This is known as the Kelvin relations. We note that the second term arises when there is a temperature difference and gives an electronic thermal contribution by $\kappa_e = \kappa_0 - S^2GT$, where $F = S^2G$ is called the power factor and plays a key role for the thermoelectric devices. Then the thermal current is expressed by $I_q = \Pi I - \kappa_e \Delta T$.

4.4.3.5 Thermoelectric relations

The electric and thermal currents on the thermoelectric properties are summarized to show the following relation as

$$\begin{aligned} I/e &= L_{11}(\Delta\mu) + L_{12}(-\Delta T/T) \\ I_q &= L_{21}(\Delta\mu) + L_{22}(-\Delta T/T) \end{aligned} \tag{4.226}$$

with $\Delta\mu = e\Delta V$. Then the thermal properties are expressed by

$$S = \frac{1}{eT}\frac{L_{12}}{L_{11}}, \quad \kappa_e = \frac{L_{11}L_{22} - L_{12}L_{21}}{L_{11}T} \quad \text{with } L_{12} = L_{21}. \tag{4.227}$$

These relations between transport coefficients are formulated in generality, called the Onsager reciprocity relations [78], which come from the requirement of time reversal invariance of the mechanical equations of motion of particles to retrace the former paths if all velocities are reversed.

In the present case, defining the function by

$$\mathcal{L}_n = \frac{2}{h}\int_{-\infty}^{\infty}\left(-\frac{\partial f(\epsilon)}{\partial \epsilon}\right)(\epsilon - \epsilon_F)^n \mathcal{T}(\epsilon)d\epsilon, \tag{4.228}$$

we have $L_{11} = \mathcal{L}_0$, $L_{12} = L_{21} = \mathcal{L}_1$, $L_{22} = \mathcal{L}_2$ and correspondingly $G = e^2\mathcal{L}_0$, $S = (1/eT)(\mathcal{L}_1/\mathcal{L}_0)$, $\kappa_e = (\mathcal{L}_0\mathcal{L}_2 - \mathcal{L}_1^2)/\mathcal{L}_0 T$. The equality $L_{21} = L_{12}$ is an example of the Onsager relations.[a]

We can check that expanding $(-\partial f/\partial\epsilon)$ and taking within the regime of $\mu \approx \epsilon_F$ as $O(k_B T)$, which produces within an accuracy of order $(k_B T/\epsilon_F)^2$, these relations are given at $\epsilon = \epsilon_F$ by

$$L_{11} = \frac{2}{h}\mathcal{T}(\epsilon),\ L_{12} = \frac{2}{h}\frac{(\pi k_B T)^2}{3}\frac{\partial\mathcal{T}(\epsilon)}{\partial\epsilon},\ L_{22} = \frac{2}{h}\frac{(\pi k_B T)^2}{3T}\mathcal{T}(\epsilon).$$

We note that these are the results from the electronic contribution to thermoelectric effects, since $\partial\mathcal{T}(\epsilon)/\partial\epsilon$ is of order $(\mathcal{T}(\epsilon_{\mathcal{F}})/\epsilon_F)$, and κ_e is dominated by the first term such as $\kappa_e \simeq \kappa_0 = L_{22}/T$. This results in the Wiedemann–Franz law, which is valid for the degenerate Fermi statistics in the metallic systems. In the semiconductor or the insulator systems, the second term of $L_{12}L_{21}/L_{11}(= S^2GT)$ for the thermoelectric field cannot be ignored.[b]

Here we consider the devices for the energy conversion using the thermoelectric effects [79]. Especially, we use the Seebeck effect to convert thermal energy to electric one, creating a thermoelectric voltage V_T. The efficiency η^c is measured by the dimensionless parameter of ZT, which is by using the Onsager relation Eq. (4.227) given by

$$ZT = \frac{S^2GT}{\kappa} = \frac{1}{e^2}\frac{L_{12}L_{21}}{L_{11}L_{22} - L_{12}L_{21} + \kappa_\ell L_{11}T}. \tag{4.229}$$

[a] When we evaluate transport properties by the Boltzmann equation, we use

$$\mathcal{K}_n = \frac{2}{(2\pi)^3}\int\left(-\frac{\partial f_0(\mathbf{k})}{\partial\epsilon_{\mathbf{k}}}\right)(\epsilon - \epsilon_F)^n\tau(\mathbf{k})\mathbf{v}_{\mathbf{k}}^2 d\mathbf{k}$$

with the matrix elements of $L_{11} = \mathcal{K}_0$, $L_{12} = L_{21} = \mathcal{K}_1$, and $L_{22} = \mathcal{K}_2$. Then we have $\sigma = e^2\mathcal{K}_0$, $S = (1/eT)(\mathcal{K}_1/\mathcal{K}_0)$, $\kappa_e = (\mathcal{K}_0\mathcal{K}_2 - \mathcal{K}_1^2)/\mathcal{K}_0 T$.

[b] For semiconductors, we note $S \approx (k_B/e)(d/2 + 1 + p - \epsilon_F/k_B T)$ for the non-degenerate Boltzmann distribution, assuming the relaxation of $\tau \propto T^p$.

[c] When we consider two materials connected with different temperatures $T_{H/C}$ as shown in the left panel of Fig. 4.18, the efficiency η of thermoelectric devices are given by the figure of merit $ZT = S^2GT/\kappa$ such as

$$\eta = \frac{T_H - T_C}{T_H}\frac{\sqrt{1+ZT} - 1}{\sqrt{1+ZT} + T_C/T_H},$$

where $T = (T_H + T_C)/2$ is the mean temperature and the Carnot efficiency is $\eta_C = (T_H - T_C)/T_H$. This relationship is derived by considering the rate of the Joule heat Q generated inside, some of which go back to the hot part, and the output power P as P/Q and by taking its maximum value.

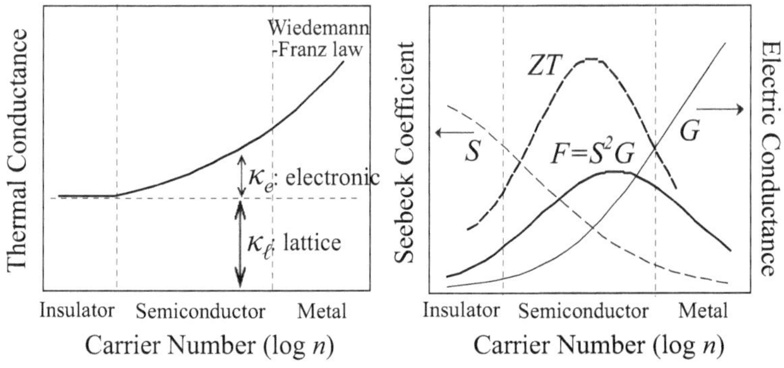

Figure 4.19 (Left) Thermal conductivity from electrons κ_e and lattice vibrations of phonons κ_ℓ. (Right) Typical behaviors of thermoelectric parameters $S, G, F = S^2G$, and ZT as a function of carrier number n.

Here $F = S^2G$ is the power factor, and $\kappa = \kappa_e + \kappa_\ell$ is the thermal conductance, which is the sum of the electronic κ_e and the lattice κ_ℓ contributions. To enhance the efficiency η, we need to find materials with (1) large Seebeck coefficient S, (2) high electric conductance G, and (3) low thermal conductance κ. Especially, the large power factor F is important to obtain electronic power and to reduce the Joule heating loss, while the low thermal conductance is important to keep the temperatures at contacts.

Figure 4.19 shows the typical behaviors of thermoelectric parameters as a function of carrier number n. In metallic systems, due to the Wiedemann–Franz law, the thermal conductance is dominated by electrons κ_e, which is proportional to the electrical conductance G. Also, the Seebeck coefficient S is relatively low; thus ZT is small and it is hard to realize high ZT in metallic systems.[a]

On the other hand, in the insulator systems, usually electric conductance G and power factor F are small. Therefore, thermoelectric devices, aiming at $ZT > 1$, are typically made of p-type and n-type doped semiconductor elements, connected so that electric current flows in series. Thermal contribution is largely from phonons κ_ℓ.

[a]We estimate from Eq. (4.223) as $S \approx (\pi^2/2)(k_B/e)(1/\epsilon_F) \approx 1 \times 10^{-5}$ V/K for $\epsilon_F = 1$ eV with $k_BT \approx 25$ meV and $ZT \approx S^2GT/\kappa_e = S^2/L \approx 0.001$.

4.4.4 *Examples of Simple Systems*

Here we present two examples by using NEGF method, electron conductance of carbon nanotubes (CNTs) and phonon conductance of SiNWs.

4.4.4.1 Conductance of carbon nanotubes

First, we treat a simple example of CNTs to show electronic states and conductance using the Green's function method with tight-binding approximation. A detailed description of the structures and electronic states of CNTs is given in Chapter 5. We mention that CNTs are characterized by two indexes (m, n) and there are two kinds of CNTs, metallic and semiconducting. The difference is whether there is a state in the Fermi energy at $\epsilon_F = 0$ or not.

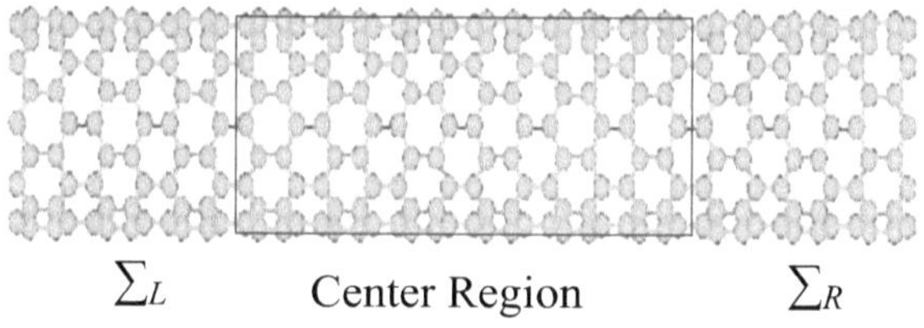

Figure 4.20 Schematic view of a structure of a CNT. We fix the center region as shown in the box and effects from the left and right outer regions are included in terms of self-energies $\Sigma_{L,R}$.

In the bulk CNTs, due to the periodic structures, we can take the center region for Green's function as shown in Fig. 4.20. Then the outer regions for the left and right sides are included in terms of self-energies. Using the surface Green's function technique, self-energies of CNT are quickly obtained. Figure 4.21 shows the density of states for the metallic (5, 5)-CNT and the semiconducting (10, 0)-CNT. We see that there are a number of van Hove singularities that originate from the one-dimensional structures.

Figure 4.22 shows the conductance as a function of energy for the metallic (left) and the semiconducting (right) CNTs. Since the bulk CNTs have no scattering, the conductance is precisely quantized

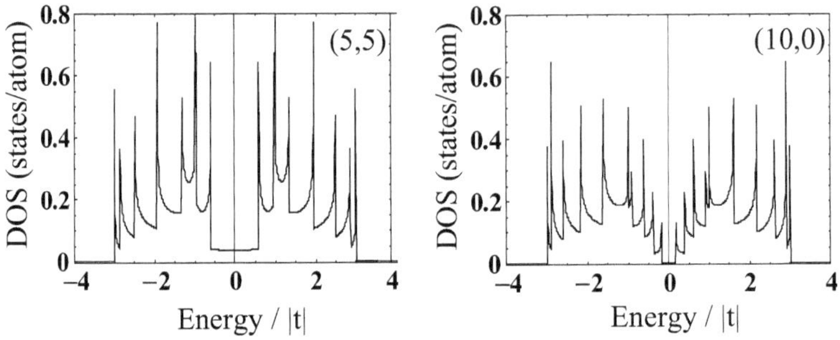

Figure 4.21 (Left) Density of states for the metallic (5, 5)-CNT. (Right) Density of states for the semiconducting (10, 0)-CNT.

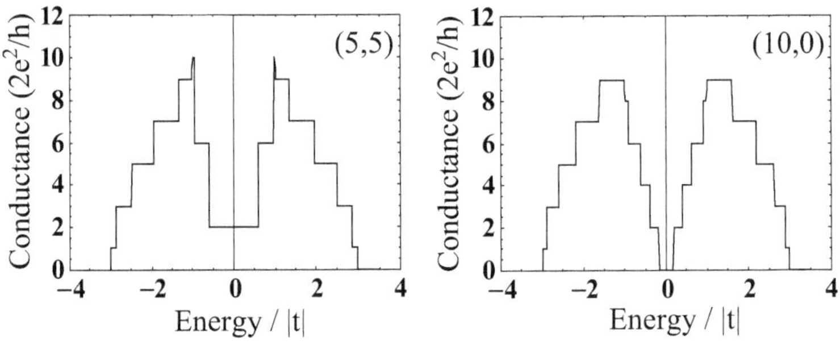

Figure 4.22 (Left) Conductance for the metallic (5, 5)-CNT. (Right) Conductance for the semiconducting (10, 0)-CNT.

as a multiple integer of $(2e^2/h)$, counting the number of bands at corresponding energies.

4.4.4.2 Thermal conductance of silicon nanowire

The second example is the phonon thermal conductance of SiNWs calculated using the NEGF formalism. Schematic pictures for the modeling is shown in the left panel of Fig. 4.23. The center panel of Fig. 4.23 shows the thermal conductance of SiNWs with and without defect vacancies. At 0 K the thermal conductance is zero since there are no phonons. As the temperature increases, a number of phonons with higher energies are excited and contribute to thermal

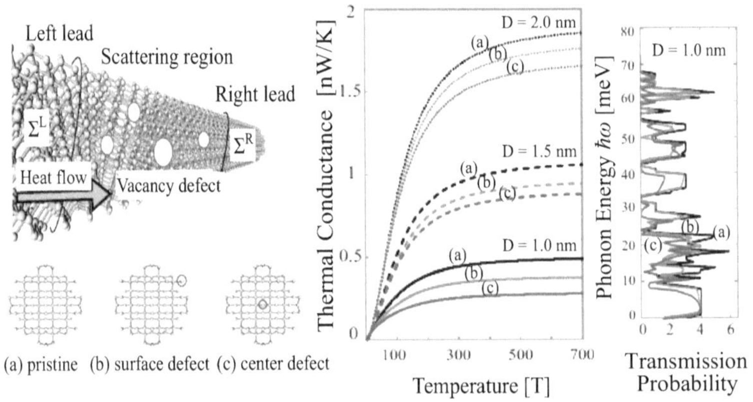

Figure 4.23 (Left) Schematic view of the atomistic model of SiNWs for ⟨100⟩ direction with a diameter of $D = 2.0$ nm. Vacancy defects are introduced in the scattering region. Atomistic configurations in diameter with and without vacancy defects are shown below. (Center) Temperature dependence of the phonon thermal conductance κ_ℓ of SiNWs with (a) no defects, (b) a surface defect, and (c) a center defect for different diameters of $D = 1.0$, 1.5, and 2.0 nm. Note that we do not take phonon–phonon scatterings into account; thus, thermal conductances show no drop at high temperatures. (Right) Corresponding transmission coefficients $\mathcal{T}_{ph}$ of SiNWs for $D = 1.0$ nm in diameter. From K. Yamamoto, H. Ishii, N. Kobayashi, and K. Hirose, *Appl. Phys. Express*, **4**, 085001 (2011) [80].

conductance with an occupation determined by the Bose–Einstein distribution. This is a very different nature from the electron conduction where the occupation is determined by the Fermi–Dirac distribution and only a small number of electrons around the Fermi level contribute to the electron conduction. We see that the thermal conductance saturates with a maximum phonon energy of $\hbar\omega \leq k_B T$ in the present ballistic limit, where $\hbar\omega \sim 70$ meV in the SiNWs as shown in the right panel of Fig. 4.23. Phonon–phonon Umklapp scattering process, which is absent here, reduces the thermal conductance significantly in the high temperature.

For SiNWs, the thermal conductance is affected from scatterings with vacancy defects. For all the diameters from $D = 1.0$ nm to $D = 2.0$ nm and for all the temperature region, pristine wire (a) has the highest thermal conductance and vacancy effect on thermal conductance is more significant in (c) a center defect than in (b) a surface defect. Transmission probability $\mathcal{T}_{ph}(\omega)$ in the right panel of Fig. 4.23 shows that $\mathcal{T}_{ph}(\omega)$ for a center defect decreases much more significantly than $\mathcal{T}_{ph}(\omega)$ for a surface defect at several specific energies, where details of phonon modes are described by the atomistic configurations.

4.5 Time-Dependent Wave-Packet Diffusion Method

In this section, we present a TD-WPD method to obtain transport properties for a huge system up to approximately 10^8 atoms from an atomistic point of views [81, 82]. The TD-WPD method utilizes wave-packet motion of an electron to obtain the diffusion constant $D(\epsilon, t)$ purely based on the quantum mechanics. Based on the Kubo formula, we can calculate transport properties of the conductance, including the effects of various scatterings with impurities and phonons. For example, the electron–phonon coupling is essential for nano-carbon materials at room temperature, for which we treat using the molecular dynamics simulation. With time-dependent wave-packet motion of electrons to combine with the molecular dynamics of atoms, we study electron transport properties with phonon scatterings, changing the behaviors from ballistic to diffusive regimes, and determine the scattering time τ and the mean free path ℓ_{mfp} from an atomistic point of view.

4.5.1 *Time-Dependent Diffusion Coefficient*

First, we explain how to obtain the electron dynamical motion for transport. Molecular dynamics simulations for the atom dynamical motion are described in the next section. We start to consider the conductivity σ based on the linear-response Kubo formula,

$$\sigma = 2\pi\hbar e^2 \sum_{n,m} \delta(\epsilon - \epsilon_n)\delta(\epsilon - \epsilon_m)\left|\langle n|\hat{v}_x|m\rangle\right|^2, \tag{4.230}$$

where $\hat{v}_x$ is the electron velocity operator along the x direction in the Schrödinger representation, $|n\rangle$ is an eigenstate of the Hamiltonian $\hat{H}$ with an eigenenergy ϵ_n, $\delta(\epsilon)$ is the delta function.

This conductivity expression is written in the time domain as

$$\sigma = 2\pi\hbar e^2 \sum_n \delta(\epsilon - \epsilon_n)\langle n|\hat{v}_x \delta(\epsilon - \hat{H})\hat{v}_x|n\rangle$$
$$= e^2 \sum_n \delta(\epsilon - \epsilon_n) \int_{-\infty}^{+\infty} \langle n|\hat{v}_x(t)\hat{v}_x(0)|n\rangle dt, \qquad (4.231)$$

where $\hat{v}_x(t)$ is the velocity operator at time t in the Heisenberg representation. Here we use the relation

$$\delta(\epsilon - \hat{H}) = \frac{1}{2\pi\hbar}\int_{-\infty}^{\infty} e^{-i\frac{(\hat{H}-\epsilon)}{\hbar}t} dt. \qquad (4.232)$$

Let us define the time-dependent diffusion coefficient $D(\epsilon, t)$ by the velocity correlation as

$$\int_{-\infty}^{+\infty} \langle n|\hat{v}_x(t)\hat{v}_x(0)|n\rangle dt = \lim_{t\to+\infty} \int_0^t 2\langle n|\hat{v}_x(s)\hat{v}_x(0)|n\rangle ds,$$
$$= \lim_{t\to+\infty} 2\frac{1}{2t}\int_0^t \int_0^t \langle n|\hat{v}_x(s)\hat{v}_x(s')|n\rangle ds ds',$$
$$= \lim_{t\to+\infty} \frac{\langle n|(\hat{x}(t) - \hat{x}(0))^2|n\rangle}{t}$$
$$\equiv \lim_{t\to+\infty} D(\epsilon_n, t), \qquad (4.233)$$

that is,

$$D(\epsilon_n, t) = \frac{\langle n|(\hat{x}(t) - \hat{x}(0))^2|n\rangle}{t}, \qquad (4.234)$$

where we use the relation of $\int_0^t \hat{v}_x(s)\, ds = \hat{x}(t) - \hat{x}(0)$ with $\hat{x}(t)$, the electron position operator along the x-axis in the Heisenberg representation. Since the density of states $\nu(E)$ is defined by

$$\nu(\epsilon) = \sum_n \delta(\epsilon - \epsilon_n), \qquad (4.235)$$

where we implicitly assume the spin degeneracy 2, the conductivity becomes

$$\sigma = e^2\nu(\epsilon) \lim_{t\to+\infty} D(\epsilon, t). \qquad (4.236)$$

We note that this formula is reduced to the well-known classical Einstein relation $\sigma = e^2\nu D$ for the diffusive transport when the time-dependent diffusion coefficient $D(\epsilon, t)$ approaches a constant value for $\lim_{t\to+\infty} D(\epsilon, t) \to D(\epsilon)$.

In the general case, we can take the maximum value as a diffusion coefficient for the conductivity,

$$\sigma = e^2 \nu(\epsilon) D_{\max}(\epsilon, t \to +\infty). \tag{4.237}$$

Then conductance, $G(L) \equiv (S/L)\sigma$, can be written as follows:

$$G(L) = \frac{S}{L} e^2 \nu(\epsilon) D_{\max}(\epsilon, t \to +\infty), \tag{4.238}$$

where the system length along x-axis is L and the cross sectional area is S, or equivalently the resistance $R(L) \equiv 1/G(L)$ is given by

$$R(L) = \frac{L}{S} \frac{1}{e^2 \nu(\epsilon) D_{\max}(\epsilon, t \to +\infty)}. \tag{4.239}$$

Here we consider how this formula derives various transport regimes. In the diffusive regime where the electrons scatter many times with impurities and phonons to converge to a constant value $D_{\max}(\epsilon, t \to +\infty) \to D(\epsilon)$. This is because the velocity correlation vanishes after the relaxation time τ with $\langle v(s)v(s')\rangle = \langle v^2\rangle \exp(-|s-s'|/\tau)$. Thus, $D(\epsilon) = \langle v^2\rangle\tau$ and the resistance is proportional to the system length L and is inversely proportional to S. This shows the classical Ohm's law. Although this formula looks quite simple, to obtain the resistance $R(L)$ from an atomistic point view is very difficult. This is because we need to evaluate the electron motions of $\hat{x}(t)$ in an environment with various scatterings. To obtain the convergent results for $D(\epsilon)$, we usually need to calculate the electron motions for a long time for huge systems, say millions of atoms. Therefore, in usual, since the diffusive regime at the Fermi energy is determined by the classical Ohm's law, the diffusion constant D is taken as a constant parameter and then its transport properties are described with it.

In the opposite limit where electrons move without any scatterings in the ballistic regime, the conductance G is determined by the number of channels N_c such that $G = (2e^2/h)N_c$. This is the well-known Landauer formula, which is also derived from the above equations as follows. Without scatterings, the electrons propagate at constant velocity v and thus the propagation length is equal to vt. The time-dependent diffusion coefficient $D(\epsilon, t)$ is therefore proportional to the time t as $D(\epsilon, t) = v^2 t = vL$. Since the density of states ν is $\nu = 2/(2\pi\hbar v/N_c)S$ where we use the density of states per one channel of length L is $2\pi\hbar v$, the conductance $G(L)$ becomes

$$G(L) = \frac{S}{L}e^2 \times \frac{2}{(2\pi\hbar v/N_c)S} \times vL = \frac{2e^2}{h}N_c, \tag{4.240}$$

independent of the length L in the ballistic regime. This limiting conductance in the ballistic regime is often calculated from an atomistic point of view by evaluating transmission coefficient T from the scattering matrix or Green's function and using the Landauer formula directly. However, the evaluation of scattering matrix or Green's function increases $O(N^3)$ with a systems size N and thus the size to be treated is limited to very small region. This prohibits us to study the larger system over mean free path (MFP) by the Landauer formula.

These observations show that if we can evaluate the time-dependent diffusion coefficient $D(\epsilon, t)$ from an atomistic point of views, we can obtain transport properties in both regimes without assuming the diffusion constant D or without using the Landauer formula directly. Furthermore, we can investigate an intermediate regime, i.e., quasi-ballistic regime, between these two regimes systematically.

4.5.2 *Calculation for Huge Atomic System*

Here we show how to evaluate the time-dependent diffusion coefficient $D(\epsilon, t)$ from an atomistic point of view. The essence of the method is that instead of calculating the numerical diagonalization of the matrix $\hat{H}$ of huge systems to obtain the eigenfunctions $|n\rangle$, which needs $O(N^3)$ calculations, we evaluate the time-dependent diffusion coefficient using *wave-packets* as follows:

$$D(\epsilon, t) \equiv \frac{\langle(\hat{x}(t) - \hat{x}(0))^2\rangle_\epsilon}{t}, \tag{4.241}$$

where we take the statistical average $\langle(\hat{x}(t)-\hat{x}(0))^2\rangle_\epsilon$ of the operator $\hat{A} = (\hat{x}(t) - \hat{x}(0))^2$ in Eq. (4.241) as an average value for several initial wave-packets $|\psi_n\rangle$, which includes information $|n\rangle$ for various energies,

$$\langle\hat{A}\rangle_\epsilon \equiv \frac{\sum_n \langle\psi_n|\delta(\epsilon - \hat{H})\hat{A}|\psi_n\rangle}{\sum_n \langle\psi_n|\delta(\epsilon - \hat{H})|\psi_n\rangle}. \tag{4.242}$$

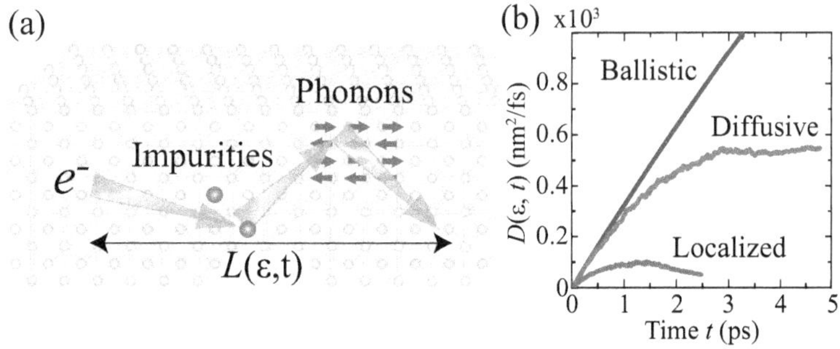

Figure 4.24 (a) Schematic view of the TD-WPD method. Electron wave packet propagates the length $L(\epsilon, t)$ in time t, under the scatterings from impurities and phonon vibrations. (b) Examples of the calculated time-dependent behaviors of diffusion coefficients in ballistic, diffusive, and localized regimes. Time-dependent diffusion coefficient $D(\epsilon, t)$ increases monotonically in the ballistic regime, while it saturates to a constant value in the diffusive regime. When the system is in the localized regime, the coefficient decreases to zero.

When we define the propagation length of an electron at the energy ϵ in time t as $L(\epsilon, t) \equiv \sqrt{\langle(\hat{x}(t) - \hat{x}(0))^2\rangle_\epsilon}$, the time-dependent diffusion coefficient $D(\epsilon, t)$ is represented by $D(\epsilon, t) = L(\epsilon, t)^2/t$.

We can show that if we use sufficient numbers of initial wave-packets, the time-dependent diffusion coefficient $D(\epsilon, t)$ approaches statistically to the one that is represented by complete sets of eigenfunctions $|n\rangle$. We will confirm the validity of this method for simple cases later. This is suitable for use of the parallel computing to perform the time-evolution calculation of each initial condition. We compute $\langle\psi_n|\delta(\epsilon - \hat{H})\hat{A}|\psi_n\rangle$ and $\langle\psi_n|\delta(\epsilon - \hat{H})|\psi_n\rangle$ separately using the Haydock's recursion method [83].

The schematic view of the TD-WPD method is shown in Fig. 4.24(a). An electron wave-packet is introduced into the system, and propagates the length $L(\epsilon_F, t)$ in time t under the scatterings by impurities and phonon vibrations. The time-dependent diffusion coefficient $D(\epsilon, t)$ as a function of time t shows various behaviors according to various transport regimes as described in the previous section and are shown schematically in Fig. 4.24(b). In the ballistic transport regime, $D(\epsilon, t)$ increases monotonically, while in the diffusive transport regime, where a number of scatterings due to impurities and energy-conserving electron–phonon couplings are present, $D(\epsilon, t)$ saturates to a constant value. When impurity scatterings are very strong, the wavefunctions show localized nature and thus $D(\epsilon, t)$ becomes zero due to the Anderson localization. The trajectories of wave-packets $\hat{x}(t)$ are calculated purely quantum-mechanically based on the time-evolution operator, which is presented in the next section.

Figure 4.25 shows the computing time of transport properties of CNTs as a function of the number of atoms. Here, we choose time step as $0.1 \times h/(1\text{eV}) \simeq 0.41$ fs and the total evolution time of a wave-packet is set as about 3 ps, corresponding to 7,200 steps. We can confirm that the $O(N)$ calculation is realized for the system of up to *80 million* atoms. The computing time of one time step can be evaluated as about 5×10^{-6} seconds per one atom. To help understand the length scale of realistic materials, the number of atoms for CNT with a 1 μm length and that for a graphene with a 1 μm×1 μm area are indicated by arrows, respectively. This indicates that we can directly compute transport properties from an atomic point of view and compare simulation results with experimental observations of samples that have the micron-order lengths. Furthermore, it is enough to obtain the convergent results for the diffusive regimes and thus the present method enables us to study electron transport properties from ballistic to diffusive regimes systematically. Note that the scattering methods or the NEGF method needs the $O(N^3)$ calculations for a matrix inversion, which limits the calculations up to about 10,000 atoms. The memory usage of the TD-WPD method again needs $O(N)$ due to the storage of non-zero components of the Hamiltonian, compared with the $O(N^2)$ for the NEGF method. Thus, the TD-WPD method requires low resources of computers.

Let us compare the time-evolution of wave packets using the traditional split-operator method instead of the Chebyshev polynomial development we explain later. The time-evolution operator is expressed in the split-operator method as $\exp(i\hat{H}_0 t/\hbar)$ for the time-independent Hamiltonian $\hat{H}_0$. Using the eigenfunctions $|n\rangle$ and the eigenenergies E_n of $\hat{H}_0$, we have the following analytical form for

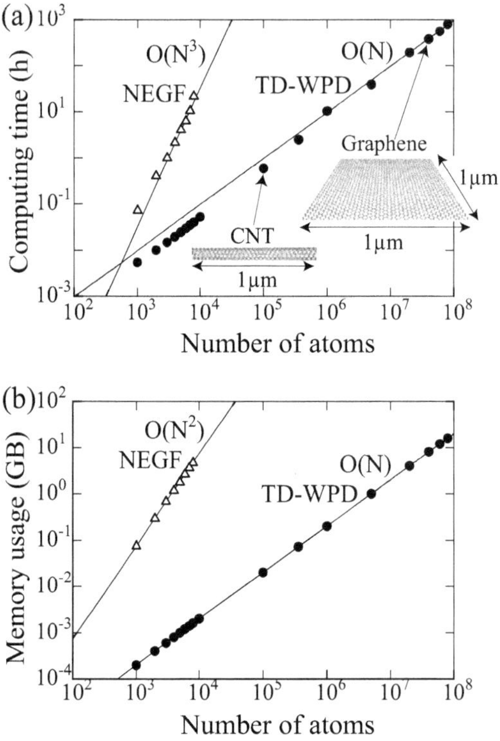

Figure 4.25 Logarithm plots of (a) computing times and (b) memory usage to calculate the time evolution of a wave packet in CNT as a function of the number of atoms using the TD-WPD method (black circles) and those by the NEGF method (white triangles). Time step is $0.1\times h/(1\text{eV}) \simeq 0.41$ fs and total evolution time of a wave-packet is set as about 3 ps, corresponding to 7,200 steps. The number of atoms for CNT with a 1 μm length and that for a graphene with a 1 μm×1μm area are indicated by arrows, respectively. Two lines in (a) represent $O(N^3)$ and $O(N)$ dependence, while $O(N^2)$ and $O(N)$ dependences are shown in (b). From H. Ishii, N. Kobayashi, and K. Hirose, *Phys. Rev. B* **82** 085435 (2010) [82].

the time-dependent diffusion coefficient defined by Eq. (4.234):

$$D(\epsilon_n, t) = \frac{\langle n|\hat{z}^2|n\rangle - \sum_m \langle n|\hat{z}|m\rangle\langle m|\hat{z}|n\rangle \cos\omega_{nm}t}{4t}, \qquad (4.243)$$

where $\omega_{nm} \equiv (\epsilon_n - \epsilon_m)/\hbar$. We can obtain transport properties from Eq. (4.243) after the maximum value of $D(\epsilon_n, t)$ is calculated. The computing time is mainly dominated by the $O(N^3)$ calculation due to the numerical diagonalization of the matrix $\hat{H}_0$, which cannot be exempted in the split-operator method. The computing time dramatically increases as the number of atom increases and the limit of split operator method is about 10,000-atom systems due to the $O(N^3)$ computation procedures; thus only the quasi-ballistic transport regime can be at most investigated.

4.5.2.1 Hall conductivity calculation

Let us extend the TD-WPD method to treat the carrier transport in a magnetic field, which enables us to obtain the magnetoconductivity and the Hall mobility [84]. The measurements of the Hall effects have been utilized to clarify the fundamental transport properties in various materials. The Hall coefficient R_H provides us information on the carrier density, the carrier type for electron and hole, and the charge of a carrier. The classical treatment of the Hall effect is described in Chapter 2 and here we formulate the Hall conductivity σ_{xy} and the Hall coefficient R_H using the TD-WPD method.

The off-diagonal part $\sigma_{xy}(\epsilon_F)$ for calculations of the Hall conductivity is obtained from an integral with respect to λ of the generalized Kubo formula with the two-velocities correlation function in the real-time domain

$$\sigma_{\zeta\xi}(\epsilon_F) = e^2 \int_0^{1/k_B T} d\lambda \int_0^{+\infty} dt \, \langle \hat{v}_\xi(0)\hat{v}_\zeta(t + i\hbar\lambda)\rangle. \qquad (4.244)$$

This yields the following expression:

$$\sigma_{xy}(\epsilon_F) = -e^2 \lim_{\eta \to +0} \int_0^{+\infty} dt \sum_{n=1}^{N} f(\epsilon_n - \epsilon_F) \times 2\mathrm{Re}\left(\langle n|\hat{v}_y e^{+i\frac{\hat{H}}{\hbar}t} \frac{1}{\epsilon_n - \hat{H} + i\eta} \hat{v}_x e^{-i\frac{\hat{H}}{\hbar}t}|n\rangle \right), \qquad (4.245)$$

where $|n\rangle$ is the eigenvector of $\hat{H}$ with the eigenvalue ϵ_n. Here $\hat{\rho}_0$ is the density matrix in equilibrium and $\hat{v}_\xi$ is the electron velocity operator along the ξ direction, which is obtained from the Heisenberg equation $i\hbar\hat{v}_\xi(t) = [\hat{\xi}(t), \hat{H}]$ where $\hat{\xi}$ and $\hat{H}$ represent the position operators and the Hamiltonian for electrons, respectively. Note that the diagonal part σ_{xx} represents the previous result of $\sigma_{xx} = e^2\nu(\epsilon) \lim_{t\to+\infty} D(\epsilon, t)$.

The first step is to eliminate the eigenvalue dependence by an insertion of energy integral of the delta function $1 = \int_{-\infty}^{+\infty} \delta(\epsilon - \epsilon_n)d\epsilon$. Then replacing the set of all eigenvectors with respect to any complete orthonormal basis set, we get eigenvector-free expression of the off-diagonal part of the Hall conductivity as

$$\sigma_{xy}(\epsilon_F) = -e^2 \lim_{\eta\to+0} \int_0^{+\infty} dt \int_{-\infty}^{+\infty} d\epsilon \; f(\epsilon - \epsilon_F) \times \sum_{n,m} 2\mathrm{Re}\Bigg(\langle\psi_n|\delta(E - \hat{H})|\psi_m\rangle \times \langle\psi_m|\hat{v}_y e^{+i\frac{\hat{H}}{\hbar}t} \frac{1}{\epsilon - \hat{H} + i\eta} \hat{v}_x e^{-i\frac{\hat{H}}{\hbar}t}|\psi_n\rangle \Bigg), \quad (4.246)$$

where $|\psi_n\rangle$ is set as an initial wave packet. This shows that $\sigma_{xy}(\epsilon_F)$ is obtained without using the direct eigenvectors. We note that the present TD-WPD method is suitable for use of the parallel computing, because we can perform the time-evolution calculation of "each" initial wave packets independently.

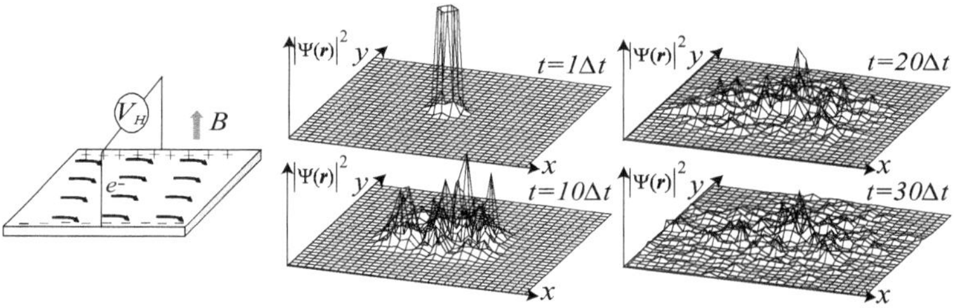

Figure 4.26 (Left) Schematic picture of the Hall effect. When B is applied perpendicular to the sample, electrons feel the Lorentz force and accumulate at the edges of the sample. As a result, there appears an electric field and a Hall voltage V_H. (Right) Snapshots of time-evolution behavior of a wave packet in the square lattice with static disorders under a magnetic field with a magnetic flux. From H. Ishii, N. Kobayashi, and K. Hirose, *Phys. Rev. B* **83** 233403 (2011) [84].

When the electric current j_x flows under a magnetic field B, the Hall coefficient R_H is defined by

$$R_H \equiv \frac{E_y}{j_x B} = \frac{\sigma_{xy}}{(\sigma_{xx}^2 + \sigma_{xy}^2)B}, \tag{4.247}$$

where the Hall field E_y appears due to transverse electromotive force. Then the diagonal conductivity $j_x = \sigma E_x$ is given by $\sigma \equiv (\sigma_{xx}^2 + \sigma_{xy}^2)/\sigma_{xx}$. In the classical model, the Hall coefficient is identical to an inverse of carrier density through $R_H = 1/nec$, where n represent the carrier density as we see in Chapter 2.

Figure 4.26 shows an example of the time-evolution behavior of a wave packet in the square lattice with static disorders of $W/\gamma^0 = 3.2$ under an external magnetic field with magnetic flux $\phi \equiv Ba^2 = (1/30) \times \phi_0$ perpendicular to the square lattice, where W, γ^0, a are parameters for disorder, transfer energy, lattice constant, and $\phi_0 \equiv ch/e$ is the magnetic flux quantum. We take snapshots of the probability density every $10\Delta t$ and see the behaviors of a wave packet in disorder systems under an external magnetic field, from which we can compute the diagonal and off-diagonal conductivities.

4.5.3 *Propagators and Molecular Dynamics*

Here we consider the procedures of time integration for the dynamics of electrons, which are used for the TD-WPD method as well as the time-dependent Kohn–Sham equation of the TDDFT and the time-dependent density matrix equation described later. Also, we consider the procedures for the calculations of atomic motions by the molecular dynamics, which is used for the time-dependent scattering potentials as phonons in the TD-WPD method.

4.5.3.1 Electron dynamics

To obtain electric current based on the time-dependent DFT formalism, we need to solve the time-dependent Kohn–Sham equation

$$i\hbar\frac{d}{dt}\psi_i(\mathbf{r}, t) = \hat{H}_{KS}(\mathbf{r}, t)\psi_i(\mathbf{r}, t), \tag{4.248}$$

where the Kohn–Sham Hamiltonian is

$$\hat{H}_{KS}(\mathbf{r}, t) = -\frac{\hbar^2}{2m}\nabla^2 + V_{\text{ext}}(\mathbf{r}, t) + e^2 \int \frac{n(\mathbf{r}', t)}{|\mathbf{r} - \mathbf{r}'|} d\mathbf{r}' + \frac{\delta E_{xc}}{\delta n(\mathbf{r}, t)}. \tag{4.249}$$

The solution to this differential equation is, taking a small time step Δt_i, expressed by the propagator

$$\hat{S}(t_i + \Delta t_i, t_i) = T \exp\left[-\frac{i}{\hbar}\int_{t_i}^{t_i+\Delta t_i} \hat{H}_{KS}(t')dt'\right] \tag{4.250}$$

and

$$\hat{S}(t, 0) = \prod_{i=0}^{N-1} \hat{S}(t_i + \Delta t_i, t_i) \tag{4.251}$$

such as

$$\psi_i(t) = \hat{S}(t, 0)\psi_i(0). \tag{4.252}$$

Here T is the time-ordering operator. By dividing the time into small intervals, we can reduce the time-dependence of the Hamiltonian $\hat{H}_{KS}$, keeping the norm and unitarity of the matrix and approximate the propagator at the value between $[t_i, t_i + \Delta t_i]$. Usually the time step is taken as a constant Δt, but it is possible to take a variable time-step. We note that the time step Δt must be smaller than the maximum frequency of the system $\Delta t_{\max} < 1/\omega_{\max}$.

If we take a smaller Δt to increase its accuracy, the computational time increases linearly with the time step of $t/\Delta t$. So we need to find the most effective way to choose $\Delta t_{\max}$ and to reduce the computational procedure for the propagator. We will show several ways to approximate the propagators. The most effective method depends on the system for which we need to choose according to the situations.

The simplest one is to use the first-order approximation

$$\hat{S}(t + \Delta t, t) = \hat{I} - \frac{i}{\hbar}\hat{H}_{KS}(t)\Delta t. \tag{4.253}$$

However, this approximation is usually unstable for any Δt, because the eigenvalues of this operator $(1 - i\epsilon_i \Delta t/\hbar)$ have moduli $(1 + \epsilon_i^2 \Delta t^2/\hbar^2)^{1/2}$ larger than the unity. This implies that the operator that have eigenvalues with moduli of $(1 + \epsilon_i^2 \Delta t^2/\hbar^2)^{-1/2}$ is always stable for any Δt, but this does not keep the unitarity of the matrix and the norm of the wavefunction continually decreases with time. The suitable way is using the second-order approximation known as Crank–Nicholson method as

$$\hat{S}(t+\Delta t, t) = \frac{\hat{I} - \frac{i}{\hbar}\frac{1}{2}\hat{H}_{KS}(t+\frac{\Delta t}{2})\Delta t}{\hat{I} + \frac{i}{\hbar}\frac{1}{2}\hat{H}_{KS}(t+\frac{\Delta t}{2})\Delta t}. \tag{4.254}$$

This propagator is unitary and preserves time-reversal symmetry. The propagation is proceeded to solve the linear equation

$$\left[\hat{I} + \frac{i}{\hbar}\frac{1}{2}\hat{H}_{KS}\left(t+\frac{\Delta t}{2}\right)\Delta t\right]\psi_i(t+\Delta t) = \left[\hat{I} - \frac{i}{\hbar}\frac{1}{2}\hat{H}_{KS}\left(t+\frac{\Delta t}{2}\right)\Delta t\right]\psi_i(t), \tag{4.255}$$

which needs propagating backward $\Delta t/2$ from $\psi(t+\Delta t)$ or forwards $\Delta t/2$ from $\psi(t)$ to obtain the state $\psi(t+\Delta t/2)$.

The second method is the split-operator method. In the present case, the Hamiltonian is split into the kinetic part $\hat{T} = -\hbar^2\nabla^2/2m$, which is diagonal in reciprocal space, and the effective potential part $\hat{V} = V_{\text{ext}} + V_H + V_{xc}$, which is diagonal in real space. The

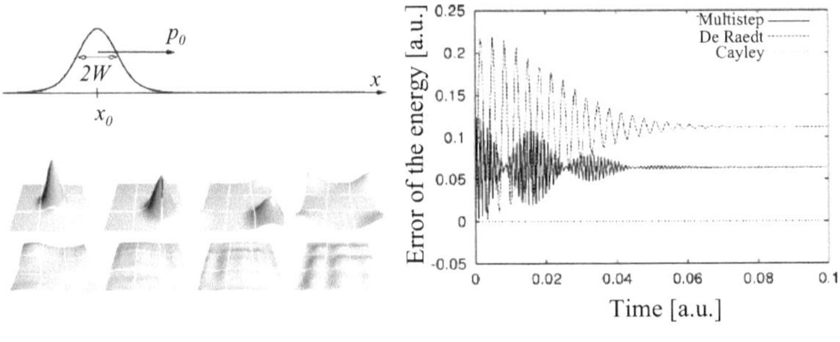

Figure 4.27 (Left) Model system for which a Gaussian wave packet is put in the system, whose initial average location x_0 and momentum p_0 are set at $x_0 = 2.0$ a.u. and $p_0 = 12.0$ a.u., respectively. The bottom panel shows evolution of the density. Time flows left to right, top row first. The Gaussian is observed to go through these areas smoothly. (Right) Time variances in energies computed by three methods, multistep, De Raedt's, and Cayley's. Energies violently oscillate in the cases of multistep method and De Raedt's method, whereas the energy is conserved exactly in the Cayley's case. From N. Watanabe and M. Tsukada, *Phys. Rev. E* **62**, 2914 (2000) [85].

approximation of the propagator in the second order is

$$\exp\left[-\frac{i}{\hbar}\hat{H}_{KS}\Delta t\right] = \exp\left[-\frac{i}{\hbar}\frac{\Delta t}{2}\hat{T}\right]\exp\left[-\frac{i}{\hbar}\Delta t\hat{V}\right]\exp\left[-\frac{i}{\hbar}\frac{\Delta t}{2}\hat{T}\right], \tag{4.256}$$

neglecting the higher order from the commutator $\left[\hat{T}, \hat{V}\right]$. These three exponentials may be computed exactly in real and reciprocal spaces and thus this propagator keeps the unitarity and stable. In the TDDFT, we use $\hat{V} = \hat{V}(t + \Delta t/2)$ as

$$\hat{V} = V_{\text{ext}}\left(\mathbf{r}, t + \frac{\Delta t}{2}\right) + e^2\int\frac{\tilde{n}(\mathbf{r}')}{|\mathbf{r} - \mathbf{r}'|}d\mathbf{r}' + \frac{\delta E_{xc}}{\delta\tilde{n}(\mathbf{r})}, \tag{4.257}$$

where $\tilde{n}$ is electron density constructed after we calculate the first kinetic exponent. This splitting technique is proposed in [85]. Figure 4.27 (left) shows an example of this method with a model system for which a Gaussian wave packet is put in the 2D system and its evolution of the density. The right panel of Fig. 4.27 shows time variances in energies computed by the three methods, multistep, De Raedt's, and Cayley's. The energies violently oscillate in the cases of the multistep method and De Raedt's method, whereas the energy is conserved exactly in the Cayley's case.

(a)

6th q=14 (p_i=p_{15-i})		8th q=15 (p_i=p_{16+i})			
p_1=p_2	0.392256805387732	p_1	0.210902950774054	p_5	0.769771783843536
p_3=p_4	0.01177866066796810	p_2	0.835657990415923	p_6	0.199415314882502
p_5=p_6	-0.5888399920894384	p_3	-0.658440728286576	p_7	0.0363152045812476
p_7=p_8	0.06575931603419684	p_4	-0.0774373402366569	p_8	-0.0892171910150694

(b)

6th q=7 (p_j=p_{8-j})	
p_1	0.784513610477560
p_2	0.0235573213359357
p_3	-1.17767998418887
p_4	1.31518632068391

Figure 4.28 Parameters set for the sixth- and eighth-order Suzuki–Trotter decomposition. Parameters determined by Suzuki satisfy $|p_i| < 1$.

The higher order approximation is known as the Suzuki–Trotter scheme. For example, the fourth approximation becomes

$$\exp\left[-\frac{i}{\hbar}\hat{H}_{KS}\Delta t\right] = \prod_{i=1}^{5} S_2(-\frac{i}{\hbar}p_i\Delta t\hat{H}_{KS}) \quad (4.258)$$

with $p_1 = p_2 = p_4 = p_5 = 1/(4 - 4^{1/3})$ and $p_3 = 1 - 4p_1$. Here S_2 is the right part of Eq. (4.256). As for the higher expansion, we give parameters for the sixth and the eighth decomposition in Fig. 4.28 [86]. With these parameters, the potential is extrapolated between t and $t + \Delta t$ properly and the Suzuki–Trotter scheme is applied to solve the time-dependent Kohn–Sham equation efficiently. These split-operator methods have an advantage when we use plane-wave basis sets, due to the FFT for $\hat{T}$ and $\hat{V}$.

The third method is to expand its exponent by the special functions. It is effective for the trajectories of wave-packets $\hat{x}(t)$ in TD-WPD method by the Chebyshev polynomial approximations, which are optimal for approximating functions such as

$$\exp\left[-\frac{i}{\hbar}\hat{H}_{KS}\Delta t\right] = \sum_{n=0}^{\infty} \exp\left(i\frac{a}{\hbar}\Delta t\right) h_n i^n J_n\left(\frac{b\Delta t}{\hbar}\right) T_n\left(\frac{\hat{H}_{KS} - a}{b}\right). \quad (4.259)$$

Here J_n is the Bessel functions and T_n is the Chebyshev polynomials of the first kind [87]. This calculation method is effective when we use the localized basis sets, which lead to the order-N ($O(N)$) and very high-speed calculation in the TD-WPD method. Energy spectrum of $\hat{H}$ has an interval $[a - b, a + b]$, $h_0 = 1$ and $h_n = 2$ ($n \geq 1$). Since the Chebyshev polynomials for an operator $\hat{x}$ obey the following simple recursive relation:

$$\begin{aligned} &T_{n+1}(\hat{x}) = 2\hat{x}T_n(\hat{x}) - T_{n-1}(\hat{x}) \qquad \text{for } n \geq 1, \\ &\text{with } \; T_0(\hat{x}) = 1, \;\; \text{and } \; T_1(\hat{x}) = \hat{x}, \end{aligned} \quad (4.260)$$

we can reduce the number of the expansion n for J_n by taking a small time step Δt in Eq. (4.259). This is because the higher order terms of J_n decreases to zero rapidly as n increases and the recursive procedure of T_n is much reduced.

We note that in case of the time-dependent QME method where we calculate the time evolution of the density matrix, these numerical propagators are important for the calculations of time evolutions.

4.5.3.2 Molecular dynamics

Here we consider the numerical method of classical molecular dynamics simulation for the atomic motions $\mathbf{R}_i$ of i-th atom in a finite temperature [88].[a]

This is based on solving the $3N$ coupled Newton equations for the motions of each atom

$$M_i \frac{d^2\mathbf{R}_i}{dt^2} = -\frac{dU(\mathbf{R}_1 \cdots \mathbf{R}_N)}{d\mathbf{R}_i} \quad (i = 1 \cdots N), \tag{4.261}$$

where M_i is a mass, $-\partial U/\partial \mathbf{R}_i (= \mathbf{F}_i(t))$ is the force acting on the i-th atom, and N is the total number of atoms.

There are various empirical forms that reduce the computational cost significantly. A simple example is the Lennard–Jones potential, two-body interatomic form represented by

$$U = \sum_{ij(i \neq j)}^{N} 4\epsilon \left[\left(\frac{\sigma}{R_{ij}} \right)^{12} - \left(\frac{\sigma}{R_{ij}} \right)^{6} \right], \tag{4.262}$$

where R_{ij} is the distance between atom i and atom j, ϵ and σ are the adjustable parameters. The Lennard–Jones potential is used for rare gases, liquids, and metals as a qualitative model.[b]

For the metallic atoms, the other empirical form is the embedded atom method (EAM) [89], which is constructed from the sum of one-body and two-body terms represented by

$$U = \sum_{i}^{N} U_i [\rho(R_i)] + \frac{1}{2} \sum_{ij(i \neq j)}^{N} U_{ij}(R_{ij}), \tag{4.263}$$

where U_i is an energy to embed electron density ρ at the i-th atom from the remaining atoms $\rho(R_i) = \sum_{j(\neq i)} \rho(R_{ij})$. This form is widely used for the metallic systems, such as Au, Pt, Cu, and W.

When we consider materials with the covalent bonds such as silicon and carbon, we need to take three-body potential $U(R_{ij}, R_{ik}, \theta_{jik})$. The Stillinger–Weber potential [90] is an example and is used for a silicon atom.

---

[a] Quantum treatment of the atom dynamics becomes essential for an atom with a small mass, such as hydrogen, deuterium, etc.

[b] The Morse potential $U = \sum_{ij(i \neq j)}^{N} D_e \left[1 - \exp(-a(R_{ij} - R_0))\right]^2$ with $a = \omega_e \sqrt{M/D_e}$ is used for the molecular systems as a qualitative model. Here D_e is the dissociation energy, ω_e is the frequency, and R_0 is the equilibrium distance.

The Tersoff potential [91] is another empirical potential for materials with the covalent bonds. It is represented by

$$U = \sum_{ij(i \neq j)}^{N} f_c(R_{ij}) \left[a_{ij} e^{-\lambda_1 R_{ij}} - b_{ij} e^{-\lambda_2 R_{ij}} \right], \tag{4.264}$$

where $f_c(R)$ is a cutoff function

$$f_c(R) = \begin{cases} 1 & \text{for } R < R^{(1)}, \\ \frac{1}{2}\left(1 + \cos \frac{\pi(R - R^{(1)})}{R^{(2)} - R^{(1)}}\right) & \text{for } R^{(1)} \leq R \leq R^{(2)}, \\ 0 & \text{for } R > R^{(2)}. \end{cases} \tag{4.265}$$

Three-body effects are included in the attractive force term $b_{ij} = \left(1 + \beta^n \xi_{ij}^n\right)^{-1/2n}$ where parameters ξ_{ij} and $g(\theta)$ are expressed by

$$\xi_{ij} = \sum_{k(k \neq ij)}^{N} f_C(R_{ik}) g(\theta_{ijk}) e^{\lambda_3^3 (R_{ij} - R_{ik})^3}$$

$$g(\theta) = 1 + \frac{c^2}{d^2} - \frac{c^2}{d^2 + (1 - \cos\theta)^2}. \tag{4.266}$$

The Tersoff potential is used for various semiconductor materials with the tetrahedral structures.

The Brenner potential [92] is a modified version of the Tersoff potential and is also used for the atom–atom interactions of semiconductor materials with the following form:

$$U = \sum_{ij(i \neq j)}^{N} f_c(R_{ij}) \left[U_R(R_{ij}) - \bar{b}_{ij} U_A(R_{ij}) \right]. \tag{4.267}$$

Here $U_{R,A}(R_{ij})$ are the repulsive Morse-type (R) and attractive (A) force terms given by

$$U_R(R) \equiv \frac{D^{(e)}}{S - 1} \exp\left(-\sqrt{2S}\beta(R - R^{(e)})\right)$$

$$U_A(R) \equiv \frac{D^{(e)} S}{S - 1} \exp\left(-\sqrt{\frac{2}{S}}\beta(R - R^{(e)})\right). \tag{4.268}$$

$D^{(e)}$, S, β, and $R^{(e)}$ are the empirical parameters for atom–atom bonds. The empirical bond-order function, $\bar{B}_{ij}$, represents a many-body coupling depending on the bond angle θ_{ijk} between i–j bond and the local environmental i–k bonds,

$$\bar{b}_{ij} = \frac{b_{ij} + b_{ji}}{2},$$

$$b_{ij} = \left\{ 1 + a_0 \sum_{k(\neq i, j)} f_c(R_{ik}) g(\theta_{ijk}) \right\}^{-\delta}. \tag{4.269}$$

Table 4.1 Empirical parameters of molecular dynamics simulation for C–C bonds with $a_0 = 0.00020813$ in thc Brenner potential

$R_{cc}^{(e)}$	$R_{cc}^{(1)}$	$R_{cc}^{(2)}$	$D_{cc}^{(e)}$	β_{cc}	S_{cc}	δ_{cc}	c^2	d^2
1.39	1.7	2.0	6.0	2.1	1.22	0.5	330^2	3.5^2

Note: $R_{cc}^{(e),(1),(2)}$ are in units of Å, $D_{cc}^{(e)}$ is in unit of eV, and β_{cc} is in unit of Å^{-1}. These reproduce the stretching force constants well.

We show the parameters for carbon–carbon bonds in Table 4.1.

Although these empirical forms of interatomic potentials are constructed for the specific kinds of atoms, the first-principles calculations with DFT is a general way for obtaining the interatomic potential function U. We note that it is possible to extract the higher-order force constants $k_{ij}, t_{ijk}, \cdots$ in Eq. (4.205) systematically from the data obtained from the first-principles total-energy calculations [93]. This enables us in general to perform the molecular dynamics simulations by using the parameters for any kind of atoms, although their accuracy must be checked carefully. This method is applicable to the thermal transport problems using the molecular dynamics or other methods.

The most accurate and reliable molecular dynamics simulations are based on the combination of the first-principles calculations for electrons with the molecular dynamics for atoms, which is called the ab initio molecular dynamics. In the Car-Parrinello (CP) method [58] widely used, electron motion is solved with the classical atomic motions simultaneously. The Lagrangian of the system is expressed in the framework of DFT by

$$L = \mu \sum_i \langle \dot{\psi}_i(t) | \dot{\psi}_i \rangle + \frac{1}{2} \sum_i M_i \dot{\mathbf{R}}_i^2 - E(\{\psi_i\}, \{\mathbf{R}_i\}) + \sum_{i \neq j} \left(\psi_i(t) \psi_j(t) - \delta_{ji} \right) \lambda_{ji}, \qquad (4.270)$$

where the Kohn–Sham Hamiltonian H_{KS} is given from the total energy functional E by $H_{KS}\psi_i = \delta E(\{\psi_i\}, \{\mathbf{R}_i\})/\delta\psi_i^*$. This

Lagrangian provides the atomic and electronic motions as

$$M_i \frac{d^2\mathbf{R}_i}{dt^2} = -\frac{dE(\{\psi_i\},\{\mathbf{R}_i\})}{dR_i}$$

$$\mu \frac{\partial^2 \psi_i(t)}{\partial t^2} = -H_{KS}\psi_i(t) + \sum_j \psi_j(t)\lambda_{j,i} \qquad (4.271)$$

with the constraint Lagrange multipliers of $\lambda_{i,j} = \langle\psi_i|\hat{H}_{KS}|\psi_j\rangle - \mu\langle\dot{\psi}_i|\dot{\psi}_j\rangle$. Here μ is a fictitious mass, which controls the dynamical response of $\psi_i(\mathbf{r})$. The integration of electron motions for the second-order differential equation in time domain is to use the Verlet method as

$$\psi_i(t+\Delta t) = -\psi_i(t-\Delta t) + 2\psi_i(t) - \frac{\Delta t^2}{\mu}\left(H_{KS}\psi_i(t) - \sum_j \psi_j(t)\lambda_{ji}\right)$$

with a normalization of $\langle\psi_i(t+\Delta t)|\psi_j(t+\Delta t)\rangle = \delta_{ij}$. Presently the CP method is computationally too demanding to apply to the transport problems for electrons and phonons.

Numerical integrations for equation of atomic motions are performed variously. Here we present the Verlet algorithm and its extension to the velocity Verlet algorithm. In the Verlet algorithm, the position $\mathbf{R}_i$ is computed on the Taylor expansion[a] as

$$\mathbf{R}_i(t+\Delta t) = 2\mathbf{R}_i(t) - \mathbf{R}_i(t-\Delta t) + \left(\frac{\mathbf{F}_i(t)}{M_i}\right)\Delta t^2 + O(\Delta t^4) \qquad (4.272)$$

with an approximate form for the velocity $\mathbf{V}_i$

$$\mathbf{V}_i(t) = \frac{(\mathbf{R}_i(t+\Delta t) - \mathbf{R}_i(t-\Delta t))}{2\Delta t}. \qquad (4.273)$$

When we give the initial positions $\mathbf{R}_i(t)$ at $t = 0$ and $t = \Delta t$, this method provides the positions at arbitrary times t. However, the accuracy of velocity $\mathbf{V}_i(t)$ becomes worse as $\Delta t \to 0$.

The velocity Verlet algorithm is an improved version to evaluate the position $\mathbf{R}_i(t)$ and the velocity $\mathbf{V}_i(t)$ on the same footing as

$$\mathbf{R}_i(t+\Delta t) = \mathbf{R}_i(t) + \mathbf{V}_i(t)\Delta t + \frac{\mathbf{F}_i(t)}{2M_i}\Delta t^2 + O(\Delta t^3) \qquad (4.274)$$

$$\mathbf{V}_i(t+\Delta t) = \mathbf{V}_i(t) + \frac{\mathbf{F}_i(t)}{M_i}\Delta t + \frac{1}{2M_i}\frac{d\mathbf{F}_i(t)}{dt}\Delta t^2 + O(\Delta t^3)$$
$$= \mathbf{V}_i(t) + \frac{\mathbf{F}_i(t+\Delta t) + \mathbf{F}_i(t)}{2M_i}\Delta t + O(\Delta t^3). \qquad (4.275)$$

[a] Here we abbreviate to write the force as $\mathbf{F}_i(t) = \sum_{j=1}^{N} F_{ij}$ at time t.

The computational procedures are as follows. First, we give the initial positions $\mathbf{R}_i(t)$ and the velocity $\mathbf{V}_i(t)$ at $t = 0$, and the force $\mathbf{F}_i(t)$, which is a function of positions. Next, we compute the position $\mathbf{R}_i(t + \Delta t)$ from Eq. (4.274) and then the force $\mathbf{F}_i(t + \Delta t)$. These are used to obtain $\mathbf{V}_i(t + \Delta t)$ from Eq. (4.275). These procedures are repeated and we obtain $\mathbf{R}_i(t)$ and $\mathbf{V}_i(t)$ for arbitrary time t. It should be noted that these calculations are stable and accurate for both $\mathbf{R}_i(t)$ and $\mathbf{V}_i(t)$ with an order of Δt^3.

To proceed with the molecular dynamics simulation, we need to impose the conditions such as constant temperature and constant pressure, which are known as the thermostats. This is because we assume the statistics for ensembles with a constant energy in the molecular dynamics simulations, while in experiments we need the canonical ensembles. For constant temperature T, the simplest way is to use a velocity scaling. We prepare initial velocities by an equilibrium of Maxwell's velocity distribution, fix the temperature T constant, and scale the velocities at each time step to satisfy

$$\frac{3}{2}Nk_B T = \sum_{i=1}^{N} \frac{1}{2} M_i V_i^2(t) \tag{4.276}$$

in the molecular dynamics simulation. We note that the Nose–Hoover thermostat [94, 95] is frequently utilized for the more sophisticated method, which introduces an auxiliary variable in the Lagrangian to construct the equation of motions for keeping the temperature.

Molecular dynamics simulations are used variously for transport problems. One example[a] is the phonon dispersion as an energy density in (ω, q)-space. Taking the 2D Fourier spectra as

$$E_{\text{ph}}(\omega, q) \propto \sum_{n=1}^{N} \sum_{\alpha=x,y,z} \left| \int dt\, V_{n\alpha}(t) \exp\left(iqZ_n^0 - i\omega t\right) \right|^2, \tag{4.277}$$

where ω and q represent the phonon frequency and wavevector, $V_{n\alpha}$ and Z_n^0 are α-component of the velocity and an equilibrium position along the z-axis from the molecular dynamics simulations.

---

[a] The other examples are the thermal conductivity described in Chapter 2 and obtaining the pair correlation function $g(r)$ for an atom at a given distance from other atoms $g(r) = \left\langle \sum_{ij(i \neq j)}^{N} \delta(r - |\mathbf{R}_i - \mathbf{R}_j|) \right\rangle / 4\pi r^2 N^2$.

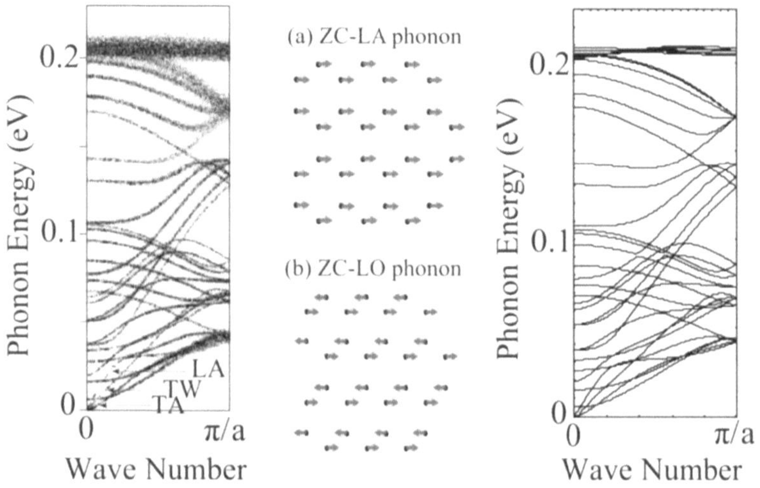

Figure 4.29 (Left) Phonon dispersion relation of the metallic (5, 5)-CNT from the molecular dynamics. The length of a unit cell is represented by a. Doubly degenerate transverse acoustic (TA) modes, twist (TW) mode, and longitudinal acoustic (LA) mode are shown by arrows. (Center) Schematic atomistic views of phonon modes corresponding to (a) acoustic and (b) optical phonons at zone-centers. (Right) Phonon dispersion relation from the direct diagonalization of dynamical matrix.

Let us show the phonon dispersion relation from molecular dynamics simulations for CNT with a 250 nm (1, 000 unit cells) with periodic boundary conditions. In the left panel of Fig. 4.29, we show the phonon dispersion relation obtained from Eq. (4.277). This dispersion reproduces very well with the calculated phonon dispersions obtained from the diagonalization of the dynamical matrix as shown in the right panel. There are four acoustic modes at a low frequency (although we can see three modes), i.e., transverse acoustic (TA) modes, which are doubly degenerate, longitudinal acoustic (LA) mode, and twisting (TW) mode, which is related to the rotation of CNT along its axis.

4.5.4 *Examples of Simple Systems*

Here we will show the validity of TD-WPD method for simple cases using the tight-binding approximation where on-site energies and nearest-neighbor hopping elements are present in the Hamiltonian. We use a metallic (5, 5)-CNT and a semiconducting (10, 0)-CNT for the first two subsections to check conductance and mean free path. For the third subsection on the check of the Hall mobility, we use 2D square lattice model with a magnetic field perpendicular to the lattice.

4.5.4.1 Conductance in ballistic limit

First, we compute the conductance in the ballistic transport limit, where all the scatterings are absent and we expect to obtain the quantized conductance. We take $\gamma^0 = 2.5$ eV for the transfer energy and $a_{cc} = 1.44$ Å for carbon–carbon bond length a_{cc}, and calculate the conductance of a metallic (5, 5)-CNT and a semiconducting (10, 0)-CNT. Energy-dependent conductances obtained from the TD-WPD method are shown by solid lines in Fig. 4.30. Charge neutral point is located at 0 eV. The calculated results show that the quantized conductance is realized, which corresponds to the number of quantum channels. For comparison, we show the conductance calculated by the NEGF method in the previous section in Fig. 4.18 by the broken lines. We see good agreements between the calculated results with the TD-WPD method and those with the NEGF method in the ballistic limit. This shows that the TD-WPD method reproduces the results by the NEGF method.

4.5.4.2 Mean free path in diffusive regime

Next, we compare the mean free path $\ell_{\rm mfp}$, which is the length to distinguish transport property from the diffusive to ballistic regimes. We compare $\ell_{\rm mfp}$ obtained from the TD-WPD method with $\ell_{\rm mfp}$ from the Fermi golden rule, which is a typical way to estimate $\ell_{\rm mfp}$. Let us consider $\ell_{\rm mfp}$ for the (5, 5)-CNTs. We use artificial short-ranged impurity potentials $\hat{V}_U$, randomly distributed with the strengths from 0 to U. We put them at the on-site energies of the Hamiltonian of (5, 5)-CNTs.

The relaxation time τ is given in the Fermi golden rule as

$$\frac{1}{\tau(\epsilon, n, U)} = \frac{2\pi}{\hbar}\left(\sum_m \left|\langle m|\hat{V}_U|n\rangle\right|^2 \nu(\epsilon)\right), \tag{4.278}$$

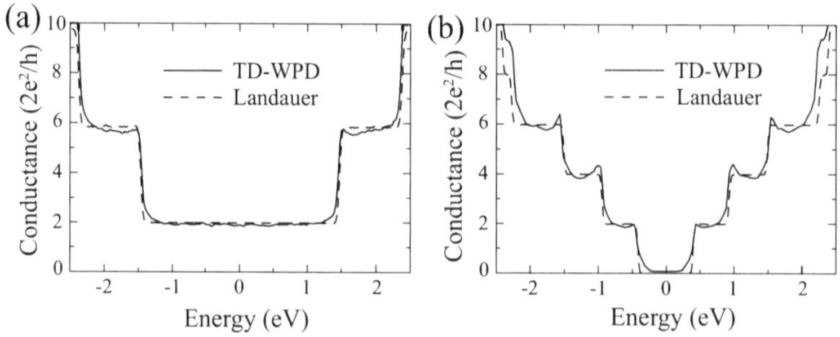

Figure 4.30 (a) Conductance as a function of energy for a metallic (5, 5)-CNT in the ballistic regime calculated by the NEGF method and the TD-WPD method, which are represented by the broken and solid lines, respectively. Charge neutrality point is located at 0 eV. (b) Conductance as a function of energy for a semiconducting (10, 0)-CNT in the ballistic regime.

where n is the ratio of impurity sites to all the sites. Here we consider τ at charge neutral point $\epsilon_F = 0$ and obtain it by averaging in the energy range from -1 eV to $+1$ eV, since the relaxation time fluctuates as a function of energy due to quantum interference effects. Figure 4.31(a) shows the relaxation time τ as a function of the energy ϵ for various parameters of U and n. Table 4.2 shows averaged relaxation times for them and mean free paths at ϵ_F obtained from $\ell_{\text{mfp}} = v_F \tau$. Here the Fermi velocity v_F is taken from an analytic form of $v_F = 3a_{cc}\gamma^0/2\hbar$, which produces the Fermi velocity of $v_F \simeq 0.82$ nm/fs.

In the TD-WPD method, the mean free path ℓ_{mfp} and relaxation time τ are obtained from the time-dependent diffusion coefficient $D(\epsilon_F, t)$ as,

$$\ell_{\text{mfp}}(\epsilon_F) = \frac{D_{\max}(\epsilon_F, t \to \infty)}{v_F} \tag{4.279}$$

and

$$\tau(\epsilon_F) = \frac{D_{\max}(\epsilon_F, t \to \infty)}{v_F^2}. \tag{4.280}$$

Figure 4.31(b) shows the behaviors of time-dependent diffusion coefficients $D(\epsilon_F, t)$ of (5, 5)-CNTs with the same parameters as in the Fermi golden rule case. The Fermi velocity v_F is obtained from the slope of $D(\epsilon_F, t)$ as a function of time t in the vicinity of $t = 0$, for

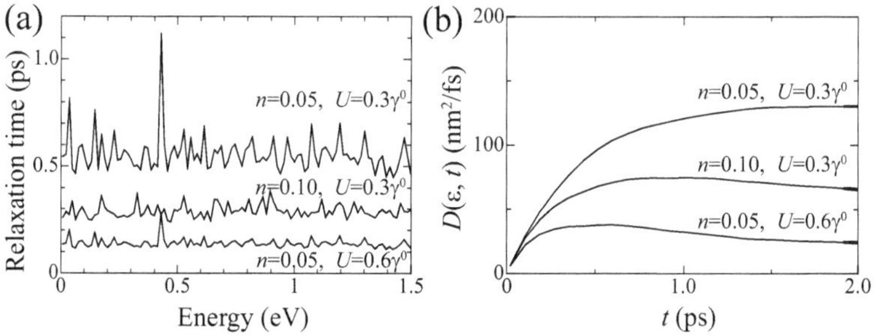

Figure 4.31 (a) Relaxation time τ as a function of energy ϵ obtained from the Fermi golden rule for impurity potentials introduced in the (5, 5)-CNTs. Charge neutrality point ϵ_F is located at 0 eV. The ratio of impurity sites to all the sites and magnitude of the perturbation potential are denoted by n and U, respectively. The energy scale is measured for an electron transfer energy, $\gamma^0 = 2.5$ eV, as a unit. (b) Behaviors of time-dependent diffusion coefficients $D(\epsilon, t)$ of (5, 5)-CNTs at ϵ_F with the same potential parameters as (a).

which we need not use the parameterized form above. Therefore, we note that mean free path $\ell_{\rm mfp}$ and relaxation time τ are all obtained using only the diffusion coefficient $D(\epsilon_F, t)$ in Eq. (4.279) and in Eq. (4.280).

The calculated results are summarized in the Table 4.2. We see that mean free paths and relaxation times agree well with those obtained using the Fermi golden rule above. We also find that the present TD-WPD method reproduces the relation satisfied in the diffusive transport regime that an inverse relaxation time is substantially proportional to the impurity ratio n and the squared

Table 4.2 Comparison of mean free paths from the Fermi golden rule and the TD-WPD method.

	Fermi golden rule		TD-WPD	
	τ (fs)	$\ell_{\rm mfp}$ (nm)	τ (fs)	$\ell_{\rm mfp}$ (nm)
n=0.05, U=0.3γ^0	500	410	470	387
n=0.10, U=0.3γ^0	280	230	280	223
n=0.05, U=0.6γ^0	130	107	140	113

Note: The ratio of impurity sites and the magnitude of impurity potentials are represented by n and U.

potential magnitude U. These results indicate the validity of the TD-WPD method in the diffusive limit.

4.5.4.3 Hall conductivity σ_{xy} and Hall mobility

Let us demonstrate how the TD-WPD formula for $\sigma_{xy}(\epsilon_F)$ works for the Hall effects and show the trajectories of wave packets and conductivities when a magnetic field is applied perpendicular to the electron systems. We consider the 2D square lattice of 30×30 sites with a lattice constant a, where the periodic boundary condition is employed. The Hamiltonian is written by

$$\hat{H} = \sum_{ij} \gamma_{ij} \hat{c}_i^\dagger \hat{c}_j + \sum_i W_i \hat{c}_i^\dagger \hat{c}_i. \tag{4.281}$$

We take $\gamma^0 = 1.0$ eV for the transfer energy and $a = 1$ Å for the lattice. The magnetic-field effect is introduced into the Hamiltonian by multiplying the transfer energy γ^0 by a phase factor

$$\gamma_{ij} = \gamma^0 \exp\left\{ \frac{ie}{c\hbar} \int_{\mathbf{R}_i}^{\mathbf{R}_j} \mathbf{A}(\mathbf{r}) \cdot d\mathbf{r}, \right\} \tag{4.282}$$

where $\mathbf{R}_i$ denotes the position vectors of the i-th site and $\mathbf{A}(\mathbf{r})$ is the vector potential. W_i is the potential energy at i-th site, which is selected randomly in the energy width $[-W/2, +W/2]$. As an initial wave packet, we employ a state localized at one site and simulate the time evolution of the wave packet with time step of $\Delta t = 0.05 \times h/\gamma^0$ up to the total evolution time of 80 time steps. In the energy integral of Eq. (4.246), we adopt an energy mesh having an interval of $9.4 \times 10^{-3}\gamma^0$ and an infinitesimal value η is $10^{-2}\gamma^0$.

First, let us compare the energy dependence of off-diagonal Hall conductivities by the TD-WPD method with those obtained from the conventional Kubo formula given by

$$\sigma_{xy}(\epsilon_F) = -i\hbar e^2 \lim_{\eta \to +0} \sum_{n,m} f(\epsilon_n - \epsilon_F) \times \frac{\langle n|\hat{v}_x|m\rangle\langle m|\hat{v}_y|n\rangle - \langle n|\hat{v}_y|m\rangle\langle m|\hat{v}_x|n\rangle}{(\epsilon_n - \epsilon_m - i\eta)(\epsilon_n - \epsilon_m + i\eta)}, \tag{4.283}$$

where $|n\rangle$ represents the n-th eigenvector of $\hat{H}$ with eigenvalue ϵ_n. Eigenvectors and eigenvalues are obtained by the direct numerical diagonalization of $\hat{H}$.

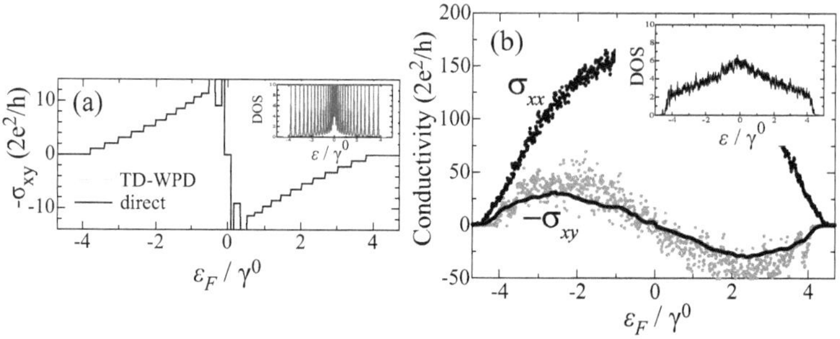

Figure 4.32 (a) Quantized Hall conductivities $-\sigma_{xy}$ calculated by the TD-WPD method as a function of ϵ_F in a disorder-free square lattice, where a magnetic field with flux $\phi/\phi_0 = 1/30$ is applied. For comparison, $-\sigma_{xy}$ by Eq. (4.283) is shown by thin line. (Inset) Landau levels of the square lattice under a magnetic field. (b) Conductivities σ_{xx} and $-\sigma_{xy}$ as a function of ϵ_F for the square lattice with the disorders $W/\gamma^0 = 3.2$ under the magnetic field of $\phi/\phi_0 = 1/30$. For comparison, we show $-\sigma_{xy}$ obtained by Eq. (4.283) by the gray dots. (Inset) DOS of the square lattice with static disorders. From H. Ishii, N. Kobayashi, and K. Hirose, *Phys. Rev. B* **83**, 233403 (2011) [84].

In the clean limit (a) where disorder potential W is absent, magnetic field induces the Landau peaks with a spacing of energy $\hbar\omega_c$ where $\omega_c = eB/m^*c$. The DOS is shown in the inset of Fig. 4.32(a) for the parameter of $\hbar\omega_c/\gamma^0 \simeq 0.1$. We can see that the DOS shows spiky peaks forming the Landau levels and the quantized σ_{xy} takes a plateau of $(e^2/h) \times N$ between two Landau levels in Fig. 4.32(a), where N is the number of Landau levels below the Fermi energy. For comparison, we plot σ_{xy} obtained from Eq. (4.283). Both of conductivities agree with each other.

In the dirty case (b) where we introduce static disorders $W/\gamma^0 = 3.2$, which is much larger than the cyclotron energy $\hbar\omega_c$, we see that the Landau peaks are completely smeared by the disorder as shown in the inset of Fig. 4.32(b). From scattering time τ estimated by the Fermi golden rule

$$\frac{1}{\tau} = \frac{2\pi}{\hbar}|W|^2\nu(\epsilon_F), \tag{4.284}$$

we evaluate $\omega_c\tau \simeq 0.1 \ll 1$, which shows the system in weak magnetic field regime. Figure 4.32(b) presents ϵ_F dependence of diagonal and off-diagonal parts of the conductivity tensor. Diagonal

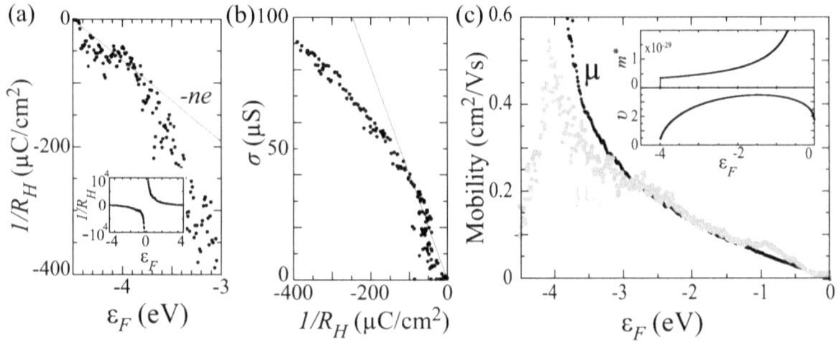

Figure 4.33 (a) Inverse Hall coefficient $1/R_H$ as a function of energy ϵ around the band edge. Asymptotic solution in the parabolic band approximation is shown by the solid line. (Inset) $1/R_H$ behavior in the whole energy region of the band. (b) Conductivity σ as a function of $1/R_H$. The solid line represents the relation of $\sigma = \mu_H \cdot R_H^{-1}$. (c) Hall mobility μ_H and mobility $\mu = e\tau/m^*$ as a function of ϵ_F. (inset) Effective mass m^* and group velocity v as a function of ϵ. From H. Ishii, N. Kobayashi, and K. Hirose, *Phys. Rev. B* **83**, 233403 (2011) [84].

part σ_{xx} has the symmetrical behavior with respect to $\epsilon = 0$, while off-diagonal part σ_{xy} shows antisymmetrical behaviors, reflecting that the square lattice with AB sub-lattice structure has a symmetrical electronic band structure. We can see that the calculated conductivities σ_{xy} by the TD-WPD method agree with those obtained from Eq. (4.283) shown by gray dots.

Next, we compare the Hall mobility μ_H with mobility $\mu = e\tau/m^*$. We evaluate the Hall coefficient by Eq. (4.247) using the conductivity tensors σ_{xx} and σ_{xy} calculated by the TD-WPD method quantum mechanically, taking a phase factor into consideration.[a] Figure 4.33(a) shows an inverse of Hall coefficient $1/R_H$ as a function of energy ϵ_F around the band bottom. The inset gives $1/R_H$ in whole energy region, where the reverse of sign of $1/R_H$ at $\epsilon = 0$ eV reveals that the type of a carrier is changed from electron to hole. In a parabolic band approximation, carrier density is given by $ne = em^*\epsilon/2\pi\hbar^2$, which is extracted from two expressions of the diagonal conductivity, $\sigma_{xx} = q^2 n\tau/m^*$ and $\sigma_{xx} = q^2 \nu v_x^2 \tau$ with the DOS of $\nu = m^*/2\pi\hbar^2$. At the bottom of band dispersion for the

[a] Here we take $c = 1$ for simplicity.

square lattice, we obtain an effective mass $m^* - -\hbar^2/2\gamma^0 a^2$ and velocity $v_x^2 = \epsilon/m^*$. Then we show the linear dependence of a carrier density ne for ϵ by the solid line in the main frame of Fig. 4.33 (a). We can see, around the band edge, the obtained $1/R_H$ is close to the carrier density in the parabolic band approximation.

From the Hall coefficient, Hall mobility defined by

$$\mu_H = \sigma R_H \tag{4.285}$$

is obtained in experiments as a linear slope of conductivity σ for $1/R_H$. In Figure 4.33(b), we see that σ increases monotonically as a function of $|1/R_H|$ in the small $1/R_H$ regime, where the parabolic-band approximation is well defined. Therefore, we can evaluate the Hall mobility μ_H from the slope of σ, which is shown as the solid line in Fig. 4.33(b). We compare the mobility $\mu = e\tau/m^*$ from the diagonal part of conductivity with no magnetic field, which is given by $\mu = eD(\epsilon)/(m^*(\epsilon)v(\epsilon)^2)$. Figure 4.33(c) shows two mobilities μ_H and μ as a function of the energy. The inset shows an effective mass $1/m^* \equiv d^2\epsilon(\mathbf{k})/\hbar^2 dk^2$ and the group velocity $v \equiv d\epsilon(\mathbf{k})/dk$. We see that both of mobilities agree with each other quantitatively for most of energy regime except for the band edge, where we cannot use m^* and v directly for the mobility. We note that Hall mobility μ_H is obtained in the TD-WPD calculations even near the band edge.

From these calculations, we can say that the TD-WPD method enables us to investigate transport properties from diffusive to ballistic transport limits, including the intermediate regime, i.e., quasi-ballistic regime, in a unified way. Furthermore, we show that the calculation of Hall mobility under a magnetic field agrees well with the mobility from the diagonal part of the conductivity tensor in the isotropic 2D square lattice case. This shows that the TD-WPD method is effective for the study of Hall effect.

4.6 Quantum Master Equation Method

To treat the electron dynamics of an open system, let us consider the total system composed of an electron subsystem (S) and a phonon bath as a reservoir (R), which are interacting each other. When we consider the systems as closed, their dynamics is reversible. However, when we assume the phonon bath is so large as an environment that it is not influenced by the interaction, the dynamics of an electron becomes irreversible, which is appropriate for electron transport with energy relaxation processes. In this section, we treat a calculation method called the QME for the dynamics of electrons in the phonon-bath dissipative systems [96].

4.6.1 *Density Matrix with Energy Relaxation*

The density matrix plays a central role to describe the systems. The irreversible description of an electron dynamics is expressed by the reduced density matrix $\hat{\rho}_S(t)$, which is obtained by tracing out the bath variables from the total density matrix as

$$\hat{\rho}_S(t) = \mathrm{Tr}_R \hat{\rho}_{\mathrm{tot}}(t). \tag{4.286}$$

Here the density matrix of the total system $\hat{\rho}_{\mathrm{tot}}(t)$ obeys the Liouville–von Neumann equation

$$\frac{\partial \hat{\rho}(t)}{\partial t} = -\frac{i}{\hbar}\left[\hat{H}_{\mathrm{tot}}, \hat{\rho}_{\mathrm{tot}}(t).\right] \tag{4.287}$$

Then the expectation value of an observable $\hat{A}(t)$ acting on the open system is determined from $\langle A(t)\rangle = \mathrm{Tr}\left\{\hat{\rho}_S(t)\hat{A}\right\}$.

In this section we treat methods to integrate the time-dependent equation for reduced density matrix in the presence of a finite electric field $\mathbf{E}$.[a] The equation deals with transport in the presence of non-energy-conserving scattering processes for relaxation to reduce the electron acceleration energy from an electric field. We derive an approximate formula for the density matrix in the form of

$$\frac{\partial}{\partial t}\hat{\rho}_S(t) = -\frac{i}{\hbar}\left[\hat{H}_S, \hat{\rho}_S(t)\right] + \hat{C}\left[\hat{\rho}_S(t)\right], \tag{4.288}$$

---

[a]We note that the TD-WPD method we describe in the previous section is based on the linear-response Kubo formula, which is derived from the density matrix using the relation of

$$\delta\hat{\rho}_S(t) = -\frac{i}{\hbar}\int_{-\infty}^{t} e^{-i\hat{H}_0(t-t')/\hbar}\left[\hat{H}_1(t'), \hat{\rho}_S(t')\right] e^{i\hat{H}_0(t-t')/\hbar} dt'.$$

Here $\hat{H}_S = \hat{H}_0 + \hat{H}_1(t)$ and an infinitesimal electric field is included in $\hat{H}_1(t)$ by $\mathbf{E} = -(1/c)\partial\mathbf{A}/\partial t$. No relaxation process is included; thus $\hat{\rho}_{\mathrm{tot}}(t) = \hat{\rho}_S(t)$. So the QME method is regarded as a generalization of the TD-WPD method (Kubo-type) to include a finite voltage (Boltzmann-type).

where the dissipative, fluctuating, and non-energy conserving phonon scattering effects are included in the last term of $\hat{C}\,[\hat{\rho}_S(t)]$. The computational approach to treat the density matrix by Eq. (4.288) is called the QME method.

It should be noted that this formula is a quantum analog of the semi-classical Boltzmann equation[a]

$$\frac{\partial}{\partial t} f(\mathbf{r}, \mathbf{p}, t) = \left\{ H, f(\mathbf{r}, \mathbf{p}, t) \right\} + \left[\frac{\partial f(\mathbf{r}, \mathbf{p}, t)}{\partial t} \right]_{\text{coll}}, \tag{4.289}$$

which is derived from

$$\frac{d}{dt} f(\mathbf{r}, \mathbf{p}, t) = \left(\frac{\partial}{\partial t} + \frac{d\mathbf{r}}{dt} \cdot \frac{\partial}{\partial \mathbf{r}} + \frac{d\mathbf{p}}{dt} \cdot \frac{\partial}{\partial \mathbf{p}} \right) f(\mathbf{r}, \mathbf{p}, t), \tag{4.290}$$

using the relations of the classical mechanics $d\mathbf{r}_i/dt = \partial H/\partial \mathbf{p}_i$, $d\mathbf{p}_i/dt = -\partial H/\partial \mathbf{r}_i$, and the Poisson bracket. The one-particle distribution function $f(\mathbf{r}, \mathbf{p}, t)$ is changed into the reduced density matrix $\hat{\rho}_S(t) = \sum_i |\psi_i\rangle\langle\psi_i|$ in the QME. In the semi-classical Boltzmann equation, the electron density and the electrical current are given by

$$n(\mathbf{r}, t) = \int f(\mathbf{r}, \mathbf{p}, t) d\mathbf{p}, \quad \mathbf{j}(\mathbf{r}, t) = e \int \frac{\mathbf{p}}{m} f(\mathbf{r}, \mathbf{p}, t) d\mathbf{p}, \tag{4.291}$$

while in the QME method, these are given by

$$n(\mathbf{r}, t) = \text{Tr}\left[\hat{\rho}_S(t)\hat{n}(\mathbf{r})\right], \quad \mathbf{j}(\mathbf{r}, t) = \text{Tr}\left[\hat{\rho}_S(t)\hat{\mathbf{j}}(\mathbf{r})\right]. \tag{4.292}$$

When we neglect non-diagonal elements, the QME formula reduces to the Boltzmann equation in an appropriate semi-classical limit.

This approach is an alternative method to quantum transport of a many-electron system for the scattering approaches using the RTM method, the LS method, and the NEGF method. We note that these methods treat finite bias effects as open boundary conditions deep in the electrodes and the non-energy-conserving processes for relaxation to reduce the acceleration energy are assumed mostly to occur outside of boundaries.

[a] This is also a quantum analog of the classical nonequilibrium molecular dynamics (NEMD) method described in Chapter 2 for electrical conductivities.

4.6.1.1 Projection operator method

First, we discuss a general formalism to deal with an open quantum system coupled to a heat reservoir bath, where we consider that electrons constitute the system while the reservoir consists of phonons (lattice vibrations). The total Hamiltonian of the system plus bath $\hat{H}_{\text{tot}}$ is written as

$$\hat{H}_{\text{tot}} = \hat{H} \otimes \hat{I}_R + \hat{I} \otimes \hat{H}_R + \hat{V}_{\text{int}}, \tag{4.293}$$

where $\hat{H}$ is the system Hamiltonian,[a] $\hat{I}_R$ is the identity operator acting on variables of the reservoir, $\hat{H}_R$ is the Hamiltonian of the reservoir, and $\hat{V}_{\text{int}}$ represents the coupling interaction between the system and reservoir. The electronic system in contact with a reservoir is not in a pure quantum state and must be described by a reduced density operator $\hat{\rho}(t) = \text{Tr}_R \hat{\rho}_{\text{tot}}(t)$ with $\hat{\rho}_{\text{tot}}(t)$ to obey

$$\frac{\partial \hat{\rho}_{\text{tot}}(t)}{\partial t} = -\frac{i}{\hbar}\left[\hat{H} + \hat{H}_R + \hat{V}_{\text{int}}, \hat{\rho}_{tot}(t)\right] = i\hat{L}_{\text{tot}}\hat{\rho}_{\text{tot}}(t), \tag{4.294}$$

where the Liouville operator is $\hat{L}_{\text{tot}} = \hat{L} + \hat{L}_R + \hat{L}_{\text{int}}$. We assume that the system and reservoir are independent at the initial time t_0 as $\hat{\rho}_{\text{tot}}(t_0) = \hat{\rho}(t_0) \otimes \hat{\rho}_R(t_0)$, where $\hat{\rho}_R(t_0) = e^{-\hat{H}_R/k_BT}/\text{Tr}_R e^{-\hat{H}_R/k_BT}$ is the density operator of the reservoir.

Let us introduce the projection operator $\hat{P}$ as

$$\hat{P}\hat{\rho}_{\text{tot}}(t) \equiv \text{Tr}_R \hat{\rho}_{\text{tot}}(t) \otimes \hat{\rho}_R(t_0) = \hat{\rho}(t) \otimes \hat{\rho}_R(t_0), \tag{4.295}$$

which projects the total system into the subspace, whereas a complementary operator $\hat{Q}$ is introduced, which projects it into the reservoir [2, 3, 5]. By definition, $\hat{P} + \hat{Q} = \hat{I}$ and we have $\hat{P}^2 = \hat{P}$, $\hat{Q}^2 = \hat{Q}$, and $\hat{P}\hat{Q} = \hat{Q}\hat{P} = 0$. For the total density matrix $\hat{\rho}_{\text{tot}}(t)$ $(= \hat{P}\hat{\rho}_{\text{tot}}(t) + \hat{Q}\hat{\rho}_{\text{tot}}(t))$, we have two coupled equations:

$$\begin{aligned}
\frac{\partial}{\partial t}\hat{P}\hat{\rho}_{\text{tot}}(t) &= \hat{P}i\hat{L}_{\text{tot}}\hat{P}\hat{\rho}_{\text{tot}}(t) + \hat{P}i\hat{L}_{\text{tot}}\hat{Q}\hat{\rho}_{\text{tot}}(t) \\
\frac{\partial}{\partial t}\hat{Q}\hat{\rho}_{\text{tot}}(t) &= \hat{Q}i\hat{L}_{\text{tot}}\hat{Q}\hat{\rho}_{\text{tot}}(t) + \hat{Q}i\hat{L}_{\text{tot}}\hat{P}\hat{\rho}_{\text{tot}}(t).
\end{aligned} \tag{4.296}$$

Putting an expression of the second equation as

$$\begin{aligned}
\hat{Q}\hat{\rho}_{\text{tot}}(t) = &\int_{t_0}^{t} e^{i\hat{Q}\hat{L}_{\text{tot}}(t-t')}\hat{Q}(i\hat{L}_{tot})\hat{P}\hat{\rho}_{\text{tot}}(t')dt' \\
&+ e^{i\hat{Q}\hat{L}_{\text{tot}}(t-t_0)}\hat{Q}\hat{\rho}_{\text{tot}}(t_0)
\end{aligned} \tag{4.297}$$

[a]We abbreviate S as $\hat{H}_S = \hat{H}$ for simplicity.

into the first equation, we obtain

$$\frac{\partial}{\partial t}\hat{P}\hat{\rho}_{\text{tot}}(t) = \hat{P}i\hat{L}_{\text{tot}}e^{i\hat{Q}\hat{L}_{\text{tot}}(t-t_0)}\hat{Q}\hat{\rho}_{\text{tot}}(t_0) + \hat{P}i\hat{L}_{\text{tot}}\hat{P}\hat{\rho}_{\text{tot}}(t)$$
$$+\hat{P}i\hat{L}_{tot}\int_{t_0}^{t} e^{i\hat{Q}\hat{L}_{\text{tot}}(t-t')}\hat{Q}i\hat{L}_{tot}\hat{P}\hat{\rho}_{tot}(t')dt'. \tag{4.298}$$

This is known as Nakajima–Zwanzig equation for the reduced system, involving the non-Markov memory effects [97, 98].

In applications, we choose the projection operator $\hat{Q}$ as $\hat{Q}\hat{\rho}_{tot}(t_0) = 0$ so that the first term depending on the initial value becomes irrelevant. Also, assuming that $\hat{P}$ commutes with $\hat{L}$, we have the homogeneous time-retarded QME

$$\frac{\partial\hat{\rho}(t)}{\partial t} = i(\hat{L}+\hat{L}_{\text{int}})\hat{\rho}(t) + \int_{t_0}^{t} K(t,t')\hat{\rho}(t')dt', \tag{4.299}$$

with the memory kernel $K(t,t') = \hat{P}i\hat{L}_{\text{int}}e^{i\hat{Q}\hat{L}_{\text{tot}}(t-t')}\hat{Q}i\hat{L}_{\text{int}}$. The first term describes the reversible motion of an electron subsystem, and the second time-retarded term expresses effects from the reservoir due to the thermal dissipation and fluctuation, which brings the irreversibility.

Although Eq. (4.299) is exact for all interactions of arbitrary (non-Markovian) systems, it is usually difficult to solve the equation. So we use the perturbation for its explicit evaluation. Up to the second order for the weak interaction, we replace $\hat{L}_{\text{tot}}$ into $(\hat{L}+\hat{L}_R)$ in the exponent, which corresponds to the Born approximation. Also, by disregarding the retardation effects, we have the required form for the quantum (Markovian) master equation as

$$\frac{\partial\hat{\rho}(t)}{\partial t} \simeq -\frac{i}{\hbar}\left[\hat{H},\hat{\rho}(t)\right] + \hat{C}\left[\hat{\rho}(t)\right] \tag{4.300}$$

with

$$\hat{C}\left[\hat{\rho}(t)\right] = \left(\int_0^{\infty}\text{Tr}_R\left\{i\hat{L}_{\text{int}}e^{i(\hat{L}+\hat{L}_R)t'}i\hat{L}_{\text{int}}\hat{\rho}_R(t_0)\right\}dt'\right)\hat{\rho}(t). \tag{4.301}$$

Here we use the Markov approximation due to coarse graining in time; the characteristic time scale with the reservoir correlation τ_R is much smaller than that of the system dynamics. Eq. (4.300) with Eq. (4.301) is called Lindblad equation [99].[a]

[a] For the interaction of the form $\hat{V}_{\text{int}} = \hat{V}_S \otimes \hat{V}_R$, the dissipation term becomes $\hat{C}[\hat{\rho}(t)] = -\sum_i\left(\hat{L}_i^{\dagger}\hat{L}_i\hat{\rho}(t) + \hat{\rho}(t)\hat{L}_i^{\dagger}\hat{L}_i - 2\hat{L}_i\hat{\rho}(t)\hat{L}_i^{\dagger}\right)$ where the Lindblad operator $\hat{L}_i$ describes effects of the reservoir on the subsystem as quantum transitions. This equation preserves the complete positivity of $\hat{\rho}(t)$ with $\text{Tr}\{\hat{\rho}(t)\} = 1$.

4.6.2 *Electron-plus-phonon bath system*

In this section, we derive the explicit formula of the Lindblad equation for the electron dynamics interacting with the reservoir of phonon bath with the coupling potential $\hat{V}_{\text{int}}$. For weak coupling that $\hat{V}_{\text{int}}$ is sufficiently small, it is convenient to use the interaction representation for perturbation theory as

$$\frac{\partial \tilde{\rho}_{\text{tot}}(t)}{\partial t} = -\frac{i}{\hbar}\left[\tilde{V}_{\text{int}}(t), \tilde{\rho}_{\text{tot}}(t)\right], \tag{4.302}$$

where

$$\begin{aligned}\hat{\rho}_{\text{tot}}(t) &= e^{-i(\hat{H}+\hat{H}_R)(t-t_0)/\hbar}\tilde{\rho}_{\text{tot}}(t)e^{i(\hat{H}+\hat{H}_R)(t-t_0)/\hbar}\\ \hat{V}_{\text{int}}(t) &= e^{-i(\hat{H}+\hat{H}_R)(t-t_0)/\hbar}\tilde{V}_{\text{int}}(t)e^{i(\hat{H}+\hat{H}_R)(t-t_0)/\hbar}.\end{aligned} \tag{4.303}$$

Here $\hat{\rho}_{\text{tot}}$ and $\hat{V}_{\text{int}}$ are in the Schrödinger representation, while $\tilde{\rho}_{\text{tot}}$ and $\tilde{V}_{\text{int}}$ are in the interaction picture and $\tilde{\rho}(t) = \text{Tr}_R\tilde{\rho}_{\text{tot}}(t)$.[a] Assuming that the reservoir is independent as $\tilde{\rho}_{\text{tot}}(t) = \tilde{\rho}(t) \otimes \tilde{\rho}(t_0)$, we obtain within the second-order in $\tilde{V}_{\text{int}}$ and by tracing out the reservoir variables as

$$\begin{aligned}\frac{\partial \tilde{\rho}(t)}{\partial t} &\approx -\frac{i}{\hbar}\text{Tr}_R\left[\tilde{V}_{\text{int}}(t), \tilde{\rho}(t_0) \otimes \tilde{\rho}_R(t_0)\right]\\ &+ \left(\frac{-i}{\hbar}\right)^2 \int_{t_0}^{t} dt'\text{Tr}_R\left[\tilde{V}_{\text{int}}(t), \left[\tilde{V}_{\text{int}}(t'), \tilde{\rho}(t') \otimes \tilde{\rho}_R(t_0)\right]\right],\end{aligned} \tag{4.304}$$

where $\tilde{\rho}_R(t_0) = e^{-\tilde{H}_R/k_BT}/\text{Tr}_R e^{-\tilde{H}_R/k_BT}$ is the mixed state [100]. Here we consider that the interaction in the Schrödinger representation is written in the form of

$$\hat{V}_{\text{int}} = \hat{V}_S \otimes \hat{V}_R = \sum_i \hat{L}_i \hat{B}_i. \tag{4.305}$$

We note that the operator $\hat{L}_i$ acts on the system and $\hat{B}_i$ acts on the reservoir bath only. Then Eq. (4.304) is integrated as [101]

$$\begin{aligned}\tilde{\rho}(t) - \tilde{\rho}(t_0) &= -\frac{i}{\hbar}\sum_i \langle \hat{B}_i\rangle_R \left[\hat{L}_i, \tilde{\rho}(t_0)\right] \int_0^{t-t_0} e^{i\omega_i t'}dt' + \left(\frac{-i}{\hbar}\right)^2\\ &\times \sum_{i,j}\int_0^{t-t_0}\left\{\left(\hat{L}_i\hat{L}_j\tilde{\rho}(t_0) - \hat{L}_j\tilde{\rho}(t_0)\hat{L}_i\right)\int_0^{t-t_0-\zeta} e^{i\omega_i t'}\langle \hat{B}_i(t')\hat{B}_j\rangle_R dt'\right.\\ &\quad \left. - \left(\hat{L}_i\tilde{\rho}(t_0)\hat{L}_j - \tilde{\rho}(t_0)\hat{L}_j\hat{L}_i\right)\int_0^{t-t_0-\zeta} e^{i\omega_i t'}\langle \hat{B}_j\hat{B}_i(t')\rangle_R dt'\right\}\\ &\quad \times e^{i(\omega_i+\omega_j)t''}dt'',\end{aligned} \tag{4.306}$$

---

[a] For the reduced density matrix, we note $\hat{\rho}(t) = e^{-i\hat{H}(t-t_0)/\hbar}\tilde{\rho}(t)e^{i\hat{H}(t-t_0)/\hbar}$.

where we use the notations for reservoir correlation functions as

$$\langle \hat{B}_i \rangle_R = \mathrm{Tr}_R\{\hat{B}_i \tilde{\rho}_R(t_0)\}, \quad \langle \hat{B}_i(t)\hat{B}_j \rangle_R = \mathrm{Tr}_R\{\hat{B}_i(t)\hat{B}_j \tilde{\rho}_R(t_0)\},$$

with $\hat{B}_i(t) = e^{i\hat{H}_R t/\hbar}\hat{B}_i e^{-i\hat{H}_R t/\hbar}$ and $\hbar\omega_i$ is the transition energy for electrons. Note $\langle \hat{B}_i \rangle_R = 0$ and the first term vanishes.

Now we make the Markov approximations. First, assuming that $\langle \hat{B}_i(t')\hat{B}_j \rangle_R = 0$ if t' is not close to zero, which means that the correlation time with reservoir τ_R is very short, we can replace the upper limit of the integral over t' with infinity. This indicates to neglect the memory effects. Introducing the spectral functions as

$$\omega^+_{i,j} = \int_0^\infty e^{i\omega_i t}\langle \hat{B}_i(t)\hat{B}_j \rangle_R dt, \quad \omega^-_{j,i} = \int_0^\infty e^{i\omega_i t}\langle \hat{B}_j \hat{B}_i(t) \rangle_R dt$$

and approximating the integral as

$$\int_0^{t-t_0} e^{i(\omega_i+\omega_j)t'}dt' \simeq (t-t_0)\delta_{\omega_i,-\omega_j}, \qquad (4.307)$$

which means that the integral time is much larger than the typical time scale of electron dynamics of transitions as $t - t_0 \gg 1/\omega_i$, we have the expression[a]

$$\frac{\tilde{\rho}(t)-\tilde{\rho}(t_0)}{t-t_0} = -\frac{1}{\hbar^2}\sum_{i,j}\delta_{\omega_i,-\omega_j} \times \left\{\left(\hat{L}_i\hat{L}_j\tilde{\rho}(t_0) - \hat{L}_j\tilde{\rho}(t_0)\hat{L}_i\right)\omega^+_{i,j} - \left(\hat{L}_i\tilde{\rho}(t_0)\hat{L}_j - \tilde{\rho}(t_0)\hat{L}_j\hat{L}_i\right)\omega^-_{j,i}\right\}.$$

Second, going back to the Schrödinger representation, we take the time derivative of $\hat{\rho}(t)$ as[b]

$$\frac{\partial\hat{\rho}(t)}{\partial t} \simeq e^{-i\hat{H}(t-t_0)/\hbar}\left\{-\frac{i}{\hbar}\left[\tilde{H},\tilde{\rho}(t)\right] + \frac{\tilde{\rho}(t)-\tilde{\rho}(t_0)}{t-t_0}\right\}e^{i\hat{H}(t-t_0)/\hbar} \quad t \to t_0. \qquad (4.308)$$

This corresponds to coarse graining in time and the scattering processes with the reservoir become instantaneous processes in the coarse-grained dynamics of the system, the Markov approximations. Putting into Eq. (4.308) with $\tilde{\rho}(t_0)$ replaced by $\tilde{\rho}(t)$, we have

$$\frac{\partial\hat{\rho}(t)}{\partial t} = -\frac{i}{\hbar}\left[\hat{H},\hat{\rho}(t)\right] - \frac{1}{\hbar^2}\sum_{i,j}\delta_{\omega_i,-\omega_j} \times \left\{\left[\hat{L}_i\hat{L}_j\hat{\rho}(t) - \hat{L}_j\hat{\rho}(t)\hat{L}_i\right]\hat{\omega}^+_{i,j} - \left[\hat{L}_i\hat{\rho}(t)\hat{L}_j - \hat{\rho}(t)\hat{L}_j\hat{L}_i\right]\hat{\omega}^-_{j,i}\right\}.$$

---

[a]We note that due to the Markov approximation, the memory effect disappears completely and the future is determined by the present time t. This is done by $\tilde{\rho}(t') \to \tilde{\rho}(t)$ in the second term of Eq. (4.304).

[b]The first term $-(i/\hbar)\left[\tilde{H},\tilde{\rho}(t)\right]$ describes some processes in the system by an external field, which is not included in the interaction picture.

This is the general equation for $\hat{\rho}(t)$ within the second-order by the reduced density operator of the system in the Schrödinger representation to have the Lindblad form of QME with $\hat{L}$ as the Lindblad operator representing electron transition.

Let us consider the electron–phonon coupling $\hat{V}_{\text{int}} = \hat{V}_{\text{e-ph}}$ for the interaction. In the coordinate representation Q, the coupling $\hat{V}_{\text{e-ph}}$ between electrons and phonons can be written as

$$\hat{V}_{\text{e-ph}} = \sum_j \int \hat{\psi}^\dagger(\mathbf{r}) \left. \frac{\partial V_{\text{ion}}(\mathbf{r} - Q_j)}{\partial Q_j} \right|_{\delta Q_j = 0} \delta \hat{Q}_j \hat{\psi}(\mathbf{r}) d\mathbf{r} \qquad (4.309)$$

in terms of ion potential V_{ion} and electron coordinate $\mathbf{r}$. Here $\delta \hat{Q}_j = \hat{Q} - \hat{Q}_j$ is displacement variables around equilibrium positions of atoms $\hat{Q}_j$ where the variable j runs all nuclei and the reservoir bath is assumed in thermal equilibrium $\langle Q_j \rangle_R = 0$. In the second quantization form, the coupling potential takes the form

$$\hat{V}_{\text{e-ph}} = \sum_{n,m,q} \gamma_{n,m,q} \hat{c}_n^\dagger \hat{c}_m a_q^\dagger + c.c., \qquad (4.310)$$

where

$$\gamma_{n,m,q} = \sum_j \sqrt{\frac{\hbar}{2 M_q \omega_q}} \left\langle n \left| \left(\frac{\partial V_{\text{ion}}}{\partial Q_j} \right) \right| m \right\rangle \qquad (4.311)$$

are electron–phonon coupling matrix elements. The first term corresponds to the phonon emission, while the complex conjugate term corresponds to the phonon absorption. From this expression, we see that the interaction energy of electron–phonon coupling in the Schrödinger representation is written in the form of

$$\hat{V}_{\text{e-ph}} = \hat{V}_S \otimes \hat{V}_R = \sum_i \hat{L}_i \hat{B}_i, \qquad (4.312)$$

where the operator $\hat{L}_i = \hat{c}_n^\dagger \hat{c}_m$, which represents a quantum transition of an electron, acts on the system only and $\hat{B}_i = \sum_q \gamma_{n,m,q} \hat{a}_q^\dagger$, which represents a phonon emission, acts on the reservoir only. Here we identify the index i with a pair of indices (n, m). Then the spectral function for $\omega_{i,j}^\pm$ of the reservoir becomes

$$\hat{B}_i(t) \hat{B}_j = \sum_{q,q'} \gamma_{i,q} \left(\gamma_{j,q'} \right)^* \times \left[e^{-i\omega_q t} \left(\hat{a}_q^\dagger \hat{a}_{q'} + \hat{a}_q^\dagger \hat{a}_{q'}^\dagger \right) + e^{i\omega_q t} \left(\hat{a}_q \hat{a}_{q'} + \hat{a}_q \hat{a}_{q'}^\dagger \right) \right]. \qquad (4.313)$$

Computing the reservoir correlation functions by using the reservoir distribution function $\hat{\rho}_R(t_0)$, we have

$$\langle \hat{B}_i(t)\hat{B}_j\rangle_R = \sum_q \gamma_{i,q}\left(\gamma_{j,q}\right)^* \left(e^{-i\omega_q t}N_q + e^{i\omega_q t}\left(N_q+1\right)\right). \quad (4.314)$$

Here $N_q = \langle \hat{a}_q^\dagger \hat{a}_q\rangle = 1/\left(e^{\hbar\omega_q/k_B T}-1\right)$ is the average occupation number of the phonon mode q at temperature T. We replace the sum over the phonon modes with an integral using the phonon density of states $\nu_{\rm ph}(\omega)$ as

$$\begin{aligned}\hat{\omega}_{i,j}^{+} &= \int_0^\infty e^{i\omega_i t}\sum_q \gamma_{i,q}\left(\gamma_{j,q}\right)^*\left(e^{-i\omega_q t}N_q + e^{i\omega_q t}\left(N_q+1\right)\right)dt \\ &= \int_0^\infty \nu_{\rm ph}(\omega)\gamma_{i,\omega}\left(\gamma_{j,\omega}\right)^* \times \left(N_\omega \int_0^\infty e^{i(\omega_i-\omega)t}dt \right. \\ &\qquad \left. +(N_\omega+1)\int_0^\infty e^{i(\omega_i+\omega)t}dt\right)d\omega. \quad (4.315)\end{aligned}$$

Using the relation of $\int_0^\infty e^{\pm i\omega t}dt = \pi\delta(\omega) \mp iP\left(1/\omega\right)$ to evaluate the integral in t, where P indicates to take the principal part and produces a small shift due to coupling with the continuum of the bath states, we have $\omega_{i,j}^{+} = \omega_{j,i}^{-}$ where we neglect this small shift. Finally we obtain

$$\frac{\partial\hat{\rho}(t)}{\partial t} = -\frac{i}{\hbar}\left[H,\hat{\rho}\right] + C\left[\hat{\rho}(t)\right] \quad (4.316)$$

with the dissipation term as

$$C\left[\hat{\rho}(t)\right] = -\sum_{n,m}\Gamma_{n,m}\left(\hat{L}_{n,m}\hat{L}_{m,n}\hat{\rho} + \hat{\rho}\hat{L}_{n,m}\hat{L}_{m,n} - 2\hat{L}_{m,n}\hat{\rho}\hat{L}_{n,m}\right). \quad (4.317)$$

This is the Lindblad form, which guarantees that the reduced density matrix $\hat{\rho}$ remains positive definite and that its trace is preserved during time evolution as $\mathrm{Tr}\{\hat{\rho}(t)\} = 1$. Here $\hat{L}_{n,m} = \hat{c}_n^\dagger \hat{c}_m$ and the transition probabilities $\Gamma_{n,m}$ are defined by

$$\Gamma_{n,m} = \begin{cases} \nu_{\rm ph}(\epsilon_m-\epsilon_n)|\gamma_{n,m}|^2\left(N_{\epsilon_m-\epsilon_n}+1\right) & \epsilon_n<\epsilon_m \\ \nu_{\rm ph}(\epsilon_n-\epsilon_m)|\gamma_{n,m}|^2 N_{\epsilon_n-\epsilon_m} & \epsilon_n>\epsilon_m, \end{cases} \quad (4.318)$$

where the phonon frequency is $\hbar\omega = \epsilon_m - \epsilon_n$ for emission and $\hbar\omega = \epsilon_n - \epsilon_m$ for absorption.

The difference between absorption and emission probability is due to the *spontaneous emission* processes. This ensures detailed balance of

$$\frac{\Gamma_{n,m}}{\Gamma_{m,n}} = e^{(\epsilon_m - \epsilon_n)/k_B T}, \tag{4.319}$$

which is essential for the thermal equilibration of the subsystem. Actually we can show that the density matrix as $\hat{\rho}_{th} = e^{-\hat{H}/k_B T}/\mathrm{Tr} e^{-\hat{H}/k_B T}$ is a stationary solution to the QME, which reveals that the subsystem relaxes to thermal equilibrium as $\hat{\rho}(t) \to \hat{\rho}_{th}$ for $t \to \infty$ for any initial state.[a,b]

Here we see the correspondence with the classical master equation. Let us consider the occupation probability as the diagonal of the density matrix $P_m(t) = \langle m|\hat{\rho}(t)|m\rangle$ where $|m\rangle$ is the eigenstate of $\hat{H}$ as $\hat{H} = \sum_m \epsilon_m |m\rangle\langle m|$. From the QME, we can show that the equation of motion for $P_m(t)$ is governed by

$$\frac{\partial P_m(t)}{\partial t} = \sum_n \left[W_{m,n} P_n(t) - W_{n,m} P_m(t) \right] \tag{4.320}$$

with the transition rates of $W_{m,n} = 2\Gamma_{n,m}\langle n|\hat{L}_{n,m}|m\rangle\langle m|\hat{L}_{m,n}|n\rangle$. This is known as the Pauli master equation, the form of the classical discrete master equation. We note that $W_{n,m}$ are real and non-negative. Due to the detailed balance of $W_{m,n}e^{-\epsilon_n/k_B T} = W_{n,m}e^{-\epsilon_m/k_B T}$, the occupation probability at equilibrium P_n satisfies the Boltzmann distribution as $P_m = e^{-\epsilon_m/k_B T}/\sum_m e^{-\epsilon_m/k_B T}$.

---

[a] A time evolution of a density matrix $\hat{\rho}(t)$ is equivalent to a time evolution of a state $|\Psi\rangle$ according to the stochastic Schrödinger equation. The density matrix is constructed by averaging $|\Psi\rangle\langle\Psi|$ over the ensemble of stochastic realizations. We note that the QME is the quantum counterpart of the classical Fokker–Planck equation, while the stochastic Schrödinger equation corresponds to the quantum counterpart of the classical Langevin equation.

[b] The irreversibility evolution of an open quantum system is associated with the entropy production, which is measured by the von Neumann entropy as $S(\hat{\rho}) = -k_B \mathrm{Tr}\,(\hat{\rho}\ln\hat{\rho})$. We have $S(\hat{\rho}) \geq 0$ where the equality holds if $\hat{\rho}$ is a pure state. Therefore, the relaxation to thermal equilibrium increases $S(\hat{\rho})$. Using the spectral decomposition for the mixed state by $\hat{\rho} = \sum_i p_i|\psi_i\rangle\langle\psi_i|$ with $p_i \geq 0$ and $\sum_i p_i = 1$, we have $S(\hat{\rho}) = -k_B \sum_i p_i \ln p_i$, the form of the Shannon information entropy. For the composite system with $\hat{H}_S \otimes \hat{H}_R$, the subadditivity condition is satisfied as $|S(\hat{\rho}_S) - S(\hat{\rho}_R)| \leq S(\hat{\rho}_{S+R}) \leq S(\hat{\rho}_S) + S(\hat{\rho}_R)$, where the equality holds for an uncorrelated state $\hat{\rho}_{S+R} = \hat{\rho}_S \otimes \hat{\rho}_R$. Here we note that $\hat{\rho}_S = \mathrm{Tr}_R \hat{\rho}_{S+R}$ and $\hat{\rho}_R = \mathrm{Tr}_S \hat{\rho}_{S+R}$. This means that quantum information on correlation is lost by tracing over the subsystems.

4.6.3 Single-Particle Approximation for DFT

Here we derive an effective single-particle formula for the QME [102] for the applications of DFT. The electrons are described by a reduced one-particle density operator

$$\hat{\rho}_{KS}(t) = \sum_{n,m} |n\rangle f_{n,m}(t) \langle m|, \qquad (4.321)$$

where $|n\rangle$ is a basis of single-particle states and the numbers $f_{n,m}(t)$ are defined by

$$f_{n,m}(t) = \mathrm{Tr}\left(\hat{\rho}(t)\hat{c}_m^\dagger \hat{c}_n\right) \qquad (4.322)$$

in terms of the many-particle density operator $\hat{\rho}(t)$.

In the absence of coupling to the phonon bath, the single-particle operator $\hat{\rho}_{KS}(t)$ obeying the equation of motion within the formalism of DFT becomes the eigenstates of $\hat{H}$ corresponding to the equilibrium density $\hat{\rho}(\mathbf{r})$ as

$$i\hbar \frac{d\hat{\rho}_{KS}(t)}{dt} = (\epsilon_n - \epsilon_m)\, \hat{\rho}_{KS}(t), \qquad (4.323)$$

where ϵ_n are eigenvalues of $\hat{H}$.[a]

When we include the coupling to the phonon bath, a single-particle expression for the dissipation term is obtained from

$$C\left[\hat{\rho}(t)\right] = -\sum_{n,m} \Gamma_{m,n} \left(\hat{L}_{n,m}\hat{L}_{m,n}\hat{\rho} + \hat{\rho}\hat{L}_{n,m}\hat{L}_{m,n} - 2\hat{L}_{m,n}\hat{\rho}\hat{L}_{n,m}\right). \qquad (4.324)$$

Using the properties of the fermion operators

$$\mathrm{Tr}\left(\hat{\rho}\hat{c}_n^\dagger \hat{c}_m^\dagger \hat{c}_p \hat{c}_q\right) = f_{q,n} f_{p,m} - f_{q,m} f_{p,n}, \qquad (4.325)$$

we obtain this term as

$$C[\hat{\rho}(t)] = (\delta_{n,m} - f_{n,m}) \sum_p \left(\Gamma_{n,p} + \Gamma_{m,p}\right) f_{p,p} - f_{n,m} \sum_p \left(\Gamma_{p,n} + \Gamma_{p,m}\right)\left(1 - f_{p,p}\right), \qquad (4.326)$$

where we neglect the exchange interactions originating from the quadratic contributions in the off-diagonal elements of $\hat{\rho}$. Therefore,

[a] We note that in the DFT formalism, electron density is the same as that of the full interacting system $\rho(\mathbf{r}, t) = \mathrm{Tr}_N\left[\rho\hat{n}(\mathbf{r})\right] = \mathrm{Tr}_1\left[\hat{\rho}\hat{n}(\mathbf{r})\right]$.

we have the QME in the single-particle approximation, which have the properties of

$$0 \leq f_{n,n} \leq 1$$
$$\sum_n f_{n,n} = N$$
$$f_{n,m} = \delta_{n,m} \frac{1}{1 + e^{(\epsilon_n - \mu)/k_B T}}. \qquad (4.327)$$

It is noted that the off-diagonal elements of $f_{n,m}$ become zero toward equilibrium.

To accelerate the electron motion of the system, we apply an external field to the effective single-particle Hamiltonian, which is included in the first term of $-(i/\hbar)[\hat{H}(t), \hat{\rho}(t)]$. In case of a uniform electric field, the standard electrostatic gauge field is related to the potential via

$$\mathbf{E} = -\nabla\Phi, \qquad (4.328)$$

where **E** is an electric field and Φ is the corresponding electrostatic potential. We note that a uniform field $\Phi = -\mathbf{E} \cdot \mathbf{r}$ is not compatible with the periodic boundary condition.[a] Here we deal with a uniform electric field using periodic boundary conditions by adopting a different *gauge* for the electric field. In particular, if we use a vector potential

$$\mathbf{A} = -c\mathbf{E}t \qquad (4.329)$$

for which the corresponding electric field becomes

$$\mathbf{E} = -\frac{1}{c}\frac{\partial \mathbf{A}}{\partial t}. \qquad (4.330)$$

Namely we use a uniform time-dependent vector potential to describe spatially uniform electric field. This is compatible with the periodic boundary conditions to have a ring geometry with a linearly varying magnetic flux traversing the surface lifted by the ring. Although this uniform electric field is a magnetically induced electric field, the situation for an electronic system subject to an electric field is the same whether we use either gauge since physics

[a] In the scattering approaches such as the RTM, the LS, and the NEGF methods, this difficulty is bypassed by adopting open boundary conditions and by treating the electron motion due to an applied electric field as a scattering problem of electrons within an applied bias voltage.

is gauge invariant. This means that we include the vector potential $\mathbf{A}$ in the kinetic energy term by replacing the momentum operator with $\hat{\mathbf{p}} - (e/c)\mathbf{A}$.

When we have the single-particle Hamiltonian $\hat{H}(t)$, including an electric field via $\mathbf{A}(t)$, the equation of motion for the single-particle density matrix $\hat{\rho}$ in the basis of the eigenstates of the equilibrium Hamiltonian H is given by

$$\begin{aligned}\frac{d}{dt} f_{n,m} = &-\frac{i}{\hbar}\sum_p \left(H_{n,p}(t) f_{p,m} - f_{n,p} H_{p,m}(t)\right) \\ &+ (\delta_{n,m} - f_{n,m}) \sum_p \left(\Gamma_{n,p} + \Gamma_{m,p}\right) f_{p,p} \\ &- f_{n,m} \sum_p \left(\Gamma_{p,n} + \Gamma_{p,m}\right)\left(1 - f_{p,p}\right). \end{aligned} \qquad (4.331)$$

Here $H_{n,m}(t) = \langle n|\hat{H}(t)|m\rangle$ are the matrix elements of the time-dependent Hamiltonian $\hat{H}(t)$ in the basis of the eigenstates of the time-independent Hamiltonian of $\hat{H}$.

A vector potential, depending linearly on time, accelerates uniformly the electrons and is counteracted by the dissipative effect of the bath. Eventually, a steady-state situation is realized when the effect of the field is balanced on average $df_{n,m}/dt = 0$. To calculate the propagation of the density matrix under the effect of an electric field, we need to perform a gauge transformation. When the vector potential changes from $\mathbf{A} = 0$ to $\mathbf{A} = -c\mathbf{E}dt$, the single-particle electronic wavefunctions acquire an additional phase by $e^{id\mathbf{k}\cdot\mathbf{r}}$ with $d\mathbf{k} = \mathbf{E}dt$. Numerically we use a finite k-point mesh, which is uniform, and model the system with a finite ring consisting of a finite number of unit cells. Then a gauge transformation is possible only when $\Delta\mathbf{k} = \mathbf{E}\Delta t$ equals to the spacing of the k-point mesh.

Let us consider the formula for the current and its conservation law. For a single-electron subsystem, the electron density at $\mathbf{r}$ is given by $n(\mathbf{r}) = \langle\mathbf{r}|\hat{\rho}|\mathbf{r}\rangle$. If both the potential in the Hamiltonian $\hat{H}$ of the system and the coupling potential $\hat{V}_{\text{int}}$ are local functions of $\mathbf{r}$, we obtain

$$\frac{dn(\mathbf{r})}{dt} = -\frac{i}{\hbar}\langle\mathbf{r}|\left[\hat{H}, \hat{\rho}\right]|\mathbf{r}\rangle = -\nabla\cdot\hat{\mathbf{j}}(\mathbf{r}), \qquad (4.332)$$

where the current density at $\mathbf{r}$ is defined as

$$\mathbf{j}(\mathbf{r}) = \mathrm{Tr}\left[\hat{\rho}\hat{\mathbf{J}}(\mathbf{r})\right] \qquad (4.333)$$

with the operator of $\hat{\mathbf{J}}(\mathbf{r}) = (1/2)\left[\hat{\mathbf{p}}\delta(\mathbf{r} - \hat{\mathbf{r}}) + \delta(\mathbf{r} - \hat{\mathbf{r}})\hat{\mathbf{p}}\right]$.

4.6.4 *Electron Transport in TDDFT formalism*

Here we present the time-dependent DFT approach, rigorous formalism to treat the dynamical problem and excited energies of electrons for interacting systems in the framework of DFT. We show two different approaches for it, the time-dependent NEGF method and the QME method.

For the scattering approaches based on the NEGF formalism for electron transport, we treat steady state in which the time-independent current is flowing between electrodes. From the self-consistent procedure based on DFT, we solve an effective single-particle equation for the static local potential by assigning different Fermi levels deep inside the left and right electrodes. This enables us to calculate the steady current. However, since electron transport occurs through excited energy levels, the direct application of DFT is not sufficient.

On the other hand, excitation energies of interacting systems are obtained using the TDDFT. In the framework of the TDDFT, an effective time-dependent local potential is calculated for a fictitious non-interacting electron system. We need to use the TDDFT to treat nonequilibrium transport problems rigorously, namely for the many-body system out of equilibrium.

4.6.4.1 Time-dependent DFT Formalism

The TDDFT is based on the Runge–Gross theorem [103], which states that *if two potentials* $v(\mathbf{r}, t)$ and $v'(\mathbf{r}, t)$ differ by more than a purely time-dependent function $c(t)$, they cannot produce the same time-dependent density $n(\mathbf{r}, t)$. This means

$$v(\mathbf{r}, t) \neq v'(\mathbf{r}, t) + c(t) \quad \Rightarrow \quad n(\mathbf{r}, t) \neq n'(\mathbf{r}, t). \tag{4.334}$$

This theorem shows a one-to-one correspondence between the time-dependent potential and the time-dependent electron density.

In the ground state, the energy satisfies the minimization principle and thus is the most important. On the other hand, in the TDDFT, the action satisfies a stationarity condition and its functional derivatives are important. Thus, the proof uses the current density $\mathbf{j}(\mathbf{r}, t)$ and the continuity equation for the current $(\partial/\partial t)n(\mathbf{r}, t) = -\nabla \cdot \mathbf{j}(\mathbf{r}, t)$.

Based on this theorem, we can construct a time-dependent single-particle effective equation

$$i\hbar\frac{\partial}{\partial t}\psi_i(\mathbf{r}, t) = \left[-\frac{\hbar^2}{2m}\nabla^2 + V_{\text{eff}}(\mathbf{r}, t)\right]\psi_i(\mathbf{r}, t), \tag{4.335}$$

where

$$V_{\text{eff}}(\mathbf{r}, t) = V_{\text{ext}}(\mathbf{r}, t) + e^2\int\frac{n(\mathbf{r}', t)}{|\mathbf{r}-\mathbf{r}'|}d\mathbf{r}' + \frac{\delta E_{xc}}{\delta n(\mathbf{r}, t)} \tag{4.336}$$

and

$$n(\mathbf{r}, t) = \sum_i |\psi_i(\mathbf{r}, t)|^2. \tag{4.337}$$

The exchange-correlation potential $\delta E_{xc}/\delta n(\mathbf{r}, t)$ depends in principle on entire history of the time and on the initial states. However, it is frequently approximated by adiabatic local density approximation (ALDA),

$$\frac{\delta E_{xc}}{\delta n(\mathbf{r}, t)} = \left.\frac{d\left[n\epsilon_{xc}(n)\right]}{dn}\right|_{n(\mathbf{r},t)}, \tag{4.338}$$

which uses electron density at time t and thus neglects the memory effect of the system.[a] The TDDFT has been used for various applications, which include excited energies, Rydberg excitation problem, atomic states in laser fields, and quantum transport [104].

In the transport applications, the time-dependent current is given by

$$I(t) = -e\int\frac{d}{dt}n(\mathbf{r}, t)d\mathbf{r}. \tag{4.339}$$

We calculate $n(\mathbf{r}, t)$ from the fictitious non-interacting system described by an effective Hamiltonian $\hat{H}_{KS}(t)$, which is given by the sum of external potential, Hartree potential, and exchange-correlation potential. We begin the calculation from thermally equilibrium electron density $n(\mathbf{r}, 0) = \sum_i f(\epsilon_i)|\psi_i(\mathbf{r})|^2$, where $\psi_i(\mathbf{r})$ is the Kohn–Sham eigenfunctions of $\hat{H}_{KS}(0)$ with the Kohn–Sham eigenvalues ϵ_i.

To proceed with electron transport using the TDDFT formalism, we will show two computational methods. One is to use the NEGF

[a]We note that the direction of improvements for the time-dependent exchange-correlation functionals have been proposed, e.g., time-dependent current density functional theory (TD-CDFT).

method in the time domain. In this framework, the current is obtained by solving a scattering problem with open boundary conditions. Electrons are injected from electrodes through these boundaries, where different chemical potentials are assigned. Then we solve the time-dependent energy-dependent scattering problems, using the NEGFs. The other method is based on the QME formalism from the density matrix, which is a generalization of the Boltzmann transport equation to the fully quantum mechanical case. In this approach, electrons are accelerated by an external driving force and the energy injected in the system is dissipated by inelastic scatterings. A steady state is achieved by the interplay of acceleration and dissipation processes where a time-independent current flows through the system.

4.6.4.2 Time-dependent NEGF formalism

To proceed with the TDDFT calculations for the many-electron case, including the Coulomb interaction within the mean field approach, we need to obtain the time-dependent Kohn–Sham orbitals for arbitrary time-dependent potentials from

$$i\hbar\frac{\partial}{\partial t}\psi_i(\mathbf{r},t) = \left[-\frac{\hbar^2}{2m}\nabla^2 + V_{\mathrm{eff}}(\mathbf{r},t)\right]\psi_i(\mathbf{r},t), \tag{4.340}$$

where

$$V_{\mathrm{eff}}(\mathbf{r},t) = V_{\mathrm{ext}}(\mathbf{r},t) + e^2\int\frac{n(\mathbf{r}',t)}{|\mathbf{r}-\mathbf{r}'|}d\mathbf{r}' + \frac{\delta E_{xc}}{\delta n(\mathbf{r},t)} \tag{4.341}$$

with

$$n(\mathbf{r},t) = \sum_i |\psi_i(\mathbf{r},t)|^2 \tag{4.342}$$

and construct the NEGFs for electric current.

The numerical calculations for the evolution of the Kohn–Sham orbitals might be proceeded when we assume the localized basis sets as follows [105]. The time-dependent Kohn–Sham equation is integrated by

$$i\hbar\frac{\partial}{\partial t}\begin{pmatrix}|\psi_L(t)\rangle\\|\psi_C(t)\rangle\\|\psi_R(t)\rangle\end{pmatrix} = \begin{pmatrix}\hat{H}_L(t) & \hat{V}_{LC} & 0\\ \hat{V}_{LC}^\dagger & \hat{H}_C(t) & \hat{V}_{CR}^\dagger\\ 0 & \hat{V}_{CR} & \hat{H}_R(t)\end{pmatrix}\begin{pmatrix}|\psi_L(t)\rangle\\|\psi_C(t)\rangle\\|\psi_R(t)\rangle\end{pmatrix}, \tag{4.343}$$

where we assume that the coupling terms $\hat{V}_{LC}$ and $\hat{V}_{CR}$ are time-independent for simplicity. Then we have

$$|\psi_{L/R}(t)\rangle = \int_0^t \hat{g}_{L/R}(t, t')\hat{V}_{L/R,C}|\psi_C(t')\rangle dt' + \hat{g}_{L/R}(t, 0)|\psi_{L/R}(0)\rangle \tag{4.344}$$

for $|\psi_{L/R}(t)\rangle$ and

$$i\hbar\frac{\partial}{\partial t}|\psi_C(t)\rangle = \hat{H}_C(t)|\psi_C(t)\rangle + \sum_{L,R}\int_0^t \hat{\Sigma}_{L/R,C}(t, t')|\psi_C(t')\rangle dt' + \left[\hat{V}_{LC}\hat{g}_L(t, 0)|\psi_L(0)\rangle + \hat{V}_{CR}\hat{g}_R(t, 0)|\psi_R(0)\rangle\right] \tag{4.345}$$

for $|\psi_C(t)\rangle$, where Green's functions $g_{L/R}(t, t')$ for electrodes are defined by

$$\left[i\hbar\frac{\partial}{\partial t} - \hat{H}_{L/R}(t)\right]\hat{g}_{L/R}(t, t') = \delta(t - t') \tag{4.346}$$

and self-energies for the couplings from the center C region to electrodes L, R are

$$\hat{\Sigma}_{L/R,C}(t, t') = \hat{V}^{\dagger}_{L/R,C}\hat{g}_{L/R}(t, t')\hat{V}_{L/R,C}, \tag{4.347}$$

respectively. Thus, we solve the time-dependent Kohn–Sham orbitals from the coupled equations of Eq. (4.344) and Eq. (4.345).

As for the initial condition, we can take the time-independent ground-state calculations,

$$\left[\epsilon\hat{I} - \hat{H}_{L/R}\right]\hat{g}_{L/R}(\epsilon) = \hat{I}, \tag{4.348}$$

which construct the static Kohn–Sham equation for the center region as

$$\left[\epsilon\hat{I} - \hat{H}_C - \hat{\Sigma}_L(\epsilon) - \hat{\Sigma}_R(\epsilon)\right]\hat{G}_C(\epsilon) = \hat{I}. \tag{4.349}$$

Here self-energies are $\hat{\Sigma}_{L/R}(\epsilon) = \hat{V}^{\dagger}_{L/R,C}\hat{g}_{L/R}(\epsilon)\hat{V}_{L/R,C}$.

In the former DFT approaches, the electric current through transmission is calculated from these static Green's functions. In the TDDFT approach, these are used only for the initial conditions and we evolve the time-dependent Kohn–Sham orbitals, constructing the time-dependent electron density $n(t, t')$ and effective potential $\hat{H}_C(t)$. Then electric current $I(t)$ is obtained using the time-dependent Green's functions.

4.6.4.3 Time-dependent quantum master equation

The alternative way for transport calculation is to use the density matrix method as the QME. This method treats in itself, as was shown in the previous section, a time-dependent equation for the density matrix of the system $\hat{\rho}(t)$

$$\frac{d}{dt}\hat{\rho}(t) = -\frac{i}{\hbar}\left[\hat{H}(t), \hat{\rho}(t)\right] + \hat{C}\left[\hat{\rho}(t)\right], \tag{4.350}$$

where the dissipation term, which brings the system to its thermal equilibrium, is expressed in the Markov approximation as

$$\hat{C}\left[\hat{\rho}(t)\right] = -\sum_{n,m}\Gamma_{n,m}\left(\hat{L}_{n,m}\hat{L}_{m,n}\hat{\rho} + \hat{\rho}\hat{L}_{n,m}\hat{L}_{m,n} - 2\hat{L}_{m,n}\hat{\rho}\hat{L}_{n,m}\right). \tag{4.351}$$

$\hat{L}_{n,m}$ represents an electron transition from eigenstates $|m\rangle$ to $|n\rangle$ and the transition strength of the coupling to the bath $\Gamma_{n,m}$ are defined by

$$\Gamma_{n,m} = \begin{cases} \nu_{\mathrm{ph}}(\epsilon_m - \epsilon_n)|\gamma_{n,m}|^2\left(N_{\epsilon_m-\epsilon_n} + 1\right) & \epsilon_n < \epsilon_m \\ \nu_{\mathrm{ph}}(\epsilon_n - \epsilon_m)|\gamma_{n,m}|^2 N_{\epsilon_n-\epsilon_m} & \epsilon_n > \epsilon_m \end{cases}, \tag{4.352}$$

where $\nu_{\mathrm{ph}}(\epsilon)$ is the density of states for phonons, $\gamma_{n,m}$ is the electron–phonon interaction strength, and N_ϵ is the thermal mean occupation number of phonons with energy $\epsilon = \hbar\omega$ as $N_\epsilon = 1/(e^{\hbar\omega/k_BT} - 1)$. The imbalance of probabilities of electronic transitions $\Gamma_{n,m}$ from the spontaneous emission of a phonon is important for the energy relaxation process toward thermal equilibrium.

To carry out the computation of many-electron systems, the TDDFT formalism is introduced. The potential is shown to be determined uniquely by electron density $n(\mathbf{r}, t)$ for a dissipative quantum system for a given dissipation term $\hat{C}$. Since the density matrix is defined by $\hat{\rho}_{KS}(t) = \sum_{n,m}|m\rangle f_{n,m}(t)\langle n|$ where $|m\rangle$ and $|n\rangle$ are the single-particle Kohn–Sham orbitals, the time evolution of the distribution coefficient $f_{n,m}$ follows:

$$\begin{aligned}\frac{d}{dt}f_{n,m} = &-\frac{i}{\hbar}\sum_p\Big(H_{n,p}(t)f_{p,m} - f_{n,p}H_{p,m}(t)\Big) \\ &+ (\delta_{n,m} - f_{n,m})\sum_p\left(\Gamma_{n,p} + \Gamma_{m,p}\right)f_{p,p} \\ &- f_{n,m}\sum_p\left(\Gamma_{p,n} + \Gamma_{p,m}\right)\left(1 - f_{p,p}\right),\end{aligned} \tag{4.353}$$

which is an analog of the Boltzmann transport equation to the quantum one with the TDDFT. Here $H_{n,m}$ are the time-dependent Kohn–Sham Hamiltonian in which Coulomb interaction effects are included within the single-particle approximation.

Electrons are accelerated by spatially constant external electric field $\mathbf{E}$, which produces a time-dependent vector potential $\mathbf{A}(t) = -c\mathbf{E}t$ with the periodic boundary conditions in a ring geometry. Electric current is obtained for the large system when the time-dependent simulation goes into a steady state. During the simulations we need a gauge transformation to change the vector potential $\mathbf{A}(t)$ into $\mathbf{A}(0)$ and to put the wavefunctions with a phase factor $\exp(-iA(\tau)z/c) = \exp(i2\pi z/L)$ at times that are integer multiple of $\tau = 2\pi/|\mathbf{E}|L$ where L is the system size along the ring geometry z. This avoids that the vector potential increases indefinitely. Electric current is obtained from

$$\mathbf{j}(\mathbf{r}, t) = \mathrm{Tr}\left[\hat{\rho}_{KS}(t)\hat{\mathbf{J}}(\mathbf{r})\right] \tag{4.354}$$

with the current operator

$$\hat{\mathbf{J}}(\mathbf{r}) = \frac{1}{2}\left[\left(\hat{p} - \frac{e}{c}\mathbf{A}(t)\right)\delta(\mathbf{r} - \mathbf{r}') + \delta(\mathbf{r} - \mathbf{r}')\left(\hat{p} - \frac{e}{c}\mathbf{A}(t)\right)\right]. \tag{4.355}$$

To satisfy the continuity equation, we need a condition

$$\langle\mathbf{r}|e^{-i\hat{H}(t-t_0)/\hbar}\frac{\tilde{\rho}(t) - \tilde{\rho}(t_0)}{t - t_0}e^{i\hat{H}(t-t_0)/\hbar}|\mathbf{r}\rangle = 0 \tag{4.356}$$

for $\Delta t = t - t_0 \to 0$ in Eq. (4.308). However, due to the coarse time-averaging procedure, this is not satisfied for a finite difference approximation $\Delta t \neq 0$ in the numerical calculations. Therefore, the additional dissipation current $\mathbf{j}_D(\mathbf{r}, t)$ is needed for $\mathbf{j}(\mathbf{r}, t)$ [106].

4.6.5 *Examples of Simple Systems*

In order to show an application of the QME, we consider a one-dimensional crystal ring in the tight-binding approximation where the band is half-filled with a metallic situation realized. We apply a small external uniform electric field by a vector potential and solve numerically the QME for the density matrix. by using the Chebyshev expansion method for the time evolution. The density of states of the

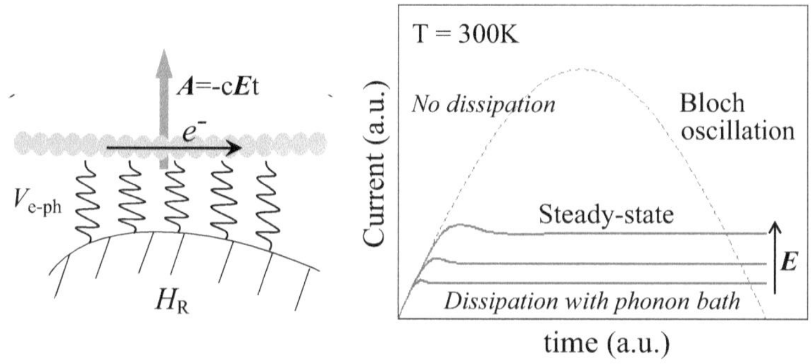

Figure 4.34 (Left) Schematic illustrations of the one-dimensional crystal model. Electron is accelerated by an electric field **E** via a vector potential **A** and is dissipated by the inelastic phonon scatterings $V_{\mathrm{e-ph}}$ coupled to the phonon bath H_R. (Right) Current as a function of time in the absence of dissipation (dashed line) and in the presence of dissipation (solid line). A steady state is realized when the effect of the field is balanced on the dissipative effect of the bath.

reservoir bath is assumed to be for the acoustic phonons with the parabolic form $\nu_{\mathrm{ph}}(\omega) = \omega^2$ with the constant coupling strength. The room temperature is set to $T = 300$ K for the dissipative phonon scatterings and the results are shown in Fig. 4.34.

In the absence of dissipation, we see that the current shows oscillatory behavior, which is known as the Bloch oscillations. On the other hand, in the presence of dissipation due to electron–phonon scatterings, a stationary situation is reached after some transient time. The current increases almost linearly as we increase an electric field. The occupation numbers of electrons, which is obtained from the diagonal elements of $f_{n,n}$, show a displacement around the Fermi energy when a steady-state condition is satisfied. This corresponds to the situation where the microscopic basis of Ohm's law is satisfied with the balance of electron acceleration due to an electric field and electron energy reduction due to dissipative inelastic phonon scatterings for relaxation. We note that the same result is obtained in this case using the Boltzmann equation, since there is no interband transition in the present model.

4.7 Problems

(1) Let us consider the interacting (3D) electron gas systems. The Hamiltonian is given by

$$H = \sum_i -\frac{\hbar^2}{2m}\nabla_i^2 + \sum_{i<j} \frac{e^2}{|r_i - r_j|^2}.$$

(a) Write the Hamiltonian in the second quantization form.

(b) Within the Hartree–Fock approximation, show that the single-particle energy becomes

$$\epsilon_k = \frac{\hbar^2 k^2}{2m} - \frac{e^2 k_F}{\pi}\left(1 + \frac{k_F^2 - k^2}{2kk_F}\ln\left|\frac{k_F + k}{k_F - k}\right|\right)$$

and that the ground-state energy per electron is given by

$$\frac{E_{\rm tot}}{N} = \epsilon_{\rm tot}(r_s) = \frac{2.211}{r_s^2} - \frac{0.916}{r_s} \quad \text{(Ryd)},$$

where $r_s a_B = (3/4\pi n)^{1/3}$ is the inter-atomic distance for the electron density n. The first term represents the kinetic energy and the second term gives the exchange energy. Note that the single-particle energy ϵ_k has a discontinuity at the Fermi wavelength $k = k_F$.

(c) The pair correlation function $g(r)$ is the probability of finding an electron at r when there is another one at $r = 0$ and is related to the two-particle density matrix. Show that $g_{\sigma\sigma'}(r)$ in the Hartree–Fock approximation are

$$g_{\uparrow\downarrow}(r) = \frac{1}{2}, \quad g_{\uparrow\uparrow}(r) = \frac{1}{2}\left\{1 - 9\left(\frac{\sin k_F r - k_F r \cos k_F r}{(k_F r)^3}\right)^2\right\},$$

and the relationship of $\int \left(g_{\uparrow\downarrow}(r) + g_{\uparrow\uparrow}(r) - 1\right) dr = -1$. Explain the meaning of this equation.

(d) Show that the ferromagnetic state where the spins are polarized for all the electrons appears for

$$r_s > 5.46$$

in the Hartree–Fock approximation.

(e) Let us consider the screening of electron gas. We define the dielectric constant $\epsilon(\mathbf{q})$ as $\tilde{\varphi}(q) = \varphi_{\text{ext}}(q)/\epsilon(q)$ where $\tilde{\varphi}(r)$ is the screened Coulomb potential due to electrons.
Show that $\epsilon(q)$ is given by

$$\epsilon(\mathbf{q}) = 1 - \frac{4\pi}{q^2}\chi(\mathbf{q}),$$

where the susceptibility is given by $\chi(\mathbf{q}) = e^2\delta n_q/\tilde{\varphi}(q)$.
Derive the induced density δn_q and then show that the susceptibility (polarizability) is given by $\chi(q) = e^2\sum_k(n_k - n_{k+q})/(\epsilon_k - \epsilon_{k+q})$ from the first-order perturbation in the random-phase approximation (RPA).
Show that

$$\epsilon_{\text{RPA}}(\mathbf{q}) = 1 + \frac{1}{2}\frac{4\pi e^2}{q^2}\nu\left(1 + \frac{(2k_F)^2 - q^2}{2q(2k_F)}\ln\left|\frac{2k_F + q}{2k_F - q}\right|\right).$$

Note that $\epsilon_{\text{RPA}}(\mathbf{q})$ has a discontinuity at $q = 2k_F$, resulting in the asymptotic behavior of the screened potential at large distance as $\tilde{\varphi}(\mathbf{r}) \sim \cos(2k_F r)/r^3$. This is called the Friedel oscillations.

(f) Derive the formula for the ground-state total energy

$$E_{\text{tot}} = E_{\text{kin}} - \sum_q\left(\int_0^\infty \frac{d\omega}{2\pi}\text{Im}\left[\frac{1}{\epsilon(q,\omega)}\right] + \frac{2\pi Ne^2}{q^2}\right).$$

Give the correlation energy per electron within the RPA.

(2) Consider the one-dimensional system with the potential V_L on the left region and V_R on the right region with a step potential at the center region.

(a) Write down the total Hamiltonian of the system.
(b) The dimension of the matrix of the center region can be taken in 2×2. Compute the self-energy matrix $\Sigma_{L/R}(\epsilon)$ for the left and right regions from the surface Green's function $g_{L/R}(\epsilon)$.
(c) Calculate the transmission $\mathcal{T}(\epsilon)$ of this system using the NEGF formalism.

(3) Derive the formula for the Hall conductivity σ_{xy} from the generalized Kubo formula

$$\sigma_{xy}(\epsilon_F) = e^2\int_0^{1/k_BT} d\lambda \int_0^\infty dt\langle\hat{v}_y(0)\hat{v}_x(t + i\hbar\lambda)\rangle,$$

where $\hat{x}$ represents the Heisenberg representation of x.

(4) On the density matrix ρ,

(a) Show that the Liouville equation for the time evolution of density matrix ρ satisfies for both pure and mixed states.

(b) Show that $\mathrm{Tr}(\hat{\rho}^2) \leq 1$ where the equality holds only for the pure states.

(c) Show that the density matrix in the interaction picture satisfies the following equation:

$$i\hbar\frac{d\tilde{\rho}(t)}{dt} = \left[\tilde{H}_1(t), \tilde{\rho}(t)\right],$$

where we define $\tilde{\rho}(t) = e^{i\hat{H}_0 t/\hbar}\hat{\rho}(t)e^{-i\hat{H}_0 t/\hbar}$ and $\tilde{H}_1(t) = e^{i\hat{H}_0 t/\hbar}\hat{H}_1 e^{-i\hat{H}_0 t/\hbar}$ for the Hamiltonian $\hat{H} = \hat{H}_0 + \hat{H}_1$.

(d) For the tensor product $|sr\rangle = |s\rangle \otimes |r\rangle$ of thc two composite subsystems $\hat{S}$ and $\hat{R}$, show that the reduced density matrix is given by $\hat{\rho}_S = \mathrm{Tr}_R\hat{\rho}_{tot}$, where $\hat{\rho}_{tot} = \sum_{ij} p_{ij}|s_i r_j\rangle\langle s_i r_j|$.

(e) When a state for two composite systems cannot be written as a tensor product, that state is called an entangled state [107]. Here let us consider a two-dimensional system with a basis set of $|0\rangle$ and $|1\rangle$. For an entangled state of $|\Psi\rangle = (|00\rangle + |11\rangle)/\sqrt{2}$ (Bell state), calculate the reduced density matrix of $\hat{\rho}_1 = \mathrm{Tr}_2\hat{\rho}_{tot}$. Show $\hat{\rho}_1$ is a mixed state.

(f) For the density matrix in the position representation $\rho(x, x') = \langle x|\hat{\rho}|x'\rangle$, using the center-of-mass coordinate $X = (x + x')/2$ and the relative coordinate $\xi = x - x'$, let us consider the following Wigner distribution function:

$$W(X, P) = \frac{1}{2\pi\hbar}\int \rho\left(X + \frac{\xi}{2}, X - \frac{\xi}{2}\right) e^{-iP\xi/\hbar} d\xi$$
$$\propto \left\langle \psi^\dagger\left(X + \frac{\xi}{2}\right)\psi\left(X - \frac{\xi}{2}\right)\right\rangle,$$

related to NEGF $-iG^<(x, x')$. Show that $\int W(X, P)dP \geq 0$, $\int W(X, P)dX \geq 0$, and $\int W(X, P)dXdP = 1$, which relates the Wigner function to the classical distribution function $f(X, P)$.

(g) Consider the density matrix of a two-level system with the Lindblad operators $\hat{L}_1 = \sqrt{\gamma}\begin{pmatrix} 1 & 0 \\ 0 & 0 \end{pmatrix}$, $\hat{L}_2 = \sqrt{\frac{\Gamma}{2}}\begin{pmatrix} 0 & 0 \\ 1 & 0 \end{pmatrix}$, calculate the density matrix $\tilde{\rho}_1(t)$ and $\tilde{\rho}_2(t)$ in the interaction picture from the Lindblad equation and explain the dephasing and decoherence for these operators.

(5) Let us obtain the time-dependent transient current though the center region (C) between left (L) and right (R) electrodes with the Fermi–Dirac distributions

$$f_{L/R} = \frac{1}{e^{(\epsilon_{L/R}-\mu_{L/R})/k_B T} + 1}.$$

Here we consider the one-dimensional system with one energy level ϵ_0 between two electrodes. The Hamiltonian is written as

$$\hat{H} = \epsilon_0 \hat{d}^\dagger \hat{d} + \sum_{k \in L,R} \epsilon_k \hat{c}_k^\dagger \hat{c}_k + \sum_{k \in L,R} V_k \left(\hat{d}^\dagger \hat{c}_k + \hat{c}_k^\dagger \hat{d} \right).$$

We assume small couplings of the center region to electrodes and the left and right regions are in thermal equilibrium.

(a) Calculate the transient current $I(t)$ for $\hat{H}$ for the sudden voltage application by the projection operator method [108].

(b) For the Liouville–von Neumann equation with

$$\hat{H} = \begin{pmatrix} \hat{H}_L & \hat{V}_{LC} & 0 \\ \hat{V}_{LC} & \hat{H}_C & \hat{V}_{CR} \\ 0 & \hat{V}_{CR} & \hat{H}_R \end{pmatrix} \quad \hat{\rho} = \begin{pmatrix} \hat{f}_L & \hat{\rho}_{LC} & 0 \\ \hat{\rho}_{LC} & \hat{\rho}_C & \hat{\rho}_{CR} \\ 0 & \hat{\rho}_{CR} & \hat{f}_R \end{pmatrix},$$

derive the reduced density matrix $\hat{\rho}_C$ in the form of

$$e\frac{d\hat{\rho}_C}{dt} = -\frac{ie}{\hbar}\left[\hat{H}_C, \hat{\rho}_C\right] + I_L + I_R,$$

using $\hat{H}_C$, $\hat{H}_{L/R}$, $\hat{V}_{LC}$, $\hat{V}_{CR}$, and $f_{L/R}$. Here $I_{L/R}$ represents an electron transfer between the center region and left/right electrode.

(c) Calculate the transient current $I(t)$ in (a) from the Liouville–von Neumann equation in (b) [109].

Chapter 5

Atomistic Nanosystems

5.1 STM Simulations and Atom Manipulation

In this section, we describe the basic operating principles of the scanning tunneling microscope (STM). STM was invented in 1982 [110] and has provided a revolutionary method for obtaining information on surfaces of metals and semiconductors with atomic resolution in real space. Now STM has become an indispensable experimental tool in surface science. Furthermore, STM has been used to manipulate atoms and molecules, and opened the possibility to create electron devices from atomistic viewpoints.

5.1.1 *Perturbation for Tunneling Current*

Figure 5.1 (left) shows a schematic view of the operation of STM. An atomic-scale probe tip is set close to the surface within a few Å separation, then there is a finite probability for an electron to tunnel between them through an overlap of wavefunctions. As the tip scans over the sample surface, a two-dimensional contour plot of the equal tunneling current is obtained and displayed as an image [111, 112].

Quantum Transport Calculations for Nanosystems
Kenji Hirose and Nobuhiko Kobayashi

ISBN 978-981-4267-32-8 (Hardcover), 978-981-4267-59-5 (eBook)
www.panstanford.com

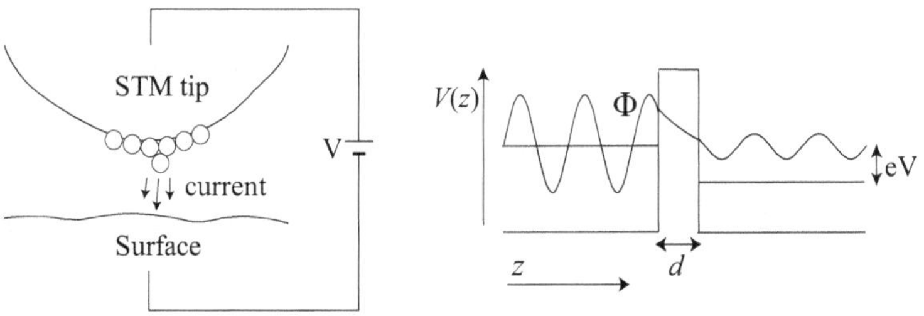

Figure 5.1 (Left) Schematic view of STM. Two-dimensional contour image with an atomic resolution is obtained through a scanning of the tip over the surface. (Right) A one-dimensional view of the tunneling current of a junction system with a separation of d and the potential barrier height Φ as a workfunction. An applied bias V corresponds to the difference of chemical potentials of the tip and the surface.

First, let us model the tip and surface simply as free-electron metals separated by a high potential barrier through a gap and apply a finite voltage (V) between them as shown in Fig. 5.1 (right). The chemical potential μ of one electrode shifts with respect to the other by eV. If the barrier height is large enough, the tunneling current from the tip to the surface can be evaluated from

$$I = \frac{2e}{h}\int_{-\infty}^{\infty} |\mathcal{T}(\epsilon)|^2 \left[f(\epsilon) - f(\epsilon + eV)\right] d\epsilon. \tag{5.1}$$

The tunneling current depends on the transmission $|\mathcal{T}(\epsilon)|^2$ for a small applied bias voltage.

When we neglect the energy dependence ϵ simply, the transmission $|\mathcal{T}|^2$ is dominated by an exponential decay of the wavefunction into the barrier as $|\mathcal{T}|^2 \propto e^{-2\kappa d}$ with a decay length of $\kappa = \sqrt{2m\Phi}/\hbar$ and the width of a potential barrier d. Since the workfunction takes typically $\Phi \sim 2$–4 eV measured from the vacuum level as a reference energy, we can assume $eV \ll \Phi$ for the tunneling regime. The tunneling current is thus evaluated from

$$I \simeq \frac{2e^2}{h} V \nu(\epsilon_F) e^{-2d\sqrt{2m\Phi}/\hbar}, \tag{5.2}$$

where $\nu(\epsilon)$ denotes the density of states (DOS). The dependence of $|\mathcal{T}|^2$ as a tunneling current, which roughly decays an order of magnitude when the barrier width d increases a few Å, leads to an atomic-scale resolution of the STM.

5.1.1.1 Bardeen's perturbation theory

Next, we derive a formula for the tunneling current using a time-dependent perturbation approach, which enables us to understand the STM image in view of the realistic surface and tip structure. We calculate electronic states of separated subsystems for the tip μ and surface ν first as shown in Fig. 5.2 (left), instead of solving the Schrödinger equation of the combined system in Fig. 5.2 (right). This is usually much easier than to obtain electronic states of the combined system. The tunneling current is calculated using a time-dependent perturbation theory by [113]

$$I = \frac{4\pi e}{\hbar} \sum_{\mu,\nu} |M_{\mu,\nu}|^2 \left[f(\epsilon_\mu) - f(\epsilon_\nu + eV) \right] \delta(\epsilon_\mu - \epsilon_\nu), \tag{5.3}$$

where the transition matrix element $M_{\mu,\nu}$ is determined from the overlap of wavefunctions of each subsystem as follows. Using the Fermi golden rule, since the transition probability per unit time from the state μ to the state ν becomes

$$\frac{|b_{\mu,\nu}(t)|^2}{t} \simeq \frac{2\pi}{\hbar} |M_{\mu,\nu}|^2 \delta(\epsilon_\mu - \epsilon_\nu) \tag{5.4}$$

with

$$\begin{aligned} M_{\mu,\nu} &= \int \psi_\mu^* (\epsilon - H) \psi_\nu d\mathbf{r} \\ &= \int \psi_\mu^* (\epsilon_\nu - H) \psi_\nu d\mathbf{r} - \int \psi_\nu (\epsilon_\mu - H) \psi_\mu^* d\mathbf{r} \\ &= \frac{\hbar^2}{2m} \iint d\mathbf{S} \left(\psi_\mu^* \nabla \psi_\nu - \psi_\nu \nabla \psi_\mu^* \right), \end{aligned} \tag{5.5}$$

we can obtain the transition matrix elements $M_{\mu,\nu}$ from ψ_μ and ψ_ν of the time-dependent wavefunction

$$\Psi(t) = a(t) \psi_\mu e^{-i\epsilon_\mu t/\hbar} + \sum_\nu b_{\mu\nu}(t) \psi_\nu e^{-i\epsilon_\mu t/\hbar}. \tag{5.6}$$

Then the STM tunneling current I is obtained from Eq. (5.3).

In this formula, we do not need to calculate the total wavefunction $\Psi(t)$ for the tunneling current I, which would be difficult and time-consuming. Since ψ_μ and ψ_ν are calculated independently once and the transition matrix elements $M_{\mu,\nu}$ are obtained through ψ_μ and ψ_ν with the tip position fixed with respect to the surface, we can

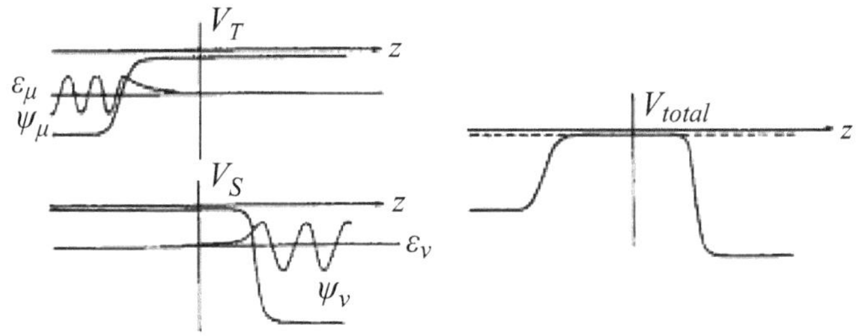

Figure 5.2 Schematic view of the Bardeen approach to tunneling current. (Left) Wavefunctions ψ_μ and ψ_ν for the tip μ and surface ν, which are obtained independently for the effective potentials of V_T and V_S at ϵ_μ and ϵ_ν. (Right) Total potential for the combined system of the tip and surface for the STM. Tunneling current is calculated perturbatively through wavefunctions of each subsystem using the Fermi golden rule.

obtain the tunneling current I for the STM image through $|M_{\mu,\nu}|^2$ quickly as a two-dimensional contour plot. We note that this formula is based on the perturbation approach for the tunneling regime where the interaction of the tip and the surface is negligible.

5.1.1.2 STM image simulation

Here we consider the physical meaning of the STM image in view of electronic states. From Bardeen's formula in Eq. (5.3), the tunneling current to the first order is obtained from

$$I = \frac{4\pi e^2}{\hbar} V \sum_{\mu,\nu} |M_{\mu,\nu}|^2 \delta(\epsilon_\mu - \epsilon_F)\delta(\epsilon_\nu - \epsilon_F). \tag{5.7}$$

As a limiting case, when we replace the tip structure to a point approximately, we then have

$$\frac{dI}{dV} \propto \sum_{\nu} |\psi_\nu(\mathbf{r})|^2 \delta(\epsilon_\nu - \epsilon_F) = \nu(\epsilon_\nu - \epsilon_F, \mathbf{r}). \tag{5.8}$$

This means that the differential conductance dI/dV measured in an STM image is proportional to the local density of states (LDOS) $\nu(\mathbf{r})$ of the surface at the bias voltage of $\epsilon_F - eV$. Here we set the Fermi energy ϵ_F at the tip side. This formula is also true when the shape of the tip is approximated by the s-wave wavefunction. This shows that the tunneling conductance corresponds approximately to the LDOS

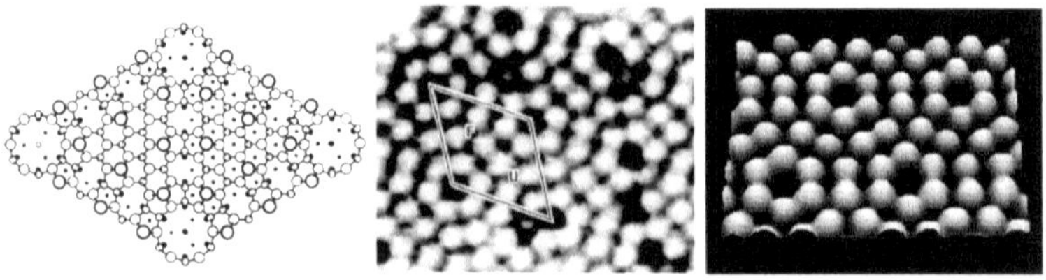

Figure 5.3 (Left) Dimer-adatom-stacking-fault (DAS) model for the Si(111)-(7× 7) surface. From K. Takayanagi, Y. Tanishiro, S. Takahashi, and M. Takahashi, *Surf. Sci.* **164**, 367 (1985) [115]. (Center) STM experimental topograph of the Si(111)-(7× 7) surface with +2 V applied to the sample. From R. J. Hamers, R. M. Tromp, and J. E. Demuth, *Phys. Rev. Lett.* **56**, 1972 (1987) [116]. (Right) Corresponding STM simulation image with the DAS model based on the ab initio DFT calculations. From K. D. Brommer, M. Needels, B. E. Larson, and J. D. Joannopoulos, *Phys. Rev. Lett.* **68**, 1355 (1992) [117].

$\nu(\epsilon, \mathbf{r})$ of the sample surface relative to the tip energy [114]. Thus, the images of STM we observe in experiments are the electronic states of surfaces.

As an example, in Fig. 5.3, we show the dimer-adatom-stacking-fault (DAS) model for the Si(111)-(7× 7) surface (left), which is the most complex and widely studied surface of a solid, experimental STM image of Si(111)-(7× 7) surface (center), and STM simulation with the DAS model based on ab initio density functional theory (DFT) calculation using Eq. (5.8) (right). We see that the simulation reproduces the STM image very well, which supports the atomistic DAS model for the reconstruction structure of Si(111)-(7 × 7) surface.

When we consider the tip structure of the STM image for more realistic STM simulation, we need to evaluate the higher order of the tunneling matrix as

$$I = \frac{4\pi e}{\hbar} \int_{\epsilon_F}^{\epsilon_F + eV} d\epsilon \, [f(\epsilon) - f(\epsilon + eV)] \times \int d\mathbf{r} \int d\mathbf{r}' V_T(\mathbf{r}) V_T(\mathbf{r}') G^T(\mathbf{r}', \mathbf{r}; \epsilon + eV) G^S(\mathbf{r} + \mathbf{R}, \mathbf{r}' + \mathbf{R}; \epsilon),$$

where $V_T(\mathbf{r})$ is the tip potential, $\mathbf{R}$ represents the position of a certain fixed point with respect to the surface. The imaginary part of Green's functions for the tip (T) and the surface (S) are

$$G^T(\mathbf{r}', \mathbf{r}, \epsilon) = \sum_{\nu} \psi_\nu(\mathbf{r}')\psi_\nu^*(\mathbf{r})\delta(\epsilon - \epsilon_\nu)$$

$$G^S(\mathbf{r}, \mathbf{r}', \epsilon) = \sum_{\mu} \psi_\mu(\mathbf{r})\psi_\mu^*(\mathbf{r}')\delta(\epsilon - \epsilon_\mu). \tag{5.9}$$

Using the matrix element of

$$J_{ij}(\mathbf{R}) = \int \psi_i^*(\mathbf{r} - \mathbf{R} - \mathbf{R}_i)V_T(\mathbf{r})\psi_j(\mathbf{r} - \mathbf{R}_j), \tag{5.10}$$

the tunneling current is approximated by

$$I = \frac{4\pi e}{\hbar} \int d\epsilon \, [f(\epsilon) - f(\epsilon + eV)] \, A(\mathbf{R}, \epsilon, \epsilon + eV), \tag{5.11}$$

where the energy density of tunneling current $A(\mathbf{R}, \epsilon, \epsilon')$ is given by

$$A(\mathbf{R}, \epsilon, \epsilon') = \sum_{i,i'} \sum_{j,j'} G^S_{i,i'}(\epsilon) G^T_{j,j'}(\epsilon') J_{ij}(\mathbf{R}) J^*_{i'j'}(\mathbf{R}). \tag{5.12}$$

We can construct $G^S_{i,i'}(\epsilon)$ and $G^T_{j,j'}(\epsilon')$ from the coefficients of the LACO representations of Eq. (5.9). The integral of Eq. (5.10) is performed over the tip side region and then $J_{ij}(\mathbf{R})$ includes the information of the tip geometry with respect to the surface. The information of the surface electronic states is included in $G^S_{i,i'}(\epsilon)$. We note that since this formula is based on the perturbative approach in Eq. (5.3), we only need to calculate electronic states of the surface and the tip once and this information can be used for any location of the tip to obtain the image of STM [118].

5.1.2 *Atomic-Scale Contacts*

In the previous section, we treat the STM system by the perturbative approach, with tip and surface independently for the tunneling current of the image. STM has been utilized not only for observing a surface as the STM image, but also for manipulating the atom and molecule on the surface [119, 120]. Atom manipulations are performed using the tip–adatom interaction with the distance much smaller than that of the STM image and by applying strong electric fields with large currents. In such situations, the perturbative

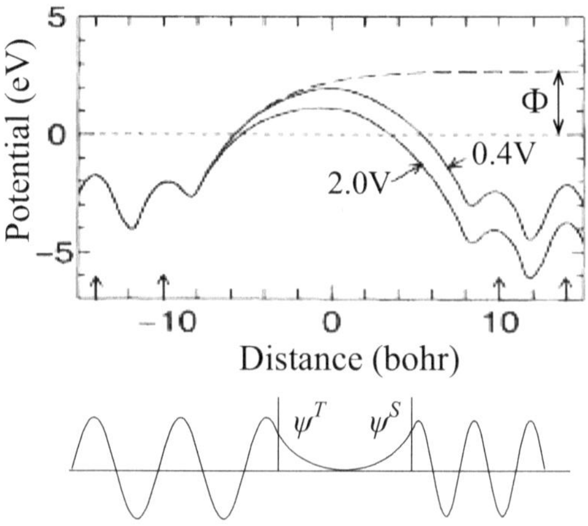

Figure 5.4 (Top) Effective potential with a short distance between the tip and the surface, the potential barrier becomes much lower than the vacuum level Φ. (Bottom) Schematic view of the wavefunctions. In the gap, the wavefunctions of the tip and the surface overlap strongly and behave completely different from those in the perturbative approach. From K. Hirose and M. Tsukada, *Phys. Rev. B* **51**, 5278 (1995) [60].

approach for the tunneling current is no more applicable. Instead, we have to treat electronic states of the tip and surface as a combined system for the simulations of atom manipulation.

As an example, the potential barrier due to overlap of wavefunctions is shown in Fig. 5.4. When the tip–surface distance is very short, due to the interaction, the effective potential barrier becomes much smaller than the workfunction Φ. The understanding of basic phenomena occurring between the tip and the surface, such as the potential barrier height and microscopic distribution of electric current, becomes more important and critical, since this leads to the construction of the atomic-scale electron devices using atomic wires and molecules bridged between electrodes.

5.1.2.1 Formation of atomic-scale point contacts

Here, we show the microscopic electronic states for the formation of atomic-scale contact between the tip and the surface, applying finite bias voltages from the first-principles calculations. Well-controlled

Figure 5.5 Electron density profiles in the (110) plane of the Na tip–surface system at the surface bias voltages of (left) $V_S = 0$ V (center) +5 V, and (right) +8 V, respectively. Tip–surface distance is fixed at $d = 14.0$ bohr. From K. Hirose and M. Tsukada, *Phys. Rev. Lett.* **73**, 150 (1994) [121].

construction of the atomic-scale contacts is essential for the operation of the atomistic electron devices. We use the RTM method described in Chapter 4 for electronic state calculations of combined nanosystems, which solves the coupled-channel matrix equation as a scattering wave with appropriate boundary conditions deep inside electrodes. A finite bias voltage is included as a difference of the chemical potentials assigned to each electrode. Self-consistency of the electron density and effective potential through scattering wavefunctions under an electric field is indispensable to include the interaction effects between the tip and the surface beyond the perturbative treatments. Note that the distance is measured by the Bohr radius a_B with 1 bohr = 0.529 Å.

Figure 5.5 shows the electron density profiles of the Na tip–surface system with finite applied voltages of $V_S = 0, +5$ V and +8 V with the distance of $d = 14.0$ bohr. This is a metallic system and thus electrons tend to extend to the gap region between the tip and surface. We see that the electron cloud swells from the tip toward vacuum gap and an atomic-scale localized region of electron accumulation appears outside the tip apex atom as an applied bias voltage increases. Apparently, such situations with the proximity effect are beyond the perturbative approach and the treatment of combined nanosystems is essential. The attractive force works

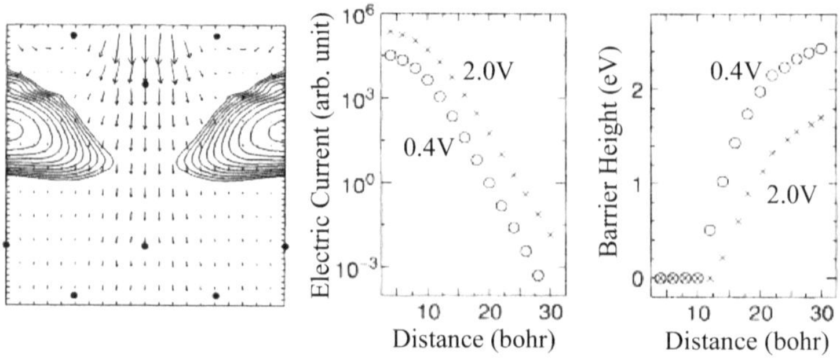

Figure 5.6 (Left) Effective potential and ballistic current distribution of the Na metallic tip–surface system with the distance of $d = 14.0$ bohr for an applied bias of $V_S = +2.0$ V. (Center) Electric current (arbitrary unit) as a function of the tip–surface distance at surface biases of $V_S = 0.4$ V and $+2.0$ V. (Right) Potential barrier height ϕ averaged in the supercell as a function of the tip–surface distance for $V_S = 0.4$ V and $+2.0$ V. From K. Hirose and M. Tsukada, *Phys. Rev. B* **51**, 5278 (1995) [60].

between the tip and surface atoms and reduces the activation barrier significantly for the atom manipulations.

The microscopic current density distribution between the tip and the surface is shown in Fig. 5.6 (left). We see that the current distribution is well collimated in the vicinity of the tip apex, reflecting the structural feature of the potential barrier. When the tip–surface distance is small enough, the ballistic current flows between the gap region in Fig. 5.6 (center) with missing of an effective potential for tunneling current in Fig. 5.6 (right). In the present Na metallic system, we see that the barrier missing occurs at $d \approx 10$ bohr for $V = 0.4$ V and $d \approx 14$ bohr for $V = 2.0$ V. This transition from the tunneling to ballistic regimes due to the disappearance of tunneling barrier and corresponding change of the current behavior provides us important information for the atomic-scale contact problem. We note that the form of the effective potential is similar to that of the mesoscopic quantum point contact in Chapter 6 where the ballistic transport regime emerges. This implies the close relationship of the atomic-scale contact problem with the formation of the quantum channel for electron transport.

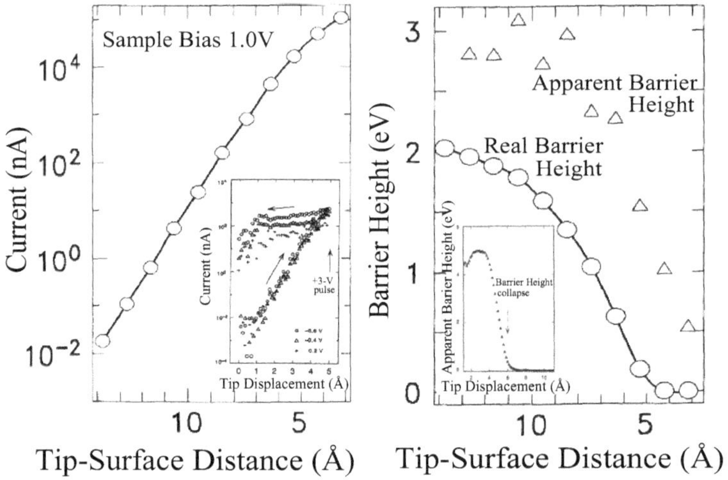

Figure 5.7 (Left) Current I as a function of tip–surface distance d at $V_S = +1.0$ V for tip-Na surface. From K. Hirose and M. Tsukada, *Phys. Rev. B* **51**, 5278 (1995) [60]. (Inset) Experimental observation of d vs. I for tip-Si surface. From I.-W. Lyo and P. Avouris, *Science* **253**, 173 (1991) [120]. (Right) Real potential barrier height Φ for electron transport from the first-principles calculations and apparent potential barrier height Φ_a obtained from the fitting to the logarithmic change of I to d. (Inset) Φ_a from experimental data of d vs. I in the left inset.

5.1.2.2 Apparent barrier height

When the tip of STM approaches sample surfaces too much, it might occur that the tip is too damaged to take the image of surfaces. Also, the missing potential barrier implies the collapse of the tunneling current. At the same time, it opens the possibility for the atom manipulation. Therefore, it is important to specify the potential barrier height. In experiments of the STM image, since $I \sim \exp(-2d\sqrt{2m\Phi_a}/\hbar)$, the apparent barrier height Φ_a is estimated from the current values I by the best fit of their logarithmic change as a function of the tip displacement d. For example, the insets of Fig. 5.7 show the current I as a function of tip displacement toward the Si surface (left) and the corresponding apparent barrier height Φ_a (right). We see that Φ_a collapses rapidly as the tip approaches the surface.

However, the formula for the apparent barrier height Φ_a assumes the tunneling regime and thus is not sure to correspond to realistic potential barrier height. Let us see in Fig. 5.7 the change of the current density I (left) and the barrier height Φ and Φ_a (right) as a decrease of the tip-Na surface distance d. Φ_a is obtained, similar to experiments, by the best fit of logarithmic change of I as a function d. First, we note that the behavior of obtained apparent barrier height Φ_a (right) is similar to the one from the experiment data (inset). It rapidly decreases as the tip approaches the surface, while electric current I increases exponentially in the tunneling regime, and disappears at certain distance to identify the contact regime. The real barrier height Φ, on the other hand, takes much smaller values than Φ_a. Correspondingly, Φ disappears at much larger distance than Φ_a. This shows that the potential barrier for the tunneling current is missing due to the tip–surface interaction to form the atomic contact even when the apparent barrier Φ_a appears to exist. We note that Φ_a is the apparent barrier that electrons feel due to the uncertainty principle. In the contact region where a potential hole is created, the current is determined by the geometrical constriction and is proportional to the incident channel number of electrons.

5.1.2.3 Formation of covalent bond at atomic contact

In the previous section, we treat the metallic systems where the electrons behave as free electrons. Effective potentials significantly reduce to form the quantum channel structure with a decrease of the tip–surface distance or with an increase of the bias voltage. Here we see how the covalent bonding states are formed for the semiconducting materials when two surfaces get close to each other. We consider the Si(111) surfaces and apply the RTM method to represent the microscopic nature of the formation of the covalent bonding in atomic scale.

Figure 5.8 shows the electronic states of Si(111) surfaces separated with a distance of d between the neighboring Si atoms. Note that $d = 2.3$ Å is the interatomic distance of the Si. We see that there is no connection between the atoms at $d = 5.2$ Å. At $d = 4.2$ Å, small overlap of an electron cloud begins to form, showing that an atomistic contact is just created. At $d = 3.2$ Å, significant

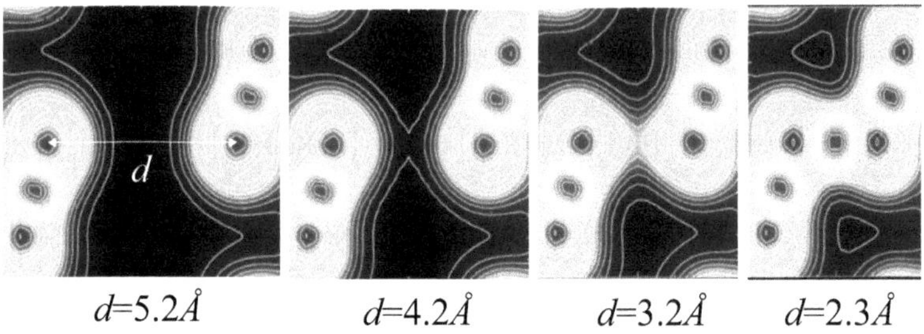

Figure 5.8 Electron density of Si(111) surfaces separated with the distance of $d = 5.2$ Å, $d = 4.2$ Å, $d = 3.2$ Å, and $d = 2.3$ Å, showing how the atomic contact is constructed as a function of d for the semiconducting material. Note that $d = 2.3$ Å is the interatomic distance of the Si crystals. The covalent bond is formed close to the interatomic distance. The overlap of wavefunctions begins to form at $d = 4.2$ Å.

connection between Si atoms is observed and finally at $d = 2.3$ Å, the complete Si crystal structure is constructed, forming a covalent bond between the Si atoms. These results show that compared with metallic Na cases, the atomic-scale covalent bond is formed with smaller distance between the surfaces of semiconducting materials.

Based on this observation, we see the atomic-scale contact formation of the STM situation for semiconducting Si(111) surfaces and the metallic Al tip, applying a finite bias voltage. For the tip, four aluminum atoms in the shape of tetrahedron cut from an ideal Al(111) surface are attached to a metallic jellium electrode corresponding to Al. Calculations using the RTM method are performed for the supercell structure with Si(111)2×2 translational symmetry in the lateral direction to the surface and we obtain the electronic states of an aluminum tip and a silicon surface in STM to discuss the microscopic nature of the potential barrier and current density distribution.

Figure 5.9 shows potential barriers and electric current distributions (top) and charge density distributions (bottom) at the surface bias of $V_S = +2.0$ V with the tip–surface distances d of 12.0 bohr, 10.0 bohr, and 8.0 bohr from left to right. In the case of $d = 12.0$ bohr, it is evident that the tunneling current flows from the apex atom of the tip, spreads in the vacuum region, and focuses on the atom just under the apex atom. Compared with the metallic

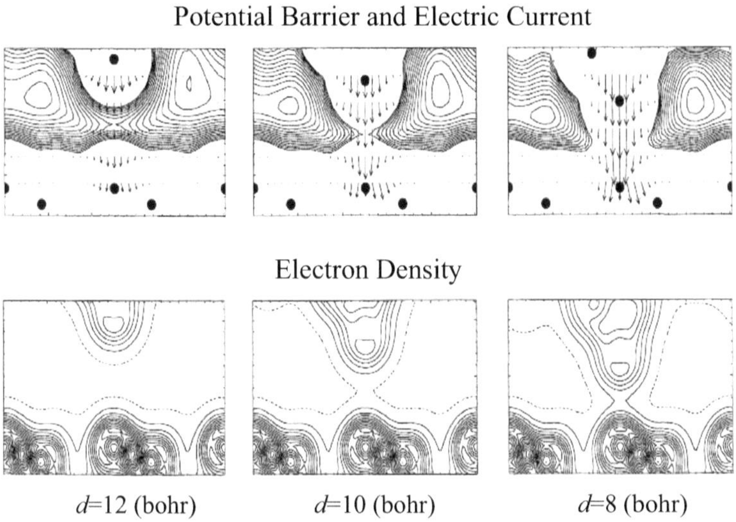

Figure 5.9 (Top) Potential barriers and current density distributions in the Si(110) plane at the surface bias of $V_S = +2.0$ V with the tip–surface distances of $d = 12.0$, 10.0, and 8.0 bohr from left to right. The effective potentials higher than the Fermi level of the tip are represented in units of 0.2 eV. The current density is shown by arrows whose lengths are proportional to the magnitude of the current density. The atomic positions are indicated by filled circles. (Bottom) Corresponding electron density distributions. The contour plot is in units of 5.0×10^{-3} electrons/bohr^3. The broken lines correspond to 1.0×10^{-3} electrons/bohr^3. From N. Kobayashi, K. Hirose, and M. Tsukada, *Jpn. J. Appl. Phys.* **35**, 3710 (1996) [122].

Na systems, due to the covalent bonding nature of Si, the distance of $d = 12.0$ bohr is too large to form the atomic-scale contact. It is interesting to note that the tunneling current flows through around the saddle point of the tunnel barrier, rather than in the direction of the apparent potential gradient. This induces the tunneling current localized in the atomic-scale space between the tip and the surface. The electron density is negligibly deformed in this tunneling regime, which suggests that the perturbative approach for the current is effective.

The potential barrier height decreases with a decrease tip–surface distance. At a distance of $d = 10.0$ bohr, a potential hole is created in the vicinity of the tip apex atom due to the proximity effect of the tip, similar to the left panel of Fig. 5.9. As the distance d becomes smaller, the potential hole in the barrier becomes larger where the ballistic electrons flow. Correspondingly electron density is deformed into a bridge region due to an enhancement of the chemical interaction between the tip and the surface. We note that even in the presence of the potential hole, the ballistic electrons are subjected to an effective barrier due to the uncertainty principle. Thus, the transmission of the electrons is reduced.

5.1.3 *Atom Manipulation*

Present technology makes it possible to manipulate individual atoms and molecules to fabricate atomic-scale structures on a surface using the tip of STM. Historically, the xenon atoms were used to position to form the shape of letters on a nickel surface and to construct an atomic-switch [119]. For the semiconductor surfaces, a single atom was removed from a silicon surface and was deposited on a predetermined position on the surface [120]. There have been a number of experiments in manipulating atoms on surfaces. Such experimental techniques have been applied to create electronic devices from atomistic viewpoints as a bottom-up approach.

5.1.3.1 Atom extraction

For an example of the atom extraction, here we present a simulation result of the tip–surface system both composed of Na atoms attached to $r_s = 2.0$ jellium electrodes first. A pyramid made up of five atoms mimics the tip and is arranged periodically in the 2D 2×2 superlattice. For an atom extraction, the topmost atom below the tip apex atom is removed. Figure 5.10 (left) shows the schematic view of the system for an atom extraction.

Figure 5.10 (center) shows the electron density profiles for the displacements of $\Delta d = 4.0$ bohr and $\Delta d = 8.0$ bohr applying the bias of $V_S = +5.0$ V. The tip–surface distance is fixed at $d = 14.0$ bohr. The displacement of $\Delta d = 4.0$ bohr corresponds to the saddle point of the adiabatic potential-energy curve shown in (b) of Fig. 5.10 (right). We can see that the electron cloud of the

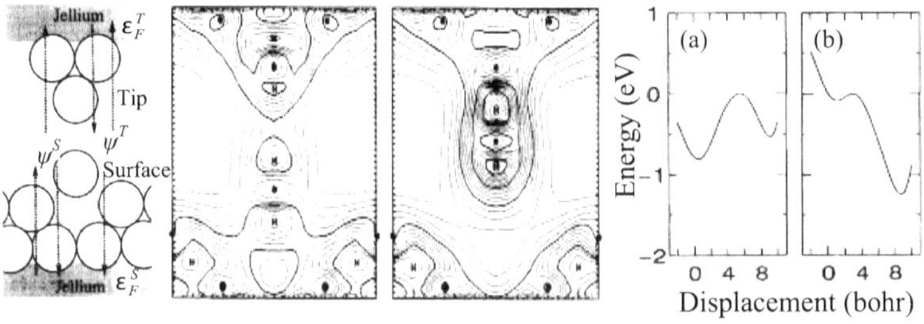

Figure 5.10 (Left) Schematic representation of the bielectrode tip–surface system. (Center) Electron density for the displacement of $\Delta d = 4.0$ bohr and $\Delta d = 8.0$ bohr from the original position applying the bias of $V_S = +5.0$ V. (Right) Adiabatic potential-energy curves of the removed atom for (a) $V_S = 0$ V and (b) $+5.0$ V. From K. Hirose and M. Tsukada, *Phys. Rev. Lett.* **73**, 150 (1994) [121].

removed atom is divided into two; one is attached to the tip and the other is attached to the surface. The displacement of $\Delta d = 8.0$ bohr is a situation where the removed atom is in the minimum of the potential-energy curve shown in (b). We see that the chemical bonding is formed between the tip apex and the removed atom, which shows the complete extraction of the atom from the surface. This chemical bonding is induced by the electric fields. Compared with the electron density in Fig. 5.5, the attractive force is enhanced when the charge density profile forms a bridge structure between the tip and the surface.

Figure 5.10 (right) shows the adiabatic potential-energy curves for the removed atom as a function of the displacement from the original position for (a) $V_S = 0$ V and (b) $+5.0$ V. There are two potential minimum in the zero bias case, the potential well close to the surface deeper than that close to the tip. The activation barrier is about 0.8 eV, which is reduced due to the tip–surface interaction compared with the sublimation energy of Na metal. With an increase of the positive applied bias voltage to the surface, the potential-energy curve near the surface is raised and that near the tip is lowered. Correspondingly the activation barrier is reduced significantly. In the case of $V_S = +5.0$ V, the activation barrier is only about 0.1 eV and it disappears more than that. Thus, the bias voltage of $V_S = +5.0$ V is close to the critical threshold voltage for an atom extraction in this system.

5.1.3.2 Atom transfer

Next, we discuss an atom transfer from a surface under electric field and current. Especially, we show the calculations regarding a Si atom transfer from the surface to the tip. As we see, Na atoms and Si atoms have quite different characters, the former forms metallic bonds, while the latter forms covalent bonds. The difference in electronic states might have significant effects on an atom transfer from the surface. The key points are how the induced electron density in a narrow space between the tip and the surface by an electric field affects the behaviors of atom transfer.

We consider an adatom transfer from a Si(111) surface. The model is shown in Fig. 5.11. We use two layers of 2×2 part of the Si(111) 7×7 DAS structure for the surface and an Al trimer for the tip, both of which are attached to semi-infinite jellium electrodes. An adatom is placed just below the center of the Al trimer tip. Under this situation, a Si adatom is transferred between the surface and the tip with an applied bias voltage. Here we consider two cases of an adatom transfer; one is for the distance of $d = 11.0$ bohr and the other is for $d = 8.8$ bohr, where d is measured from the

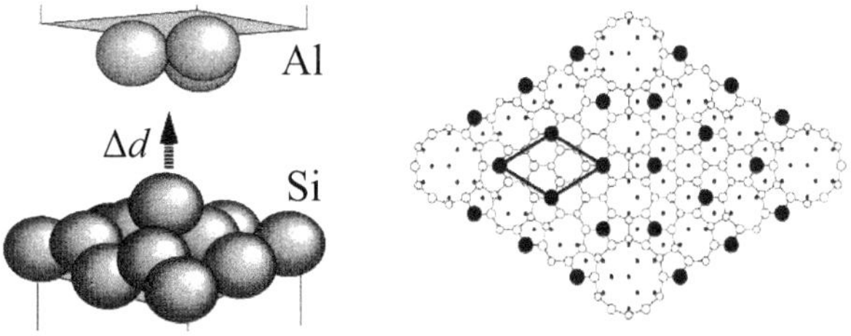

Figure 5.11 (Left) Schematic representation of the atomic structure corresponding to a unit of the supercell. Two layers of Si(111) with 2×2 part of the DAS structure are used for a surface and an Al trimer for a tip. An adatom of Si is transferred from a surface to a tip. (Right) The DAS model of Si(111) surface with an area surrounded by the thick lines corresponding to the supercell. From N. Kobayashi, K. Hirose, and M. Tsukada, *Jpn. J. Appl. Phys.* **36**, 3791 (1997) [123].

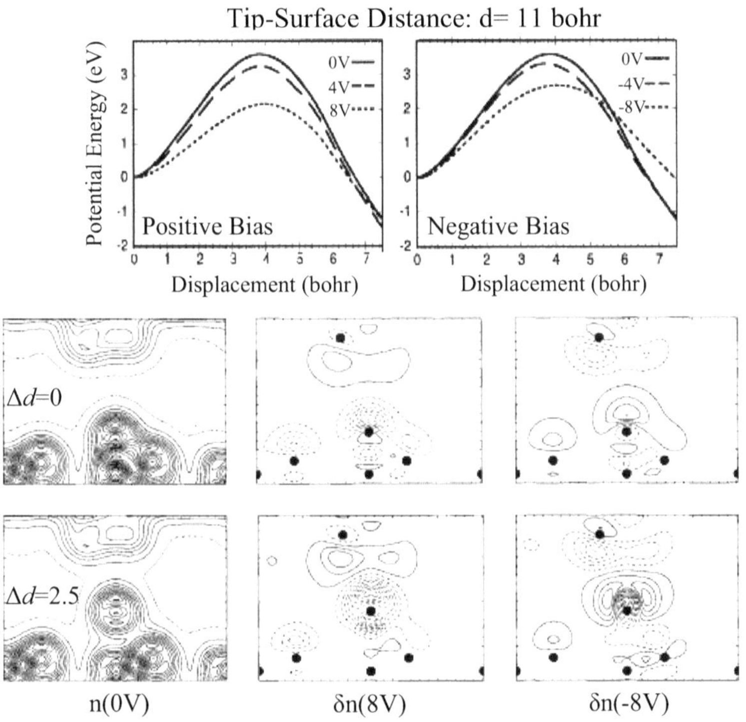

Figure 5.12 (Top) Adiabatic potential surfaces with $d = 11.0$ bohr as a function of the displacement Δd of a Si adatom for positive (left) and for negative (right) biases. (Bottom) Electron density $n(\mathbf{r})$ and induced electron density $\delta n(\mathbf{r})$ for the displacement of $\Delta d = 0$ and $\Delta d = 2.5$ at surface biases of $V_S = 0, \pm 8$ V. The solid and broken lines correspond to electron accumulation and deficiency for $\delta n(\mathbf{r})$. Atomic positions are indicated by filled circles. From N. Kobayashi, K. Hirose, and M. Tsukada, *Jpn. J. Appl. Phys.* **36**, 3791 (1997) [123].

stable position on the surface without bias to the center of the trimer atoms. We note that in these narrow spaces, the tail of wavefunctions between the tip and the surface hardly overlaps for the distance of $d = 11.0$ bohr, but overlaps significantly for the distance of $d = 8.0$ bohr.

Figure 5.12 (top) shows the adiabatic potential surfaces of an adatom transfer from the Si(111) surface as a function of the displacement Δd for the distance of $d = 11.0$ bohr. The adiabatic potential is obtained by integrating the force acting on the adatom along the displacement. The maximum value of the potential surface corresponds to the activation barrier height, which is about 3.6 eV in the present system. We see that while the potential surfaces do not change much up to the biases of ± 4 V, they change significantly at the biases of ± 8 V. The decrease of the potential barrier for both polarities of the bias voltage in the case of an Si atom transfer was observed in experiments [124]. This cannot be explained simply from an electric field effect and we need to study electronic states of the Si adatom in a narrow space under electric fields more carefully.

Figure 5.12 (bottom) shows the electron density distributions $n(\mathbf{r}, 0\ \text{V})$ and the induced electron density distributions $\Delta n(\mathbf{r}, \pm 8\ \text{V})$ for the surface bias voltages of $V_S = \pm 8$ V. The displacements of a silicon adatom from the surface toward the tip are $\Delta d = 0$ and $\Delta d = 2.5$ bohr. When a Si adatom is located $\Delta d = 0$ at the stable point on the surface, the effect of an applied bias on the induced electron density distribution $\Delta n(\mathbf{r})$ is qualitatively similar for both polarities, that is, the induced electron density is reversed from positive (electron increase) to negative (electron decrease) or *vise versa* by changing the bias polarities. On the other hand, when a Si adatom is displaced $\Delta d = 2.5$ bohr from the surface, the effect of the bias polarity on the induced electron density changes considerably. We see that electron density $n(\mathbf{r})$ around the adatom *decreases* when the bias is positive $V_S = 8$ V. In this case, an accumulation of electrons appears in front of the transfer adatom below the tip and the adatom is subject to an attractive force from the accumulated electrons. This reduces the activation barrier height significantly. In contrast, for the negative bias $V_S = -8$ V, the electron density $n(\mathbf{r})$ around the adatom *increases* toward the tip in the lateral directions, while decreases in the vicinity of the center of the adatom. The polarization of this complicated electron distribution, where the electron accumulation appears below the tip, acts for a significant attractive force and reduces the activation barrier height.

Next, let us see the cases for $d = 8.0$ bohr where we see different behaviors. Figure 5.13 shows the adiabatic potential surfaces of the Si transfer atom. Since the tip–surface distance decreases, even without bias voltage, the activation barrier takes a much smaller value of about 1.3 eV due to the proximity effect of the tip and the

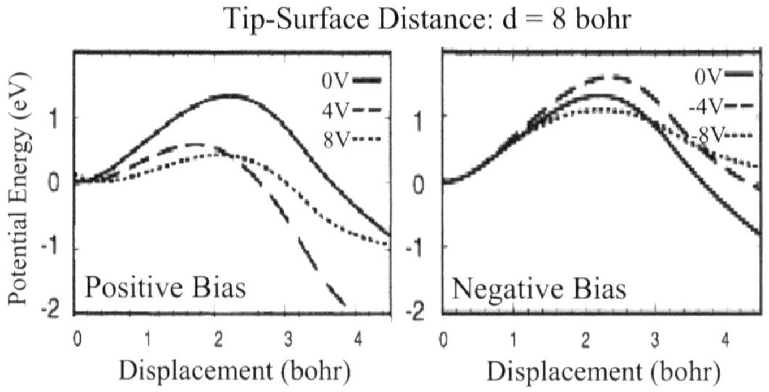

Figure 5.13 Adiabatic potential surfaces with $d = 8.0$ bohr as a function of the displacement Δd of a Si adatom for positive biases (left) and for negative biases (right). From N. Kobayashi, K. Hirose, and M. Tsukada, *Jpn. J. Appl. Phys.* **36**, 3791 (1997) [123].

surface. We note that the wavefunctions overlap in this narrow space significantly even at zero displacement.

The effect of bias voltages on an increase of induced electron density becomes much larger for this small tip–surface separation, since the electric field becomes much larger. Furthermore, the bias effect on the electron density distribution is not limited around the adatom below the tip apex but spreads over much larger regions for $d = 8.0$ bohr cases. This is completely different from the $d = 11.0$ bohr cases, where the electric field effect is limited almost around the adatom below the tip. The potential barrier decreases further by the imposition of a *positive* bias at the surface. We see that at the bias of $V_S = 4$ V, the activation barrier reduces less than 0.5 eV and reduces slightly more at $V_S = 8$ V. However, in the *negative* bias case, the potential does not change much and correspondingly the barrier height does not change monotonously. The activation potential barrier increases for $V_S = -4$ V case, but it decreases slightly for $V_s = -8$ V case.

From these observations for the Si atom transfer, we can say that when the tip–surface distance is large as $d = 11.0$ bohr, the activation barriers decrease for both positive and negative bias voltages and the charge density distributions change due to the effect of electric fields primarily around the atom below the tip. On the other hand, when the tip–surface distance decreases as d = 8.0 bohr, the activation barrier decreases monotonously only for the positive bias case, while the activation barrier does not change much in the negative bias case. The monotonic decrease of the activation barrier in the positive bias case is due to the proximity effect as in the d = 11.0 bohr case. These interesting behaviors for the different polarities show the complicated nature of the response of the Si adatom to an applied electric field and the atom transfer is much influenced by the sensitivity of the electronic states between the tip and the surface under strong electric fields. The present calculations suggest that according to the distance between the tip and the surface, the behavior of adatom changes considerably for the bias polarity, which explains the experimental observations [125].

5.1.3.3 Atom within nanospace in electric field

In this section, we consider the influence of strong electric fields on an adatom on the surface when it confines within a tiny nanospace between the tip and the surface. We expect that the adatom in a nanospace changes its behavior of electronic states for the electric fields differently as that in the free space. Thus, it is very important to understand the fundamental properties of the adatom behaviors in a nanospace, since atom manipulations are performed in such situations.

Here we consider a simple situation where an adatom, situated on the surface, is put between jellium electrodes with a nanometer-scale spacing under an influence of electric fields with applications of bias voltages. When an atom is in an electric field in a free space, the atom changes its electronic states with a polarization to screen the external electric field. The dipole moment thus formed is a fundamental physical value of an atom. However, it is not sure that we can use the polarizability of an atom in a free space for the atom manipulation, since it is confined within a nanospace. Let us calculate the electronic states by the RTM method and see the different behaviors of polarizability for Na, Al, and Si atoms.

Figure 5.14 shows the DOS (left) and the electron density distributions of adatoms (right) for Na (top), Al (center), and Si (bottom) between the jellium electrodes with r_s = 2.0

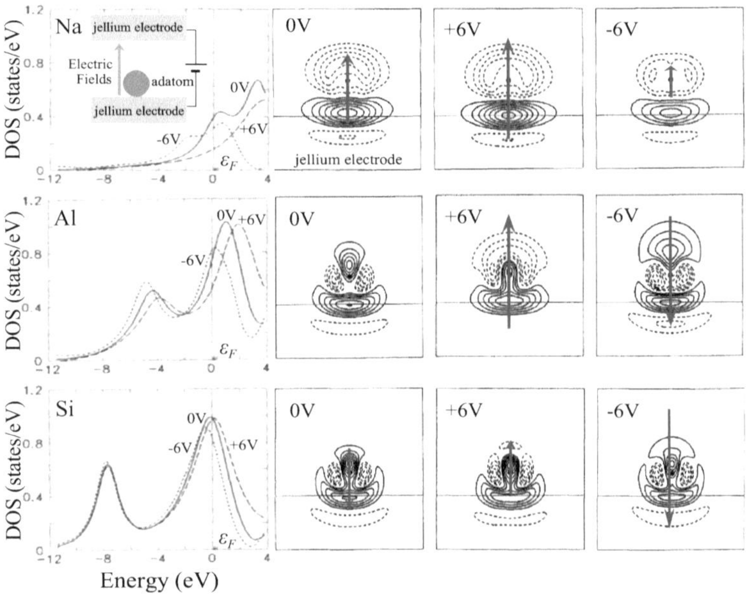

Figure 5.14 (Left) Density of states for the adatom of Na (top), Al (center), and Si (bottom) with and without bias voltages between the jellium electrodes with $r_s = 2.0$, corresponding to Al electrodes. The applied bias is ± 6 V with a space of $d = 10$ bohr. The DOS from jellium electrodes are subtracted. (Inset) Schematic picture of the situation is illustrated. (Right) Electron density distributions with and without bias voltages. Electron density of an atom in a free space and that of jellium electrodes are subtracted. Electron affinity is plotted by the solid lines, while electron deficiency is plotted by the dashed lines in $5 \times 10^{-5} e/\text{bohr}^3$ for Na, in $1 \times 10^{-4} e/\text{bohr}^3$ for Al, and in $2 \times 10^{-4} e/\text{bohr}^3$ for Si. Horizontal lines show the surface edges of electrode and the arrows represent the magnitude and direction of polarization of an adatom.

corresponding to Al electrodes. The DOS from the jellium electrodes (left) and the electron density of atom in a free atom and that of jellium electrodes (right) are subtracted to see the polarized behaviors clearly. Electric fields are applied to the adatom by applying finite bias voltages of ± 6 V between electrodes with the spacing of $d = 10$ bohr; thus the electric fields are $E \approx \pm 11.3$ V/nm. The schematic picture of the situation is shown in the inset of the left panel.

First, we see the Na adatom case (top). In the zero bias without electric fields, the DOS of Na adatom (left) has a broad peak at the Fermi energy ϵ_F. This state is constructed from the $3s$ orbital and thus should be symmetric and tend to spread out. The electron density distribution (right), the solid lines representing the electron accumulation and the dashed lines representing the electron deficiency, shows that there is an electron accumulation at the edges of jellium electrodes, which form the chemical bond of Na atom with the semi-infinite electrodes on the surface, and correspondingly an electron deficiency in the vacuum region. This electron-polarized distribution produces the apparent dipole moment even without electric fields. When the positive electric fields are applied to the Na adatom, the peak of the DOS from $3s$ orbital goes up to a higher energy regime and the redistribution of electron density occurs to enhance the electron accumulation and electron deficiency, increasing the polarizability and thus dipole moment. On the other hand, when the negative electric fields are applied, the peak of the DOS from $3s$ orbital shifts to a lower energy regime and the behavior of the polarization is reversed. The Na atom resembles to the free state with a little dipole moment.

In the Al (center) and Si (bottom) atom cases, the Fermi energy ϵ_F is situated just below the peak of $3p$ states for Al and at the peak of $3p$ states for Si, depending on the occupation of $3p$ states, while the peak of $3s$ states situate much below ϵ_F. Electron density distributions (right) show the polarized behaviors due to $3p_z$ states. We see the electron accumulation at the edges of jellium electrodes, which form the covalent bonds of Al and Si atoms with the semi-infinite electrodes on the surface; however, the corresponding electron deficiencies are much complicated than Na atom. When the positive electric fields are applied to the Al and Si adatoms, the peak of the DOS from $3p$ orbitals goes up, and when the negative electric fields are applied, it shifts to the lower energy regime. We see that the movement of the peak is small for Si atom than that for Al atom. Especially for Si atom, while the peak of $3s$ state rarely changes, the shifts of $3p$ states at ϵ_F look more complicated. The corresponding redistributions of electron densities for Al and Si atoms show much more complicated behaviors due to the polarization of $3p_z$ orbital compared with the Na atom.

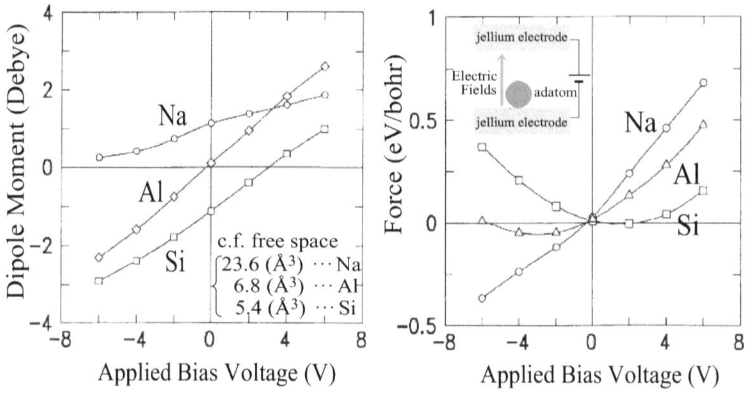

Figure 5.15 (Left) Dipole moments μ (Debye) as a function of applied bias voltages (V) for three kinds of adatoms, Na, Al, Si on the electrode. The atomic configurations are the same as in Fig. 5.14. (Right) Force acting on the adatom in the perpendicular direction as a function of V. (Inset) Schematic representation of the present system with an adatom in a nanospace with an electric field.

Figure. 5.15 shows the dipole moment μ (left) obtained from

$$\mu = -\int (z - z_0)\delta n(\mathbf{r})d\mathbf{r}, \tag{5.13}$$

and the force F_z (right) acting on the adatom in the perpendicular direction to the surface as a function of applied bias voltage V. In the Na atom case, it has a finite dipole moment $\mu \simeq 1.0$ Debye even without the electric fields as we see. From the slope of an increase of dipole moments for E, we can obtain the polarizability α from the formula as

$$\mu \simeq \mu_0 + \alpha E, \tag{5.14}$$

which becomes $\alpha \approx 2.1$ Å^3. This polarizability of Na atom is much smaller than that in the free space, which takes[a] $\alpha_0 = 23.6$ Å^3. This is due to the effects of the confinement in a narrow nanospace where the Na atom is put. The screening of metallic jellium electrodes prevents the 3s state of Na spreading out, and eventually results in a small polarizability of Na atom. The force F_z becomes reversed when we change the polarity of bias voltage from the positive to negative.

[a]From the data of *CRC Handbook of Chemistry and Physics.*

Thus, the atom transfer of Na adatom in a nanospace is possible only for the positive bias.

In Al atom case, the dipole moment in the absence of an electric field is $\mu = 0$ Debye, since the Al adatom is put on the Al surface. The polarizability α for Al adatom becomes $\alpha \approx 6.3$ Å^3. This is not so different from α_0 in the free space, which takes $\alpha_0 = 6.8$ Å^3. This shows that the spreading out of the $3p_z$ state of Al adatom due to an electric field is much reduced within this nanospace. The force F_z works for the positive bias, but becomes almost zero for the negative bias. In Si atom case, the Si adatom has a negative dipole moment with $\mu \simeq -1.0$ Debye when there is no electric field. The polarizability α for Si adatom becomes $\alpha \approx 5.1$ Å^3, which is also close to α_0 in the free space with $\alpha_0 = 5.4$ Å^3. This again shows that the spreading out of the $3p_z$ state of Si adatom due to an electric fields is much reduced. However, the force F_z shows completely different behaviors. The Si atom has positive values for both polarities, with slightly small values in the positive bias. This result is in consistent with the Si transfer in the previous section for $d = 11$ bohr, where the decrease of potential barrier is observed for both positive and negative bias. Therefore, the complicated Si transfer behaviors in the tip–surface system is due to the complicated polarized nature of Si adatom for an electric field in a tiny nanospace.

5.1.3.4 Atom sliding

Here, we consider the atom sliding on the surface induced by the STM tip. This is also an example of atom manipulations. We note that atom sliding technique has been used for the movement and positioning of atoms at arbitrary points on the surfaces and thus is very important. Let us consider the following situation, where one adatom is on the jellium surface for sliding and the other atom to mimic the STM tip is placed on the opposite surface with applying a finite bias voltage between them. Figure 5.16 (left) shows the electron density (solid lines), current distribution (arrows), and the electric fields (dashed lines) for the Al tip and Al adatom case with tip–adatom separation of $d = 10$ bohr. Figure 5.16 (right) shows the resistance as a function of tip–adatom separation d bohr for

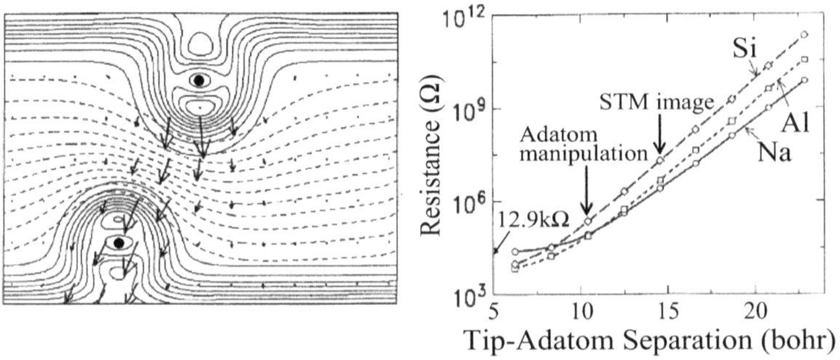

Figure 5.16 (Left) Tip–adatom system where an Al tip atom is placed on the jellium surface and the Al adatom is on the opposite jellium surface, applying a finite bias voltage with a tip–adatom distance of $d = 10$ bohr. Electron density (solid lines), current distribution (arrows), and electric fields (dashed lines) are shown. (Right) Resistance of tip–adatom system for Na, Al, and Si adatom with Al tip as a function of the distance d. The image of STM is taken in the tunneling regime, while atom manipulation is performed close to the contact regime.

Na, Al, and Si adatoms with the Al tip. The STM image is usually taken within tunneling regime around $d = 15$ bohr, while atom manipulation is performed with much smaller distance close to contact regime of $d = 10$ bohr.

Figure 5.17 shows the adiabatic potential surfaces for the atom sliding on the jellium surface induced by the Al tip. These are obtained from the force F_x acting on the adatom, which is integrated along the surface direction. The atomic configuration is schematically shown in the inset of the right panel. We note that for all the cases, the position just below the Al tip becomes the most stable and thus the adatom is induced to move in that direction.[a]

We see that the adiabatic potential surface for Na adatom becomes larger for the positive bias than that for the negative bias. In the case of Al adatom, since both tip atom and adatom are the same element, the adiabatic potential surfaces are the same for both polarity, while in the Si adatom case, the negative bias is more

[a]In realistic surfaces, there are additional potential barriers for adatoms to move between the atomic sites on the surfaces. Therefore, the adiabatic potential energy for an adatom to move on the surface induced by the tip becomes larger than that in the present calculations.

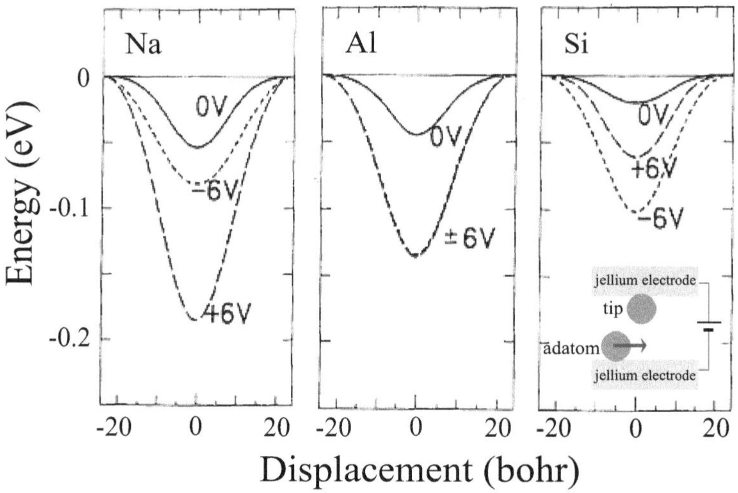

Figure 5.17 Adiabatic potential surfaces for the atom sliding on the jellium surface with tip–surface distance of $d = 10.0$ bohr with applied bias voltages of ± 6 V for the Al tip and Na (left), Al (center), and Si (right) adatoms. The atomic configuration is shown in the inset.

favorable for atom sliding of Si adatom on the surface. We note that these behaviors of adiabatic potential surfaces are qualitatively understood in view of the dipole moment of adatoms Eq. (5.13) in electric fields induced by the tip as

$$U \approx -\int \mu dE = U_0 - \mu_0 E - \frac{1}{2}\alpha E^2, \tag{5.15}$$

where the positive (Na) and negative (Si) values of μ_0 produce the opposite atom sliding behaviors for the different polarity of applied bias voltages.

5.1.4 *Field Emission*

Electron emission from nanometer-scale systems under an electric field has become important from scientific and also technological points of view. Field emission microscopy (FEM) images of nanometer-scale systems observed have been used to study the effects of microscopic electronic states. As for an example of

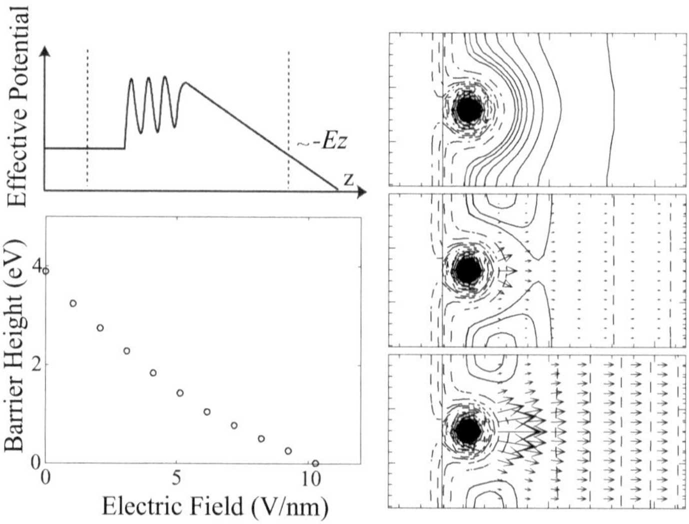

Figure 5.18 (Left) Average effective potential around an Al apex atom on a jellium electrode (top) and tunneling barrier height as a function of applied electric fields (bottom). (Right) 2D representation of electron current flow and effective potential. Solid lines are the effective potential above the Fermi energy corresponding to a tunneling barrier, while dashed lines are the effective potential below the Fermi energy for applied electric fields of $E = 0$ (top), $E = 10.3$ (middle), and $E = 20.6$ (bottom) V/nm. From N. Kobayashi, K. Hirose, and M. Tsukada, *Appl. Surf. Sci.* **237**, 568 (2004) [126].

technological applications, field emission from carbon nanotubes (CNTs) is utilized for a field emission display as a sharp field emitter. Therefore, to analyze the microscopic nature of field emission from the theoretical viewpoints is an important topic. In this section, we present a method for calculating field emission current using the RTM method and show microscopic distributions of the electron emission current from an apex atom adsorbed on a metallic electrode and the behavior of barrier height to the strength of an applied electric field. The boundary condition of the RTM method shown in Chapter 4 is modified to express a constant electric field deep into a vacuum region corresponding to the field emission.

In the field emission system, the gradient of an effective potential becomes constant deep into the asymptotic vacuum region, which corresponds to the slope of an applied electric field E as

$$V_{\text{eff}}(\mathbf{r}) = -Ez + C \qquad (z \to \infty), \tag{5.16}$$

where C is a constant. Under this boundary condition, the wavefunction deep into the vacuum region applying an electric field of E is expressed by

$$\psi(z) = \frac{A}{2\xi^{\frac{1}{4}}} \exp\left[i\left(\frac{2}{3}\xi^{\frac{3}{2}} - \frac{\pi}{4}\right)\right], \tag{5.17}$$

where ξ is the Airy function

$$\xi = \left(z + \frac{\epsilon - C}{E}\right)(2E)^{1/3}. \tag{5.18}$$

In the RTM calculation scheme, the ratio matrices between two subsequent mesh points are handled to solve the wavefunctions. Since the wavefunctions are diagonalized in the asymptotic vacuum region, by discretizing the z coordinate into fine meshes $z = z_p \quad (p = 0, 1, \ldots, l+1)$, we can construct the ratio matrix $S_{mn}(z_l) = \delta_{mn}\psi(\mathbf{G}_{||}^n, z_{l+1})/\psi(\mathbf{G}_{||}^n, z_l)$ at the boundary of the vacuum region z_l to connect to the Airy function of

$$S_{mn}(z_l) = \delta_{mn}\left[\frac{\epsilon_\perp(z_l)}{\epsilon_\perp(z_{l+1})}\right]^{1/4} \times \exp\left[i\frac{2\sqrt{2}}{3E}\left\{\epsilon_\perp(z_{l+1})^{\frac{3}{2}} - \epsilon_\perp(z_l)^{\frac{3}{2}}\right\}\right]. \tag{5.19}$$

Here the energy for the perpendicular direction is given by

$$\epsilon_\perp(z_p) = \epsilon - \frac{1}{2}|\mathbf{k}_{||} + \mathbf{G}_{||}^n|^2 - V_{\text{eff}}(\mathbf{G}_{||}^0, z_p). \tag{5.20}$$

The Hartree potential of the $\mathbf{G}_{||}^0$ components has a constant gradient deep into the vacuum region corresponding to the electric field E. Therefore, we solve the Poisson equation

$$\frac{d^2}{dz^2}V_H(\mathbf{G}_{||}^0, z) = -4\pi n(\mathbf{G}_{||}^0, z) \tag{5.21}$$

with the boundary condition of

$$\begin{aligned} V_H(\mathbf{G}_{||}^0, z) = &-\int_{z_0}^{z}\int_{z_0}^{z} 4\pi n(\mathbf{G}_{||}^0, z')dz' \\ &+ (z - z_0)\left(\frac{d}{dz}V_H(\mathbf{G}_{||}^0, z)\bigg|_{z_{l+1}} + \int_{z_0}^{z_{l+1}} 4\pi n(\mathbf{G}_{||}^0, z')dz'\right) \\ &+ V_H(\mathbf{G}_{||}^0, z_0). \end{aligned} \tag{5.22}$$

The electronic states and effective potential are obtained self-consistently under the condition that the electric field is constant deep into the vacuum region.

The top panel of Fig. 5.18 (left) shows the effective potentials for a system consisting of an Al apex atom adsorbed on a jellium electrode. Current density distributions are shown in the right panels for applied electric fields of $E = 0$ (top), $E = 10.3$ (middle), and $E = 20.6$ (bottom) V/nm. Here the solid lines indicate the effective potentials above the Fermi energy, which corresponds to a tunneling barrier, and the dashed lines show the effective potentials below the Fermi energy. The current density distribution is shown as the arrows. The current flows from the adsorbed atom and spreads out in the vacuum region. With an increase of the applied electric field, the hole created in the tunneling barrier increases and correspondingly the current from the adsorbed Al atom is enhanced significantly.

The barrier height ϕ shown as a function of electric field E in the bottom panel of Fig. 5.18 (left), which is defined as the height of effective potential at the saddle point above the Fermi energy, decreases with an increase of electric fields E. Corresponding tunneling current I increases exponentially with E, following the Fowler–Nordheim tunneling expression for the triangular barrier by

$$I \propto E^2 \exp\left(-A\frac{\phi^{3/2}}{E}\right) \tag{5.23}$$

with A constant. When an electric field E exceeds $E = 10.3$ V/nm ($\sim 1 \times 10^{10}$ V/m), we see that the barrier height reduces and is missing. This value is close to that obtained by the other calculation [127]. This indicates that the potential barrier exhibits a hole for an electron to emit into the vacuum region directly through this hole. We see that the current distribution becomes considerably large after a hole is created in the potential barrier.

Figure 5.19 shows the current density of states (CDOSs) for the field emission from Na, Al, and Si single atom on the jellium surface. The applied electric field is $E \approx 10^{10}$ V/m. We see that the CDOSs increase gradually as a function of energy. This shows that not only the current density close to the Fermi energy ϵ_F contributes to the total current. We note that the resonant state ϵ for $3s$ orbital, which

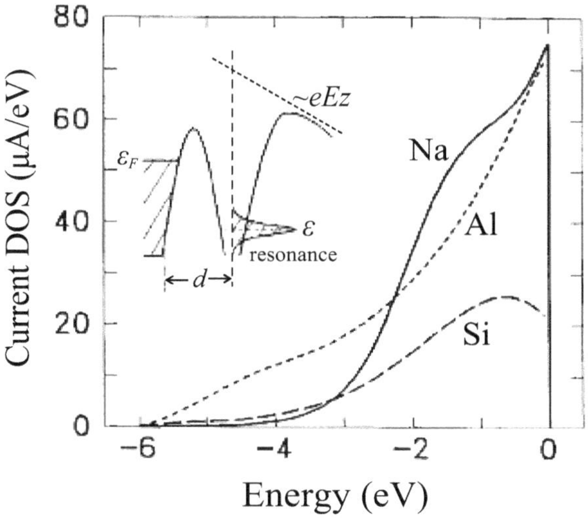

Figure 5.19 Current density of states for the field emission from Na, Al, and Si single atom on the jellium surface with an applied electric field of approximately $E \approx 10^{10}$ V/m. The inset shows the schematic representation of the effective potential with $-eEz$ in the vacuum region, with d the separation of a single atom from the jellium surface, and ϵ the energy of the resonance of a single atom.

shifts down to lower energy due to the electric field E, is observed for the Na case in the CDOS.

5.2 Atomic Wires

The atomic-scale nanostructures have been created using very sophisticated experimental techniques. Among them, the simple but very important nanostructures would be atomic-scale point contacts and atomic wires. Historically, atomic-scale point contact structures were created after the invention of the STM and measured the resistance about 10 kΩ [128, 129].

After that, a new technique called the mechanically controllable break junction (MCBJ) was developed to create an atomic-scale point contact [130]. The conductance of an order of $2e^2/h$ was

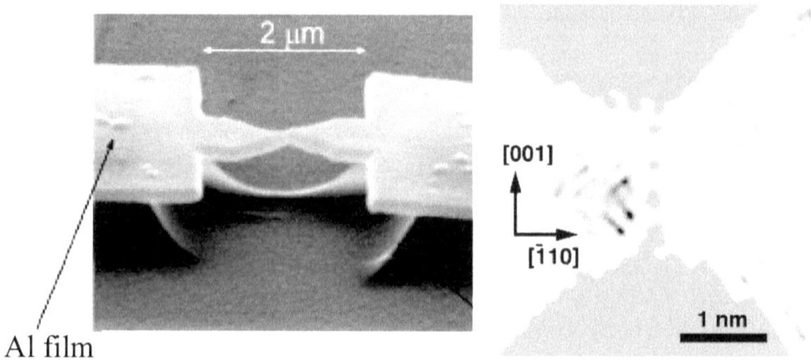

Figure 5.20 (Left) Micrograph of a suspended Al microbridge fabricated by MCBJ techniques. From E. Scheer, P. Joyes, D. Esteve, C. Urbina, and M. H. Devoret, *Phys. Rev. Lett.* **78**, 3535 (1997) [131]. (Right) Electron microscope image of a gold atomic wire. From H. Ohnishi, Y. Kondo, and K. Takayanagi, *Nature* **395**, 780 (1998) [132].

measured and the connection with the quantized conductance has been discussed. In Fig. 5.20, we show the atomic-scale point contact and the atomic wires created by the sophisticated experiments. The conductance data was recorded for atomic-sized contacts while the contacts are stretched to the point of breaking. The conductance in such experiments show to decrease approximately with an integer step of order of the quantum unit of conductance $2e^2/h$, although they fluctuate much since various atomic configurations are created. The histograms of conductance values from a large number of individual conductance curves demonstrate that the conductance has finite values for multiples of $2e^2/h$. An example of theoretical and experimental observations of conductance for the atomic wire is shown in Fig. 5.21.

In this section, we show the transport properties of atomic wires. The calculation methods for the electron transport are based on the scattering wave approach with the density functional theory. Since the atomic wires change their atomic configurations, we consider the bending effect for the conductance here. Eigenchannel decomposition technique plays an important role to understand the transport from an atomic-orbital point of view.

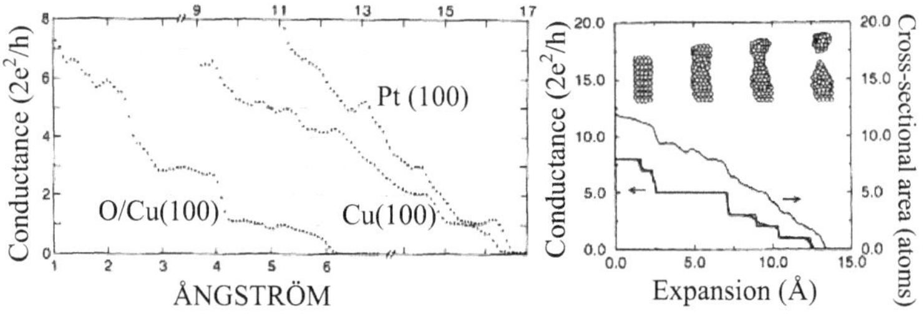

Figure 5.21 (Left) Conductance curves for atomic-scale point contact. (Right) Calculation of conductance curve and cross-sectional area for the elongation. From L. Olesen, E. Lægsgaard, I. Stensgaard, F. Besenbacher, J. Schiøtz, P. Stoltze, K. W. Jacobsen, and J. K. Nørskov, *Phys. Rev. Lett.* **72**, 2251 (1994) [133].

5.2.1 *Eigenchannel Decomposition Analysis*

5.2.1.1 Al atomic wire

Let us consider the transport properties of Al atomic wires here. We use a model consisting of three-atom Al wires connecting to the triangular atomic bases sandwiched between two semi-infinite metallic jellium electrodes. We also see the transport properties of bending Al wires later, which are expressed by the displacement of the electrodes spacing Δd. As for an experiment using the MCBJ technique for Al junctions, the number of channels through a single atomic contact and the channel transmission during the elongation processes were measured and analyzed [131]. According to the experiment, just before the contact is broken, three channels contribute to the transport through the contact, and the transmission of one of these channels becomes much larger than the other ones. The positive slopes were observed clearly in the MCBJ and the STM experiments.

We use the RTM method for the electronic states and the electric current, which are decomposed into a sum over conduction channels using the eigenchannel decomposition (ECD) technique. In particular, the conductance, the LDOS, and the current density are decomposed into eigenchannels. These are effective in clarifying the details of transport through various atomic wires, which enables us to elucidate the number of conduction channels, the relation between atomic orbitals and the channels, and their dependency on the geometry of the various atomic wires.

5.2.1.1.1 Straight Cases First, we show the conductance of straight Al atomic wire. The atomic configuration is schematically shown in the inset of Fig. 5.22. Figure 5.22 (top) shows the channel DOS and channel transmission as a function the energy measured from the Fermi energy. The conductance is expressed by the sum of individual channel transmissions in the quantized unit of $2e^2/h$. The channel DOS for the first channel shows the one-dimensional feature, with a long tail on the higher energy side $1/\sqrt{\epsilon}$ from the resonant energy close to $\epsilon_0 = -5.8$ eV. The transmission from this channel is completely open at the Fermi energy where it is located well above ϵ_0 and contributes to one quantum unit $2e^2/h$. For the second and third channels, which are degenerate due to the symmetry of atomic wire structure, the Fermi energy is located close to the resonant energies. Then the transmissions from these channels are partially open and take fractional values of ~ 0.25. As a result, the total conductance becomes close to $G \simeq 1.5 \times (2e^2/h)$.

The character of each channel is well seen in the channel LDOS and channel current density shown in Fig. 5.22 (bottom). The first channel (left) has s, p_z, $d_z^2 \ldots$ orbital components with $l_z = 0$ of the Al atoms in the wire. The second and third channels (center and right), which are degenerate, have p_x,$d_{zx} \ldots$, and p_y, $d_{zy} \ldots$ orbital components with $l_z = \pm 1$. Therefore, we can rename these channels as s-p_z for the first channel, p_x and p_y for the second and third channels, since the contributions from s and p orbitals are dominant below the Fermi energy for the Al atomic wires. Note that higher angular momentum components are automatically included since we expand wavefunctions with plane-wave basis sets.

We point out that the relation between the resonant energies and the Fermi energy is important when we understand the changes of transport properties due to geometrical distortions. Since the Fermi energy is located around the resonant energies of the second and third channel p_x and p_y for the Al atomic wires, the conductance would be sensitive to geometrical changes, which are the subjects in the next section.

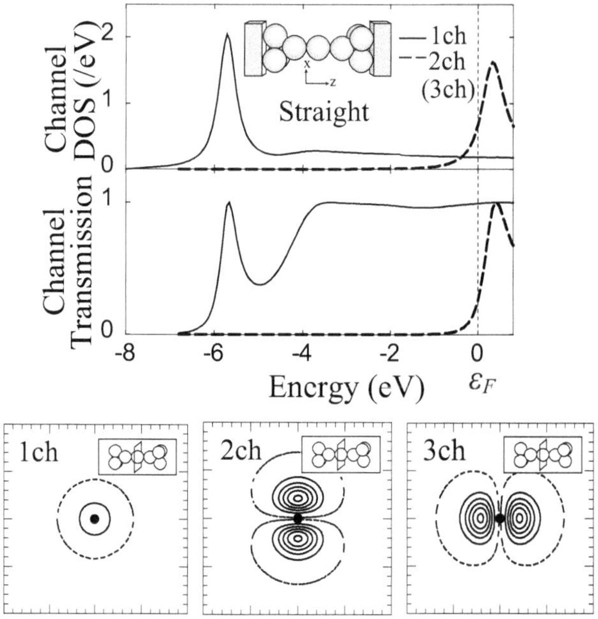

Figure 5.22 (Top) Channel DOS around the center atom (upper) and channel transmission (lower) for the straight Al atomic wire as a function of the energy measured from the Fermi energy. The atomic configuration is schematically shown in the inset. (Bottom) Channel LDOS ($n^{L}_{\epsilon,i}(\mathbf{r})$) at the Fermi energy. The electronic states incident from the left electrode are shown. The left, center, and right panels show the first, second, and third channels. From N. Kobayashi, M. Brandbyge, and M. Tsukada, *Phys. Rev. B* **62**, 8430 (2000) [134].

5.2.1.1 Bending Cases Next, we consider the effects of bending of Al atomic wire on conductance. The bending is represented by the displacement Δd of the center atom in the x direction. The bond length is kept constant for the variation of Δd. We take $\Delta d < 0$ to represent the bending case.

In the top panel of Fig. 5.23 (left), we show the channel DOS around the center atom and channel transmissions for $\Delta d = -2.0$ bohr. We realize that the degenerate channels become lifted at the Fermi energy due to the breaking of the symmetry. The conductance and channel transmission at the Fermi energy are shown in Fig. 5.23 (right) as a function of displacement Δd. We

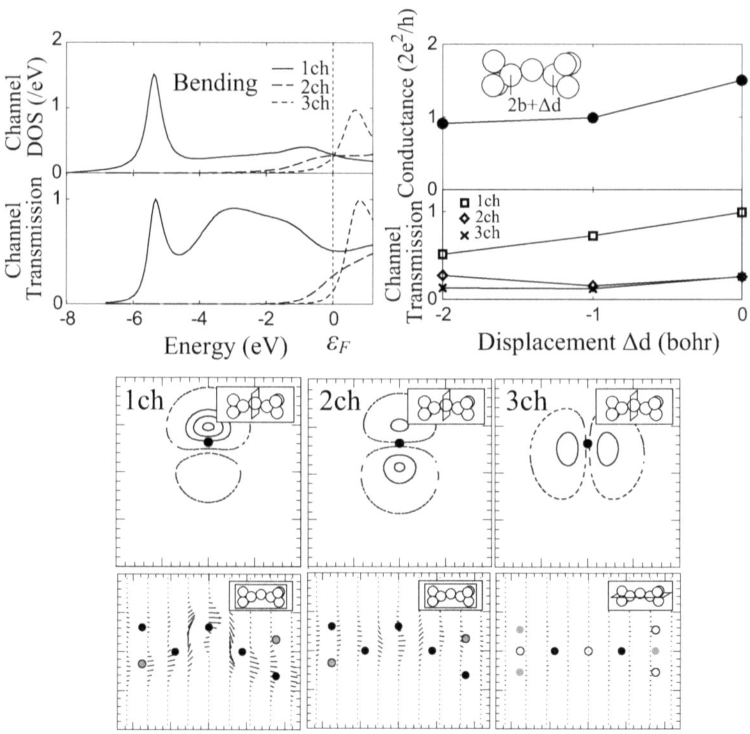

Figure 5.23 (Top) (Left) Channel DOS around the center atom and channel transmission (right) Conductance and channel transmission at the Fermi energy as a function of displacement of Δd for the bending. (Bottom) Channel LDOS ($n^L_{\epsilon,i}(\mathbf{r})$) and channel current density ($\mathbf{j}^L_{\epsilon,i}(\mathbf{r})$) distributions for the first, second, and third channels. The length of the arrows indicates the magnitude of current density. The solid black circles indicate atomic positions. From N. Kobayashi, M. Brandbyge, and M. Tsukada, *Phys. Rev. B* **62**, 8430 (2000) [134].

see that the conductance takes $G \simeq 1.5(2e^2/h)$ for the straight wire $\Delta d = 0$ and it decreases with increasing the displacement $|\Delta d|$ for bending, which becomes less than the quantized unit of $2e^2/h$ for $|\Delta d| \geq 1$. In more detail, the first channel of the straight wire is mixed with its second channel and the one-dimensional character is weakened. Therefore, the first channel transmission at the Fermi energy has a smaller value than in the straight case. On the other hand, the second channel goes through a minimum for

$\Delta d = -1$ bohr, but has almost the same transmission for $\Delta d = -2$ bohr as in the straight configuration. The third channel decreases because of the slight upward shift of the channel onset energy relative to the Fermi energy. Thus, the conductance of the bent wire is smaller than that of the straight wire, and decreases with an increase of the bending $|\Delta d|$. Here we note that even though the conductance decreases monotonously, not all channel transmissions decreases as bending.

Figure 5.23 (bottom) shows the channel LDOS distributions at the Fermi energy for $\Delta d = -2$ bohr. Due to the breaking of symmetry, we can no longer observe the s-p_z or the p_x character of the first two channels around the central atom. Only the structure of the third channel, which has the p_y orbital character, is similar to that of the straight wire. It is interesting to note that the current distribution $\mathbf{j}^L_{\epsilon,i}(\mathbf{r})$ for the first channel constitutes a loop current circulating the center atom, which induces a magnetic moment.

5.2.1.2 Si atomic wire

Here, we consider the electron transport in a Si atomic wire at a finite bias voltage, focusing on the covalent bond nature of Si atoms and its response to a finite bias. The calculations are performed by the Lippmann–Schwinger equation method with the plane-wave Laue representation. The atomic wire is composed of three Si atoms between semi-infinite metallic jellium electrodes. The electron density and current density distribution for a bias voltage of 3 V in the left of Fig. 5.24 (top) shows that high-electron-density regions corresponding to the covalent bonds between the Si atoms, which are slightly polarized due to bias voltage. The bonding natures are also seen at contacts between Si atom and metals. The corresponding potential drop for a bias voltage of 3 V extends over a broad region in the wire, reflecting a small screening of Si covalent bonds.

Let us see the electronic states and the contact effects of metallic electrodes on the Si atomic wires under a finite bias voltage. Figure 5.24 (bottom left) shows the DOS and Fig. 5.24 (right) shows the LDOS and effective potentials for the bias voltages of 0 V and 3 V. In the zero bias case, we see a localized state around -8 eV

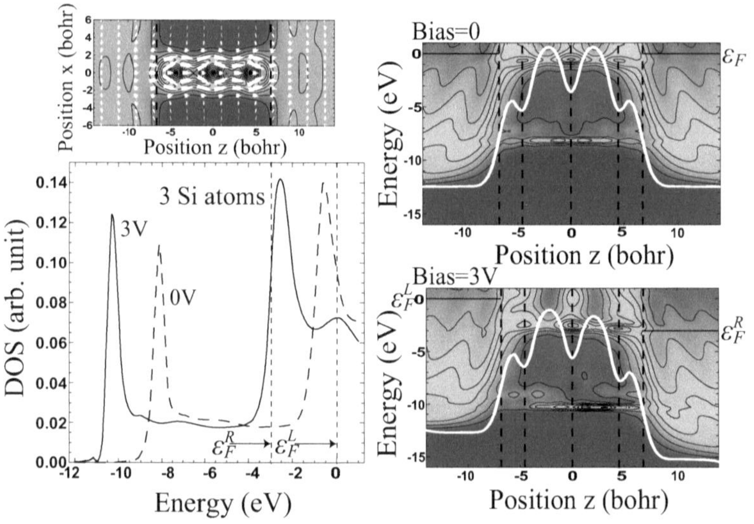

Figure 5.24 (Left) (Top) Electron density and current density distribution for a bias voltage of 3 V. The white arrows show the current density and the black circles indicate the Si atomic positions. (Bottom) Density of states (DOS) in the center region of $-2.2 < z < 2.2$ in bias voltages of 3 V (solid) and 0 V (dotted) with $\epsilon_F^L=0$ (vertical dashed line). The vertical dashed line at -3 V is ϵ_F^R for a bias of 3 V. (Right) Local density of states and effective potentials averaged in the x–y plane. The vertical dashed-dotted lines and dashed lines indicate the positions of Si atoms and the edges of the electrodes, and horizontal lines indicate the Fermi energies in the electrodes. From H. Kusaka and N. Kobayashi, *Appl. Surf. Sci.* **258**, 1985 (2012) [135].

from the s orbitals in the LDOS. This is also observed in the DOS as a distinct peak, showing a localized singular character of a 1D system $\sim 1/\sqrt{\epsilon}$ with a longer tail to the higher energy. On the other hand, the LDOS shows a strong mixing between the p orbitals and electrodes near the Fermi energy 0 eV where the electronic states of Si atoms at contacts are spread out to electrodes, while the Si atom in the center has a clear bonding nature. Note that due to the existence of metallic electrodes, there is no energy gap in the Si atomic wire. This is completely different from the Si atomic wire with infinite length, where band structure shows an energy gap slightly below the Fermi energy. For a finite bias of 3 V, we see that the DOS shifts to lower energies. The LDOS shows that the p state of the Si atom connected to the right electrode changes its character significantly at the Fermi energy, which reflects in the potential drop. On the other hand, the s orbital apart from the Fermi energy at ~ -8 eV in 0 V case only shifts the energy level to ~ -10 eV, since the mixing of the s atomic orbitals with electrodes is small.

5.2.2 *Effects of Atomic-Scale Contacts*

As we see, when atomic wires are directly connected to electrodes, the conductance is well described by the concept of eigenchannel decomposition. Namely, the number of conducting channels in the atomic wire and their energy dependence relative to the Fermi energy determine the magnitude of conductance. This conductance nature is based on the observation that electrons are inserted to the atomic wire smoothly, which indicates that the potential modulation is adiabatic compared with the Fermi wavelength. However, when the contact to electrodes is strongly modified, the conductance should be determined not only by the nature of atomic wire but also by the nature of the contact. Here, let us show the effect of contact on electron transport in atomic wires.

Figure 5.25 (left) shows electron densities and current distributions of Al atomic wires with a single Al (top), Na (center), and Cl (bottom) atom at contact adjacent to the left electrode. Effective potential is strongly modified at contact according to the kinds of a substitution atom. Figure 5.25 (right) plots the conductance of Al atomic wires composed of six atoms with a different atom (from Na to Cl) connecting to the left electrodes. Conductance is obtained from $G = \int I(\mathbf{r})d\mathbf{r}_{||}/V$ with a tiny bias voltage of 10 mV in the linear-response regime. We see that conductance is strongly dependent on the bonding nature of the atom at contact. Since an applied bias is very small, conductance is determined by the states close to the Fermi level, which leads to the net current. In the clean Al atomic wire case, the magnitude of conductance is close to $2 \times (2e^2/h)$, which results from the partially occupied $3p$ states of Al atoms at the Fermi level, as we see in the previous section. Conductance decreases considerably as the number of valence electrons decreases from Mg to Na. Notably, the conductance

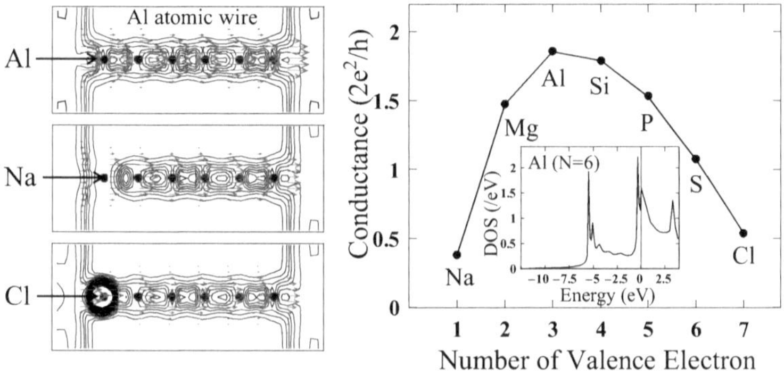

Figure 5.25 (Left) Electron densities and current distributions of Al atomic wires composed of six atoms. Different kinds of a single atom, Al (top), Na (center), and Cl (bottom) are placed at contact with the left semi-infinite Al jellium electrode ($r_s = 2$). (Right) Conductance of Al atomic wires with a different kind atom (from Na to Cl) connecting to the left electrode with an applied bias of $V = 10$ mV. (Inset) Density of states of clean Al atomic wires. Density of states from the left and right electrodes are subtracted. From K. Hirose, N. Kobayashi, and M. Tsukada, *Phys. Rev. B* **69**, 245412 (2004) [62].

decreases less than $0.5 \times (2e^2/h)$ when we put a single Na atom at contact. This is mainly due to the deficiency of charge density in the conducting channel from the partially occupied 3s state of the Na atom, and the atomic wire is almost disconnected from the electrode. On the other hand, conductance also decrease monotonically as the number valence electrons increases from Si to Cl. This behavior is attributed to off resonance in the conducting channel of the Al atomic wire. This is because the energy levels of the valence 3p states of connecting atoms at contact shift to a deeper energy and exhibit a localized nature as the number of valence electrons increases, which reduces conductance significantly.

Next, we consider where applied bias drops in the Al atomic wires. In the present inhomogeneous atomic-wire systems connected to electrodes, which are different from purely one-dimensional systems, we expect nonlinear behaviors in bias drops due to atomic-scale local polarization. Figure 5.26 (left) plots the differences in electrostatic potentials with +5 V bias and with zero bias along the Al atomic wire. The dotted curve plots the difference

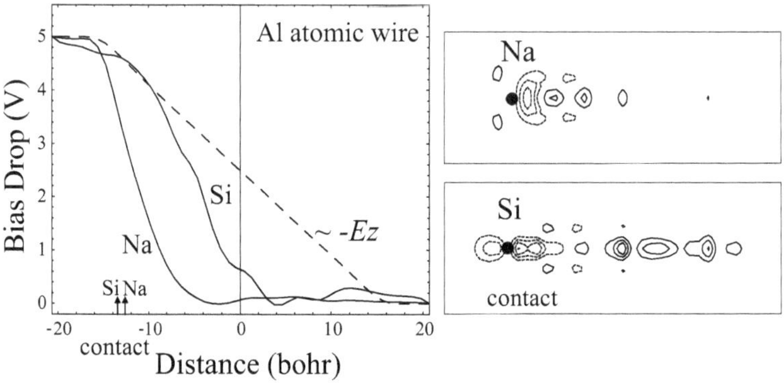

Figure 5.26 (Left) Differences in electrostatic potentials of Al atomic wires with a single Na and Si atom at the contact for an electrode biased at +5 V and the potentials of the same system at zero bias plotted along the atomic wire. Dotted line represents the difference in electrostatic potential of a bare electrode without atomic wires at the same finite bias voltage. (Right) Differences in the charge densities of Al atomic wires with a single Na (top) and Si (bottom) atom at the contact biased at +5 V and the charge densities of the same system at zero bias. Solid lines represent charge affinity and dotted lines represent charge deficiency. Large dots show the positions of Na and Si atoms at contact connecting to electrode.

for bare electrodes without atomic wire, which reveals linear drop between electrodes since bare electrodes act as a planar capacitor. We note that the slope of linear drop corresponds to an applied electric field. When a single Na atom is placed at the contact in the Al atomic wire, we can see that the drop in potential occurs in the vicinity of Na atom. This shows that the main resistance in the present atomic wire comes from the local position of Na atom at the contact. Figure 5.26 (top right) shows a difference in the charge density of Al atomic wires with a single Na atom at the contact biased at +5 V and those from the same system at zero bias. We see that local polarizations occur in the vicinity of the Na atom, which induces rapid drop in electrostatic potential. This strong polarization is attributed to the ionic nature of the 3*s* orbital of the Na atom and the electric field is completely screened out in the vicinity of the Na atom. On the other hand, when a single Si atom is placed at contact in Al atomic wire, bias drop in potential is much more slow, which shows a different behavior from the Na case. The difference in the charge density in Fig. 5.26 with a single Si atom at the contact shows that the polarization of the Si atom does not occur only in the vicinity of the Si atom, but spreads out to the whole region of the atomic wire. This character reflects the covalent bond nature of the 3 *p* orbital of Si atom coupled with other Al atoms. Since the potential drop is not rapid, local electric field at contact in the Si atom does not become so strong.

These observations of bias drops in the atomic wires reveal that the local electric field depends strongly on the nature of the atom at the contact, whose polarization behaviors determine the electronic states and correspondingly the transport properties, and open the possibility of changing the local potential profile by mixing atoms in the atomic wire locally, which would lead to decrease the potential barrier at the contact for the electron transport. The atomic-scale contact problem for electron transport is much more important and essential when we put molecules between electrodes, where strong nonlinear $I-V$ characters emerge in the electron transport due to the HOMO–LUMO energy gap. This problem will be presented in the section on the transport of single molecule later.

5.3 Spin-Dependent Transport of Magnetic Materials

In this section, we describe the transport in atomic-scale dots, point contacts, and wires, including magnetic materials. When the dots, contacts, or wires are made of a magnetic-metal such as Fe, Co, Ni and Mn, the exchange energy removes the spin degeneracy of conduction electrons and an atomic scale domain wall is created. It is then intriguing to ask how the exchange energy and the domain wall affect the electron transport. We show transport properties for such atomic-scale spin-dependent phenomena.

5.3.1 *Diluted Magnetic Semiconductor Dot*

First, we describe the spin-dependent transport of diluted magnetic dots in the influence of electric fields and magnetic fields. With the rapid progress in semiconductor growth techniques, fabrication of various diluted magnetic semiconductor (DMS) dot structures is now possible. Here we treat the spin-dependent transport for $Cd_{1-x}Mn_xTe$ DMS dot systems.

The novel properties of DMS arise from the strong sp–d exchange interaction between the carriers and the magnetic ions Mn^{2+} in DMS structures. Of interest in the spin-dependent transport for semiconductor DMS systems is whether the phenomenological model can account for the experiments of DMS dots. Using empirical parameters, we apply the RTM method for the spin-dependent transport through 3D nanostructures of DMS dots. We note that the spin-dependent transport of GaAs quantum dots will be described in Chapter 6, where we use the spin-density functional theory for electronic state calculations to treat the Coulomb interaction, including Hartree and electron correlation energies in the single electron tunneling (SET) regime.

We consider a DMS dot embedded in another semiconductor material. External magnetic fields B and electric fields E are applied to the system. The Hamiltonian can be written as

$$H = \frac{1}{2m^*}\left(\mathbf{p} + \frac{e}{c}\mathbf{A}(\mathbf{r})\right)^2 \pm g^*\mu_B B - e\mathbf{E}\cdot\mathbf{r} + V_{\text{ext}}(\mathbf{r}) + V_x(\mathbf{r}), \quad (5.24)$$

where $\pm g^*\mu_B B$ is the Zeeman splitting with $g^* = -0.7$ and $\mathbf{E} = (0, 0, E)$ is the external static electric field. We use the symmetric gauge $\mathbf{A} = \mathbf{B} \times \mathbf{r}/2 = (B/2)(-y, x, 0)$ for vector potential $\mathbf{A}$ and take confining potentials of electrons $V_{\text{ext}}(\mathbf{r})$ simply as

$$V_{\text{ext}}(\mathbf{r}) = \begin{cases} 0 & \text{in the dot} \\ \epsilon_c & \text{otherwise.} \end{cases} \quad (5.25)$$

As for the Coulomb interaction, we neglect the electron correlation and consider mean-field exchange potential $V_x(\mathbf{r})$ only to account for the spin-dependent electron transport,[a] which is expressed by

$$V_x(\mathbf{r}) = -J_{sd}\sigma_z\langle S_z\rangle. \quad (5.26)$$

Here $J_{sd} = N_0\alpha x_{\text{eff}}$ and $\langle S_z\rangle = S_0 B_J\,[S g_{Mn}\mu_B B/k_B(T + T_0)]$, where $S = 5/2$ corresponds to the spins of the localized $3d^5$ electrons of the Mn^{2+} ions, B_J is the Brillouin function, and N_0 is the number of cations per unit volume. The phenomenological parameters x_{eff} (reduced effective concentration of Mn) and T_0 account for the

[a]This corresponds to the mean-field (single-particle) treatment of the s–d Hamiltonian described for the many-body Kondo effect in Chapter 6.

reduced single-ion contribution due to the antiferromagnetic Mn-Mn coupling, $g_{Mn} = 2$ is the g factor of the Mn^{2+} ion, and $\sigma_z = \pm 1/2$ is the electron spin. For a spin-dependent transport through a single $Cd_{1-x}Mn_xTe/Cd_{1-y}Mg_yTe$ spherical DMS dot, which is located at the middle of the system, we take the width of the system and the radius of the dot as $2L$ and R. The empirical parameters used are $\epsilon_c = 0.8(\epsilon_g^2 - \epsilon_g^1)$, where $\epsilon_g^1 = (1.586 + 1.51x)$ eV and $\epsilon_g^2 = (1.586 + 1.705y)$ eV are the bandgaps of $Cd_{1-x}Mn_xTe$ and $Cd_{1-y}Mg_yTe$. Other parameters are $m^* = 0.096m$, $x_{\text{eff}} = 0.045$, $N_0\alpha = 0.22$ eV, $N_0\beta = -0.88$ eV, $S_0 = 1.32$, and $T_0 = 3.1$ K.

Let us consider magnetic and electric field effects of spin splitting in a DMS dot. We set $x = 0.07$ and $y = 0.062$ so that $\epsilon_c = 0$ for the $Cd_{0.93}Mn_{0.07}Te$/ $Cd_{0.938}$ $Mg_{0.062}Te$. Figure 5.27 (a) shows the spin-dependent transmission coefficient $\mathcal{T}_{\uparrow\downarrow}$ as a function of the Fermi energy ϵ_F for $L = 20$ nm and $R = 5$ nm with $B = 10$ T and $E = 0$. $\mathcal{T}_{\uparrow}$ is almost restrained if $\epsilon_F < 10$ meV, while $\mathcal{T}_{\downarrow} \approx 1$ as long as $\epsilon_F > 1$ meV. This shows that a 100% polarized electric current is possible by injecting electrons with 1 meV $< \epsilon_F <$ 10 meV for the structure with parameters given above. Figure 5.27(b,c) shows $\mathcal{T}_{\uparrow\downarrow}$ as a function of ϵ_F for (b) $B = 30$ T and (c) $E = 10^4$ V/cm. The axial magnetic field increases the transverse confined energy levels, making the resonant peaks move to the high Fermi energy side. The difference between $\mathcal{T}_{\uparrow}$ and $\mathcal{T}_{\downarrow}$ is enhanced as the magnetic field B become large mainly due to exchange interaction and partly due to Zeeman splitting.

On the other hand, the electric field E on $\mathcal{T}_{\uparrow\downarrow}$ moves the transmission peak positions with reduction of the heights. The movements of the peak positions to higher or lower Fermi energy depends on the direction of E. Therefore, the polarized electric current is present when E is applied for the DMS dot with $B =$ 10 T. Change of the parameter into $y = 0.4$ produces the band offset energy as $\epsilon_c = 416$ meV for $Cd_{0.93}Mn_{0.07}$ $Te/Cd_{0.6}Mg_{0.6}Te$. Figure 5.27(d) shows the variation of $\mathcal{T}_{\uparrow\downarrow}$ with ϵ_F. The solid and dashed lines are $\mathcal{T}_{\uparrow}$ and $\mathcal{T}_{\downarrow}$. The difference between $\mathcal{T}_{\uparrow}$ and $\mathcal{T}_{\downarrow}$ decreases as the band offset ϵ_c increases. This is because the split energy between the spin-up and spin-down electrons is much smaller than the large ϵ_c and thus can be treated as a perturbation.

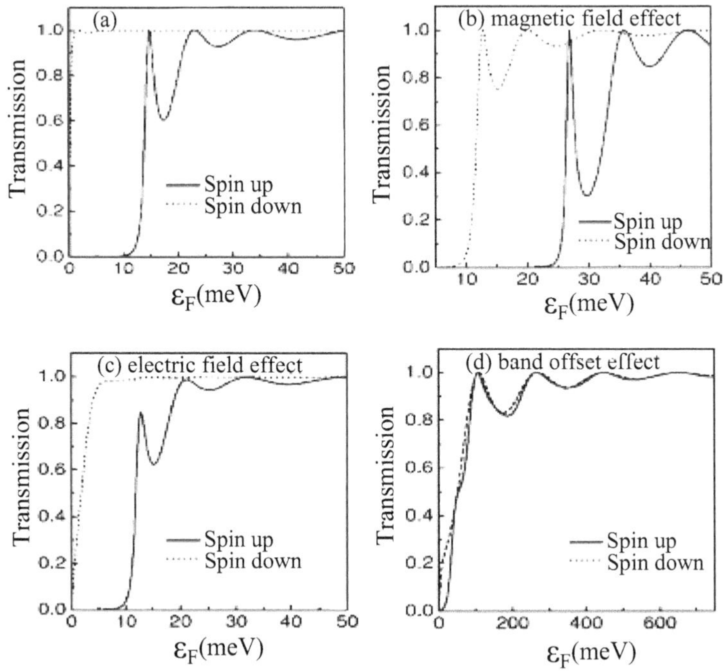

Figure 5.27 (a) Spin-dependent transmission coefficients $\mathcal{T}_{\uparrow\downarrow}$ as a function of the Fermi energy ϵ_F in $Cd_{0.93}Mn_{0.07}Te/Cd_{0.938}Mg_{0.062}Te$ with the parameters of $R = 5$ nm, $L = 20$ nm, $B = 10$ T, and $E = 0$. The solid line represents $\mathcal{T}_{\uparrow}$ and the dashed line for $\mathcal{T}_{\downarrow}$. (b) Magnetic-field effect on $\mathcal{T}_{\uparrow\downarrow}$ with $B = 30$ T. (c) Static electric field effect on $\mathcal{T}_{\uparrow\downarrow}$ with $E = 10^4$ V/cm. (d) Structural change on $\mathcal{T}_{\uparrow\downarrow}$ due to the band offset in $Cd_{0.93}Mn_{0.07}Te/Cd_{0.6}Mg_{0.4}Te$. For (b)-(d) the other parameters are the same as (a). From S. S. Li, K. Chang, J.-B. Xia, and K. Hirose, *Phys. Rev. B* **68**, 245306 (2003) [136].

These results indicate that for the spherical DMS dots, (1) we can get a spin-polarized electric current by using suitable structure parameters; (2) the spin-polarized effect is slightly increased for larger magnetic fields; (3) the external static electric field moves the transmission peaks to higher or lower Fermi energy depending on the direction of the applied field; and (4) the spin-polarized effect decreases as the band offset increases. We note that the present spin-dependent transport calculations based on the phenomenological treatments are effective when the two spin channels are separated and the parameters are determined from the experiments well. Extension to include the spin-orbit coupling, for example by the Rashba-type $H_{RSO} = (\alpha/\hbar)\sigma \cdot (\mathbf{p} \times \mathbf{z}) = (\alpha/\hbar)(p_y\sigma_x - p_x\sigma_y)$ is possible, where the single-particle treatments are still effective.

5.3.2 *Magnetic Atomic Point Contact*

Next, we treat the magnetic point contact. When the contact is made of a magnetic-metal, the conductance depends on the relative orientation of magnetizations between left and right electrodes, like magnetic tunnel junctions. The spin-dependent transport such as the tunnel magnetoresistance (TMR) in magnetic tunnel junctions and giant magnetoresistance (GMR) in magnetic multilayers is of current interest in both fundamental physics and application in spin-electronics. In Ni point contact, experimental results show the magnetoresistance (MR) in excess of 200% [137], the spin-dependent conductance quantization [138], and the $2e^2/h$ to e^2/h switching of the quantized conductance [139].

Let us consider the electron transport through a magnetic point contact shown in Figs. 5.28(a) and (b) by using the RTM method. When a magnetic field is applied, the magnetizations of left and right electrodes are parallel. In this ferromagnetic (F) alignment, the quantized conductance of odd integer multiples of e^2/h appears, since the spin-up and spin-down electrons contribute to the conductance in the different way. On the other hand, in the absence of the magnetic field, the system is in the antiferromagnetic (A) alignment, where the magnetizations of left and right electrodes are antiparallel. The domain wall is created inside the constriction. The spin precession of conduction electrons is forbidden in such an atomic-scale domain wall. The contribution to the conductance from spin-up and spin-down electrons are the same and the unit of the conductance quantization is $2e^2/h$. The MR is strongly enhanced for the narrow point contact and oscillates with the conductance. These explain the experimental results and provides a new direction for spin-electronic devices with an atomic-scale domain wall.

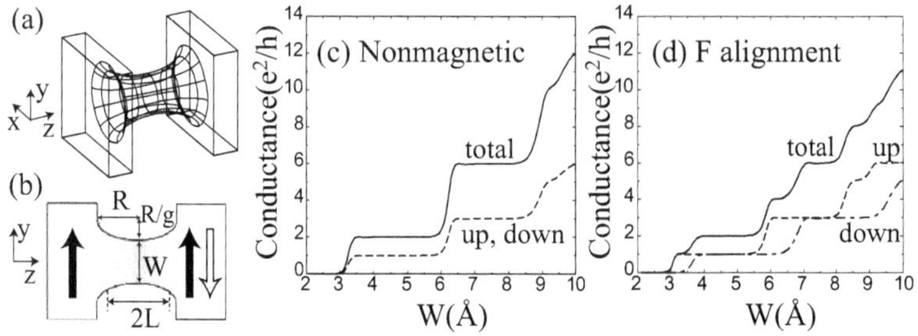

Figure 5.28 (a) Schematic picture of the geometry of constriction potential $V(\mathbf{r}_{||}, z)$. (b) Cross-section of the constriction potential. Arrows represent the magnetization vectors with the filled (hollow) arrow for the F (A) alignment. The shaded region in the constriction is the domain wall with thickness $2L$ for the A alignment. (c) Conductance curves $h_0 = 0$. The total conductance is plotted by the solid line, while those for spin-up (spin-down) electrons by the dotted lines. (d) Conductance curves for the F alignment with $h_0 = 0.5$ eV. Conductance for spin-up (spin-down) electrons is plotted by the dashed (dot-dashed) line. From H. Imamura, N. Kobayashi, S. Takahashi, and S. Maekawa, *Phys. Rev. Lett.* **84**, 1003 (2000) [140].

The effective one-electron Hamiltonian is given by

$$H = -\frac{\hbar^2}{2m}\nabla^2 + V(\mathbf{r}_{||}, z) - \mathbf{h}(z)\cdot\sigma. \tag{5.27}$$

Here $V(\mathbf{r}_{||}, z)$ is the constriction potential defined by

$$V(\mathbf{r}_{||}, z) = V_0 \begin{cases} \theta(R+z)\theta(R-z) \\ \qquad\qquad \text{for } |\mathbf{r}_{||}| \geq R/g + W/2 \\ \theta(R - \sqrt{z^2 + (R + gW/2 - g|\mathbf{r}_{||}|)^2}) \\ \qquad\qquad \text{for } |\mathbf{r}_{||}| < R/g + W/2, \end{cases} \tag{5.28}$$

where V_0 is the height of the potential, R is the radius of the elliptic envelope with a deformation parameter g, and W is the width of the constriction. $\mathbf{h}(z)$ is the exchange field, σ is the Pauli matrix, and $\mathbf{r}_{||} = (x, y)$ is the position vector in the x–y plane. The x-axis is taken parallel to the exchange field in the left electrode. For the F alignment, the exchange field is constant $\mathbf{h}(z) = (h_0, 0, 0)$. For the A alignment $\mathbf{h}(z)$ is not constant inside the domain wall. In the classical theory, $\mathbf{h}(z)$ is given by

$$h_x(z) = 0, \quad h_y(z) = h_0\theta(L - |z|)\cos(Qz),$$
$$h_z(z) = h_0\{\theta(-L-z) - \theta(-L+z) - \theta(L - |z|)\sin(Qz)\}, \tag{5.29}$$

where L is half the thickness of the domain wall and $Q = \pi/2L$. The domain wall is about the size of the point contact. For such an atomic-scale domain wall, we have to derive an exchange field $\mathbf{h}(z)$ on the basis of the quantum theory. As we see later, the y component of the exchange field vanishes $h_y(z) = 0$ and the spin of conduction electrons cannot rotate in atomic-scale domain wall.

In order to obtain the stationary scattering states, we employ the RTM method, including the spin degree of freedom. We choose the commonly accepted values of the material parameters for typical ferromagnetic metals of Ni, Co, and Fe. The Fermi energy and the height of the constriction potential are taken as $\epsilon_F = 3.8$ eV, ($k_F = 1.0$ Å^{-1}), and $V_0 = 2\epsilon_F$. The length of the constriction and the thickness of the domain wall are taken as $2R = 2L =$ 10 Å corresponding to 3$\sim$5 atoms and the deformation parameter is $g = 10$. We choose the magnitude of the exchange field $h_0 =$ 0.3 $\sim$ 0.7 eV and replace the step function θ in Eq. (5.28) by the Fermi–Dirac function with a width of 0.25 Å to make the constriction smooth. In Fig. 5.28(c), we show the conductance curve for a nonmagnetic point contact. The missing of the plateau at $4e^2/h$ and $8e^2/h$ is due to the rotational symmetry of the constriction potential. For a magnetic point contact, the degeneracy of the spin-up and spin-down conductances is removed by the exchange energy and plateaus of odd integer multiples of e^2/h appear for the F alignment as shown in Fig. 5.28(d).[a] The key point is that the width of the constriction W, at which the number of transmitting channels changes, becomes spin-dependent, because the Fermi wavelength is different between spin-up and spin-down electrons. The width W at which the new transmitting channel opens for spin-up (spin-down) electrons decreases (increases) as the exchange filed increases.

Let us move on to the effect of the domain wall for the A alignment. First, we show how the spin of a conduction electron rotates without the constriction potential. We consider a 1D classical domain wall. The analytical results of the probability current density is obtained by connecting wavefunctions at the boundary of the domain wall. We rotate the spin-quantization axis parallel to the

[a]We note that the spin-dependent conductance quantization was observed in semiconductor point contacts under high magnetic fields [141], where the spin degeneracy is removed by the Zeeman energy, which is shown in Chapter 6.

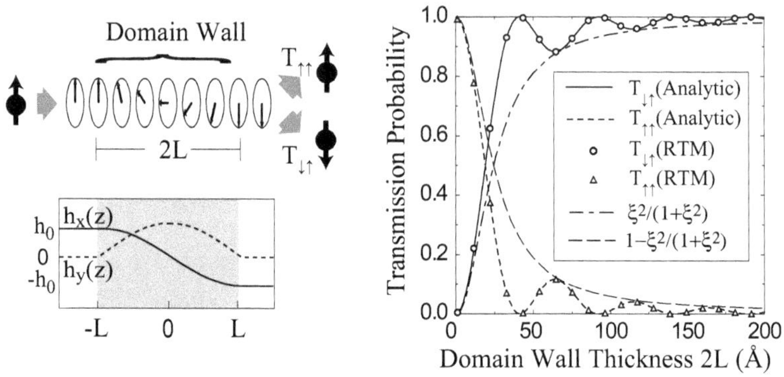

Figure 5.29 (Left) (Top) Magnetization vector of 1D classical domain wall with $2L$. (Bottom) Exchange field $\mathbf{h}(z)$. (Right) Transmission probabilities for the incident electron. Analytical results of $\mathcal{T}_{\downarrow\uparrow}$ and $\mathcal{T}_{\uparrow\uparrow}$ with solid and dotted lines and numerical results by the RTM method of $\mathcal{T}_{\downarrow\uparrow}$ ($\mathcal{T}_{\uparrow\uparrow}$) with circles (triangles). The dashed (dot-dashed) line is the spin conservation (flip) probability. From H. Imamura, N. Kobayashi, S. Takahashi, and S. Maekawa, *Mat. Sci. Eng. B* **84**, 107 (2001) [142].

z-axis. In this case, the eigenstates of the Hamiltonian in the domain wall region are the same as those of a "spin spiral". Assuming that the spin-up electron ($s_z = 1/2$) is incident from the left electrode as shown in Fig. 5.29(a), the wavefunctions in the left (L) electrode, domain wall, and right (R) electrode are given by

$$\psi_L = \begin{pmatrix} e^{ik_L^\uparrow z} \\ e^{ik_L^\uparrow z} \end{pmatrix} + r_{\uparrow\uparrow} \begin{pmatrix} e^{-ik_L^\uparrow z} \\ e^{-ik_L^\uparrow z} \end{pmatrix} + r_{\downarrow\uparrow} \begin{pmatrix} e^{-ik_L^\downarrow z} \\ -e^{-ik_L^\downarrow z} \end{pmatrix}$$

$$\psi_{DW} = \sum_{j=1}^{4} D_j \begin{pmatrix} e^{ik_D^j z} \\ \alpha_j e^{i(k_D^j + Q)z} \end{pmatrix}$$

$$\psi_R = t_{\uparrow\uparrow} \begin{pmatrix} e^{ik_R^\uparrow z} \\ e^{ik_R^\uparrow z} \end{pmatrix} + t_{\downarrow\uparrow} \begin{pmatrix} e^{ik_R^\downarrow z} \\ -e^{ik_R^\downarrow z} \end{pmatrix}, \tag{5.30}$$

where $k_L^\uparrow = k_R^\downarrow = \sqrt{2m(\epsilon_F - h_0)}/\hbar, k_L^\downarrow = k_R^\uparrow = \sqrt{2m(\epsilon_F + h_0)}/\hbar$, k_D^j is the four solutions to the equation $(\varepsilon_{k_D^j} - \epsilon_F)(\varepsilon_{k_D^j+Q} - E_F) = h_0^2$ and the coefficients $\alpha_j = (\epsilon_F - \varepsilon_{k_D^j})/ih_0$ with $\varepsilon_k = \hbar^2 k^2/2m$.

The spin precession is often related in analogy with the magnetic resonance. In Fig. 5.29(b), the dot-dashed line is the probability that the spin $s_z = \hbar/2$ at $t = 0$ have $s_z = -\hbar/2$ at

$t = 2L/\bar{v}_F$ in a magnetic field h_0 rotating with a frequency $\pi\bar{v}_F/2L$ with $\bar{v}_F = [\sqrt{2m(\epsilon_F + h_0)} + \sqrt{2m(\epsilon_F - h_0)}]/2$. Let us introduce the spin precession length $l_s \equiv h\bar{v}_F/4h_0$. For the exchange field $h_0 = 0.5$ eV, l_s is estimated to be 24 Å. The spin flip probability is given by $\xi^2/(1+\xi^2)$ with $\xi = 2L/l_s$. $\mathcal{T}_{\downarrow\uparrow(\uparrow\uparrow)} = |t_{\downarrow\uparrow(\uparrow\uparrow)}|^2 k_R^{\downarrow(\uparrow)}/k_L^{\uparrow}$ for $h_0 = 0.5$ eV is plotted by the solid (dotted) line in Fig. 5.29(b). The probability $\mathcal{T}_{\downarrow\uparrow}$ oscillates between a lower envelope given by $\xi^2/(1+\xi^2)$ and unity, while $\mathcal{T}_{\uparrow\uparrow}$ oscillates between zero and an upper envelope $1 - \xi^2/(1+\xi^2)$. The lower and upper envelopes represent the spin-flip and spin-conservation probabilities in an oscillating magnetic field. The probability current density for the reflection, $|r_{\uparrow\uparrow}|^2 k_L^{\uparrow}/k_L^{\uparrow} + |r_{\downarrow\uparrow}|^2 k_L^{\downarrow}/k_L^{\uparrow}$, is less than 0.005 even for the 1 Å thick domain wall and decreases as the thickness of the domain wall increases. We also plot the numerical result of $\mathcal{T}_{\downarrow\uparrow}$ and $\mathcal{T}_{\uparrow\uparrow}$ obtained by the RTM method in Fig. 5.30(b). The excellent agreement between the analytic and numerical results indicates that the RTM method can accurately describe the spin precession of conduction electrons. As shown in Fig. 5.29(b), the transmission probability $\mathcal{T}_{\downarrow\uparrow}$ to the spin-down state is as small as 0.16 at $2L = 10$ Å. This means that the spin of the conduction electron hardly rotates through 1D classical domain wall of the size of the point contacts that show the conductance quantization.

In a point contact, the geometrical constriction plays a crucial role in the spin precession through the classical domain wall. In the adiabatic picture, the velocity for channel n is given by $v_z^n = \sqrt{(2/m)(\epsilon_F - \epsilon_{n\parallel})}$, where $\epsilon_{n\parallel}$ is the energy eigenvalue of the transverse mode. $\epsilon_{n\parallel}$ is a decreasing function of W and the channel n opens when $\epsilon_{n\parallel}$ is smaller than ϵ_F. Electrons transmitting through the channel n have the small velocity $v_z^n \ll v_F$. Electrons track the local exchange field of the domain wall adiabatically and feel the constant exchange field as if they were in the F alignments. Therefore, the conductance curve for the A alignment is similar to that for the F alignment as shown in Fig. 5.30(a). However, experimentally the sequence of the quantized conductances is different between F and A alignments: the conductance plateaus of odd integer multiples of e^2/h appear only for the F alignment. This discrepancy is resolved considering an atomic-scale domain wall on the basis of the quantum theory. The spin-mixing term vanishes for

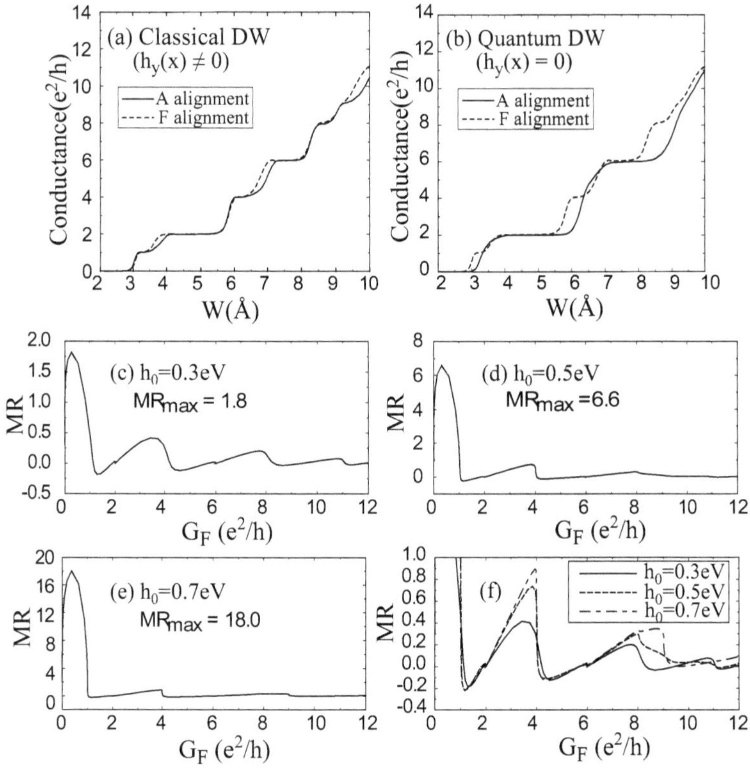

Figure 5.30 The conductance curves for the A (F) alignment with (a) classical domain wall and (b) quantum domain wall. The spin precession of conduction electron in the quantum domain wall is forbidden. In both panels, parameters are those in Fig. 5.28(b). (c–f) Magnetoresistances for the point contact with h_0 = (c) 0.3, (d) 0.5, and (e) 0.7 eV. (f) Enlarged views of MR. From H. Imamura, N. Kobayashi, S. Takahashi, and S. Maekawa, *Phys. Rev. Lett.* **84**, 1003 (2000) [140].

the atomic-scale quantum domain wall. The spin of the conduction electrons cannot rotate in the domain wall.

The conductance curve for the A alignment with a quantum domain wall is plotted in Fig. 5.30(b). The sequence of the quantized conductances is clearly different between F and A alignments. In the adiabatic picture, the number of transmitting channels is the same for spin-up and spin-down electrons, because the exchange energy

-$h_z(z)\sigma_x$ is an odd function of z. The sequence of the quantized conductances is the same as that for a non-magnetic point contact as shown in Fig. 5.28(c). The conductance curve shows clear plateaus at $G = 2e^2/h$ and $6e^2/h$ and the plateaus at odd integer multiples of e^2/h disappear. Comparing Figs. 5.30(a) and (b), we conclude that the experimental results [138, 139] that the sequence of the quantized conductances is different between A and F alignments is because the spin of conduction electrons cannot rotate in the atomic-scale quantum domain wall.

In Figs. 5.30(c–f), we show the MR calculated by the formula: MR$= (G_F/G_A) - 1$, where $G_{F(A)}$ is the total conductance for the $F(A)$ alignment. We find the strong enhancement of the MR at $G_F = 1$ ($W \simeq 3$ Å) where the first transmitting channel opens for the F alignment but channels for the A alignment are hardly transmittive. The point is that the domain wall makes the number of transmitting channels different between F and A alignments as shown in Fig. 5.28(d). Note that the conductance plateau at integer multiples of e^2/h means that the scattering intrinsic to the domain wall is negligible. Magnetoresistance, i.e., the difference of conductances between F and A alignments increases as an exchange field h_0 increases. The maximum values of MR, MR_{max}, for $h_0 = 0.3$, 0.5, and 0.7 eV are respectively 1.8, 6.6, and 18.0. The enhancement of MR is expected to be large if we use magnetic-metals with large exchange field such as Co and Fe. We also find that the MR oscillates with G_F as shown in Fig. 5.30(f). Since the difference of the conductances or the number of transmitting channels between the A and F alignments does not increase with G_F as shown in Fig. 5.30(d), the magnitude of the oscillation decreases with G_F. The oscillation of MR was experimentally confirmed in Co contacts later.

5.3.3 *Magnetic Atomic Wire*

Finally, we treat the spin-dependent transport and magnetoresistance of Ni atom wires. Although empirical parameters are used for the exchange energy in the former topics, here we use the nonequilibrium Green's function (NEGF) method in ab initio spin-density functional theory. We show spin-dependent transmission spectra and MR of Ni atomic wire, which shows a value of 250% due to the scattering of d orbital channels. We consider 6 Ni atoms in the conductor region coupled with semi-infinite atomic wire electrodes. The interatomic distance is 2.1 Å and pseudo-atomic orbitals of Ni-$s2p2d2$ are used for numerical atomic basis sets. From the realistic atomic calculations based on the spin-density-functional theory (SDFT), we can specify the atomic orbitals that contribute to the spin-dependent conductance.

When the magnetizations of the electrodes are parallel, the transmission shows integer values corresponding to the number of bands of infinite atom wires formed by the s-d_{z^2}, d_{xz}, d_{yz}, d_{xy}, and $d_{x^2-y^2}$ orbitals as shown in the left panel of Fig. 5.31. The z-axis is taken to be parallel to the atomic wire. At the Fermi energy, all orbitals contribute to the transport for the spin-up state, while only the s-d_{z^2} orbital contributes to the transport for the spin-down state. Therefore, the total conductance is $7e^2/h$.

When the magnetizations of the electrodes are antiparallel, the domain wall is included in the conductor region as shown in the inset of the right panel of Fig. 5.31, which depicts the Mulliken charges for the spin states of the 4s, 3p, and 3d orbitals at each atom. The change of the Mulliken charges due to the domain wall is seen around only two atoms. Figure 5.31 (right) shows the transmission spectrum for the antiparallel alignment. The transmission is the

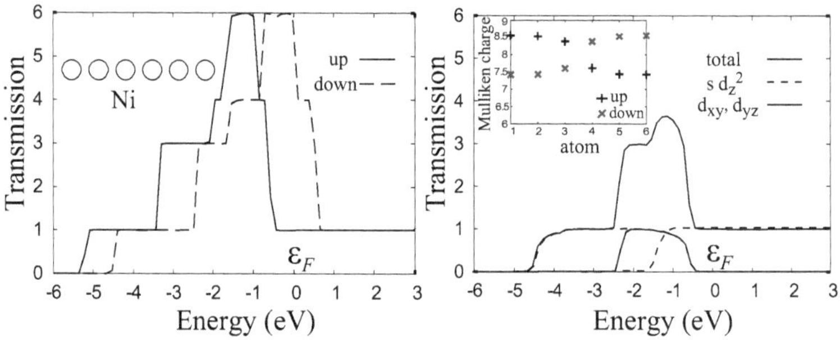

Figure 5.31 (Left) Transmission spectrum of Ni atom wire between parallel magnetizations of electrodes. (Right) Total transmission and channel transmission spectra of Ni atom wire between antiparallel magnetizations of electrodes. (Inset) Mulliken charges of Ni atom wire between antiparallel magnetizations of electrodes. From N. Kobayashi, T. Ozaki, and K. Hirose, *J. Phys, Conf. Ser.* **38**, 95 (2006) [143].

same for the spin-up and spin-down states for the symmetry. In this case, the d states with a small momentum in the direction of atomic wires do not transport owing to the scattering at the domain wall, and only the s-d_{z^2} states contribute to the transport. Thus, the conductance is $2e^2/h$, which corresponds to the plane-wave calculation.

The MR defined by $MR = (G_P/G_A) - 1$, where $G_{P(A)}$ is the conductance for the parallel (antiparallel) alignment of the electrode magnetization, is calculated to be 250%, which is consistent with the magnetic point contact experiment. This shows the MR in excess of 200%. It should be noted that the transport properties are sensitive to the geometries of the contact and the electrodes, and thus the small quantitative difference is expected to be due to the difference between the theoretical and experimental atomic configurations.

5.4 Nano-Carbon Materials

Nano-carbon materials such as CNTs [144] and graphene [145] have attracted much attention. If we consider these nano-carbon materials as single molecules, they are categorized as the most widely studied molecules whose electrical conductivity have been measured in a variety of ways as shown in Fig. 5.32. For example, since CNTs have long length for contacts to metal electrodes fabricated by the electron beam lithography, their unique conductivity properties of metallic and insulating behaviors have been measured in a number of experiments.

In the present section, we describe first the basic electronic structures of nano-carbon materials and their unique characteristics for the transport briefly. Then we show how the electron transport properties change of CNTs from ballistic to diffusive regimes due to phonon scatterings. Temperature dependence and energy dependence on the mean free path (MFP) will be shown and are compared with experiments. Then, we present how the phase coherence of CNTs is preserved for various temperatures due to inelastic vibrations. Finally, we treat the graphene nanoribbons (GNRs) and show their transport properties focusing the edge scattering effects.

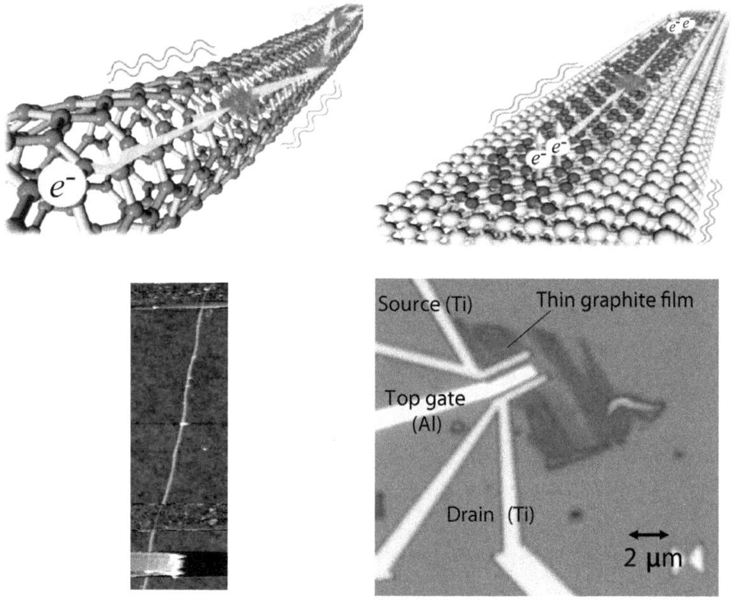

Figure 5.32 (Top) Schematic views of the transport through CNTs and graphenes. (Bottom) Examples of experiments for the CNT transistors and graphene transistors. Experiments from (left) A. Javey, J. Guo, Q. Wang, M. Lundstrom, and H. Dai, *Nature* **424**, 654 (2004) [146] and (right) H. Miyazaki, S. Okada, T. Ssato, S. Tanaka, H. Goto, A. Kanda, K. Tsukagoshi, Y. Ootuka, and Y. Aoyagi, *Appl. Phys. Express* **1**, 034007 (2008) [147].

5.4.1 *Anomalous Band Structure*

5.4.1.1 Electronic structures of graphene

The graphene is a two-dimensional sheet of carbon atoms in the honeycomb structure. The CNT is constructed from the graphene by rolling up. The graphite consists of layers of graphene sheets by weak van der Waals forces. Therefore, we consider the band structure of graphene first [148–150].

When the carbon atoms are placed onto the graphene hexagonal lattice, the electronic wavefunctions from different atoms overlap. The $2s$, $2p_x$, and $2p_y$ orbitals comprise the so-called sp^2 hybridization in the plane to form the strong covalent σ and σ^* bonds. The remaining $2p_z$ orbital, whose overlap with σ and σ^* states is

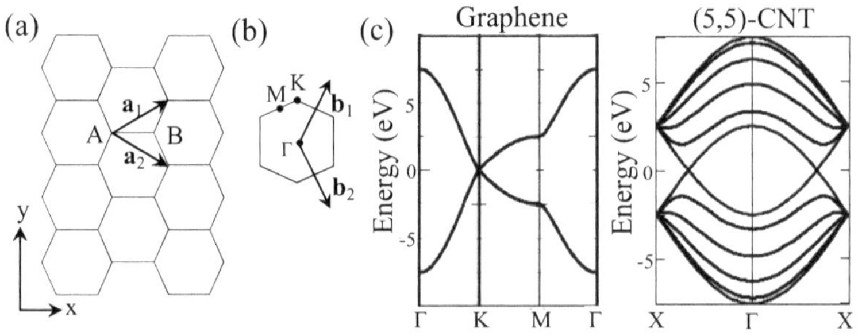

Figure 5.33 (a) Unit cell with lattice vectors of $\mathbf{a}_1$ and $\mathbf{a}_2$ of two-dimensional graphene honeycomb network. (b) Brillouin zone and reciprocal space with corresponding reciprocal vectors. (c) Energy dispersion of graphene and (5,5) CNT.

strictly zero, creates delocalized π and π^* bonds. The π bonds are perpendicular to the plane and are responsible for the transport, crossing the Fermi energy. On the other hand, the σ bonds are far away from the Fermi energy and are responsible for the binding energy and elastic properties.

Here we derive the band structure of graphene using the simple tight-binding model. The unit cell of graphene contains two atoms (A and B) and is characterized by a basis defined as $\mathbf{a}_1$ and $\mathbf{a}_2$ as

$$\mathbf{a}_1 = a\left(\frac{\sqrt{3}}{2}, \frac{1}{2}\right), \qquad \mathbf{a}_2 = a\left(\frac{\sqrt{3}}{2}, -\frac{1}{2}\right), \tag{5.31}$$

where the $a = \sqrt{3}a_{cc}$ is the length of the basis vector and $a_{cc} =$ 1.42 Å. From the condition of $\mathbf{a}_i \cdot \mathbf{b}_j = 2\pi\delta_{ij}$, we can obtain the reciprocal-lattice vectors $\mathbf{b}_1$ and $\mathbf{b}_2$ as

$$\mathbf{b}_1 = \frac{4\pi}{\sqrt{3}a}\left(\frac{1}{2}, \frac{\sqrt{3}}{2}\right), \qquad \mathbf{b}_2 = \frac{4\pi}{\sqrt{3}a}\left(\frac{1}{2}, -\frac{\sqrt{3}}{2}\right). \tag{5.32}$$

These are shown in Fig. 5.33(a,b). Note that the reciprocal space becomes the hexagonal lattice.

To derive the electronic structure of graphene, the wavefunctions are expanded using the Bloch theorem as

$$\Psi_k(\mathbf{r}) = c_k^A \varphi_k^A(\mathbf{r}) + c_k^B \varphi_k^B(\mathbf{r}) \tag{5.33}$$

with

$$\varphi_k^{A,B}(\mathbf{r}) = \frac{1}{\sqrt{N}} \sum_{\ell} e^{i\mathbf{k}\cdot\ell} p_z(\mathbf{r} - \mathbf{r}_{A,B} - \ell), \tag{5.34}$$

where $\mathbf{k}$ is the electron momentum and ℓ is the position vectors. The spectrum ϵ as a function of k is obtained by the diagonalization of 2×2 matrix of

$$H - \epsilon = \begin{pmatrix} H_{AA}(k) - \epsilon & H_{AB}(k) \\ H_{BA}(k) & H_{BB}(k) - \epsilon \end{pmatrix} \tag{5.35}$$

with the matrix element of

$$H_{AB}(k) = \frac{1}{N} \sum_{\ell, \ell'} e^{i\mathbf{k}\cdot(\ell' - \ell)} \langle p_z^A | H | p_z^B \rangle. \tag{5.36}$$

Here we abbreviate $p_z^{A,B} = p_z(\mathbf{r} - \mathbf{r}_{A,B} - \ell)$. The other matrix elements $H_{AA}(k)$, $H_{BA}(k)$, and $H_{BB}(k)$ are calculated in the same way. Restricting the interactions to the first nearest neighbor atoms only, we get

$$\begin{aligned} H_{AB}(k) &= -t_0 f_0(k) = -t_0 \left(1 + e^{-i\mathbf{k}\cdot\mathbf{a}_1} + e^{-i\mathbf{k}\cdot\mathbf{a}_2}\right) \\ &= -t_0 \left(1 + 2e^{-ik_x(\sqrt{3}a/2)} \cos(k_y a/2)\right), \end{aligned} \tag{5.37}$$

where t_0 is the transfer integral between first neighboring π orbitals of carbon atoms and $f_0(k) = 1 + e^{-i\mathbf{k}\cdot\mathbf{a}_1} + e^{-i\mathbf{k}\cdot\mathbf{a}_2}$.

When we take the energy reference as $\langle p_z^A | H | p_z^A \rangle = \langle p_z^B | H | p_z^B \rangle = 0$, the dispersion relation becomes

$$\begin{aligned} \epsilon^{\pm}(k) &= \pm t_0 |f_0(k)| \\ &= \pm t_0 \sqrt{1 + 4\cos^2\left(\frac{k_y a}{2}\right) + 4\cos\left(\frac{\sqrt{3}k_x a}{2}\right)\cos\left(\frac{k_y a}{2}\right)}. \end{aligned} \tag{5.38}$$

The energy band for graphene is presented in Fig. 5.33(c, left panel) along the high symmetry directions of the Brillouin zone defined by the Γ, M, and K points. The positive sign gives the conduction band from π^* anti-bonding states and the negative sign the valence band from π bonding states. These two bands intersect at $\epsilon = 0$ and the conduction and valence bands of graphene sheet have mirror images to each other. This nature works advantageous to construct complementary electron devices. At the Fermi energy $\epsilon = 0$ as a charge neutral point, the occupied and unoccupied bands cross at the six K points. We can show easily that at two K point $k_K = (0, \pm 4\pi/3a)$, which are the independent Fermi points, the energy $\epsilon^{\pm}(k_K)$ becomes zero.

For the transport, the dispersion relation $\epsilon(k)$ near the Fermi energy is important. Expanding $f_0(k)$ around the K point at $k_K = (0, \pm 4\pi/3a)$

$$f_0(k) \approx \delta k_x \left(\frac{\partial f_0}{\partial k_x}\right)_{k_K} + \left(\delta k_y \mp \frac{4\pi}{3a}\right)\left(\frac{\partial f_0}{\partial k_y}\right)_{k_K}$$
$$= -\frac{\sqrt{3}}{2} ia(\delta k_x - i\delta \bar{k}_y), \quad (5.39)$$

the energy dispersion around the K points shows a linear energy-momentum relation

$$\epsilon = \pm t_0 |f_0(k)| \approx \pm t_0 \frac{\sqrt{3}}{2} a|\delta k| = \pm \hbar v_F |\delta k| \quad (5.40)$$

with the Fermi velocity of $v_F = (1/\hbar)(\partial\epsilon/\partial k) = \sqrt{3}at_0/2\hbar$, typically given by $v_F = 8 \times 10^7$cm/s, and $|\delta k| = \sqrt{\delta k_x^2 + \delta \bar{k}_y^2}$. The linear band dispersion close to the Fermi energy will have important consequences for the transport properties.

5.4.1.2 Formation of carbon nanotubes

The formation of CNTs is viewed as the rolling up of graphene sheet. As shown in Fig. 5.34, the chiral vector $\mathbf{C}_h$ specifies the direction of the roll-up by a pair of integers (m, n),

$$\mathbf{C}_h = n\mathbf{a}_1 + m\mathbf{a}_2. \quad (5.41)$$

Using $\mathbf{C}_h$ the diameter of the nanotube becomes

$$d = \frac{|\mathbf{C}_h|}{\pi} = \frac{a}{\pi}\sqrt{n^2 + nm + m^2}. \quad (5.42)$$

If $\mathbf{C}_h$ lies along the x-axis, we have an armchair nanotube. While if $\mathbf{C}_h$ lies along the y-axis, we have a zigzag nanotube. The chiral angle θ, defined by the angle between $\mathbf{C}_h$ and $\mathbf{a}_1$ with values of $0^\circ \le |\theta| \le 30^\circ$, denotes the tilt angle of the hexagons with respect to the direction of the nanotube axis:

$$\cos\theta = \frac{\mathbf{C}_h \cdot \mathbf{a}_1}{|\mathbf{C}_h||\mathbf{a}_1|} = \frac{2n + m}{2\sqrt{n^2 + nm + m^2}}. \quad (5.43)$$

Note that $\theta = 30^\circ$ corresponds to the armchair nanotube (n, n) and $\theta = 0^\circ$ corresponds to the zigzag nanotube $(n, 0)$. For other angles, we have chiral nanotubes.

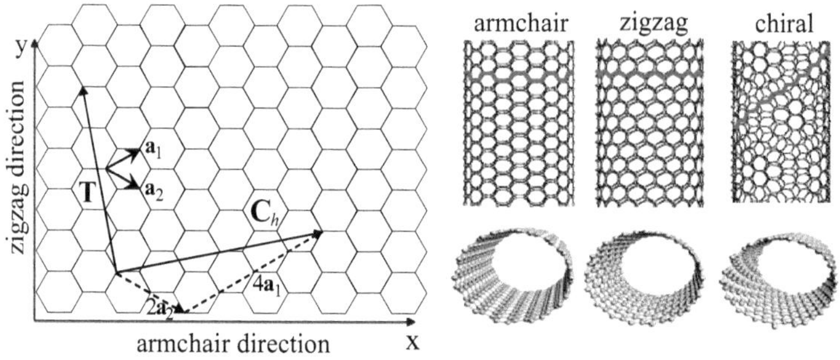

Figure 5.34 (Left) Honeycomb lattice of the graphene with lattice vectors $\mathbf{a}_1$ and $\mathbf{a}_2$. The chiral vector $\mathbf{C}_h = (4, 2)$ shows the wrapping of the two-dimensional graphene sheet into a tube form. The direction $\mathbf{T} = (4, -5)$ perpendicular to $\mathbf{C}_h$ is the tube axis. If $\mathbf{C}_h$ lies along the x-axis, we have an armchair nanotube. If $\mathbf{C}_h$ lies along the y-axis, we have a zigzag nanotube. (Right) Atomic structures of armchair, zigzag, and chiral nanotubes, respectively.

The smallest lattice vector $\mathbf{T}$ perpendicular to $\mathbf{C}_h$ defines the translational period along the tube axis from the condition of $\mathbf{C}_h \cdot \mathbf{T} = 0$ as

$$\mathbf{T} = t_1\mathbf{a}_1 + t_2\mathbf{a}_2 \quad \text{with} \quad t_1 = \frac{2m+n}{N_R}, \quad t_2 = -\frac{2n+m}{N_R}. \tag{5.44}$$

Here N_R is the greatest common divisor of $(2m + n)$ and $(2n + m)$. The length of $\mathbf{T}$ is given by

$$|\mathbf{T}| = \frac{\sqrt{3}\pi d}{N_R} = \frac{\sqrt{3}a\sqrt{n^2 + nm + m^2}}{N_R} \tag{5.45}$$

and the number of C atoms in a unit cell is given by

$$N_C = \frac{4(n^2 + nm + m^2)}{N_R}. \tag{5.46}$$

5.4.1.3 Electronic states of carbon nanotubes

The electronic states of CNTs are obtained from those of graphene by applying a periodic boundary condition along the circumference direction of the nanotube. Since the wavefunctions follow the Bloch theorem

$$\Psi_k(\mathbf{r} + \mathbf{C}_h) = e^{i\mathbf{k}\cdot\mathbf{C}_h}\Psi_k(\mathbf{r}) = \Psi_k(\mathbf{r}), \tag{5.47}$$

which results in

$$\mathbf{k} \cdot \mathbf{C}_h = 2\pi M. \tag{5.48}$$

Here M is an integer. The allowed values of **k** are quantized by the boundary condition as

$$\frac{\sqrt{3}a}{2}(n+m)k_x + \frac{a}{2}(n-m)k_y = 2\pi M. \tag{5.49}$$

This comprises an equation for a set of parallel lines. There are two possibilities, either the allowed k states do or do not pass through one of the K points (two are independent) where the valence and conduction bands meet. In the first case, we have a metallic nanotube, while in the second case we have a semiconductor nanotube.

The condition that the equation (5.49) to pass through the K point at $k_K = (0, \pm 4\pi/3a)$ becomes

$$\pm\frac{2\pi}{3}(n-m) = 2\pi M. \tag{5.50}$$

This means that when $(n-m)$ is a multiple of 3, that is $\mathrm{mod}(n-m, 3) = 0$, the nanotube becomes metallic. Otherwise, $\mathrm{mod}(n-m, 3) \neq 0$, the nanotube becomes semiconducting.

Let us first consider the armchair nanotubes here, which are specified by (n, n). This corresponds to the condition of $\mathrm{mod}(n-m, 3) = 0$, so all armchair nanotubes are metallic. We show the band structure of the (5, 5) armchair CNT in Fig. 5.33(c, right panel). We see that the valence and conduction bands cross at $k = \pm 2\pi/3a$, which corresponds to the points located at two-thirds of $\Gamma - X$ line and where the original K points of the graphene are folded. The energy dispersion keeps the linear band relation around the Fermi energy

$$\epsilon = \pm t_0 |f_0(k)| \approx \pm t_0 \frac{\sqrt{3}}{2} a |\delta k| = \pm \hbar v_F |\delta k|. \tag{5.51}$$

In Fig. 5.35, we show the DOS for (5, 5), (10, 10), (20, 20), and (200, 200) armchair nanotubes by increasing the diameters. When the diameter is small, the DOS diverges as $1/\sqrt{\epsilon}$ close to the band edges. These spiky structures are called the van Hove singularities, characteristics of the 1D systems described in Chapter 2. There are finite states at $\epsilon = 0$, showing that the armchair nanotubes are

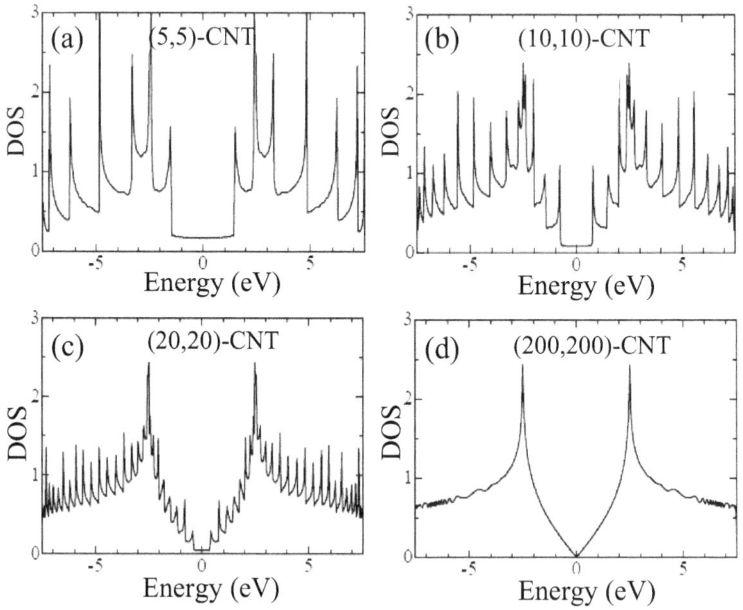

Figure 5.35 Density of states of the CNTs with various diameters. (a) (5,5)-CNT, (b) (10,10)-CNT, (c) (20,20)-CNT, and (d) (200,200)-CNT. As the diameter becomes large, the 1D characters of CNT approaches the 2D characters of graphene.

metallic. As the diameter becomes large, the spiky structures are reduced and electronic structures approach the 2D characteristics of the graphene. At the charge neutral point $\epsilon = 0$, the states decrease to show the semi-metallic properties with no bandgap of the graphene.

Next, we consider the zigzag nanotubes specified by $(n, 0)$. When $\mathrm{mod}(n - m, 3) \neq 0$, the nanotubes become semiconducting. Since $e^{i\mathbf{k}\cdot\mathbf{C}_h} = \pm e^{2i\pi/3}$, the relation in the vicinity of the Fermi energy is obtained from the similar Taylor expansion as $\epsilon \approx \pm t_0(\sqrt{3}a/2)\sqrt{\delta k^2 + (2/3d)^2}$, where d is the diameter of the nanotubes. From this, we have the bandgap of $\Delta\epsilon = t_0(2a/\sqrt{3}d)$. This shows that the bandgap $\Delta\epsilon$ is inversely proportional to the diameter d of the nanotubes. The bandgap disappears in the limit of $d = \infty$, which corresponds to the semi-metallic graphene.

5.4.1.4 Absence of backscatterings

Due to the linear dispersion relation in the vicinity of the charge neutral point, the electron transport properties of graphenes and metallic nanotubes have strong ballistic natures. Here following the reference [151], we show the anomalous properties of an absence of backscatterings by the effective-mass description based on the $k \cdot p$ approximation. Since the Bloch states having the wavevector $k = K + \delta k$ can be written as $\Psi_{K+\delta k}(\mathbf{r}) = e^{i\delta\mathbf{k}\cdot\mathbf{r}}\psi_K(\mathbf{r})$, where $\psi_K(\mathbf{r})$ satisfies the Bloch condition, the Schrödinger equation becomes

$$\left(-\frac{\hbar^2}{2m}\nabla^2 + V(\mathbf{r}) + H'\right)\psi_K(\mathbf{r}) = \epsilon_{K+\delta k}\psi_K(\mathbf{r}) \qquad (5.52)$$

with $H' = (\hbar/m)\delta k \cdot p + \hbar^2\delta k^2/2m$. The $k \cdot p$ approximation extracts information by the expansion of $\psi_K(\mathbf{r})$ around the Fermi energy.

Expanding the wavefunction near the K points as

$$\Psi_k(\mathbf{r}) = F^A(\mathbf{r})\psi_K^A(\mathbf{r}) + F^B(\mathbf{r})\psi_K^B(\mathbf{r}), \qquad (5.53)$$

where $F^{A,b}(\mathbf{r}) = \int c_{\delta k}^{A,B} e^{i\delta\mathbf{k}\cdot\mathbf{r}} d\delta k$ are the envelope functions, the $k \cdot p$ equation becomes

$$\begin{pmatrix} \epsilon_{\delta k} & t_0\sqrt{3}(\delta k_x + i\delta k_y)/2 \\ t_0\sqrt{3}(\delta k_x - i\delta k_y)/2 & \epsilon_{\delta k} \end{pmatrix} \begin{pmatrix} F^A(\mathbf{r}) \\ F^B(\mathbf{r}) \end{pmatrix} = \begin{pmatrix} 0 \\ 0 \end{pmatrix},$$

which is reduced to $\epsilon_{K+\delta k} = \pm\hbar v_F|\delta k|$ near the K point. Note that $\epsilon_K = 0$. The eigenvectors satisfy $F(\mathbf{r}) = e^{i\delta\mathbf{k}\cdot\mathbf{r}}R^{-1}(\theta_{\delta k})|s = \pm 1\rangle$, where $R(\theta_k)$ is the rotation matrix

$$R(\theta_k) = \begin{pmatrix} e^{i\theta_k/2} & 0 \\ 0 & e^{-i\theta_k/2} \end{pmatrix}, \quad |s = \pm 1\rangle = \frac{1}{\sqrt{2}}\begin{pmatrix} -is \\ 1 \end{pmatrix}$$

with $s = -1$ for $\epsilon < 0$ and $s = 1$ for $\epsilon > 0$.

When the scattering from the impurity potential $V(\mathbf{r})$ is slowly varying on the scale of the interatomic distance a_{CC}, $V(\mathbf{r})$ can be treated perturbatively to the $k \cdot p$ Hamiltonian. Then $\langle\Psi_{k,s}(\mathbf{r})|V(\mathbf{r})|\Psi_{k',s'}(\mathbf{r})\rangle$ describe scattering amplitude $k \to k'$ due to a single impurity. A simple picture shows that

$$\langle s'|V(\mathbf{r})|s\rangle \propto (-is', 1)\begin{pmatrix} V(\mathbf{r}) & 0 \\ 0 & V(\mathbf{r}) \end{pmatrix}\begin{pmatrix} -is \\ 1 \end{pmatrix} = 0 \qquad (5.54)$$

for the case of ($s = s' = \pm 1$). This observation is shown to be generalized to all higher-order terms in the perturbation series for the multiple-scatterings of the T matrix

$$T(\epsilon) = V + V\frac{1}{\epsilon - H_0}V + V\frac{1}{\epsilon - H_0}V\frac{1}{\epsilon - H_0}V + \cdots . \quad (5.55)$$

The matrix elements of the scattering $\langle s|T(\epsilon)|s\rangle$ are constructed by the multiplications of the rotation matrix of $\langle s_p|R(\theta_{k_p})R^{-1}(\theta_{k_{p'}})|s_{p'}\rangle$. Due to the symmetry of eigenstates, time-reversal terms cancel out and the total amplitude is proportional to $\langle s|R(\theta_k)R^{-1}(\theta_{-k})|s\rangle = \cos[(\theta_k - \theta_{-k})/2]$, which is zero since $\theta_k - \theta_{-k} = \pi$. This shows that the backscattering is suppressed in the low-energy regime around the charge neutral point to all orders in the multiple scatterings. Thus, the transport properties of the nanotubes and graphenes are expected to have large conductivities due to the linear dispersion relation in the low energy regime when the impurity potentials are slowly varying.

5.4.2 *From Ballistic to Diffusive Regime*

In the last section, we see that backscattering is greatly suppressed for graphene and metallic nanotubes. Thus, strong ballistic nature is expected for metallic CNTs with fewer disorders. Phonon scatterings are the main scattering effects for the conduction of electrons. Generally, electron transport changes its character from ballistic to diffusive regimes according to the scatterings as shown in Fig. 5.36. The elastic MFP ℓ_{mfp} distinguishes the length from ballistic to diffusive regimes.

To obtain the MFP, we employ the TD-WPD method, suitable to simulate quantum transport in huge and complex systems with up to approximately 10 millions of atoms. Various electron scattering effects such as impurities and electron–phonon interactions for the computation of transport properties are included from an atomistic point of views, which enables us to compare with experimental observations. The resistivity $\rho(T)$ is obtained from the Kubo formula as

$$\frac{1}{\rho(T)} = \lim_{t\to+\infty}\int_{-\infty}^{+\infty} 2e^2\left(-\frac{df(\epsilon)}{d\epsilon}\right)\mathrm{Tr}\left[\frac{\delta(\epsilon - \hat{H}_e)}{L}\frac{\left(\hat{x}(t) - \hat{x}(0)\right)^2}{t}\right]d\epsilon, \quad (5.56)$$

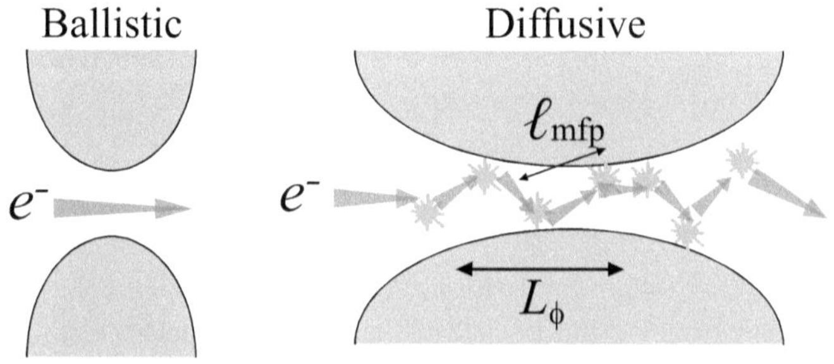

Figure 5.36 Schematic pictures of ballistic (left) and diffusive (right) transport regimes. $\ell_{\rm mfp}$ is the MFP and L_ϕ is the phase coherence length.

where $\mathrm{Tr}[\delta(\epsilon - \hat{H}_e)/L]$ corresponds to the DOS per unit and $\hat{x} = \hat{U}^\dagger(t)x\hat{U}(t)$ is the position operator along the nanotube in the Heisenberg representation. The quantum diffusion coefficient is given as the long-time limit of the time-dependent diffusion coefficient $D = \lim_{t\to+\infty} D(t)$ by

$$D(t) = \frac{1}{t}\frac{\int d\epsilon f(\epsilon)\mathrm{Tr}\left[\delta(\epsilon - \hat{H}_e)\left(\hat{x}(t) - \hat{x}(0)\right)^2\right]}{\int d\epsilon f(\epsilon)\mathrm{Tr}\left[\delta(\epsilon - \hat{H}_e)\right]}. \tag{5.57}$$

We use the time-evolution operator

$$\hat{U}(t) = \prod_{n=0}^{N-1} \exp\left(i\frac{\hat{H}_e(n\Delta t)}{\hbar}\Delta t\right) \tag{5.58}$$

with $\Delta t = t/N$ employing the Chebyshev polynomial expansion.

The Hamiltonian is the sum of the electronic states described in the previous section and the phonon scatterings

$$\hat{H}_e(t) = \hat{H}_0 + \hat{H}_{\text{e-ph}}(t) + \hat{H}_{\rm imp}. \tag{5.59}$$

The π-effective Hamiltonian based on Eq. (5.35) is employed for $\hat{H}_0$ where the fluctuating carbon-carbon bond lengths due to the realistic phonon modes are taken into account from the molecular dynamics simulations at a finite temperature from atomistic viewpoints. Randomly selected disorder potentials are included at on-site energies of π orbitals with the interval width of $\left[-W\gamma^0/2, W\gamma^0/2\right]$ for $\hat{H}_{\rm imp}$, where $\gamma_{ij}^0 = 2.5$ eV (for C–C bond) is the transfer energy at equilibrium with the bond length of 0.14 nm. Electron–phonon scattering effects for $\hat{H}_{\text{e-ph}}(t)$ from atomic displacements with phonon vibrations are included in the off-diagonal elements through a time-dependent electron transfer energies $\gamma_{ij}(t)$. Here we use the empirical rule [152]

$$\gamma_{ij}(t) = \frac{\gamma_{ij}^0}{\Delta R(t)^2}, \quad \text{with} \quad \Delta R(t) = \frac{\mathbf{R}_i(t) - \mathbf{R}_j(t)}{\mathbf{R}_i^0 - \mathbf{R}_j^0}. \tag{5.60}$$

We note that this is semi-classical, since the displacement of atomic positions is not quantized; thus, the detailed balance for phonon emission and absorption is not correctly taken into account. This approach is justified in the temperature not so low as we need quantization of phonon and in the linear-response regime where the spontaneous emission of phonon is negligible. Heating due to inelastic phonon emission is absent when the source-drain bias voltage is small.

Since all the phonon modes are taken into account from the molecular dynamics simulation, we expect the simulations close to the experiments and obtain the resistance from diffusive to ballistic regime. We use two initial electron wave packets for the electron propagation, which are localized at one π orbital on A and B sublattice around the center part of the CNT as well as eight Maxwell velocity distributions of initial atoms are employed in the molecular dynamics simulations, thereby averaging totally by 2×8 initial conditions. The transport calculations are performed up to $t \approx 15$ ps with $\Delta t \equiv 0.1 \times h/(1 \text{ eV}) = 0.41$ fs for up to 20 μm (80,000 unit cells) CNTs with open boundary conditions, while the MD simulations are done with periodic boundary conditions, sufficiently large to achieve convergence transport calculations.

5.4.2.1 Mean free path

First, we see how the transport properties of the CNTs change according to the channel length L from the ballistic to diffusive regimes due to the electron–phonon coupling ($W = 0$). Figure 5.37 shows the calculated length-dependent resistance of the (5, 5)-CNTs at 300 K. The length-dependent resistance exhibits a crossover from a diffusive behavior where the increase of resistance almost scales linearly with length to a ballistic-like one where the resistance is almost independent of length.

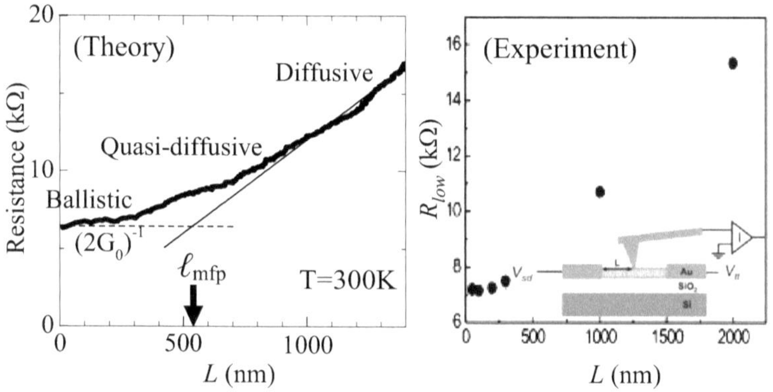

Figure 5.37 (Left) $R(L)$ of metallic (5, 5)-CNTs from ballistic to diffusive regimes. Thin solid line represents the diffusive limit and the thin broken line shows the ballistic limit of $(2G_0)^{-1} = 6.45$ kΩ. These two lines intersect at the value of the MFP ℓ_{mfp} indicated by the thick arrow. (Right) Experimental values for low-bias resistance $R = dV/dI$ as a function of L for the single-walled nanotubes (SWNTs). Experiment from J. Y. Park, S. Rosenblatt, Y. Yaish, V. Sazonova, H. Üstunel, S. Braig, T. A. Arias, P. W. Brouwer, and P. L. McEuen, *Nano Lett.* **4**, 517 (2004) [153] and theory from H. Ishii, N. Kobayashi, and K. Hirose, *Appl. Phys. Express* **1**, 123002 (2008) [81].

When L is large enough, $R(L)$ shows a monotonically increasing behavior for L. Thin solid line to fit to an asymptotic behavior with a linear equation of $R(L) = \rho \cdot L + R_0$, where ρ is the resistivity, converges to $R_0 = 0$ as $L \to 0$. This shows the classical resistance and thus asymptotic behavior for large L manifests a typical transport property in the diffusive regime $L \gg \ell_{\mathrm{mfp}}$. Note that this situation, illustrated schematically in Fig. 5.36 (right), is realized when an electron loses its velocity correlation $\langle v(t)v(0)\rangle = \langle v_F^2\rangle \exp(-t/\tau)$ due to many scatterings; thus the diffusion coefficient is saturated as $D = \langle v_F^2\rangle\tau$ independent of both the time and the length. On the other hand, when L becomes small to zero, $R(L)$ is independent of L and converges to $(2 \times 2e^2/h)^{-1} = (2G_0)^{-1}$. This is a typical character in the ballistic transport limit and an asymptotic behavior is shown by the thin broken line. The crossing points of the two asymptotic lines for diffusive and ballistic regimes allow for an estimation of ℓ_{mfp} shown in Fig. 5.37 by the

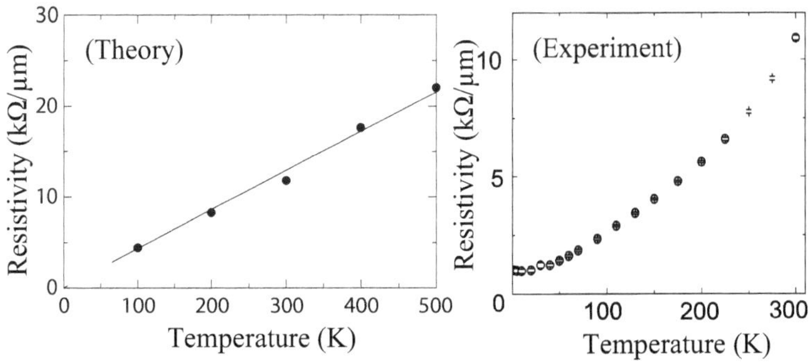

Figure 5.38 (Left) Resistivity of the (5, 5)-CNTs as a function of temperature. The resistivity is proportional to T in the high-temperature regime. (Right) Experimental data of the temperature dependence of CNT resistivity. Experiment from M. S. Purewal, B. H. Hong, A. Ravi, B. Chandra, J. Hone, and P. Kim, *Phys. Rev. Lett.* **98**, 186808 (2007) [154] and theory from H. Ishii, F. Triozon, N. Kobayashi, and K. Hirose, *Appl. Phys. Express* **1**, 123002 (2008) [81].

thick arrow, which is very close to the length of $\ell_{\rm mfp} = D(t \to \infty)/v_F$ with $\ell_{\rm mfp} = 515$ nm at room temperature 300 K. Thus, $\ell_{\rm mfp}$ from $D(t \to \infty)$ corresponds to the MFP, which separates the regimes from ballistic to diffusive one. Around $\ell_{\rm mfp}$, say 300–800 nm as the quasi-ballistic regime, $R(L)$ deviates from the linear curve in the diffusive regime to take larger values. This leads to the quasi-ballistic regime to deviate from both the diffusive and ballistic asymptotic lines.

Next, we see the temperature-dependent resistivity. Figure 5.38 shows the temperature versus resistivity characteristic with the solid line representing the linear slope of $\rho(T)$. The resistivity is almost proportional to the temperature as $R(T) \propto T$, which agrees with the experimental observation in the right panel and shows that the dominant mechanism in the high-temperature regime is from the phonon scatterings. Here the resistivity, which does not approach zero at low temperature in the experiment, shows the residual resistance from impurity scatterings ($W \neq 0$). The obtained temperature T- and diameter d-dependent resistivity in the diffusive regime as $\rho \simeq 0.032T$ (K)$/d$ (nm) is in good agreement with experiments, which varies from 0.024 to 0.121 [153, 154]. It is

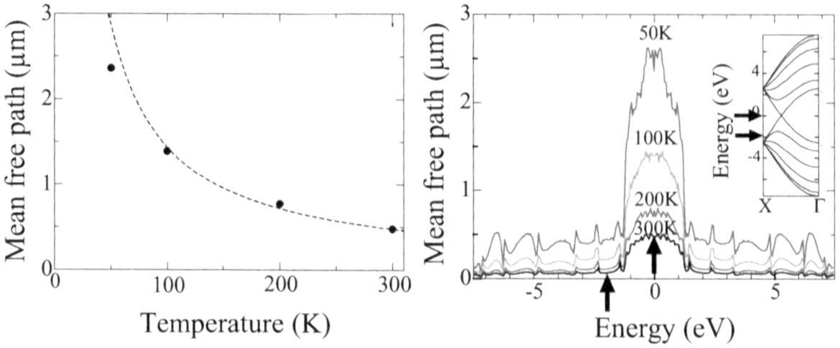

Figure 5.39 (Left) Mean free path $\ell_{\rm mfp}$ of the (5, 5)-CNTs from 50 K to 300 K. The dashed curve represents the inverse temperature dependence, which indicates that the acoustic phonons dominate the transport property. (Right) $\ell_{\rm mfp}$ as a function of μ at temperatures ranging from $T = 50$ K to 300 K. Arrows indicate $\mu = 0$ (charge neutral point) and $\mu = -2$ eV where the new bands appear. (Inset) Electron band structure of the (5, 5)-CNT. From H. Ishii, N. Kobayashi, and K. Hirose, *Phys. Rev. B* **82**, 085435 (2010) [82].

also consistent with the result based on the effective mass theory with a continuum model [155], which clarifies that only TW mode contributes to the backscattering for the armchair CNTs.

Let us study the MFP $\ell_{\rm mfp}$ of (5, 5)-CNT. Figure 5.39 (left) shows $\ell_{\rm mfp}$ as a function of the temperature. It is seen that $\ell_{\rm mfp}$ exceeds larger than a few hundred nanometers even at room temperature and shows $\ell_{\rm mfp} \propto T^{-1}$ dependence due to the scattering by acoustic phonons. Indeed, $\ell_{\rm mfp}$ increases remarkably as the temperature decreases up to about 3 μm at $T = 50$ K. The inversely temperature dependence is consistent with the result from the effective mass theory as $\ell_{\rm mfp} = \pi\mu_0(3a_{\rm cc}\gamma^0/4g_2)^2 d/k_B T \propto T^{-1}$. Figure 5.39 (right) shows $\ell_{\rm mfp}$ as a function of the chemical potential μ at several temperatures. The behavior of extremely long $\ell_{\rm mfp}$ around the charge neutral point 0 eV is due to the suppression of backscattering from the linear dependent electronic band structure shown in the inset of Fig. 5.39. However, $\ell_{\rm mfp}$ drops rapidly down in the higher subband region ($|\epsilon| \geq 1.5$ eV), since new scattering channel bands are open for electrons. For example, $\ell_{\rm mfp}$ with 3 μm around $\epsilon = 0$ is reduced to 500 nm around $\epsilon = 2$ eV at

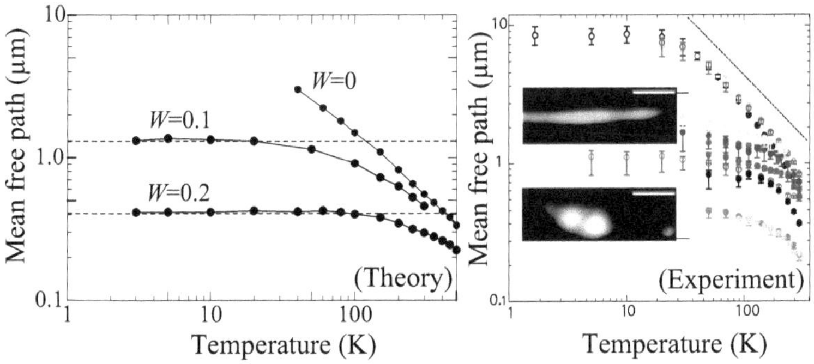

Figure 5.40 (Left) Temperature dependence of MFP ℓ_{mfp} for a metallic (5, 5)-CNT with both dynamical disorder and static disorder potential W. ℓ_{mfp} is proportional to T^{-1} for $W = 0$, while ℓ_{mfp} saturates to a T-independent value tuned by the disorder strength $W \neq 0$. In high T regime, ℓ_{mfp} approaches the value in the $W = 0$ limit, evidencing that phonon scatterings dominate the transport properties. (Right) Corresponding experiment. Experiment from M. S. Purewal, B. H. Hong, A. Ravi, B. Chandra, J. Hone, and P. Kim, *Phys. Rev. Lett.* **98**, 186808 (2007) [154] and theory from H. Ishii, S. Roche, N. Kobayashi, and K. Hirose, *Phys. Rev. Lett.* **104**, 116801 (2010) [159].

$T = 50$ K and 500 nm to 100 nm at $T = 300$ K. Similar energy dependences of the transport behaviors have been obtained in the experiment [156].

By tuning the static disorder parameter W to include residual resistance, we reproduce experimental observations of the T dependence of ℓ_{mfp} very well. In Fig. 5.40, the logarithm plot of temperature-dependent ℓ_{mfp} is shown for $W = 0$, 0.1, and 0.2. One clearly notices that ℓ_{mfp}s are determined by the static disorder in low-temperature regime ($T < 50$ K), while they tend to merge in the high-temperature regime, when phonon scatterings contributions to the transport properties prevail. For $W = 0$, $\ell_{\mathrm{mfp}} \propto T^{-1}$ is fully dominated by phonon-driven scatterings. The obtained temperature dependence of ℓ_{mfp} from the numerical results are in full agreement with experimental data in Fig. 5.40 (right). By setting $W = 0.2$, we can reproduce the typical elastic MFP of 400 nm found in the experimental data, which is used to discuss the behavior of L_{ϕ} later.

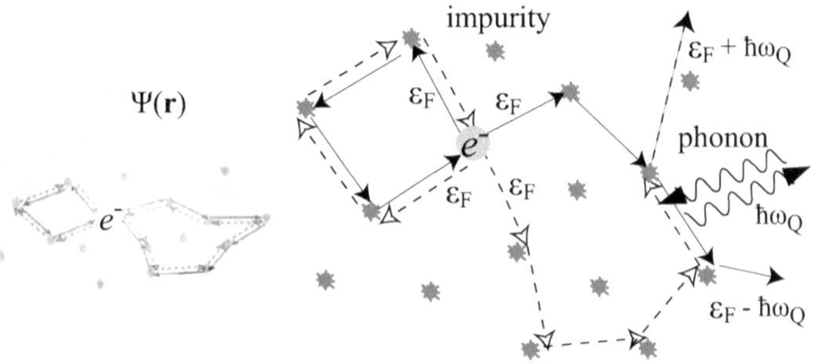

Figure 5.41 (Left) Conceptual picture of localization phenomena due to QIE between clockwise and counter-clockwise back-scattering paths. (Right) Delocalization process induced by the inelastic electron–phonon scattering in disordered systems.

5.4.2.2 Phase coherent length

Here, we consider the phase coherence length of the CNTs, especially at high-temperature regime where the effects of phonon scatterings on the phase coherence are dominant. The phase coherence length has been investigated in view of the localization in disordered mesoscopic systems based on the quantum interference effects (QIE) on charge transport. As we see in Chapter 2, the QIE between clockwise and counter-clockwise backscattering paths develops in the coherent regime and yields an increase of the probability for propagating wave-packets to return to the origin, resulting in localization. The conductance is related to the probability distribution $P(\mathbf{r}, \mathbf{r}')$ for an electron at the point $\mathbf{r}$ to reach the point $\mathbf{r}'$ as shown in Fig. 5.41. The quantum nature to the conductance appears due to the interference

$$P(\mathbf{r}, \mathbf{r}') = \sum_{p} |A_p|^2 + \sum_{p \neq p'} A_p A^*_{p'}, \qquad (5.61)$$

where A_p is a probability amplitude. In general, the interference contribution terms disappear after averaging on disorder, except for the interference between clockwise path A_p and counter-clockwise backscattering path A_{-p}. This enhances by a factor of 2 compared with the classical value and gives a negative correction to the conductance due to the localization [157, 158].

Decoherence mechanisms in mesoscopic systems take place as soon as dephasing processes start to reduce the quantum coherence of propagating wavepackets. The negative weak localization correction depends on the phase coherence time τ_ϕ and the dimensionality d through

$$\frac{\Delta\sigma}{\sigma} \sim -\langle P(\mathbf{r}, \mathbf{r}')\rangle = -\int_{\infty}^{\tau_\phi} C(t)dt, \qquad (5.62)$$

where the probability $C(t)dt$ for a diffusing electron to follow the path between t and $t + dt$ is given in the diffusive regime by

$$C(t)dt = \frac{v_F \lambda_F^2 dt}{(Dt)^{d/2}}. \qquad (5.63)$$

This is derived from the ratio of the volume of a tube of section λ_F^2 and the length $v_F dt$ with the total volume of all possible diffusive trajectories during the time t. The correction of the conductance for $d = 1$ case is evaluated by

$$\Delta\sigma = -\frac{e^2}{2\pi\hbar} \times 2\pi(D\tau_\phi)^{1/2} \qquad (5.64)$$

and thus

$$\Delta G = \frac{\Delta\sigma}{L} = -\frac{2e^2}{h}\frac{L_\phi}{L} \quad \text{with} \quad L_\phi = \sqrt{D\tau_\phi}. \qquad (5.65)$$

L is the system length and L_ϕ is derived from the temperature-dependent dephasing mechanisms taking place in the materials.

At low temperatures, the main decoherence mechanism is the electron–electron (e–e) interaction. A usual way to obtain the phase coherence length is to evaluate the MR in a transport experiment. When a magnetic field B is applied to a 1D sample with the width W and the length L, phase coherent time τ_ϕ is given from the quantum correction as

$$\tau_\phi = \left(\frac{\sqrt{DmW}}{2\pi k_B T}\right)^{2/3}. \qquad (5.66)$$

This approach was successfully used for the analyses of multiwall CNTs, where the experiments are performed at low temperatures up to about 60 K [156]. At such low temperatures,

$$L_\phi \sim T^{-1/3} \qquad (5.67)$$

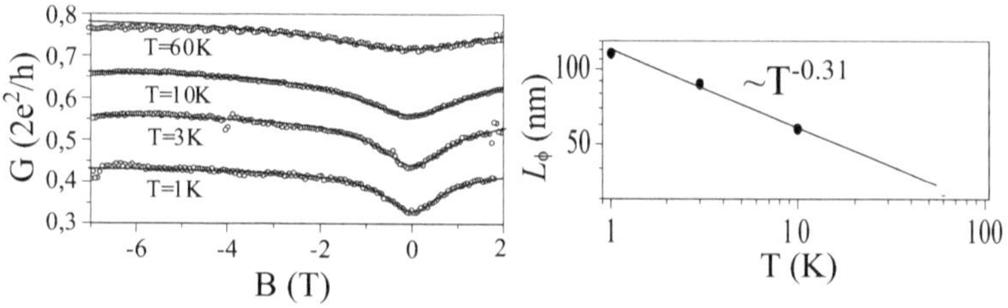

Figure 5.42 (Left) Averaged magnetoconductance of multiwalled CNTs up to $T = 60$ K. (Right)Temperature dependence of the phase coherence length L_ϕ.The line corresponds to a power law fit with an exponent of -0.31. From B. Stojetz, C. Miko, L. Forró, and C. Strunk, *Phys. Rev. Lett.* **94**, 186802 (2005) [156].

was roughly found as shown in Fig. 5.42 and agrees with the predictions of weak localization theory, in which the *e–e* scattering predominates over the electron–phonon (*e–ph*) interaction in the one-dimensional systems. While the *e–e* interaction dominates at very low temperature, the *e–ph* coupling is usually found to dominate at higher temperatures. Actually, both the *e–e* and *e–ph* interaction effects on decoherence are observed in experiments using the rope of single-wall CNTs, which are more ideal 1D systems than multiwall CNTs.

Here we show the phase coherence length L_ϕ of the CNTs evaluated due to the *e–ph* scatterings and its temperature dependence from low to high-temperature regimes. The weak localization regime is induced by introducing a static source of elastic scatterings. The phase coherence length L_ϕ is derived using the weak localization theory as follows. In presence of static disorder, beyond the scale defined by the elastic MFP, quantum interferences produce a quantum correction to conductance $\Delta G_{\mathrm{WL}} = -(2e^2/h)L_\phi/L$. The resistance at zero-temperature is thus entirely monitored by static impurity scattering and localization effects. Since the resistance is defined as the inverse of conductance, $R = 1/(G_{\mathrm{Imp}} + \Delta G_{\mathrm{WL}}) \simeq 1/G_{\mathrm{Imp}} - (1/G_{\mathrm{Imp}})^2 \Delta G_{\mathrm{WL}} \equiv R_{\mathrm{Imp}} + \Delta R_{\mathrm{WL}}$, the quantum correction ΔR_{WL} is given by

$$\frac{\Delta R_{\mathrm{WL}}}{R_{\mathrm{Imp}}^2} = \frac{2e^2}{h}\frac{L_\phi}{L}. \tag{5.68}$$

The correction is reduced by the decoherence event due to inelastic scattering such as e–ph scatterings. Therefore, we can extract the coherence length from the resistance reduction due to the inelastic e–ph scatterings. The coherence length is deduced from

$$L_\phi(T) = \frac{h}{2e^2}\left\{\frac{\rho(T) - \rho_{\rm Imp} - \rho_{\rm ph}(T)}{\rho_{\rm Imp}^2}\right\}, \tag{5.69}$$

where $\rho(\equiv R/L)$ is the 1D resistivity in the diffusive regime. $\rho(T)$ and $\rho_{\rm ph}(T)$ are obtained from $\rho(T) = (h/2e^2)(v_F/D(W, T))$ and $\rho_{\rm ph}(T) = (h/2e^2)(v_F/D(W = 0, T))$. Since $\rho(T \to 0)$ includes the weak localization effects, we extract $\rho_{\rm Imp}$ from the high-temperature limit as

$$\begin{aligned}\rho_{\rm Imp} &= \lim_{T\to+\infty}\Big\{\rho(T) - \rho_{\rm ph}(T)\Big\}\\ &= \lim_{T\to+\infty}\Big[\Big\{\rho_{\rm Imp} + \rho_{\rm ph}(T) + \Delta\rho_{\rm WL}(T)\Big\} - \rho_{\rm ph}(T)\Big],\end{aligned} \tag{5.70}$$

assuming that the localization effect vanishes in high-temperature limit, i.e., $\Delta\rho_{\rm WL}(T \to +\infty) = 0$. We evaluate the temperature dependence of resistivity from 3 K to 550 K, averaged within [−0.25 eV, +0.25 eV] energies. $(\rho(T) - \rho_{\rm ph}(T))$ saturates to 9.38×10^{-3} kΩ/nm over 400 K. $L_\phi(T)$ is evaluated from $\rho(T)$, $\rho_{\rm ph}(T)$, and $\rho_{\rm Imp}$ using Eq. (5.69). We note that in this approach non-energy conserving scattering processes are neglected and only dephasing mechanisms driven by dynamical disorder are retained. This computational methodology can be rationalized following the picture of quantum decoherence as described in [160]. Dynamical disorder will generate many energy-conserving wavepackets and corresponding loss of quantum coherence. This loss accumulates in time as the wavepacket diffuses through the disordered system and once the uncertainty on the quantum phase is in the order of 2π, the full suppression of coherence allows to extract L_ϕ.

Figure 5.43 gives the inverse phase coherence length $L_\phi{}^{-1}$ as a function of T. We see that the temperature dependence of L_ϕ is markedly changed at 250 K. The data from 20 to 100 K are fitted very well by $L_\phi \propto T^{-0.558}$, while $L_\phi \propto T^{-1}$ fits the data from 250 to 350 K. This observation is well understood from the temperature dependence of relaxation times τ. The inelastic relaxation time due

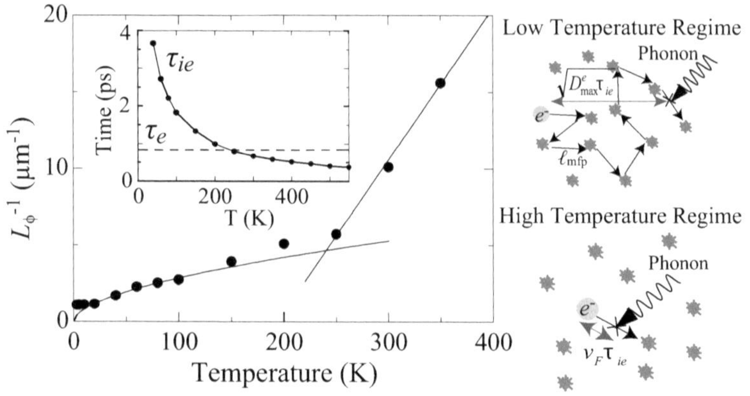

Figure 5.43 (Left) L_ϕ^{-1} vs. T curve for (5, 5)-CNT with $W = 0.2$. The T dependence of L_ϕ changes at 250 K. Data from 20 K to 100 K are well fitted to $L_\phi^{-1} \propto T^{0.558}$, while data from 250 K to 350 K are fitted to $L_\phi \propto T$. (Inset) T dependence of elastic and inelastic scattering times. The elastic scattering time equals the inelastic one at 250 K. (Right) (Top) Schematic picture of dephasing in low-temperature regime. Diffusive transport is realized due many elastic scatterings with impurities. Electrons lose their phase memories by inelastic scatterings with phonons. (Bottom) In high-temperature regime case. Inelastic phonon scatterings occur before elastic scatterings with impurities. From H. Ishii, S. Roche, N. Kobayashi, and K. Hirose, *Phys. Rev. Lett.* **104**, 116801 (2010) [159].

to phonon scatterings is given by $\tau_{ie}(T) = D(W = 0, T)/v_F^2$ and shows T^{-1} dependence. On the other hand, the elastic relaxation time $\tau_e(\equiv \ell_{mfp}/v_F)$ due to impurity scatterings is temperature independent; thus we see that $\tau_{ie}(T)$ and τ_e cross each other at 250 K (Fig. 5.43, inset). This indicates that when T is less than 250 K, elastic scattering with impurities dominates and *e–ph* mediated scattering occurs after the system reaches the diffusive transport regime as shown in the left panel of Fig. 5.40. In this regime, using the diffusion coefficient $D^e_{max} \equiv v_F \ell_{mfp}$, the inelastic scattering length is determined by $\sqrt{D^e_{max}\tau_{ie}}$ and therefore L_ϕ has the $T^{-0.5}$ dependence since τ_{ie} is proportional to T^{-1}. It should be noted that $L_\phi(T \to 0)$ has no meaning in the present calculation, since we treat the phonons semi-classically and since the *e–e* scattering effect (which is important in low-temperature regime) is neglected.

In high-temperature regime ($T \geq 250$ K), on the other hand, electrons propagation is fully dominated by e–ph scatterings, as shown in Fig. 5.43 (right). The inelastic scattering length is given by $v_F \tau_{ie}$, which is proportional to the inverse temperature T^{-1}. It is important to note that the coherence length is not determined from $L_\phi \sim \sqrt{D^e_{\max} \tau_{ie}}$ close to room temperature. These observations on the power-law behaviors of T-dependent inelastic scattering length agree qualitatively well with the present simulation results of $L_\phi(T)$ in Fig. 5.43. The coherence length does not equal the inelastic scattering length, because the quantum decoherence takes place before non-energy conserving processes dominate conduction. Here we note the W dependence on $L_\phi(T)$. Since the elastic scattering time increases as $\tau_e \propto W^{-2}$, when the disorder W decreases from 0.2 to 0.1, the intersection of the temperature dependence of $\tau_{ie}(T)$ and τ_e is decreased from 250 K to 50 K. In such temperature range, L_ϕ shows T^{-1} of $T^{-0.5}$ behavior depending on the static disorders of samples. We note that experimental data on the energy dependence of both elastic MFP ℓ_{mfp} and coherence length L_ϕ are sensitive to onsets of new subbands. Comparing L_ϕ with ℓ_{mfp}, weak localization theory becomes inapplicable in high-temperature regime. This underlines the strong decoherence effects driven by vibrations in CNTs.

5.4.3 *Transport of CNTs and GNRs*

Here we show the transport properties of nano-carbon materials for the applications to the channel materials of field-effect transistor (FET) devices.

5.4.3.1 Field-effect transistor devices from CNTs

The first is the CNTs. In addition to the nature of good conductor for the ballistic transport, CNTs have a number of fascinating properties, such as the bandgap engineering by the doping, free from dangling bonds to use the high-k materials for the gate controls, no expensive lithograph for the fabrication, the large current density, etc. These might be useful for the FET and other quantum devices. Here we consider the CNT-FET devices, for which we need to select the

Table 5.1 Comparison for the typical values of silicon and CNT

	Silicon	Carbon Nanotube
Fabrication	Lithography	Self-assembly
Bandgap	Fixed at 1.12 eV	Flexible 0.2 $\sim$ 1.2 eV
Doping (n,p)	B, P by sputtering	O, K contamination
Fermi velocity	10^7 cm/s	$2 \sim 8 \times 10^7$ cm/s
Resistivity	$3\text{–}4 \times 10^{-4}$ Ω cm	1.67×10^{-5} Ω cm
Current density	$\sim 10^7$ A/cm^2	10^8 A/cm^2
Thermal conduct.	150 W/mK	3, 000–5, 500 W/mK

semiconducting CNTs for channel materials. Table 5.1 shows the comparison of the fundamental properties of Si and CNT.

In the experiments of CNT-FET, the device operation is mostly determined by the contact, where potential barriers are constructed from the heterostructure of metallic electrode and semiconducting CNT. This is in contrast to the Si-MOSFETs, where the contacts of channel with source and drain are made of heavily doped Si and the contacts are ohmic. In CNT-FET, the contact of semiconducting CNT with metallic electrodes such as Al, Au, and Ti produces a potential barrier called Schottky barrier. This is due to the charge transfer from the metal to the CNT at interface, which is determined by the different workfunctions, and the dipole moment at the interface produces potential barrier. In the CNT, due to weak screening for the 1D system, the tail of the Schottky barrier becomes very long (several tens of nm) for both source and drain electrodes. Then both electrons and holes are injected as carriers and the CNT-FET shows an ambipolar nature. Therefore, the understanding and the control of the contact are a crucial problem for the better performance of CNT-FET devices.

5.4.3.2 Experiments on CNT-FETs

In general, FET has a structure with a channel made of a semiconductor connected to a source (S) and a drain (D) electrodes. An insulator separates the channel from the third gate (G) electrode. By applying a voltage at the gate electrode, we can modulate the conductance in semiconductor channel. The first realizations of

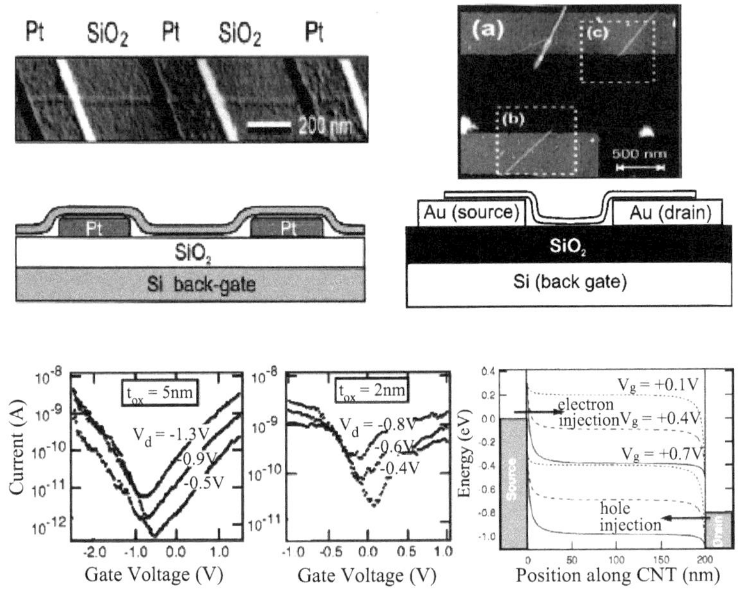

Figure 5.44 (Top) Examples of the realization of the CNT-FET devices. (Left) S. J. Tans, A. R. M. Verschueren, and C. Dekker, *Nature* **393**, 49 (1998) [161]. (Right) R. Martel, T. Schmidt, H. R. Shea, T. Hertel, and Ph. Avouris, *Appl. Phys. Lett.* **73**, 2447 (1998) [162]: (Bottom) (left and center) Experimental I_d–V_g characteristics for ambipolar CNT-FETs at different drain voltage V_d. The length of the CNT is $L = 300$ nm. (Right) Schematic band diagram of the Schottky barrier transistors. Both electrons and holes are injected into the CNT channel through the Schottky barriers for ambipolar operations. From M. Radosavljević, S. Heinze, J. Tersoff, and Ph. Avouris, *Appl. Phys. Lett.* **83**, 2435 (2003) [163].

the CNT-FET devices were reported in 1998 [161, 162], where the semiconducting CNT plays the role of the channel. The key points to use the CNT for the FET channel are its high mobility and low scatterings (Fig. 5.44 (top)).

Figure 5.44 (bottom) shows I_d–V_g characteristics of ambipolar CNT-FET devices. For both negative and positive gate voltages V_g, holes and electrons are injected where at the minimum the two currents are equal. The efficiency of the switching characteristics is measured by the parameter $S = (\mathrm{dlog_{10}I_d/dV_g})^{-1}$, called subthreshold slope S, which shows how sensitive an applied gate voltage drives the source-drain current. The current in the "off" state I_{off} is also important for the low leakage current. When the insulator thickness t_{ox} is small or the source-drain voltage V_d is large, the "off" current is significant. Typically $I_{\text{on}}/I_{\text{off}} \gg 10^4$ is desired for the device operation. The behavior of the Schottky barrier is controlled by the selectively doping of the CNT-FET. Different from the traditional dopant such as B or P atoms, chemical doping of O, K, or molecules to absorb on the CNT induces the charge transfer at contacts. Also, to use the high-k materials for the gate insulator is effective to increase the coupling of the CNT channel and the gate. When we make the ohmic contacts, the currents of short-channel CNT-FETs should show the ballistic nature.

5.4.3.3 Empirical potential for Schottky barrier contact

Here we show a phenomenological planar gate model for the Schottky potential close to the contact to electrodes [164]. Consider the CNT with radius R surrounded uniformly by the insulating materials with κ, which is attached to metallic electrodes along the z-axis and is separated from the gate substrate with a distance of R_s (Fig. 5.45, left). The potential $\phi(z)$ under a gate voltage V_g is obtained from self-consistent calculations of the Poisson equation and charge density $n(z)$.

The Poisson equation relates the total potential $\phi(z)$ to charge density $n(z)$ of the CNT under an applied gate voltage V_g,

$$\phi(q) = U(q)n(q) + M(q)V_g(q), \tag{5.71}$$

where $\phi(q)$, $n(q)$, and $V_g(q)$ are the Fourier transformation of $\phi(z)$, $n(z)$, and the z-dependent gate bias $V_g(z)$, respectively. Using the modified Bessel functions I_0 and K_0, the coefficients $U(q)$ and $M(q)$ are given by the analytic forms of

$$U(q) = \frac{2}{\kappa}\{I_0(qR)K_0(qR) - K_0(2qR_s)\} \tag{5.72}$$

and

$$M(q) = \exp(-|q|R_s). \tag{5.73}$$

Due to the conservation of the total energy of electrons, the charge neutrality level $\epsilon_0(z)$ of the CNT measured from the Fermi energy $\epsilon_F = 0$ eV is related to the total potential $\phi(z)$ as

$$\epsilon_0(z) + e\phi(z) = \Delta W, \tag{5.74}$$

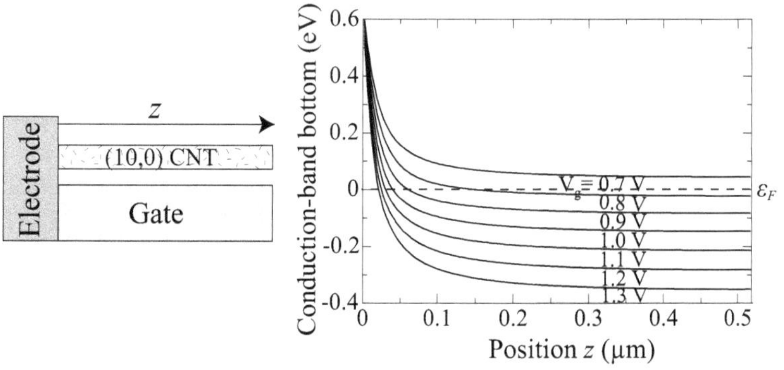

Figure 5.45 (Left) Schematic illustration of the planar gate model of the CNT-FET. (Right) Bottom of the conduction band of the semiconducting (10, 0)-CNT with about 1.0 μm channel-length for various gate voltages V_g. The Schottky barrier is formed at the right-side contact. ϵ_F is represented by the broken line.

where ΔW is one of the fitting parameters independent of z. Since the charge neutrality level is given, the z-dependent charge density can be obtained from

$$n(z) = n_0(z) + e\frac{8\pi R}{3\sqrt{3}a_{cc}^2}\left(D_0\epsilon_0(z)\exp\left(-\frac{z}{d}\right)\right)\text{sgn}(z), \qquad (5.75)$$

where

$$n_0(z) = e\int N(\epsilon)\text{sgn}(\epsilon)f[\{\epsilon - \epsilon_0(z)\}\text{sgn}(\epsilon)]d\epsilon. \qquad (5.76)$$

Here D_0 is the pinning strength and an effective decay constant d of the surface states is taken to be 2 nm. $N(\epsilon)(= \nu(\epsilon)S)$ represents the DOS per unit length of the CNT and $f(\epsilon)$ is the Fermi–Dirac distribution function. Equations (5.71), (5.74), and (5.75) are solved self-consistently to obtain $\phi(z)$. The boundary conditions at $\phi(z = 0)$ are imposed according to symmetric sgn $= 1$ (or anti-symmetric sgn $= -1$) condition with $V_g(z) = V_g\text{sgn}(z)$ and $\epsilon_0(z) + e\phi(z) = \Delta W\text{sgn}(z)$.

Let us consider the semiconducting (10, 0)-CNT, which is surrounded by a dielectric material of uniform dielectric constant of $\kappa = 5.0$. The distance between the CNT and the gate substrate is set as $R_s = 19$ nm. The experimental values are fitted by the parameters of $D_0 = 0.1$ and $\Delta W = 0.65 \times \epsilon_G/2$, where the energy bandgap ϵ_G

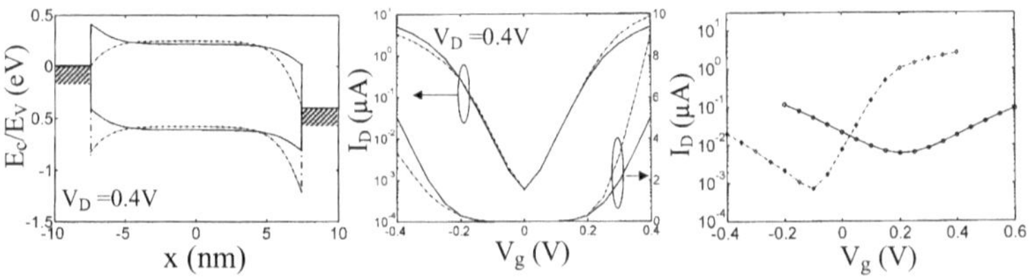

Figure 5.46 (Left) Schematic diagram for the conduction and valence band profiles. (Center) Shifted I_d vs. V_g characteristics for the thin gate oxide ($t_{ox} = 2$ nm and $\kappa = 25$) CNT-FET. The channel length L of 15 nm for (13,0)-CNT is used as channel. The minimal leakage current is shifted to $V_g = 0$ by adjusting the flat band voltage. The left axis shows the I_d–V_g on log-scale and the right axis shows the same curves on linear scale. (Right) I_d vs. V_g for the thick gate oxide ($t_{ox} = 40$ nm and $\kappa = 25$). The channel length L of 100 nm for (25,0)-CNT is used as channel. From J. Guo, S. Datta, and M. Lundstrom, *IEEE Trans. Electron. Devices* **51**, 172 (2004) [165].

of the (10, 0)-CNT is equal to 0.8 eV. The right panel of Figure 5.45 shows effective potentials, conduction-band bottom of the 1.0 μm channel-length CNT, in the vicinity of the contact as a function of the position z for several gate voltages ranging from 0.7 to 1.3 V. The Schottky barriers are clearly observed to form close to the contact to electrode, whose effects range up to 100 nm inside the CNTs. This anomalous long-range tail with logarithmic length dependence is due to the weak screening of CNTs as one-dimensional materials.

5.4.3.4 Carrier transport calculations for CNT-FET

Here we see the I_d–V_g characteristics of the CNT-FET devices obtained from the NEGF atomistic calculations. The coaxially gated CNT-FET with the channel length of L is used at room temperature, for which the ballistic conduction is assumed.

The calculations are performed using the NEGF formalism. The electron density from the nonequilibrium Green's functions is obtained with self-consistently solving the Poisson equation. The Schottky barrier at the metal–CNT interface is treated by the phenomenological model described in the previous section. After the self-consistency is achieved, the source-drain current I_d is computed from

$$I_d = \frac{4e}{h} \int \mathcal{T}(\epsilon) \left[f(\epsilon) - f(\epsilon + eV_d) \right] d\epsilon, \tag{5.77}$$

where two channel for the CNT bands is taken into account for transmission and $\mathcal{T}(\epsilon)$ is the surface-drain transmission obtained from the NEGF formalism. Figure 5.46 (left) shows the schematic effective potential for the conduction and valence bands of CNT-FET devices with the Schottky barriers.

Figure 5.46 shows the I_d as a function of gate voltage V_g for the thin insulator $t_{ox} = 2$ nm (center) and thick insulator $t_{ox} = 40$ nm (right) cases. The transistor shows the ambipolar nature, symmetric for electron and hole conduction. The minimum current I_{off} is achieved when $V_g = V_d/2$ where the electron and hole currents become the same. We see that for the thin insulator case, the I_d–V_g characteristics remains symmetric. This indicates that the Schottky barrier becomes thin when the gate insulator is thin and the current is mostly determined by the thermionic emission, close to the conventional FET. On the other hand, when the insulator is thick, I_d–V_g characteristics are quite different. For these devices, the current is determined by the tunneling through the thick Schottky barriers. Since the barriers are thick, an asymmetric barrier height leads to quite asymmetric electron and hole conduction. The on-off current ratio $I_{\text{on}}/I_{\text{off}}$ becomes much better. These observations show that the device efficiency is determined by the Schottky barriers, which behave differently by the other parameters such as the gate insulator.

5.4.3.5 Mobility of semiconducting CNT

In the previous section, the current I_d is calculated assuming the ballistic nature. Here we consider the transport properties of the semiconducting CNT for the performance of CNT devices at room temperature by using the TD-WPD method [166]. The mobility and cutoff frequency of semiconducting (10, 0)-CNT FET devices are investigated from ballistic to diffusive regimes.

Let us study how the mobility changes according to the length of the CNT from the ballistic to the diffusive regimes. Here we define the pseudo-mobility, μ^*, by the following equations within the effective mass approximation:

$$\mu^* \equiv \frac{e\tau}{m^*}, \quad \tau = \frac{D_{\max}(V_g)}{v^2}, \tag{5.78}$$

where τ represents the scattering time defined in Eq. (4.280). The velocity v is obtained as 0.59 nm/fs for $V_g = 1.0$ V and we approximate the bottom of conduction band by a parabolic curve and the effective mass of electrons, which is defined by $1/m^* \equiv d^2\epsilon(k)/\hbar^2 dk^2$ and is evaluated as $m^* \simeq 0.296 \times 10^{-31}$ kg. The parameters for the Schottky barrier are used as shown in Fig. 5.45 (right) ($\kappa = 5.0$, $R_s = 19$ nm).

Here, we call the calculated μ^* as the pseudo-mobility, since the mobility is generally defined in only the diffusive regime, not in the (quasi-)ballistic regime. The important point here is that there are two factors to determine the time τ. One is the relaxation time we expect in the diffusion regime due to the scatterings with phonons. The other is the time at which electrons reach the contact where the potential barrier is formed. Therefore, if the channel length L is short enough, we expect that the pseudo-mobility μ^* will be proportional to L and approaches zero for the limiting $L = 0$ case. We may call such a regime as a (quasi-)ballistic regime since the electron mobility is determined by the device structure itself. Here we put localized wave packets at the right edge of CNT and then $D_{\max}$ and μ^* are obtained from the time-dependent calculations of $D(\epsilon, t)$ with the molecular dynamics simulations for phonon-scatterings.

Figure 5.47 (left) shows μ^* vs. L for the semiconducting (10, 0)-CNT FET devices applying $V_g = 1.0$ V at 300 K. We study the transport properties as a function of L ranging from 200 to 1.4 μm. It is seen that the obtained μ^* changes its character gradually from $L \approx 600$ nm. Since $\mu^* \approx 30{,}000$ cm^2/Vs independent of L as shown by the straight line, the diffusive regime is realized for $L > 600$ nm and μ^* is determined by the electron–phonon scatterings there. This corresponds to the intrinsic mobility of the CNT observable in experiments, not related to the device structure such as contacts to electrodes. Note that the reported experimental values of mobility range from 1,000 to 79,000 cm^2/Vs [167], depending on the various samples. On the other hand, μ^* decreases as L decreases for $L <$ 600 nm. This clearly shows that electron motions are significantly affected by the Schottky barrier in the quasi-ballistic and the

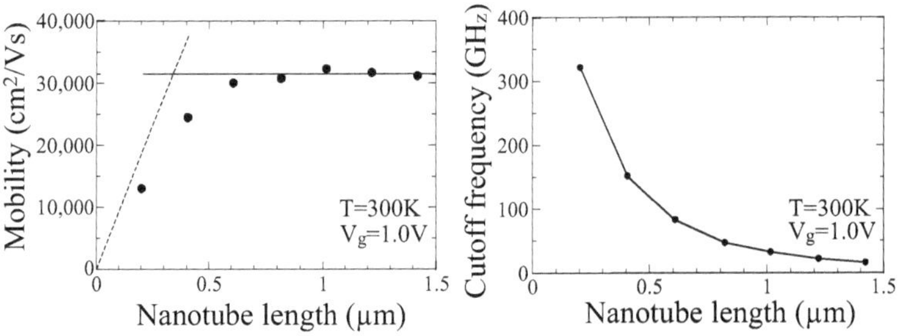

Figure 5.47 (Left) Channel-length dependence of pseudo-mobility of the CNT-FET at 300 K. A gate voltage 1.0 V is applied to the CNT. The thin solid line represents the diffusive limit, while the ballistic limit is shown by the thin broken line. (Right) Cutoff frequency versus length of CNT. From H. Ishii, N. Kobayashi, and K. Hirose, *Phys. Rev. B* **82**, 085435 (2010) [82].

ballistic regimes. The MFP $\ell_{\text{mfp}}(V_g) = D_{\max}(V_g, t \to \infty)/v$, including the electron–phonon coupling and the Schottky barrier is 343 nm, which is shorter than that of metallic (5,5)-CNTs, reflecting that opening bandgap disrupts the linear band relation around the Fermi energy. Thus, for the Schottky barrier parameters considered, the transport properties of (10,0)-CNT change from diffusive to quasi-ballistic regime at $L = 600$ nm and from quasi-ballistic to ballistic regimes at $L = 350$ nm. Since the observed mobility would change dramatically, careful treatment is needed for the analyses of the transport properties in the different transport regimes.

Next, we see the cutoff frequency f_T for a criterion of high-speed performance transistors. High f_T is essential to realize the high-speed response of devices. Experiments demonstrate f_T up to 50 GHz using CNT-FETs with a channel length of few hundred nanometers. Though f_T of CNT-FETs with a 2 nm gate is estimated to reach as high as 1.6 THz using the NEGF method, the length dependence of f_T raging from ballistic to diffusive regimes is not well understood. Here let us show f_T as a function of the channel length by the TD-WPD method. f_T is related to the time τ_T in which electron propagates from the source to the drain electrodes as $f_T \equiv 1/2\pi\tau_T$, where τ_T is obtained using D in a low source-drain bias limit by $\tau_T = L^2/D_{\max}(\epsilon_F)$ for the channel length L. We show f_T vs. L for the (10,0)-CNT FET at 300 K in Fig. 5.47 (right) for $V_g = 1.0$ V. We

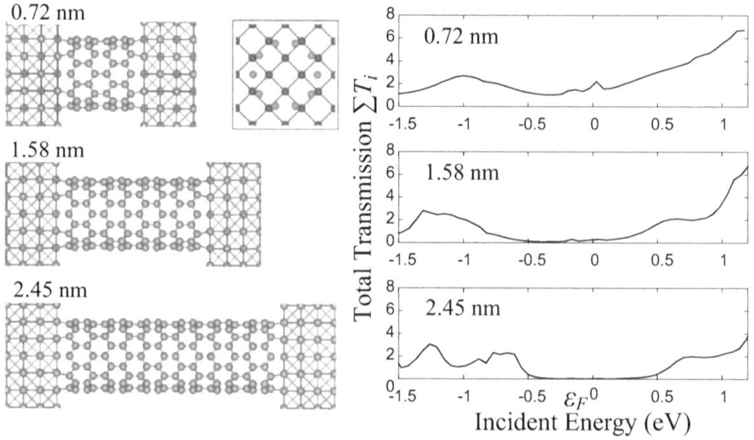

Figure 5.48 (Left) Atomic configuration for (10,0)-semiconducting CNT with $L = 0.72$ nm (top), 1.58 nm (middle), and 2.45 nm (bottom) between semi-infinite Al(100) electrodes. Inset shows aluminum atoms on the outermost surface of the electrode. (Right) Transmission spectra for the corresponding lengths. From N. Kobayashi, T. Ozaki, and K. Hirose, *Surf. Sci.* **601**, 4136 (2007) [168].

see that f_T increases with decreasing L, showing that $f_T = 300$ GHz is achieved for the 0.2 micron-long CNTs. f_T reaches at $f_T = 1$ THz for $L = 100$ nm with $v = 0.59$ nm/fs,

5.4.3.6 Contact effects in short-channel CNT devices

For electron transport in CNT devices, the contacts to electrodes sensitively influence the transport properties. When the CNT is long enough to screen the effects of metallic electrodes, the empirical model describing the Schottky barrier is effective. However, as the length of the channel of CNTs becomes short, the phenomenological treatment of the contact potential is not appropriate and we need the approach from the atomistic viewpoint to describe the effects of electrode contact on the transport of CNTs.

Electron transport in short-channel semiconducting CNTs between metallic electrodes is obtained using the NEGF method with first-principles calculations. Electronic states are calculated using a numerical atomic orbital basis set in the framework of

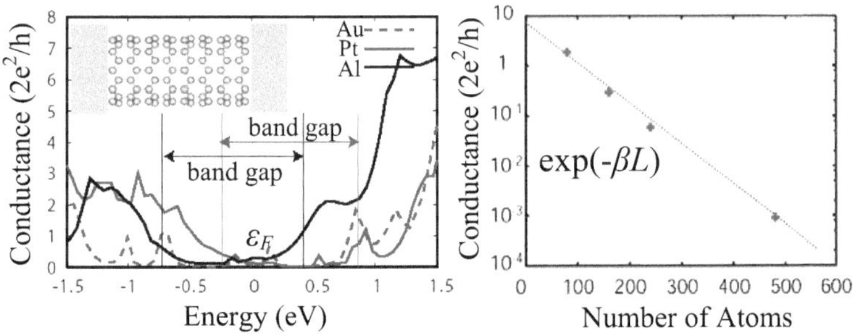

Figure 5.49 (Left) Conductance as a function of Fermi energies of (10,0)-semiconducting CNTs for Al, Au, and Pt metallic electrodes. (Right) Conductance as a functional of the channel length L with Al electrodes. As the effective potential barriers are formed, electron transport becomes tunneling process with a decay length of β.

the DFT, which include screening effects between metallic and semiconductor interfaces self-consistently, though we do not take the device structure of the gate into account. We consider here the models consisting of a finite size of a (10,0)-semiconducting CNT with the channel length L of 0.72, 1, 58, and 2, 45 nm sandwiched between Al(100) electrodes as shown in Fig. 5.48 (left). The spacing from the edge of the nanotube to the Al surface is taken as 1.6 nm.

The right panels of Fig. 5.48 show the total transmission spectra for each CNT. For the shortest CNT, the metallization of the CNT due to the mixing with metallic electrode contact is significant and correspondingly the bandgap becomes missing in the energy region. Thus, the FET operations will not work less than this ultra-short channel length of $L = 0.72$ nm. With an increase of the channel length of CNTs, the strong mixing effect reduces and a gap-like structure appears. Then the hybridization between the carbon atoms and metallic states of the electrode appears and the evanescent tunneling state from the electrode becomes dominant in the transport. The conductance decreases exponentially with an increase of the channel length of the CNTs.

Figure 5.49 shows the conductance of CNTs for various metallic electrodes (left) and tunneling behavior as a function of CNT length (right). We see that use of different metallic electrodes changes the relative Fermi energy in the bandgap. Since the CNT device operation depends strongly on the coupling to metallic electrodes, large-scale ab initio calculations for the transport properties become more and more important.

5.4.3.7 Transport of graphene nanoribbons

The second is the graphene nanoribbon (GNR). The transport properties of graphene, a single atomic layer of graphite with a fascinating two-dimensional material, have attracted special attention [169]. Graphene has unconventional transport properties of Dirac fermions, such as the Klein tunneling and the anomalous Hall effect, due to the linear dispersion relation in the electronic band structure. Also, since the backscattering is absent around the charge neutral point, graphene shows a very large mobility up to 200,000 cm^2/Vs [170], which fascinates to realize extremely high speed of electron devices operating at room temperature. A key issue to use graphene as a channel material of electron devices is how to create a bandgap from the gapless semiconductor materials. There are several proposals for the creation of the bandgap such as the confinement effect of electrons due to reduction of dimensionality, an application of electric fields perpendicular to the graphene, the use of bilayer graphene, to break the equivalence of the A and B sub-lattices of graphene by interacting with the substrate, and so on.

Graphene nanoribbons, fabricated from graphene sheet into one-dimensional structures, have finite bandgap by the confinement effects of electrons. The bandgap is proportional to the inverse of the width of GNRs and becomes smaller as the width becomes wider. GNRs with 10 nm width are fabricated using the lithographic pattering of graphene sheet [171] as shown in Fig. 5.50 (left), and the energy bandgaps reach close to 0.1 eV. From the calculation, the bandgap close to 1 eV is only possible with the width of a few nanometers.[a] However, experimental observations of the transport show that the mobility of sub-10 nm-wide GNRs dramatically decreases to 100 cm^2/Vs from the mobility of 100,000 cm^2/Vs

---

[a] For the CNT with the same circumference as the GNR width, the energy bandgap of CNT takes twice of that of GNR, reflecting the difference of boundary conditions along the circumference directions in Fig. 5.50 (right).

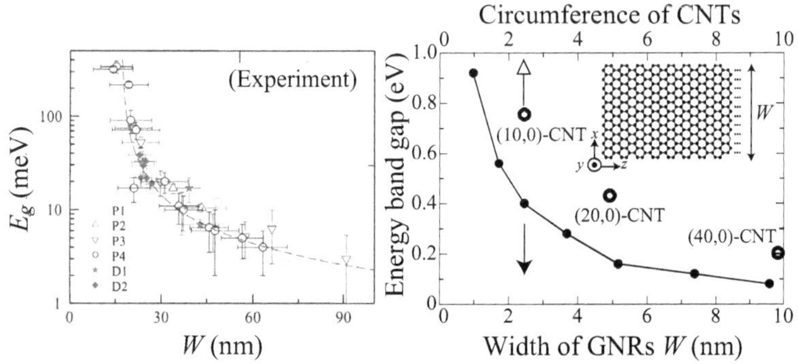

Figure 5.50 (Left) Experimental observation of the bandgap vs. the width of nanoribbons. From M. Y. Han, B. Özyilmaz, Y. Zhang, and P. Kim, *Phys. Rev. Lett.* **98**, 206805 (2007) [171]. (Right) Calculated energy bandgap as a function of the width of armchair GNRs. The bandgap is inversely proportional to the width. For comparison, the circumference dependence of the bandgap of the semiconducting zigzag CNTs is shown. From H. Ishii, N. Kobayashi, and K. Hirose, *Appl. Phys. Express* **3**, 095102 (2010) [172].

for the 2D graphene sheets [173]. This is because the opening bandgap disturbs the linear energy dispersion of graphene, which weakens the tolerance to various scatterings. Thus, the engineering of opening bandgap with keeping the high mobility is desired.

Different from the CNTs, the GNRs have edge structures. The edge of GNRs have a fascinating nature as to construct the polarized spin states, predicted to appear due to the flat band structure at the edge of GNRs [174]. Also, the control of edges is essential for the realization of high mobility of GNRs. Here let us see the effects of edge doping on the transport of GNRs. We attach the nitrogen (N) atom (n-type) and the boron (B) atom (p-type) randomly to the edges of the GNRs with the width of 1 nm. Figure 5.51 shows the conductance as a function of energy from the tight-binding NEGF calculations. For both polarities, conductance increases almost linearly from the linear-dispersion band relation of the band structure. Additional zigzag structures are due to the confinement effect and due to the scatterings with doping atoms at the edges. For the n-type doping, the conductance for the electron transport through the conduction band slightly increases, while for the p-type doping the conductance for the hole transport through

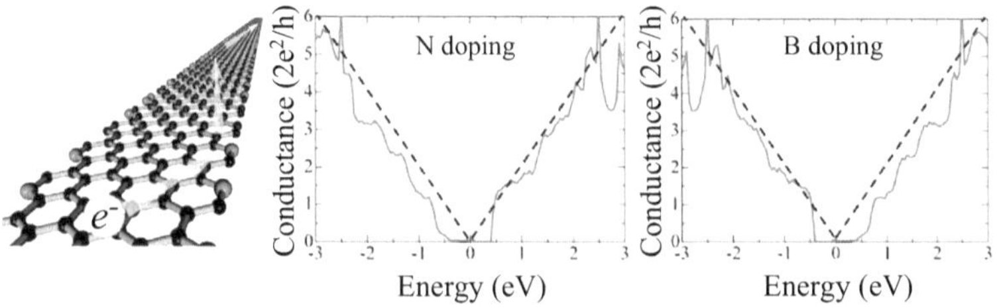

Figure 5.51 (Left) Schematic view of the transport of GNR. We fix the width of GNR as 1 nm. (Right) Conductances as a function of energy when the nitrogen atoms are attached randomly to the edges (n-type doping), whereas when the boron atoms are attached randomly to the edges (p-type doping).

the valence band increases, indicating the possibility to construct the *p*–*n* junctions inside the GNRs. It is important to understand the effects of edges on the transport properties of GNRs. We consider the edge-phonon and edge-roughness effects here.

5.4.3.8 Edge-phonon effect of GNR

There appear localized phonon modes at the edge in low-frequency regime for the GNRs. Such phonon modes are thermally excited at room temperature and might influence the transport properties of the thin GNRs. Figure 5.52 (left) shows the phonon dispersion of armchair GNR with $W = 2.5$ nm from Eq. (4.277) by the MD simulations. Since the vibrations of carbon atoms around the edge are different from those of the center ones due to the symmetry, the edge-phonon modes appear in the low-frequency regime as indicated by the arrow. The phonon branch is categorized into two out-of-plane modes, i.e., in-plane vibration mode and anti-phase vibration mode from the edge carbon-carbon dimers. In the TD-WPD method, the charge mobility $\mu(\epsilon)$ is obtained from $\mu(\epsilon) = e\nu(\epsilon)D(\epsilon)/n(\epsilon)$ where the DOS is $\nu(\epsilon) \equiv \mathrm{Tr}[\delta(\epsilon - \hat{H})]/(L \cdot W)$ and L, W are the length and width of GNR, $n(\epsilon)$ is the carrier number given by $n(\epsilon) = \int_{\epsilon_{\mathrm{CNP}}}^{\epsilon} \nu(\epsilon')d\epsilon'$.

Figure 5.52 (right) shows the mobility $\mu(\epsilon)$ and MFP ℓ_{mfp} as a function of the Fermi energy ϵ_F for the armchair GNRs with width of $W = 2.5$ nm. When ϵ_F is close to the charge neutral point, MFP

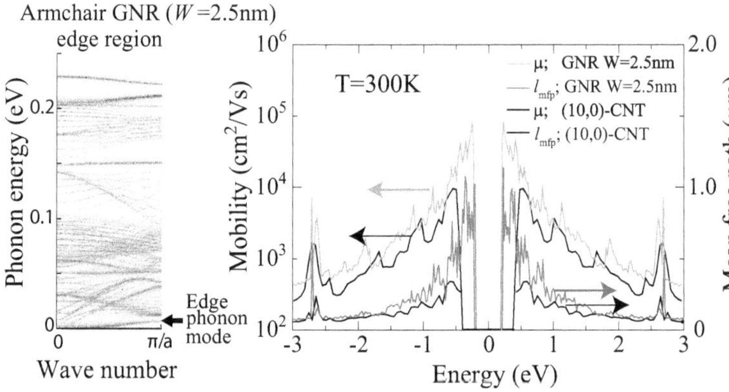

Figure 5.52 (Left) Phonon-dispersion in (ω,q) space near the edge of armchair GNRs with $W = 2.5$ nm width. Edge-localized phonon modes appear, indicated by the arrow. (Right) Mobility $\mu(\epsilon)$ and MFP $\ell_{\rm mfp}$ at $T = 300$ K as a function of ϵ_F for the armchair GNRs with $W = 2.5$ nm and the (10,0)-CNT. From H. Ishii, N. Kobayashi, and K. Hirose, *Appl. Phys. Express* **3**, 095102 (2010) [172].

exceeds 1 μm at room temperature. This is due to the suppression of backscattering of electrons in the linear band dispersion regime. $\mu(\epsilon)$ reaches $10^3 - 10^5$ cm^2/Vs around the charge neutral point beside the bandgap. For comparison, $\mu(\epsilon)$ and $\ell_{\rm mfp}$ of CNTs with no edge-phonon modes are shown. There is no distinct difference for the transport properties of CNTs and GNRs, except for the magnitude of the bandgap. This shows that transport properties of GNRs are not affected by the edge-phonon modes at room temperature in a small applied bias voltage regime.

5.4.3.9 Edge-roughness effect of GNR

The edge roughness of GNRs with fluctuations of a few nanometers in the process of lithographic pattering might become the source of scatterings of the GNRs. Here we evaluate the effects of edge-roughness scatterings on the mobility of GNRs to reveal the relation between energy bandgap and the mobility for the width W. In the inset of 5.53 (top), we show an example of the atomic configuration for the edge roughness of GNRs modeled by the generating random numbers. Here the edge shape between $z = Z_n$ and Z_{n+1} is simply

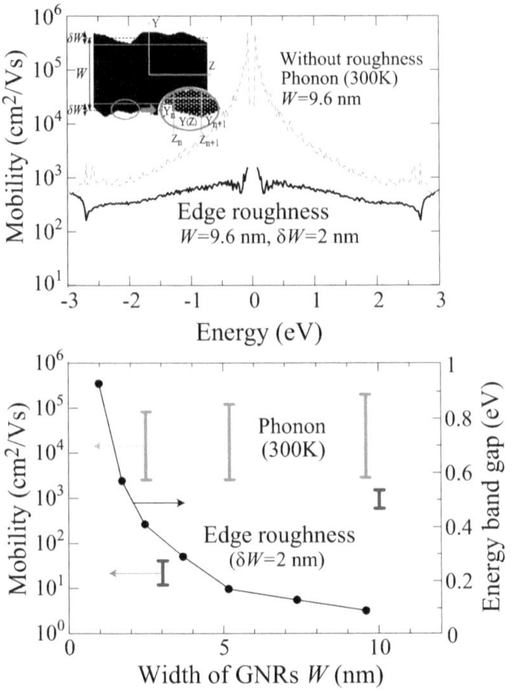

Figure 5.53 (Top) Mobility μ of the GNRs as a function of ϵ for $L = 8.5$ μm with the width $W = 9.6$ and the edge roughness of $\delta W = 2$ nm. For comparison, μ of the armchair GNRs without edge roughness is shown by the broken line. (Inset) An example of the part of atomistic configurations of micron-long GNR. (Bottom) W dependence of μ and energy bandgap E_g of GNRs. Edge-roughness effect induces large W dependence of μ, while electron–phonon effect is independent of W.

described by the straight line given by

$$Y(Z) = \frac{Y_{n+1} - Y_n}{Z_{n+1} - Z_n}(Z - Z_n) + Y_n, \tag{5.79}$$

where $Y_n = -W/2 + \delta W \times p_n$ and p_n is given by random numbers ranging from $-1/2$ to $1/2$, assuming $(Z_{n+1} - Z_n) = \delta W$. Recently experiments estimate the edge roughness of GNRs of a few nanometers. Here we set the edge roughness δW to 2 nm.

First, we see the edge-roughness effect on transport properties with $W = 9.6$ nm. Figure 5.53 (top) shows the ϵ dependence of mobility μ, where the charge neutral point is located at $\epsilon = 0$ eV.

For comparison, μ of the GNR without edge roughness is shown by the broken line, where μ is determined only from the electron–phonon coupling at 300 K. While the GNRs with ideal edges achieve the high electron- and hole-mobility ranging from 10^3 to 10^5 cm^2/Vs, the edge-roughness scattering decreases the mobility considerably up to 10^2 to 10^3 cm^2/Vs. Thus, scatterings at edge roughness with the order of a few nanometers determine the transport properties over the electron–phonon scattering. This result qualitatively agrees with experimental observations for graphene sheets. We note that the 9.6 nm-wide GNR still keeps the mobility comparable to that of silicon materials but the energy bandgap is less than 0.1 eV.

Next, we see W dependence of the energy bandgaps E_g and μ. Figure 5.53 (bottom) shows E_g by the solid line with μ. E_g is consistent with the analytical one $\pi\gamma^0 a_{CC}/W$. The obtained mobility dramatically decreases down to 15–40 cm^2/Vs for the GNR with only $W = 3$ nm, because the edge roughness become a major scattering source as the ratio of edge roughness to the width increases. Experiments show that the mobility of 3 nm-wide GNRs is restricted to less than 10^2 cm^2/Vs. Assuming the relation $\mu \propto W$ for the mobility by the edge-roughness scatterings, we can estimate that the edge-roughness effect on the mobility becomes comparable to the electron–phonon coupling effect when W becomes longer than 20 nm. We note that for the two-dimensional graphene sheet, the MFP increases more than 500 nm in experiments. These results show that the transport properties of GNR are strongly influenced by the scattering effects at the edge roughness.

5.5 Single Molecule

The molecular electronics has recently attracted much attention. The idea of molecular electronics is to utilize single molecules to operate as functional electron devices. Presently various applications such as sensors, diodes, switches, circuits, the next generation of transistors, and others are proposed. Furthermore, to operate the electron devices based on purely quantum mechanics will open possibilities for the new types of quantum devices.

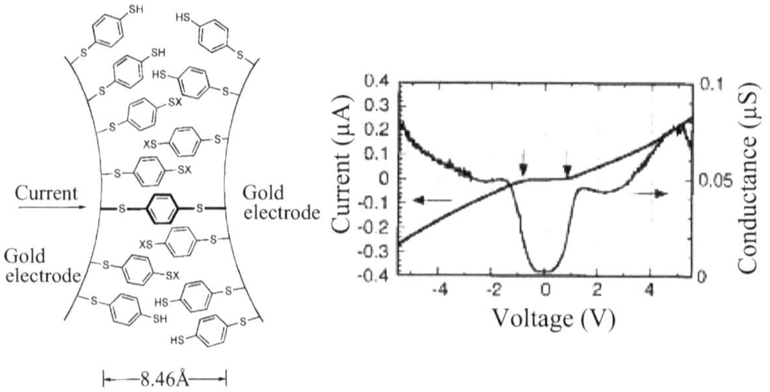

Figure 5.54 Realization of measurements of electron transport through a single molecule between atomic-scale nanogap. (Left) Schematic view of a benzene-1,4-dithiolate between proximal gold electrodes formed with an MCBJ technique. (Right) Typical I–V characteristics and the differential conductance $G(V)$. From M. A. Reed, C. Zhou, C. J. Muller, T. P. Burgin, and J. M. Tour, *Science* **278**, 252 (1997) [175].

Although the concept to use molecules as components in electron devices dates back to the theoretical proposals in the 1970s, the recent rapid progress of the nanometer-size technologies drives much interest in this field. In such situations, the basic technology and the understanding how to fabricate the atomic-scale contacts, to attach a single molecule between electrodes, and to control the single molecules for the useful electron transport properties become more and more important in order to realize electron devices for the molecular electronics. In this section, we see transport properties through single molecules. Especially, we focus on the contact problem of single molecules to electrodes, sensitivity of terminal structure, and detection of vibration spectroscopy of single molecules.

5.5.1 *Experiments toward Molecular Devices*

In order to realize a molecular device, the following two problems have to be solved. The first is the formation of nanometer-scale gap, and the second is a way to arrange single molecules between the gap for creation of a nanometer-scale structure with use of the chemical interactions of molecules with electrodes.

5.5.1.1 Nanogap formation

Presently two experimental methods are technically utilized to make nanometer-scale gap. One is the MCBJ technique as we presented in the previous section briefly, and the other is by using the electromigration phenomena. Both techniques can be used to create nanometer-scale gap for the single molecular junctions.

In the break-junction technique, a thin metal wire is prepared on top of the flexible substrate. The wire is subsequently broken by bending the substrate to form a very small gap in the wire. The advantage for this technique is that the substrate is bent to adjust the size of the gap under an accurate control. The formed nanojunction is well reproduced. We show an example of the nanogap thus formed in Fig. 5.55 (left) [175]. The measurements of the I–V characteristics and the conductance through the single molecule were performed to place benzene-1,4-dithiol molecules between the nanogap formed in the gold wire as shown in Fig. 5.54 [176].

In an alternative technique to use the electromigration phenomena, when electric current is large enough, electron scattering events in a wire move the atoms along the wire and the missing

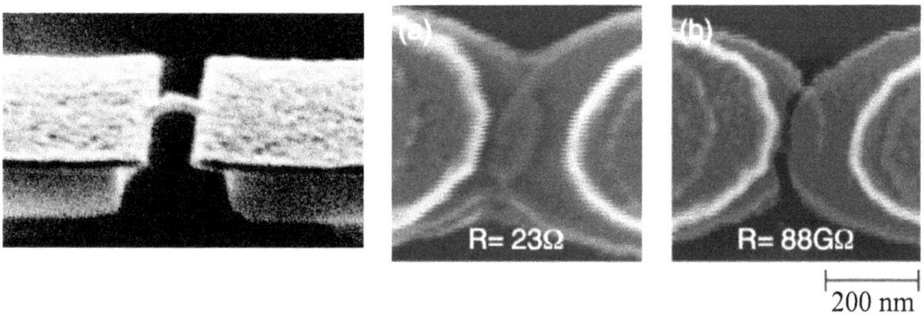

Figure 5.55 (Left) SEM photograph of the connecting gold wire just before the elongation to break by the MCBJ technique. From C. Zhou, C. J. Muller, M. R. Deshpande, J. W. Sleight, and M. A. Reed, *Appl. Phys. Lett.* **67**, 1160 (1995) [176]. (Right) SEM photograph of the gold nanowire before and after the breaking by the electromigration technique. From H. Park, A. K. L. Lim, A. P. Alivisatos, J. Park, and P. L. McEuen, *Appl. Phys. Lett.* **75**, 301, (1999) [177].

atoms create holes. Note that although this phenomenon is a serious problem for the circuits in micro-electronics, it can be used advantageously in the molecular electronics. With a control of electric current to form the onset of electromigration, a nanometer-gap can be created in the wire with highly reproducible. We show an example of the nanogap thus formed in Fig. 5.55 (right) [177]. The single-molecule transistors was created using the individual C_{60} molecules connected between the nanogap in the gold wire formed by the electromigration technique, and the nanomechanical oscillation behaviors were detected in the transport measurement [178].

5.5.1.2 Transport measurement

By putting single molecules between the nanometer-scale gap, nanometer-scale bridge structures are created and the transport measurements are performed for the molecular electronics. There have been a number of experiments on transport through single molecules so far. Let us show two examples of I–V characteristics in Fig. 5.56 (top) [179, 180], where we observe similar non-linear behaviors for the I–V characteristics. Namely, we see that the I–V curves exhibit very small up to several bias voltages. When the applied bias exceeds the threshold bias voltages, the electric current flow increases. Correspondingly, the differential conductance dI/dV have peaks at these voltages and shows a valley structure as a function of the bias voltage.

These I–V characteristics are often explained as follows. In general, a single molecule has electronic states with an energy gap of several eV between the highest occupied molecular orbital (HOMO) and the lowest unoccupied molecular orbital (LUMO). When the molecule makes contact with electrodes, these states are broadened and the Fermi level lies between the HOMO and the LUMO. Applying a bias voltage, electric current flows with peaks when one of the electronic states of the molecule aligns with the left or right chemical potentials through resonant tunneling. The electronic states and electron transport process are schematically shown in Fig. 5.56 (bottom). Therefore, we observe peak structures in the differential conductance in the transport experiments through the resonance of molecular states and the valley structure around the zero bias

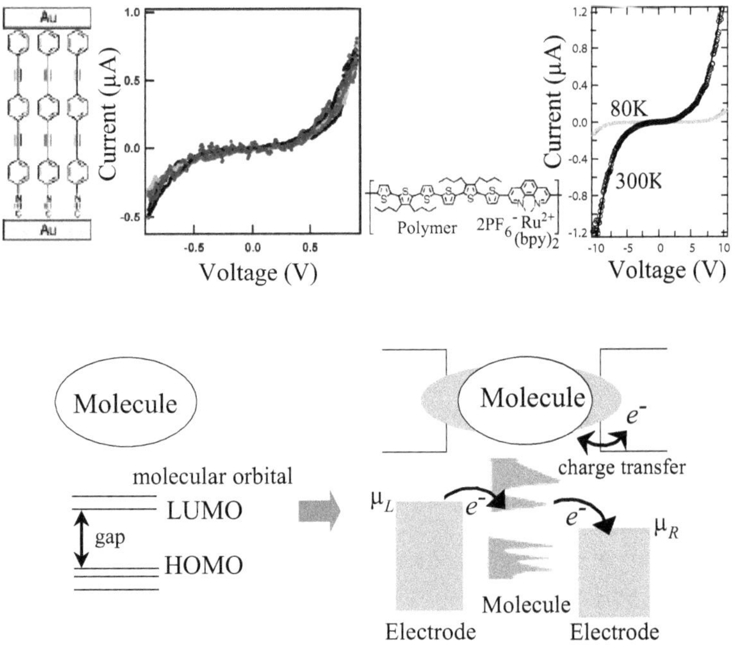

Figure 5.56 (Top) Examples of I–V measurements through molecules. (Left) From J. G. Kushmerick, J. Naciri, J. C. Yang, and R. Shashidhar, *Nano Lett.* **3**, 897 (2003) [179]. (Right) From K. Araki, H. Endo, H. Tanaka, and T. Ogawa, *Jpn. J. Appl. Phys.* **43**, L634 (2004) [180]. (Bottom) Schematic representations of electronic states of a single molecule. There is an energy gap of several eV between the HOMO and LUMO states. These states are broadened due to the contact to electrodes. Electric current flows between the left and right electrodes with peaks when the molecular states align with one of the chemical potentials. Peak and valley structures appear in the differential conductance due to these resonant tunneling phenomena.

reflects the energy gap of single molecule. This looks to explain the I–V characteristics observed in experiments.

5.5.2 *Transport Calculations of Single Molecule*

In this section, we show some examples of transport properties of single molecules. First, we consider the linear-response transport of the infinite-length molecular wires with no connection to electrodes and see the effects of geometrical configuration change on the transport. Second, we consider the relative Fermi energy dependence of the electrodes to the HOMO and LUMO states of molecules in mixing of molecular states with those of electrodes. As an example, we use the Bucky ball C_{60} for a single molecule. Third, we consider tunneling current in the HOMO–LUMO energy gap in the long molecular wires, where the Fermi energy position lies in the energy gap.

5.5.2.1 Poly-thiophene molecular wires

First, we see the behavior of linear-response conductance for the infinite 1D molecular wires. Molecular wires have flexible structures and we expect that small change of structural configurations would give significant influences on the transport properties. Here we choose poly-thiophene molecular wires, organic materials for various applications such as light-emitting diodes (LEDs), solar cells, and others.[a]

Figure 5.57 shows the atomic configurations (left) and corresponding conductances (right) for various geometrical configurations. Here we consider molecular wires with infinite chain lengths and use the NEGF method with localized basis sets with the self-energies $\Sigma_{L/R}$ constructed from the component of molecular wires. In this situation, since contact effects to electrodes are absent, the conductance shows plateau structures without rotation (0°), which is a characteristic feature for the transport of 1D structure. It should be noted that these plateaus count the number of bands of 1D wires. We see that the first plateaus for HOMO and LUMO states correspond to π-electron single bands with a gap of 1.8 eV.

Let us rotate one component of poly-thiophene wires and see how the transport properties are modified by the conformal change. As center component of thiophene molecule rotates with 30°, 60°, and 90° the conductances decrease significantly. Especially, the plateau structure for HOMO states is completely missing for the rotation with 90°. The total energy shown in the inset shows that the energy difference for the rotation is small $\sim$1 eV up to

[a]Here we do not take the phonon effects on the electronic states and transport properties into account.

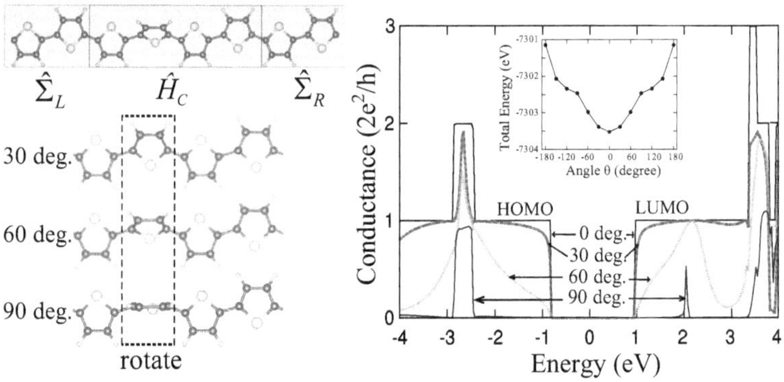

Figure 5.57 (Left) Schematic pictures of atomic configurations for poly-thiophene molecules. The center component shown in the dotted line rotates. (Right) Total transmissions for various angles θ of rotation in the center component of poly-thiophene molecular wires. (Inset) Total energy as a function of the angle θ.

90°, but larger than the room temperature. This means that by applying the local electric fields to one component of thiophene wires, the conductances can be changed significantly. This suggests the possibility to construct ultimately small switching devices by the conformal changes of molecular wires if we apply the atomic-scale gate voltage locally to poly-thiophene molecular wires.

5.5.2.2 Bucky ball

Second, as an example of single molecule, we present transport properties of Bucky ball C_{60} sandwiched between Al electrodes. Figure 5.58 (left) shows the atomic configuration for C_{60} between semi-infinite Al(111)4×4 electrodes with protrusions, which mimic microprobes. The electrode spacing is taken to be 10 Å and the centers of the five-membered ring of C_{60} are contacted to the electrodes. We use the NEGF method with localized basis set.

We see that there are number of peaks in the total transmissions shown in Fig. 5.58 (right), which indicate sharp resonant tunneling behaviors. The splitting in the degenerate states is clearly seen because of the breaking of symmetry due to the connection to the electrodes. The isolated C_{60} has the fivefold degenerate h_u HOMOs

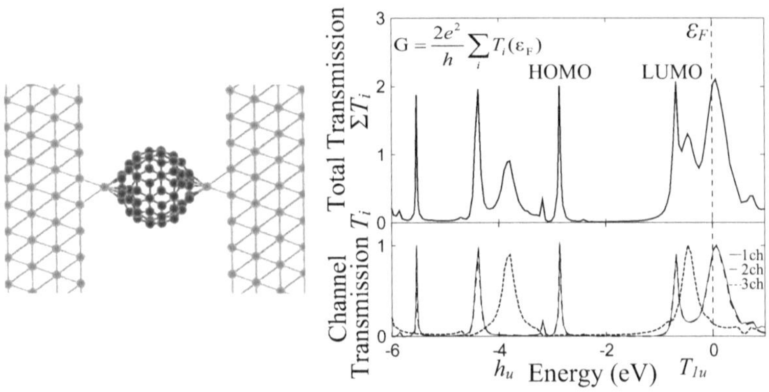

Figure 5.58 (Left) Schematic picture of atomic configuration for a Buck ball C_{60} sandwiched between semi-infinite Al electrodes. (Right) Total transmission (top) and eigenchannel transmission (bottom) of C_{60} between Al electrodes. From N. Kobayashi, T. Ozaki, and K. Hirose, *Jpn. J. Appl. Phys.* **45**, 2151 (2006) [181].

and the threefold degenerate t_{1u} LUMOs, but the degeneracy is lifted into two states. From the channel decomposition analysis, we see that three eigenchannels contribute to the transport at ϵ_F. Also, there are sharp peaks originating in individual molecular orbitals. The resonant widths at a higher energy are wider than those at a lower energy, since the wavefunctions of LUMOs spread out around the five-membered ring and the hybridization between the C_{60} molecule and the electrode is large.

It is important to note that the electron transfer occurs from Al electrodes to C_{60}, and thus the relative Fermi energy position is not in the energy gap but is located inside the conduction band (LUMOs). Therefore, C_{60} acts as a conducting molecule and electrons move through the states originating from the unoccupied molecular orbitals of C_{60}. The situation is close to the schematic picture in Fig. 5.56 (right). This indicates that the Bucky ball C_{60} acts as an acceptor for Al electrodes and works for an n-type material for electron transport. We note that C_{60} is an important material as an acceptor in the organic devices. The molecular orbitals contributing to the carrier transport depend on the interaction and the charge transfer between molecules and electrodes.

5.5.2.3 Tunneling of molecule wire

Next, we consider the transport through molecular wires between electrodes. When the Fermi energy ϵ_F of electrodes is located in the large insulating HOMO–LUMO gaps as characterized in organic molecules, the transport behavior should be governed by the tunneling process. This produces different characteristics from the metallic transport nature of the Al atomic wires in the previous section. The signature of off-resonant tunneling transport has an exponential dependence of $G = G_c e^{-\beta L}$ for the conductance G on the length L of the molecular wires. The tunneling exponent β is related to the complex band structure of the molecules.

The calculations are based on the supercell containing the phenyl-chain molecular wires sandwiched by the 12 layers of gold electrodes with the periodic boundary conditions, then the exponent β is derived. Figure 5.59 (left) shows the band structures

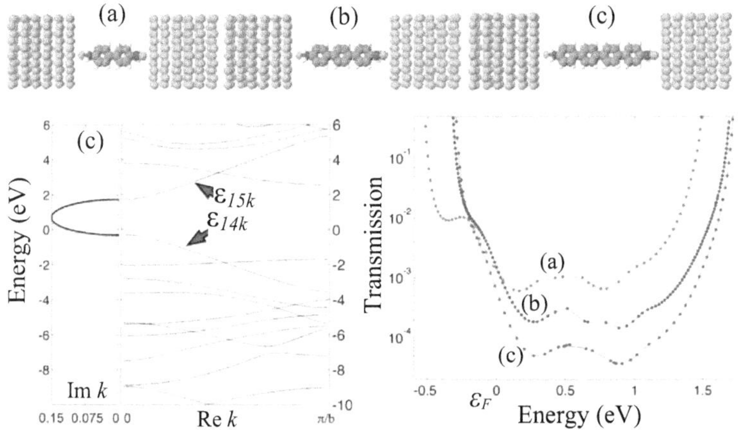

Figure 5.59 (Top) Schematic views of atomic configurations of the molecular wires between electrodes. (Bottom) (left) Real and complex band structures corresponding to the periodic potential for (c). Only the complex band with smallest Im[k] is shown. The unit for Im[k] is Å and the Fermi energy ϵ_F is set to zero The arrows indicate the energy bands, which determine the asymptotic value of the tunneling conductance. (Right) Plots of the transmission as a function of energy for (a), (b), and (c). From E. Prodan and R. Car, *Phys. Rev. B* **80**, 035124 (2009) [182].

corresponding to the four phenyl rings linked to gold electrodes via amine groups. We see the energy gap about 2 eV and the Fermi energy, which is set to zero, is located close to the edge of the valence band (HOMOs). Thus, the conducting states of electrodes decay rapidly to zero inside the phenyl chain in the tunneling regime. Figure 5.59 (right) shows the transmission as a function of energy. The conductances of the molecules composed of two, three, and four phenyl rings linked to gold electrodes at the Fermi energy are $G = 1.5 \times 10^{-3} G_0$, $G = 1.5 \times 10^{-3} G_0$, and $G = 4.3 \times 10^{-4} G_0$ where $G_0 = (2e^2/h)$. The tunneling exponent β, computed from the imaginary complex band $\beta = 2b\text{Im}[k_F]$ where b is the length of one phenyl ring, are equal to $\beta = 1.15$, $\beta = 0.98$, and $\beta = 0.98$ for two, three, and four ring cases. This shows that the tunneling regime is reached when the number of phenyl rings is more than three. The conductance for the two-phenyl-ring case is very close to the experimental value of $G = 1.16 \times 10^{-3} G_0$, but that for the three-phenyl-ring case is 8.3 times larger than the experimental value of $G = 1.8 \times 10^{-4} G_0$ [183].

We note that since the Fermi energy is located close to the band edge of the valence band, the magnitude of conductance is very sensitive to the alignment of the molecular level. This indicates that tunneling conductance for the molecular wire is determined not only from the bandgap but also from the band alignment relative to the electrodes. This sensitivity of transport properties to the contact conditions makes the desired molecular devices extremely difficult to construct.

5.5.3 *Contact Problem to Electrodes*

In the previous sections, we see that where the Fermi level of electrodes lies in the molecular states determines the carrier transport (n-type, p-type) or tunneling transport through the molecular energy gap. This is determined by the charge transfer between molecules and electrodes, which is very sensitive to the alignment of the molecular levels, terminal structures of molecules, and the geometry and conditions of contacts to electrodes. When we consider the situations of experiments, the control of the spacing of nanogap with atomic-scale accuracy is difficult, and thus it is not sure how the single molecules bridge the nanogap to form the atomic-scale junction. It might occur that the length of nanogap is not suitable for attaching molecules to both electrodes. Then we cannot know how well the chemical bonds are formed between them. Since the characteristics of electron transport might change significantly as the condition of contacts, this is known as the contact problem.

Ideally, we expect the resonant transport of currents through molecular states, where the measured currents should increase approximately by the quantized unit of $2e^2/h$ times an applied bias voltage ($\simeq 77.5 \times V$ μA) when the Fermi levels of electrodes ($\mu_L - \mu_R = eV$) align the molecular states.[a] However, the observed currents are usually much smaller than this value. This suggests that large resistances might exist between molecules and electrodes, probably at contacts. These observations lead us to the question of what determines the resistance in the nanometer-scale junction with single molecules between electrodes. Since the device performance as the operation speed is strongly affected by large resistance, this is an important problem.

5.5.3.1 Sensitivity of terminal structure

Here let us see how the terminal structures of the molecules affect the conductance through molecular bridges even when molecules have good contacts to both electrodes. Here we use the tight-binding method with parameters obtained from the electronic states based on the density functional theory (TB-DFT). We show the transmission spectra of the phenalenyl-like molecules as a prototype here.

Figure 5.60 (left) shows the molecular bridge structures where the phenalenyl-like molecules are attached to semi-infinite gold electrodes through the mercapto-vinyl groups. The difference between two systems in Fig. 5.60(a,b) lies in the positions of the connecting sites of the leads (mercapto-vinyl groups). Here α and β denote the second and third nearest sites measured from the center site of the molecule. As a candidate of the center atom X, we consider three atomic species of carbon (C), nitrogen (N), and boron (B)

[a] We neglect the energy dissipation inside the molecule for simplicity.

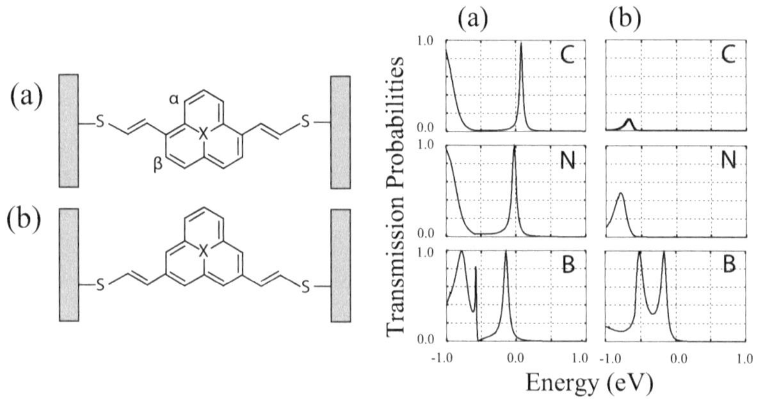

Figure 5.60 (Left) Schematic views of structures of phenalenyl-based molecular bridge, where $X = C$, N, or B. The mercapto-vinyl groups are attached to (a) α site and (b) β site. (Bottom) (Right) Transmission spectra when both leads attached to (a) α site and (b) β site. From M. Tsukada, K. Tagami, K. Hirose, and N. Kobayashi, *J. Phys. Soc. Jpn.* **74** 1079, (2005) [184].

atoms. Note that $X = C$ indicates that the unpaired electron exists in the molecule, but this does not mean that this electron is localized on the center site. As for the connection between mercapto-vinyl groups and gold electrodes, we set the S–Au bonds with the length of 2.55 Å.

Figure 5.60(a) (right) shows the transmission spectra when both leads are attached to the α sites. The origin of an electron incident energy ϵ is set to the Fermi level of the gold electrodes. In the case of $X = C$ and N, the transmission peaks appear at $\epsilon = 0.08$ eV and $\epsilon = -0.02$ eV close to the Fermi level, while three transmission peaks are observed at $\epsilon = -0.78, -0.58, -0.14$ eV between $-1.0 < \epsilon < 0.0$ eV in the case of $X = B$. On the other hand, when both leads are connected to β sites shown in Fig. 5.60(b)(right), the prominent peaks at around the Fermi level disappear in the case of $X = C$ and N. In contrast, there are strong peaks close to the Fermi level at $\epsilon = -0.52$ and -0.17 eV in the case of $X = B$. The origin of such a strong dependence of the transmission on the center atom or the connecting sites of leads is explained from the energy diagram of the free phenalenyl-like molecule. In the case of

$X = C$ and N, the transmission peaks close to the Fermi level is caused by the orbital in high-energy, which has amplitudes only on the α atomic sites. Therefore, if the leads are connected to β sites, this orbital does not contribute to the transmission and shows an absence of transmission peak. On the other hand, in the case of $X = B$, the transmission is carried by the orbitals in lower-energy, which has amplitudes both on the α and β sites. Thus, the system is conductive irrespective of the sites connected to the electrodes. Such a sensitivity of the conductance on the terminal sites has also been found in the tape-porphyrin molecules. These indicate that the matching of molecular states to the electronic states of electrodes at contact gives significant influence on the transport.

5.5.3.2 $I-V$ characteristics in the absence of molecules

Next, we consider the $I-V$ characteristics. Since we cannot observe directly the single molecule bridged between electrodes, we are not sure that the observed $I-V$ reflects the molecular states. Thus, we confirm here the characteristics of $I-V$ curve when a single molecule is missing between electrodes. Let us consider the junction system composed of Na atoms with an apex atom for each electrode to construct atomic-scale contacts and use the RTM method for electron transport based on the DFT formalism. The inset of Fig. 5.61 (top left) shows the schematic picture of the system with d the distance between electrodes measured from the equilibrium positions of the apex atoms. The current is dominated in the narrow region between the apex atoms, which creates the effective channel for electron transport. The linear-response conductance as a function of the distance d in Fig. 5.61 (top left) distinguishes the tunneling regime from the contact regime, as we see in the STM system. We note that $I-V$ characteristics show the linear Ohmic behavior in the contact regime due to the formation of effective channels. Thus, the current contribution should be proportional to the difference of chemical potentials, applied bias voltage, in each electrode. On the other hand, non-linear behaviors might appear in the $I-V$ curves in the tunneling regime.

Figure 5.61 (bottom) shows the $I-V$ behaviors for $d = 10$ bohr (left) and $d = 14$ bohr (right). We see that strong non-linear

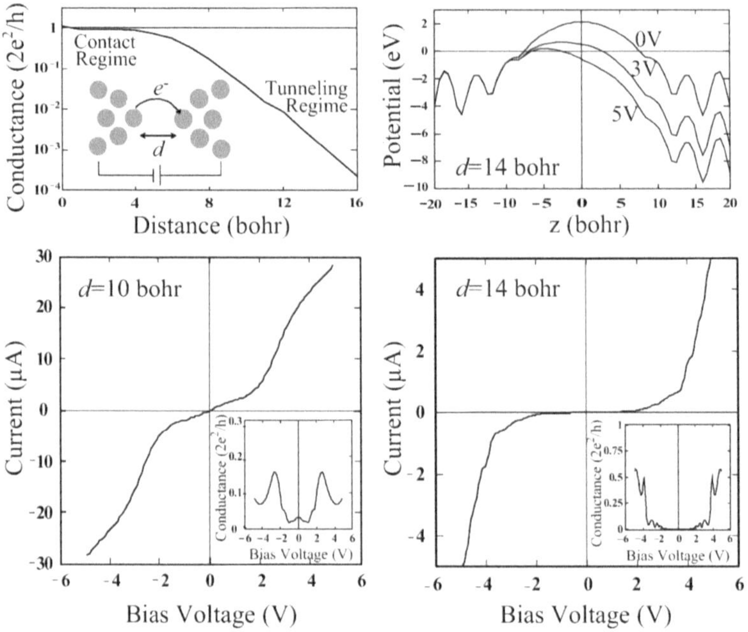

Figure 5.61 (Top) (Left) Linear-response conductance as a function of the distance d between electrodes measured from the equilibrium positions of the apex atoms. (right) Effective potential for $d = 14$ bohr for the applied bias voltages of 0, 3, and 5 V. (Bottom) (left) I–V characteristics with $d = 10$ bohr. The upper left inset shows the schematic picture of the system with d the distance measured from the equilibrium positions of the apex atoms. (right) I–V characteristics with $d = 14$ bohr. The upper left inset shows the effective potentials along the z direction are shown for the bias voltages of 0, 3, and 5 V cases. For both cases, the lower right insets show the differential conductance dI/dV as a function of applied bias voltage. From K. Hirose and N. Kobayashi, *Physica E* **29**, 515 (2005) [185].

behaviors appear as a distance d becomes large with the apex atoms at each electrode well separated. The differential conductances dI/dV shown in the insets show clearly valley structures around the zero bias regime. It is important to note that the non-linear I–V characteristics and the valley structures in the dI/dV behaviors are very similar to those observed in the conductance measurements for molecular-bridged systems. It is sometimes argued that the valley structures observed in the differential conductance dI/dV reflects the HOMO–LUMO gap of the molecule between electrodes. However, this calculation implies that the non-linear I–V characteristics and corresponding gap structure of dI/dV emerge even without a molecule between electrodes. The mechanism to appear the non-linear characteristics in the I–V curves and corresponding valley structures in the dI/dV curves even without the molecule with HOMO–LUMO gap is understood when we see the effective potentials along the z direction for several bias voltages. We see in Fig. 5.61 (top right) for $d = 14$ bohr that the effective energy barrier disappears when the applied bias exceeds 3 V. Correspondingly, the current changes its behavior from exponential to linear one around the bias voltages of ± 3 V. This shows that the non-linear behavior of I–V characteristics appears when the transport properties change from tunneling to ballistic regime. In other words, the energy level in the vacuum (the workfunction of electrodes), which is usually close to the HOMO–LUMO gap with several eV, makes it difficult to argue that we measure the conductance of single molecules only from the I–V characteristics, since we cannot see the molecules bridging between electrodes.

5.5.3.3 Contact effects on I–V characteristics of molecules

Here, we see the transport properties of single molecules sandwiched between electrodes. As we see in the previous section, the non-linear behaviors of I–V similar to the experimental observations appear even if the molecules are missing in the nanogap. We adopt the benzene dithiol (BDT) molecule as a prototype of single molecules and calculate its transport properties. Especially, we focus on the I–V characteristics when the single molecule has a chemical bond only to one side of the electrode and the other side of the contact is not well formed. For simplicity, we use the metallic jellium model as electrodes [186] and apply the RTM method for the transport properties of molecular junction systems.

Figure 5.62 (top left) shows the charge density profile of the benzene dithiole molecule connected to the jellium electrodes when the molecules have good contacts to both electrodes. Figure 5.62 (bottom) shows the I–V characteristics through the single molecule between electrodes. For the perfect contact case (left), we see that the I–V characteristic shows non-linear behavior, symmetric due to the geometrical symmetric structure of molecules. The differential conductance dI/dV in Fig. 5.62 (top right) shows the valley structure around the zero bias regime. Two peaks appear in conductance, which correspond to the resonant tunneling through π orbital state of the molecule. This shows that when the contacts to both electrodes are very good, we observe the molecular states in the differential conductance.[a] Next, we see how the atomic-scale contacts affect transport properties of single molecules. We consider the situations in which BDT molecule has good contact to one of the contacts, while the other contact to electrode is separate from the electrode. The distance from the edge of the jellium electrode is d as schematically shown in Fig. 5.62 (top center). Figure 5.62 (bottom right) shows the I–V characteristics with $d = 8$ bohr, where strong non-linear behavior is observed in the I–V curve. This looks a similar behavior for the perfect contact case in the left, thought the magnitude is of two orders smaller. Correspondingly, its differential conductance shows a valley structure around the zero bias.

It is important to note that this non-linear behavior of I–V curve has a different origin. When we look at the effective potential for the case of $d = 8$ bohr, a potential barrier is formed between the single molecule and the electrode, which dominates the transport process as tunneling. The mechanism for the appearance of valley structure in the differential conductance is the same as the previous section, where the effective potential barrier disappears as we increase bias voltages. This means that the resistance is determined not by the molecular states but by the contact to electrodes. We note that intriguing behavior of transport appears in I–V when the distance $d = 5$ bohr as shown in Fig. 5.62 (bottom center). We see that the current decreases even as an applied bias increases around 2 V. This type of anomalous I–V characteristic is sometimes observed in experiments [188, 189].

[a]The conductance at zero bias is larger than the experimental observation, which attribute to the contact condition [186] or due to the single-particle treatment based on the DFT, where the wavefunctions tend to expand as metallic nature.

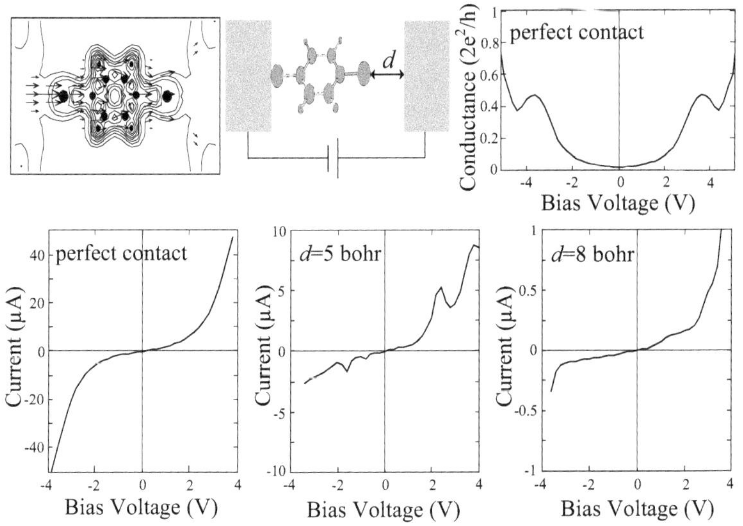

Figure 5.62 (Top) (Left) Electron density profile of the benzene dithiol (BDT) single molecule connected to jellium electrodes. Current density distribution is also shown. (Center) Schematic illustration of the molecular junction system. BDT molecule has a strong chemical bond only to the left electrode. The distance d is measured from the edge of the right jellium electrode. (Right) Differential conductance dI/dV when the single molecule has strong chemical bonds to both electrodes forming perfect contacts. (Bottom) $I-V$ characteristics through the single molecule between electrodes with perfect atomic-scale contacts (left), with $d = 5$ bohr (center), and with $d = 8$ bohr (right). From K. Hirose and N. Kobayashi, *Surf. Sci.* **601**, 4113 (2007) [187].

5.5.3.4 Detecting molecules in the nanogap

In the previous section, we show that the $I-V$ characteristics take similar non-linear curves even in the absence of single molecules between electrodes. This shows that the magnitude of the current I, which is usually small due to tunneling process through the HOMO–LUMO energy gap of molecules, is sensitive to the contact condition. Thus, it is difficult to convince whether the measured current I reflects the molecular states only from the $I-V$ data. The peaks of differential conductance dI/dV do not always correspond to the molecular states as resonances and the gap structure in the data of dI/dV curve does not always reflect the HOMO–LUMO energy gap of single molecules. In such situations, how can we confirm that the single molecules are bridged between electrodes. One approach to detect individual molecules in the nanogap is to observe the vibration signatures of molecules.

This experimental technique, called inelastic electron tunneling spectroscopy (IETS), has been developed after the STM was invented to observe the molecular states from the STM tunneling current. Tunneling electrons are excited and induce the energies due to the absorption of phonons from single molecules. The energies and intensities of the spectral features are characteristics of the chemical bonds and their interactions of single molecules with the surrounding. When the bias is increased and exceeds the threshold for the excitation energies of a vibrating mode, the inelastic tunneling channel opens with the change of the differential conductance dI/dV. Since this change is too small to detect, we need to observe the second derivative of the current d^2I/dV^2 as a function of the bias voltage, which exhibits peaks or dips corresponding to the single-molecule vibration spectra. Schematic picture of the emergence of inelastic tunneling current and its characteristics of I, dI/dV, and d^2I/dV^2 are shown in Fig. 5.63. It should be noted that the detection of the vibrating properties of molecules by IETS does not mean that the molecule is connected to both electrodes to form bridge structures. Figure 5.64 (top)

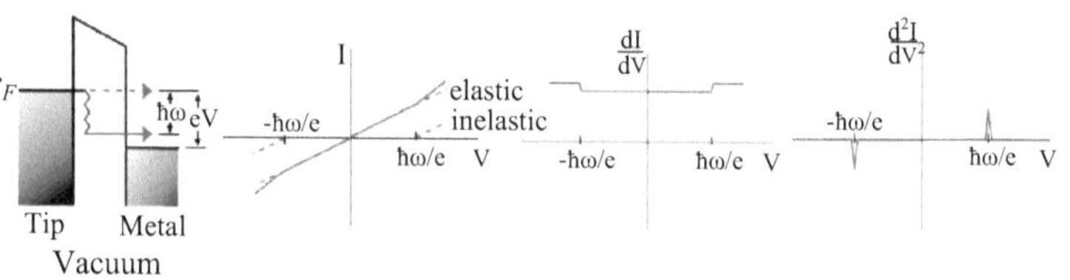

Figure 5.63 (Left) Schematic illustration of the emergence of inelastic tunneling at the threshold for vibrational excitation. (Right) Characteristics of the tunneling current I, dI/dV, and d^2I/dV^2. The change of tunneling current I due to vibrational excitation is too small to be measured from the $I-V$ curve in experiments. The vibrational features of molecules are extracted from the second derivative d^2I/dV^2 of the tunneling current. From W. Ho, *J. Chem. Phys.* **117**, 11033 (2002) [190].

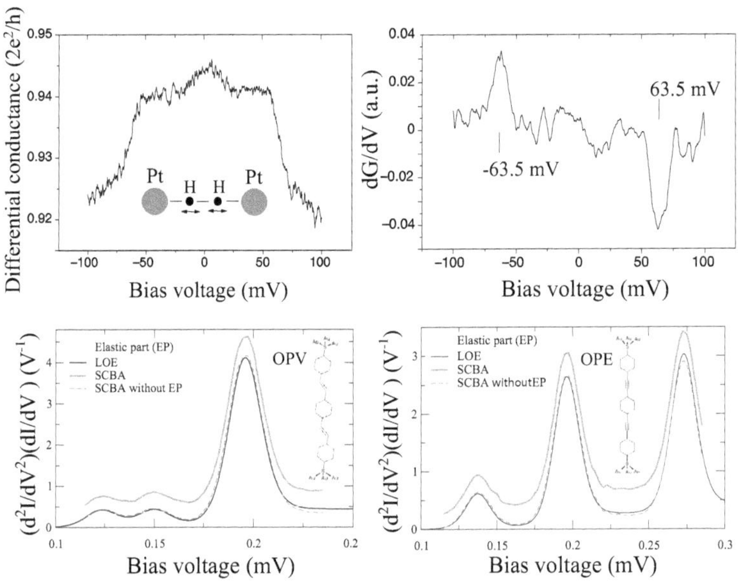

Figure 5.64 (Top) Experiments of (left) differential conductance and (right) its derivative for a Pt/H_2 contact taken at a conductance plateau close to $2e^2/h$. (Bottom) Calculations of IETS spectra for (left) an OPV molecule and (right) an OPE molecule with various approximation schemes. The chemical structures are shown in the insets. Experiments from R. H. M. Smit, Y. Hoat, C. Untiedt, N. D. Lang, M. C. van Hemert, and J. M. van Ruitenbeek, *Nature* **419**, 906 (2002) [191] and calculations from T. Frederiksen, M. Paulsson, M. Brandbyge, and A.-P. Jauho, *Phys. Rev. B* **75**, 205413 (2007) [193].

shows an example of experimental observations of (left) differential conductance and (right) its derivative for the hydrogen molecule H_2 between Pt electrodes, taken at a $2e^2/h$ plateau, constructed by the mechanically controllable break junction technique. This is based on the point contact spectroscopy, similar to IETS. Compared with the isotope molecules D_2 and HD, the observed spectra are confirmed from the vibration of the hydrogen molecules. Since the hydrogen molecule is bonded to the Pt surface, this shows that the stable molecular bridge is formed between the metallic electrodes.

The computational schemes for the observed IETS spectra of molecules have also been developed [192, 193]. Figure 5.64

(bottom) shows the calculations of the IETS spectra for the hydrocarbon molecules of (left) OPV and (right) OPE between Au electrodes based on the NEGF formalism, including electron–phonon coupling. Since the number of molecules in the nanogap is unknown, the computational spectra are shown for the renormalized values as IETS $= (d^2I/dV^2)/(dI/dV)$. Note if the current I scales with the number of molecules, if the molecule is present in the nanogap junction, the IETS spectra is independent of the number of molecules. Compared with experiments, the peaks of IETS spectra of an OPE molecule (right) are specified to correspond to C–S breathing mode (first peak), ring mode (second peak), C $\equiv$ C mode (third peak) and reproduce the positions and relative heights of IETS experiments. The inexpensive scheme in the lowest-order expansion (LOE) captures the inelastic current for the IETS spectra, compared with the more elaborate scheme in self-consistent Born approximation (SCBA). This shows that the electron–phonon interactions of organic molecules connected to electrodes are weak, except for the sharp resonance case weakly coupled to electrodes.

5.6 Organic Semiconductors

5.6.1 *Weakly Bonded Molecular Systems*

Organic semiconductors are crystals composed of organic materials such as organic oligomers, polymers, and small organic molecules (typically two to 10 benzene rings).[a] The characteristics of these organic materials are their flexibility, since they are formed in a solid by weak van der Waals interaction. Recently much attention has rapidly been paid to the development of the field of organic electronics, which are expected to grow as one of the candidates for future electronic-device applications with a structural flexibility and large area coverage using inexpensive printing processes. For example, representative applications include FETs, LEDs, electronic paper, and solar cells. The recent technological developments

[a]It can be regarded as the assemblies of π-conjugated single molecules with weak van der Waals interaction.

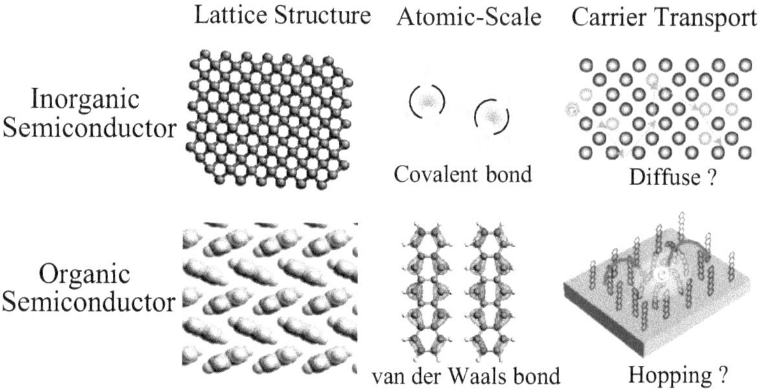

Figure 5.65 Difference of inorganic semiconductors such as silicon and GaAs from organic semiconductors in view of lattice structures (left), atomic bondings (center), and carrier transport (right).

in the organic electronics require us to obtain much better fundamental understanding of the nature of their charge-transport mechanisms [194].

Figure 5.65 represents the typical characteristics of the inorganic semiconductors such as silicon and the organic materials. Different from inorganic materials, which are held together with strong covalent bonds, organic materials are formed with very weak van der Waals forces between molecules. For typical organic semiconductors, the intermolecular bonding energies, called transfer integrals, are small in the range $10 - 10^2$ meV to form the narrow electronic bands, which is comparable to the depth of carrier-trap potentials by the static structural disorders, impurities, and grain boundaries. Therefore, the charge carriers are strongly scattered and are localized in a molecule, creating a quasi-particle state called polaron as a result of the strong interaction with intramolecular vibrations of the local lattice environment.

5.6.1.1 Experiments on electron transport measurements

To consider carrier transport properties of organic semiconductors, the understanding of polaronic state is very important since the binding energy of polaron (reorganization energy) has similar energetic orders of transfer integrals and thermal excitation energy $k_B T \simeq 25$ meV at room temperature. In fact, the incoherent molecular-to-molecular hopping of polaron, which is a slow moving electron coupled with lattice distortion, is observed in experiments as a thermally activated behavior of temperature-dependent mobility, where the carrier mobility is generally below 0.1 cm^2/Vs at room temperature. In such a low mobility regime, the charge-transport properties have been investigated theoretically using the Marcus equation, taking polaron effects by the Holstein model, which describes the strong coupling between the intramolecular vibration and the electron.

Recent rapid progress of the technology enables us to fabricate the single-crystal organic thin films and to construct the flexible thin-film transistor (TFT) devices with high carrier mobility up to ~ 40 cm^2/Vs, which presently exceeds the mobility of amorphous silicon. This shows that structural disorders are almost excluded from the organic TFTs. In Fig. 5.66 (bottom), we see that the temperature dependence of mobility is remarkably different from the disordered materials. In single crystals in Fig. 5.66 (right), the mobility decreases with increasing temperature as $\mu \propto T^{-n}$ [198]. Such power-law dependence is a typical character of coherent band transport due to delocalized carriers and originates from the scattering processes by the coupling between carriers and intermolecular vibrations, i.e., the lattice phonons. Moreover, recent measurements of Hall effects on the organic TFTs provide us with the evidence of the possible coherent charge transport in single-crystal organic materials [199]. On the other hand, a problem in the coherent band picture arises, that is, the estimated MFP is comparable to or shorter than the distance between adjacent molecules, which imply a breakdown of the coherent band transport.

5.6.1.2 Large and small polaron problems

The molecules in organic materials are coupled to each other by the weak Van der Waals interactions, which make these materials much more flexible than inorganic materials. The carrier motions in the organic materials are affected strongly by the lattice distortions, which form the so-called polaronic state, quasi-particle state

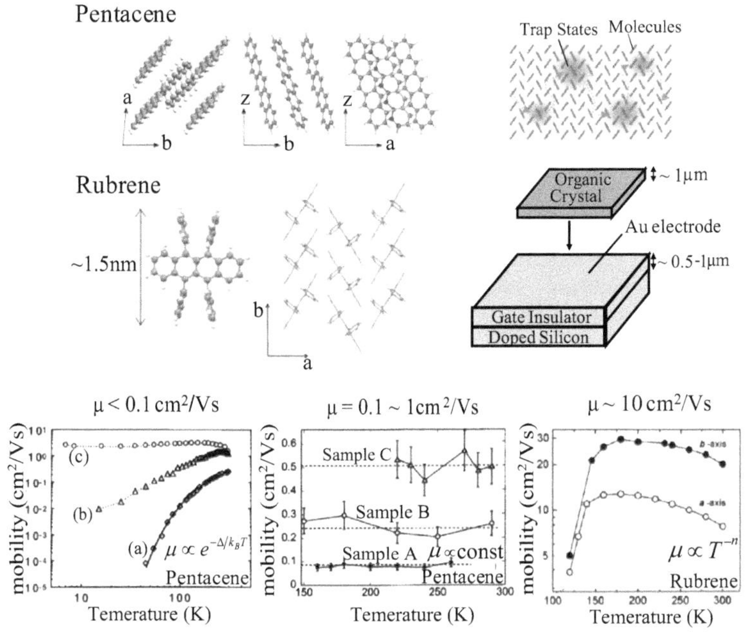

Figure 5.66 (Top) (Left) Molecular and crystal structures of pentacene and rubren. (Right) Schematic illustration of thin-film transistors (TFTs). From T. Hasegawa and J. Takeya, *Sci.Technol.Adv.Mater.* **10**, 024314 (2009) [195]. (Bottom) Comparison of temperature-dependent behaviors of carrier transport from slow ($\mu < 0.1$ cm/Vs with $\mu \propto e^{-\Delta/k_BT}$) to fast ($\mu \sim 10$ cm/Vs with $\mu \propto T^{-n}$) systems. From (left) S. F. Nelson, *et al.*, *Appl. Phys. Lett.* **72**, 1854 (1998) [196], (center) J. Takeya, *et al.*, *J. Appl. Phys.* **84**, 5800 (2003) [197], and (right) V. Podzorov, *et al.*, *Phys. Rev. Lett.* **93**, 086602 (2004) [198].

coupled with electron and molecular vibrations. Therefore, the charge-carrier transport must be described using different models from those for covalently bonded semiconductors. Theoretical studies tackle these transport problems on the single-crystal organic materials with high mobility from an atomistic viewpoint.

For the intermolecular coupling (Peierls model as large polaron), the time-dependent Schrödinger equation for electron is solved combined with the classical equation for molecular motion to take the polaron effects, which describes the strong coupling between the intermolecular vibration and the electron [200]. It is shown that large thermal fluctuations of molecular motion are sufficient to destroy the translational symmetry of the molecular lattice and that the carrier state becomes sometimes a delocalized band-like and sometimes a strongly localized one. The simultaneous presence of band carriers and incoherent states is also found in the Peierls model [201]. The investigations of polaron dynamics in the molecular lattice applied by a finite magnitude of electric field within the Peierls model show the change of transport process from an adiabatic polaron drift process to a combination of sequences of adiabatic drift and non-adiabatic hopping events [202].

On the other hand, the intramolecular coupling (Holstein model as small polaron) cannot be ignored to account for the transport properties of single-crystal organic materials because the Holstein-type polaron has a large binding energy and describes the charge transport in the localized hopping regime due to static disorders, which inevitably exist in the molecular crystals. The analytical evaluations of mobility by using the canonical transformation method for the mixed Holstein–Peierls model show the contribution of both the coherent and incoherent scattering events on the transport properties [203].

5.6.2 *Transport with Polaronic States*

To investigate the carrier transport properties of single-crystal organic semiconductors around the room temperature, let us extend the time-dependent wave-packet diffusion (TD-WPD) method and apply it for the mixed Holstein–Peierls model to study the temperature dependence of carrier transport. With this approach, since we perform the quantum-mechanical calculations of electron wave-packet dynamics with a combination of the classical molecular-dynamics simulation, we can evaluate the carrier mobilities, MFPs, and diffusion coefficients, including both the Holstein-type and the Peierls-type polaronic states on the same footing based on the Kubo formula. Here we study the pentacene organic semiconductors and show their temperature dependence of transport properties, especially focusing on the competition among thermal-fluctuation of molecular motion, polaron formation, and static disorders.

5.6.2.1 TD-WPD methodology for organic materials

To clarify the charge transport properties from an atomistic viewpoint, we adopt a semi-classical approach for phonons assuming that the nuclear motion of the lattice is treated classically via Born-Oppenheimer approximation, whereas the charge carrier dynamics is evaluated purely based on the quantum mechanical approach. This assumption has been frequently taken in the studies of conductive polymers. The electron motion in materials is described directly by coupling the quantum-mechanical time evolution calculation of an electron wave packet with the classical molecular lattice dynamics simulations as the TD-WPD approach. We note that in the previous sections, we have applied the TD-WPD method to inorganic materials, such as CNTs and graphene nanoribbons, which are formed by the covalent bondings.

We extract the time-dependent form of the electron conductivity σ of materials with the volume Ω from the Kubo formula

$$\sigma = \lim_{t\to+\infty} \int_{-\infty}^{+\infty} e^2 \left(-\frac{df(\epsilon)}{d\epsilon}\right) \mathrm{Tr}\left[\frac{\delta(\epsilon - \hat{H}_e)}{\Omega} \cdot \frac{(\hat{x}(t) - \hat{x}(0))^2}{t}\right] d\epsilon, \tag{5.80}$$

where $\mathrm{Tr}[\delta(\epsilon - \hat{H}_e)/\Omega] \equiv \nu(\epsilon)$ is the DOS. The dynamical change of electronic states by the lattice distortion due to the polaron formation is included in the time-dependent Hamiltonian $\hat{H}_e(t)$.

The mobility of a carrier with the elementary charge q is defined as the conductivity of Eq. (5.80) divided by the charge density

$$\mu = \frac{\sigma}{qn}, \tag{5.81}$$

where the charge density is defined as $qn = q\int d\epsilon f(\epsilon - \epsilon_F)\nu(\epsilon)$. When the diffusion coefficient D is evaluated as the long-time limit of the time-dependent diffusion coefficient

$$D = \lim_{t\to+\infty} \frac{1}{t} \frac{\int d\epsilon f(\epsilon)\mathrm{Tr}[\delta(\epsilon - \hat{H}_e)\,(\hat{x}(t) - \hat{x}(0))^2]}{\int d\epsilon f(\epsilon)\mathrm{Tr}[\delta(\epsilon - \hat{H}_e)]}, \tag{5.82}$$

we can extract the well-known Einstein relation in Chapter 2:

$$\mu = \frac{qD}{k_B T}. \tag{5.83}$$

Here we approximate the Fermi–Dirac distribution function as $f(\epsilon - \epsilon_F) \simeq \exp\{-(\epsilon - \epsilon_F)/k_B T\}$, since we consider the low carrier states of the semiconductor. The MFP is also obtained from the diffusion coefficient $\ell_{\mathrm{mfp}} = D/v$, where v is the carrier velocity defined by $v = \lim_{t\to 0}\sqrt{D(t)/t}$.

Using the real-time molecular dynamics simulations, we describe the dynamical lattice distortions both by the thermal fluctuation and by reorganization upon ionization of charge carriers. When we employ the generalized coordinate system $\{R\}$ to describe the lattice distortion, the equation of motion for the n-th site with the mass M is derived from the following canonical equation, $Md^2R_n/dt^2 = -\partial E_{tot}/\partial R_n$, where the total energy E_{tot} is defined by the sum of the electron and molecular vibration energies. Here, for instance, R_n represents the displacement of the n-th molecular position ΔR_n and the intramolecular distortion Δu_n. In general, the total energy is divided into four terms,

$$E_{tot} = E_e(\{\rho\}) + E_l^{\mathrm{pot}}(\{R\}) + E_l^{\mathrm{kin}}(\{\dot{R}\}) + E_{el}(\{\rho \cdot R\}), \qquad (5.84)$$

where E_e, E_l^{pot}, E_l^{kin} and E_{el} represent the electron energy, elastic potential energy, kinetic energy of the lattice vibration, and electron–phonon coupling energy. Then we obtain the following equation of motion for the lattice vibration and distortion:

$$M\frac{d^2R_n}{dt^2} = -\frac{\partial E_l^{\mathrm{pot}}(\{R\})}{\partial R_n} - \frac{\partial E_{el}(\{\rho \cdot R\})}{\partial R_n}, \qquad (5.85)$$

where the first term in the right-hand side corresponds to the elastic force and the second term induces the lattice distortion due to the polaron formation. The temperature T is fixed by normalizing the kinetic energy of lattice vibration at each time step on the condition of $\sum_{n=1}^{N} M\dot{\mathbf{R}}_n^2/2 = dNk_BT/2$ in the d-dimensional systems. Solutions to the coupled equations (5.80) and (5.85) allow us to describe the electron motion strongly coupled with both the thermal fluctuating lattice and the distortion by the polaron formation. The flowchart of the calculations is shown in Fig. 5.67.

5.6.2.2 van der Waals interaction from ab initio calculation

We extract the parameters for the transport properties of organic semiconductors, such as electron–phonon coupling constants and elastic constants from the calculations based on the DFT, including

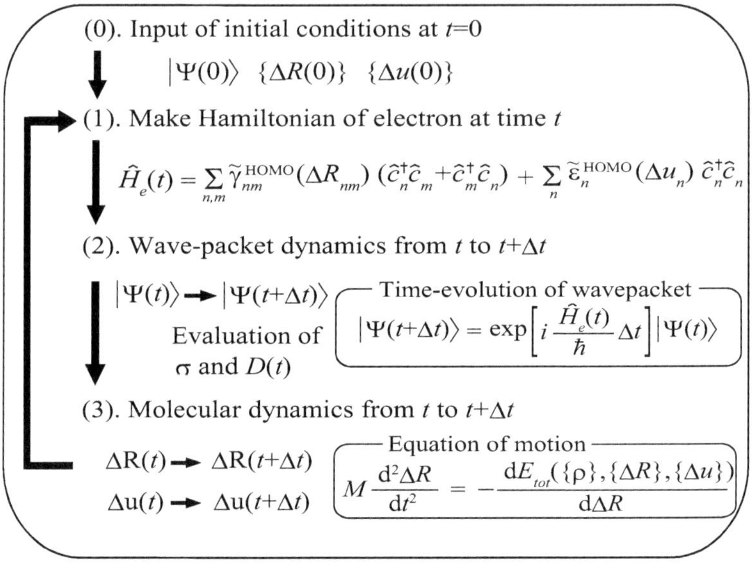

Figure 5.67 Flowchart of numerical computation according to the TD-WPD methodology in case of the Holstein–Peierls model.

van der Waals interactions for the energy. Here we adopt the functional form called the DFT-D approach [204]. The recent developments for introducing van der Waals interaction energy due to the dipole-dipole interaction $\sim -1/R^6$ is well described by the formula as

$$E_{\text{disp}} = -s_6 \sum_{i=1}^{N_{at}-1} \sum_{j=i+1}^{N_{at}} \frac{C_6^{ij}}{R_{ij}^6} f_{\text{dmp}}(R_{ij}), \qquad (5.86)$$

which are shown to be accurate for the molecular systems. Here N_{at} is the number of atoms in the systems, C_6^{ij} are the dispersion coefficients for atom pair ij, s_6 is the scaling factor, and $f_{\text{dmp}}(R) = 1/(1 + e^{-\alpha(R/R_0-1)})$ is the sum of atomic van der Waals radii.

One of the ways to extract the effective transfer integrals γ is to use the Lödin's symmetric transformation method from the DFT-D calculations. When we consider $\gamma^{\text{HOMO(LUMO)}}$, we have HOMO (LUMO) and HOMO-1 (LUMO+1) orbitals Ψ_i from the interaction of two molecules with a distance d and an angle θ, including the effects from other environmental molecules, the effective orbital energies are described from the 2 × 2 secular equation of $\det|\bar{H} - \varepsilon\bar{S}| = 0$. Here $\bar{H}$ and $\bar{S}$ are expressed by

$$\bar{H} = \begin{pmatrix} e_1 & J_{12} \\ J_{12} & e_2 \end{pmatrix} \qquad \bar{S} = \begin{pmatrix} 1 & S_{12} \\ S_{12} & 1 \end{pmatrix} \tag{5.87}$$

with the matrix elements of $e_i = \langle\Psi_i|\bar{H}|\Psi_i\rangle$, $J_{ij} = \langle\Psi_i|\bar{H}|\Psi_j\rangle$, and $S_{ij} = \langle\Psi_i|\Psi_j\rangle$. The matrix elements of e_i and J_{ij} have the same physical meaning of the parameters of the on-site energy and the transfer integral, although these two sets are not identical. According to Lödin's symmetric transformation, we have orthogonal basis sets, which maintain the initial character of the orbitals as much as possible. Then the effective Hamiltonian becomes

$$\bar{H}^{\text{eff}} = \begin{pmatrix} e_1^{\text{eff}} & \gamma \\ \gamma & e_2^{\text{eff}} \end{pmatrix} \tag{5.88}$$

with

$$e_{1,2}^{\text{eff}} = \frac{1}{2}\frac{(e_1 + e_2) - 2J_{12}S_{12} \pm (e_1 - e_2)\sqrt{1 - S_{12}^2}}{1 - S_{12}^2},$$
$$\gamma = \frac{J_{12} - \frac{1}{2}(e_1 + e_2)S_{12}}{1 - S_{12}^2}. \tag{5.89}$$

In this way, the secure equation becomes the standard form and the resulting energy splitting becomes $\Delta\varepsilon_{12} = \sqrt{(e_1^{\text{eff}} - e_2^{\text{eff}})^2 + 4\gamma^2}$. The data for the evolution of γ would be useful for evaluating the charge transport of organic semiconductors, where the molecular structures are flexible and change variously.

Here we obtain the transfer integrals between pentacene molecules as $\gamma^{\text{HOMO}} = 75$ meV and $\gamma^{\text{LUMO}} = -125$ meV. The HOMO and LUMO energy levels of an isolated pentacene molecule are taken as $\varepsilon^{\text{HOMO}} = -4.60$ eV and $\varepsilon^{\text{LUMO}} = -2.384$ eV. The Peierls-type elastic constant can be estimated as $K_P = 2.19$ eV/Å^2 by fitting the computed bond length-dependent total energy $U(\Delta R)$ with the parabolic form $U = K_{\text{P}}(\Delta R)^2/2$ around the equilibrium molecular position. The intermolecular vibration has the continuous phonon-band structure, $\hbar\omega_P(q) = \hbar\sqrt{2K_P(1 - \cos qa)/M}$ with the bandwidth of 3.8 meV with the bond length $a = 5.16$ Å. The phonon dispersion is in good agreement with an acoustic phonon branch of oligo-acene crystals. By coupling the wave-packet dynamics with the classical molecular dynamics, we can take various phonon modes from $q = 0$ to $\pm\pi/a$ thermally excited at a finite temperature to study the realistic phonon-scattering effects on the charge transport properties. The Holstein-type electron–phonon coupling and the elastic constant are $\alpha_{\mathrm{H}}^{\mathrm{HOMO}} = -93$ meV and $K_{\mathrm{H}} = 92$ meV, where we employ the dimensionless-effective-scalar coordinate Δu_n as the intramolecular distortion. When the effective coordinate Δu is equal to 1, the relaxation energy defined by $K_H \cdot (\Delta u)^2$ and the binding energy of Holstein-type polaron are in good agreement with those reported by other theoretical studies.

5.6.2.3 Application to the Peierls–Holstein model

To investigate the charge-carrier transport of organic semiconductors, we employ the mixed Holstein–Peierls model for the carriers interacting tightly with molecular distortion. The Holstein model is based on the local electron–phonon coupling, which acts purely as intramolecule, i.e., at the single molecule ionized by the charge of carrier. On the other hand, intermolecular vibration influences the time dependence of transfer integrals between adjacent molecules. The resulting nonlocal coupling leads to the Peierls model. In Fig. 5.68(b), we show the schematic picture of the hole transport in a single-crystal organic semiconductor.

The Hamiltonian for a hole motion in the highest occupied molecular orbital (HOMO) band is written as

$$\hat{H}_e(t) = \sum_{n,m} \tilde{\gamma}_{nm}^{\mathrm{HOMO}}(\Delta R_{nm}(t)) \cdot \left(\hat{c}_n^{\dagger}\hat{c}_m + \hat{c}_m^{\dagger}\hat{c}_n\right) + \sum_{n} \left\{\tilde{\varepsilon}_n^{\mathrm{HOMO}}(\Delta u_n(t)) + W_n\right\} \hat{c}_n^{\dagger}\hat{c}_n, \tag{5.90}$$

where $\hat{c}_n^{\dagger}$ and $\hat{c}_n$ are creation and annihilation operators of a hole at the HOMO with the orbital energy $\varepsilon_n^{\mathrm{HOMO}}$. The transfer integral $\tilde{\gamma}_{nm}^{\mathrm{HOMO}}$ and the orbital energy $\tilde{\varepsilon}_n^{\mathrm{HOMO}}$ coupled with the inter- and intramolecular lattice distortions are defined by

$$\begin{aligned} \tilde{\gamma}_{nm}^{\mathrm{HOMO}}(\Delta R_{nm}) &= \gamma_{nm}^{\mathrm{HOMO}} + \alpha_{\mathrm{P}}^{\mathrm{HOMO}} \cdot \Delta R_{nm} \\ \tilde{\varepsilon}_n^{\mathrm{HOMO}}(\Delta u_n) &= \varepsilon_n^{\mathrm{HOMO}} + \alpha_{\mathrm{H}}^{\mathrm{HOMO}} \cdot \Delta u_n, \end{aligned} \tag{5.91}$$

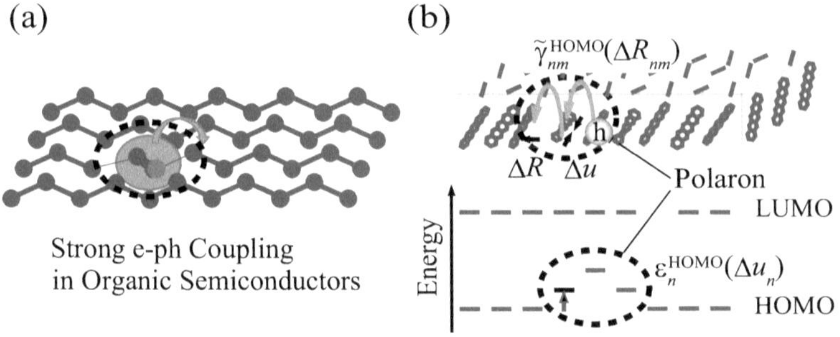

Figure 5.68 (a) Schematic picture of electron–phonon coupling in organic semiconductor systems to form a polaronic state due to the deformation of the molecular structures. (b) The hole transport in organic semiconductors with the Holstein-type and the Peierls-type couplings. (top) Transfer energy between two molecules is modulated from γ^{HOMO} to $(\gamma^{\mathrm{HOMO}} + \alpha_{\mathrm{P}}^{\mathrm{HOMO}} \Delta R)$ by the bond-length change ΔR due to both the thermal fluctuation and the polaron formation induced by the Peierls-type coupling. (Bottom) HOMO and LUMO levels of each molecule in the organic semiconductor. The intramolecular deformation by the reorganization, which is represented by Δu in the upper panel, changes the HOMO level through the Holstein-type coupling.

where $\gamma_{nm}^{\mathrm{HOMO}}$ is the transfer integral between n-th and m-th molecules. Here $\alpha_{\mathrm{P}}^{\mathrm{HOMO}}$ is the Peierls-type electron–phonon coupling constant. When we represent the displacement of n-th molecule as ΔR_n, the change of intermolecular distance is given by $\Delta R_{nm} = |(\mathbf{R}_m^e + \Delta\mathbf{R}_m) - (\mathbf{R}_n^e + \Delta\mathbf{R}_n)| - |\mathbf{R}_m^e - \mathbf{R}_n^e|$ where $\mathbf{R}_n^e$ is the equilibrium position of the molecule. On the other hand, $\alpha_{\mathrm{H}}^{\mathrm{HOMO}}$ is the Holstein-type electron–phonon coupling constant. When we replace the intramolecular displacement $\Delta\mathbf{u}_n$ with the effective scalar coordinate Δu_n, we can express the shift of orbital energy by $\{\alpha_{\mathrm{H}}^{\mathrm{HOMO}} \cdot \Delta u_n\}$. In addition to these polaron effects, we also consider the effects of static disorder, which inevitably exists in the molecular crystals. We introduce the disorder potentials W, which modulate the on-site orbital energies randomly within the width $[-W/2, +W/2]$.

As for the MD simulations for the dynamical lattice distortions, the elastic energies of both the inter- and the intramolecular deformations are approximated by a harmonic form. Then we obtain the energies of molecular lattice as

$$E_{tot} = \sum_{n} \frac{1}{2} M \left\{ \frac{d\Delta R_n(t)}{dt} \right\}^2 + \sum_{n,m} \frac{1}{2} K_{\mathrm{P}} \left\{ \Delta R_{nm}(t) \right\}^2 + \sum_{n} \frac{1}{2} K_{\mathrm{H}} \left\{ \Delta u_n(t) \right\}^2, \tag{5.92}$$

where the inter- and intramolecular elastic constants represent K_{P} and K_{H}. The intramolecular distortion Δu_n is determined by the variational principles for the minimum total energy $\partial E_{tot}/\partial \Delta u_n = 0$, resulting in $\Delta u_n(t) = (\alpha_{\mathrm{H}}/K_{\mathrm{H}}) \cdot \{\rho_{nn}(t)/\sum_m \rho_{mm}(t)\}$ with the density matrix ρ_{nm}.

5.6.3 *From Localized to Extended Regime*

5.6.3.1 Polaron formation energy

Let us see the electron–phonon coupling effect on the polaron formation of single-crystal organic semiconductors without both thermal fluctuations and static disorders. The polaronic state is achieved by self-consistent calculations to minimize the total energy E_{tot} with respect to the intra- and the intermolecular displacements, i.e., Δu_n and ΔR_n, under the condition to keep the length of the molecular lattice fixed.

5.6.3.1.1 Holstein-type coupling effects Here, we focus on the Holstein-type electron–phonon coupling, which describes the interaction between the electron and intramolecular distortion inducing the energy shift of the HOMO levels. A discrete polaronic state is observed in the energy spectrum of DOS located above the HOMO band whose binding energy E_B is evaluated as 14.3 meV, the same order of transfer integral γ^{HOMO} and the thermal excitation energy $k_B T$. This implies that the Holstein-type coupling affects the carrier transport properties through the polaron formation. In Table 5.2, we present the calculated polaron binding energies for three different models; the Holstein–Peierls model, the Peierls model with $\alpha_P \neq 0$ and $\alpha_H = 0$, and the Holstein model with $\alpha_P = 0$ and $\alpha_H \neq 0$. The Holstein–Peierls model gives rise to a much larger binding energy, which indicates that both the Holstein-type and the

Table 5.2 Polaron binding energies E_B by various models

	Peierls	Holstein	Peierls–Holstein
Binding energy	1.3 meV	7.5 meV	14.3 meV

Peierls-type electron–phonon coupling should be taken into account simultaneously.

5.6.3.1.2 Peierls-type coupling effects and thermal fluctuation Then we consider the Peierls-type electron–phonon coupling effect on the transfer integrals, which describes the interaction between the electron transfer integral and the intermolecular vibration called the lattice phonon. We can observe the formation of the Peierls-type large polaron from the shrink of the molecular bond. On the other hand, since the organic semiconductor devices are operated at room temperature, it is essential to consider how the thermal fluctuations of molecular motions, as well as the polaron formation, affect the transfer integrals. It is found that the thermal lattice vibrations give rise to a large dynamic disorder in the intermolecular transfer integrals [205]. These results indicate that the charge-transport properties of organic semiconductors are determined in the subtle balance of energies among the thermal-fluctuation of molecular motion, the thermal excitation, and polaron formation.

5.6.3.2 Polaron effects on intrinsic charge transport

First, we study intrinsic carrier transport for single-crystal organic semiconductors using the time evolution of wave-packets. The charge transport is determined by the coupling with inter- and intra vibration of polaron characters. This calculation would provide us the possible maximum value of the mobility.

Figure 5.69 shows the carrier mobility μ as a function of temperature T. The mobility decreases monotonically with increasing temperature approximately by the power-law dependence, which shows an apparent evidence of the band-like transport. This indicates that the intrinsic mobility reaches about 100 cm^2/Vs around the room temperature and increases up to 300 cm^2/Vs with decreasing

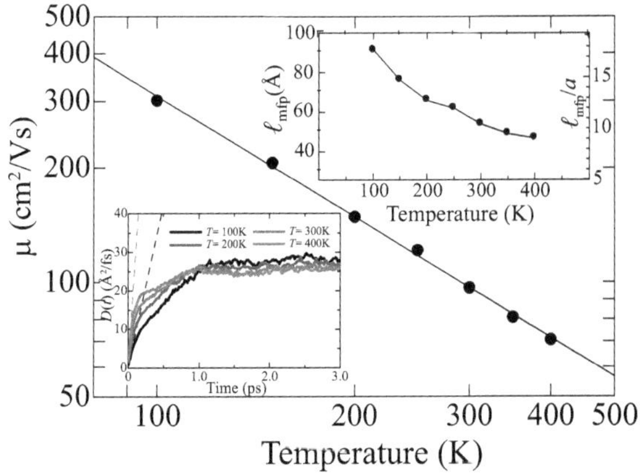

Figure 5.69 Intrinsic carrier mobility μ as a function of temperature in case of no static disorders. (Lower left) Time-dependent diffusion coefficients $D(t)$ for several temperatures from 100 to 400 K. (Upper right) ℓ_{mfp} as a function of T. The ratio of ℓ_{mfp} to a is shown in the right axis. From H. Ishii, K. Honma, N. Kobayashi, and K. Hirose, *Phys. Rev. B* **85**, 245206 (2012) [205].

temperature to 100 K. From $D(t)$ shown in Fig. 5.69(lower left), we see that the intrinsic carrier motion in the single-crystal organic semiconductors has a typical diffusive transport character. Figure 5.69(upper right) shows the temperature dependence of the MFP ℓ_{mfp}, which decreases monotonically as the temperature increases $\ell_{\mathrm{mfp}} = v\tau \propto v/\left\langle(\Delta R_{nm})^2\right\rangle \propto v/(k_B T)$. This is due to the carrier scatterings induced by the thermal fluctuation of molecular motion through the Peierls-type coupling. If ℓ_{mfp} is equivalent to or shorter than the lattice constant a (distance between adjacent molecules) as $\ell_{\mathrm{mfp}} < a$, then the carrier takes the hopping transport process, while the band-like transport process is characterized as ℓ_{mfp} much longer than a such as $\ell_{\mathrm{mfp}} \gg a$. The ratio of ℓ_{mfp} to a is plotted in the right axis. We see that the MFP reaches 10 times longer than the lattice constant around the room temperature, supporting the possibility of the band-like behavior in the intrinsic transport when the static disorders are absent.

Let us see the polaronic behavior for the herringbone structure with the anisotropy of transfer integrals. Here, for simplicity, we approximate van der Waals interaction by the Lennard–Jones

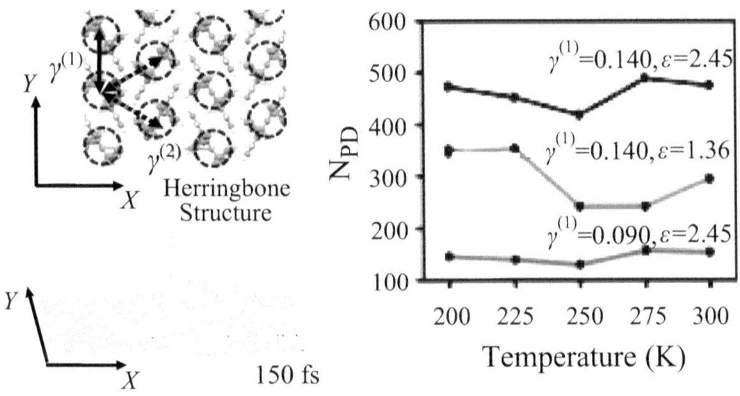

Figure 5.70 (Left) (Top) Schematic picture of the herringbone structure. (Bottom) Example of wave-packet dynamics at $t = 150$ fs. (Right) Polaron delocalization number at the initial condition of $N_{PD} = \pi\langle R(0)^2\rangle/S_{\text{unit}}$. From H. Tamura, M. Tsukada, H. Ishii, N. Kobayashi, and K. Hirose, *Phys. Rev. B* **85**, 245206 (2012) [206].

potential $E_{disp} = 4\epsilon\left\{\left(\sigma/R_{ij}\right)^{12} - \left(\sigma/R_{ij}\right)^{6}\right\}$ with two parameters of $\sigma = 12.12$ bohr and $\epsilon = 1.36$ eV or 2.45 eV. Two kinds of transfer integrals are taken as $\gamma^{(1)} = 0.09$ eV or 0.14 eV and $\gamma^{(2)} = 0.22\gamma^{(1)}$. The polaron behavior is characterized by the polaron delocalization number of $N_{PD} = \pi\langle R(0)^2\rangle/S_{\text{unit}}$, where $\langle R(0)^2\rangle$ is the mean square displacement averaged with the Boltzmann factor, S_{unit} is the area of the Wigner–Seitz cell, and we take 100 samples of lattice configurations for an average of $\langle R(0)^2\rangle$. Thus, N_{PD} indicates the extent to which the polaron is delocalized over molecules. Figure 5.70 (right) shows N_{PD} as a function of the temperature. We note that the carrier mobility shows the band-like behavior, decreasing monotonically with increasing temperature approximately by the power-law dependence. We see that N_{PD} increases with increasing the transfer integral and that the polaron can be delocalized over few hundreds of molecules for a large transfer integral. The localization of polaron originates from the inhomogeneity of intermolecular coupling due to thermal fluctuations.

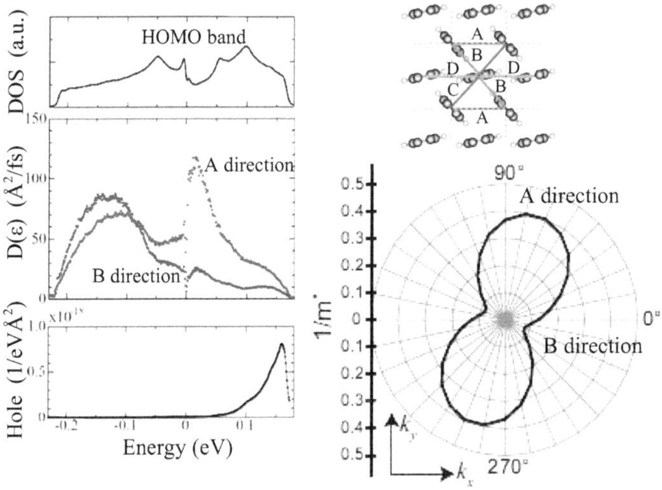

Figure 5.71 (Left) (Top) DOS of HOMO band, which spreads approximately 50 meV. (Middle) $D(\epsilon)$ for the A and B directions. (Bottom) Hole density appearing near the edge of the HOMO band. (Right) (Top) Schematic picture of pentacene layers with four independent directions A–D. (Bottom) Anisotropy of inverse of angle-resolved hole effective mass of pentacene for each direction.

5.6.3.3 Anisotropic effects on intrinsic charge transport

Next, we consider the effective mass in the different directions of geometrical configurations, which produces anisotropic field effects on the charge transport. For the computations we prepare three layers of pentacene with 300×200 unit cells, including 3.6×10^5 pentacene molecules.

The left panels of Fig. 5.71 show the DOS of HOMO band, which spreads approximately 50 meV (top) and the hole density (bottom). The diffusion coefficient $D(\epsilon)$ in the center shows that the mobility becomes anisotropic in the upper energy and isotropic in the lower energy. The anisotropic behavior of mobility in various directions is clearly observed in the right panel, where the ratio of the inverse of effective mass in the A direction with the lowest mobility in the B direction becomes $m^*_{\mathrm{max}}/m^*_{\mathrm{min}} \sim 5$. The anisotropic behaviors of mobilities are determined by the competition of transfer integrals γ^{HOMO} in the A and B directions and thermal fluctuations of γ^{HOMO}, which moderate the anisotropy $\mu_{max}/\mu_{min} \sim 4$ at room temperature due to intermolecular vibrations.

5.6.3.4 Static disorder effects on the carrier transport

Here, we study how the transport properties are affected by the static disorders W, which inevitably exist in the molecular crystals in terms of static defects such as crystal imperfections, the presence of impurities, and static dipole disorder caused by interaction of the charge with induced dipole moments in the gate dielectric. Several experimental evaluations show that the potential depths of static disorders are estimated at about 50 meV. Thus, we change the parameter W for the random distribution from 50 meV to 200 meV, which is comparable to the HOMO bandwidth. Therefore, the charge-transport properties are expected to be strongly disturbed by the static disorders in the various temperature regimes.

Figure 5.72 shows the carrier mobility μ as a function of temperature T for several strengths of static disorders $W = 200$, 100 and 50 meV. As the static disorders decreases, carrier mobility increases significantly from 2 cm^2/Vs ($W = 200$ meV) up to 100 cm^2/Vs (no static disorder) around 300 K, and more importantly their temperature dependences are changed completely. In case of $W = 200$ meV, the magnitude of μ is about 1 cm^2/Vs, the sign of $d\mu/dT$ is positive in whole temperature regime. The carrier is trapped tightly by the disorders and ℓ_{mfp} is shorter than the molecule-to-molecule distance, which renders the concept of band-like transport meaningless. The carrier transport shows typical thermally activated behaviors, as observed with low quality.

When W decreases to 100 meV, $d\mu/dT$, μ changes its behavior from the thermal activated in the low T regime to band-like in the high T regime. In this situation, the carrier is localized by the static disorders and induces the intramolecular distortions, resulting in a low-mobility polaron formation. At temperatures more than 150 K, the mobility reduction by the polaron formation becomes negligible. The polaronic localized state is destroyed with an increase of T because the thermal excitation energy beyond the polaron binding energy is given to the carriers. When the static disorders decrease up to $W = 50$ meV, μ becomes larger than 50 cm^2/Vs around

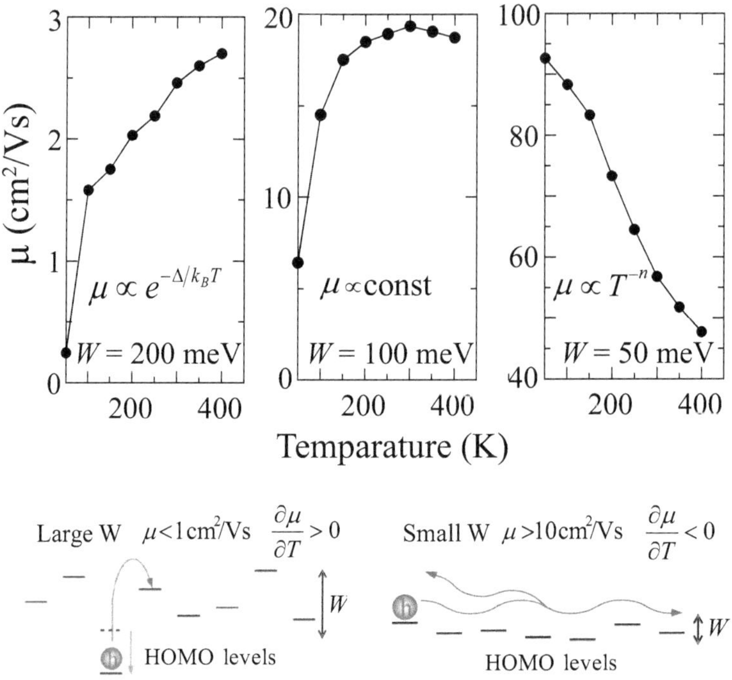

Figure 5.72 (Top) Carrier mobility μ of pentacene as a function of temperature T for the static disorders $W = 200$ meV (left), 100 meV (center), and 50 meV (right). (Bottom) Schematic pictures of charge transport in the static disorder systems. For large W ($W = 200$), the carriers are trapped in the molecules and form the polaronic state. For small W ($W = 50$), the carriers are extended over the crystals and are scattered from thermal fluctuations due to the dynamical disorder effects. From H. Ishii, K. Honma, N. Kobayashi, and K. Hirose, *Phys. Rev. B* **85**, 245206 (2012) [205].

the room temperature and the $d\mu/dT$ takes negative values in whole temperature regime, showing the monotonic increase of μ as a decrease of T with the power-law dependence. The time dependences of $D(t)$ indicates the realization of the diffusive transport rather than the hopping transport of a localized carrier. These behaviors are described in Fig. 5.72 (bottom). For $W = 200$, the carriers are trapped in the molecules and form the polaronic state with thermal activated hopping transport behavior $d\mu/dT > 0$. As W decreases, the carriers become extended and the coupling with intramolecular distortions to form the polaronic state becomes small. For $W = 50$, the carriers are extended over the crystals and are scattered from thermal fluctuations by the dynamical disorder, showing the band-like transport behavior $d\mu/dT < 0$. These results indicate that thermal fluctuation, polaron formation, and static disorders play important contributions to understand transport mechanisms of organic semiconductors.

As we note, organic semiconductors are used to construct the TFT. Figure 5.66 (right) shows a schematic illustration of a flexible TFT device with carrier mobility up to $\sim$40 cm^2/Vs at room temperature using printing processes. This is in contrast to the 2D electron gas (2DEG) systems constructed from the inorganic semiconductor heterostructures such as Si/SiO_2, which exhibit larger mobility and electron density, but with no flexibility. In the next chapter, we see the transport properties of traditional MOSFETs at the interface of $GaAs/Al_xGa_{1-x}As$.

Chapter 6

Artificial Nanosystems

6.1 Overview of 2D-Electron Gas Systems

6.1.1 *2D-Electron Gas in Semiconductor Heterostructures*

6.1.1.1 Effective mass approximation

Two-dimensional electron gas (2DEG) systems are constructed at interfaces between two different materials, where electrons are confined vertically within thin spaces and their energy levels are quantized in the vertical direction. Since the energy spacings become larger than the level broadening and the thermal energy $k_B T$, the electronic states in the vertical direction is discrete and has a standing-wave form. Therefore, electrons cannot move in that direction and the transport is limited to two-dimensional layer.

Examples of the interfaces for 2DEG systems are an interface between Si and SiO_2 insulating layer and an interface between GaAs and AlGaAs heterostructures. These systems are important for a number of technological applications of electron devices such as a metal–oxide–semiconductor field-effect transistor (MOSFET) and electron transistor with high mobility. Thus, their transport properties have been studied extensively [207, 208].

Quantum Transport Calculations for Nanosystems
Kenji Hirose and Nobuhiko Kobayashi

ISBN 978-981-4267-32-8 (Hardcover), 978-981-4267-59-5 (eBook)
www.panstanford.com

The 2DEG systems are also important to create further lower-dimensional systems by the subsequent lateral confinement. These are done by applying an electrostatic field from top or bottom gate electrodes. Correspondingly the one-dimensional systems (called quantum wire [QW]) and the zero-dimensional systems (quantum dot [QD]) are created from the 2DEG systems [209].

Typically, the vertical confinement at the interface is on the order of the Fermi wavelength λ_F, which, in the 2D systems, is given by

$$\lambda_F = \frac{2\pi}{k_F} = \sqrt{\frac{2\pi}{n}}. \tag{6.1}$$

Here n is an electron carrier density and we use the relation for 2D electron gas $n = (2/(2\pi)^2) \times \pi k_F^2$. In GaAs/AlGaAs heterostructures, a typical electron carrier density is $n \sim 3 \times 10^{11}$ cm^{-2} and we have $\lambda_F \sim 40$ nm. Confining the carriers to a 2D plane is made possible through the bandgap engineering of two different materials as shown in Fig. 6.1. When a layer of AlGaAs is grown on top of a layer of GaAs, due to the discontinuities of conduction and valence bands, electron carriers are trapped and thereby they form a 2D sheet of carriers at the interface of AlGaAs and GaAs. Since the bottom of the conduction band of GaAs is approximated well by the parabolic effective potential with an effective mass of m^*, the density of states $\nu(\epsilon)$ for the GaAs/AlGaAs 2D electron systems is given by

$$\begin{aligned}\nu(\epsilon) &= \frac{2}{(2\pi)^2} \int \delta\left(\epsilon - \frac{\hbar^2 k^2}{2m^*}\right) d\mathbf{k} \\ &= \frac{2}{(2\pi)^2} \int 2\pi k \delta\left(\epsilon - \frac{\hbar^2 k^2}{2m^*}\right) dk = \frac{m^*}{\pi \hbar^2},\end{aligned} \tag{6.2}$$

which takes a constant value. This should be compared with the $\sqrt{\epsilon}$ behavior for the 3D case and $1/\sqrt{\epsilon}$ behavior for the 1D case, which we see for the CNT case in the previous chapter. Schematic views of $\nu(\epsilon)$ for low-dimensional systems are shown in Chapter 2.

Let us consider the mean free path of electrons for the transport. To reduce the electron scatterings from the impurities and to accelerate an electron motion in a 2D system for the high electron mobility, carrier doping is performed in the n-AlGaAs layer, away for

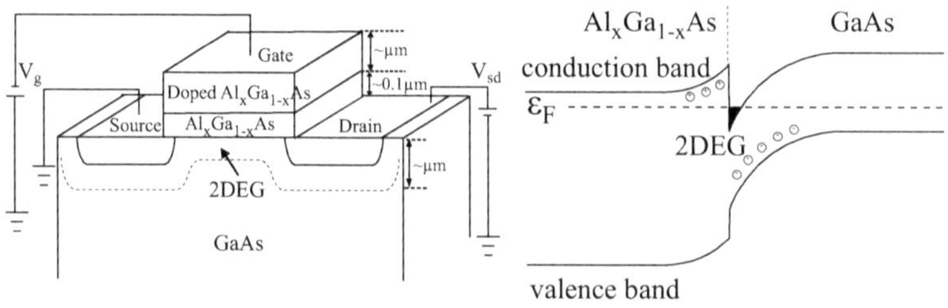

Figure 6.1 (Left) Schematic illustration of n-type MOSFET. (Right) Band diagram of a GaAs/$Al_xGa_{1-x}As$ heterostructure with n-type doping layer forming a 2DEG at the interface.

some distance from the interface in which the 2DEG is formed. With use of this modulation doping technique, electrons can tunnel from these donors to the lower energy state at the interface and leave the ionized impurity states behind. Since conducting 2D electrons are spatially removed from these ionized donors, the scattering from these impurities is much reduced and the mean free path increases significantly. To decrease the temperature, the effects from the phonon scatterings are also much reduced, and thus we have a 2DEG system whose mean free path is as much larger than that of bulk materials for the same carrier density. This type of transistor with high electron mobility is very important not only for the fundamental research of transport but also for a number of practical applications.

Figure 6.2 shows the behaviors of electron mobility μ for the GaAs 2DEG system as a function of the temperature T. The mean free path is calculated from

$$\ell_{\mathrm{mfp}} = v_F \cdot \tau = \frac{\hbar k_F}{m^*}\frac{m^*}{e}\mu = \frac{\mu\hbar\sqrt{2\pi n}}{e}. \tag{6.3}$$

For the typical electron density of $n \sim 3 \times 10^{11}$ cm^{-2}, when we have a sample with the mobility of $\mu \sim 1 \times 10^6$ cm^2/Vs at low temperature, the mean free path becomes $\ell_{\mathrm{mfp}} \sim 10\ \mu$m, which is much larger than the typical channel length of the electron devices, and thus the ballistic regime is readily accessible in GaAs devices at low temperature. We summarize the typical values of the various quantities for the GaAs 2DEG in Table 6.1.

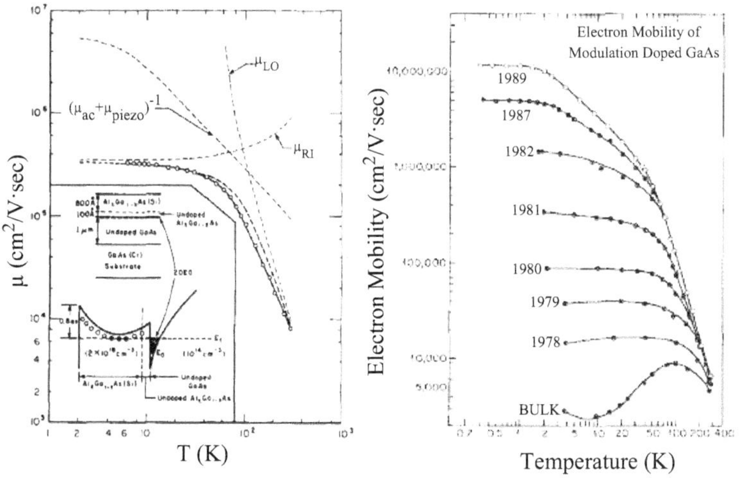

Figure 6.2 (Left) Temperature dependence of the GaAs 2DEG mobility with $n_{2D} = 5.35 \times 10^{11}\text{cm}^{-2}$. Mobilities from LO phonon scatterings μ_{LO}, remote ionized impurities μ_{RI}, acoustic phonon scatterings μ_{ac}, and the piezoelectric couplings μ_{piezo} are also shown. The heavy dashed line is the combined mobility. The inset shows the structure and band diagram. From B. J. F. Lin, D. C. Tsui, M. A. Paalanen, and A. C. Gossard, *Appl. Phys. Lett.* **45**, 695 (1984) [210] . (Right) Temperature dependence of the mobility with some of the samples in the history of modulated-doped GaAs. From L. Pfeiffer, K. W. West, H. L. Stormer, and K. W. Baldwin, *Appl. Phys. Lett.* **55**, 1888 (1988) [211].

Next, we consider the range of Coulomb interaction strengths fabricated in 2DEG systems. Coulomb energy is determined by the electron density n, which can be controlled by an applied gate voltage, typically changed in the range from $n = 1 \times 10^{10}$ cm^{-2} up to $n = 1 \times 10^{12}$ cm^{-2}. The Coulomb energy of an electron is screened by the other electrons, and thus the behaviors of electrons are as if they are in the space with the dielectric constant κ. In such a case, the motion of an electron i is described approximately by the effective-mass Hamiltonian

$$H^* = \sum_i -\frac{\hbar^2}{2m^*}\nabla_i^2 + \sum_{i<j} \frac{e^2}{\kappa|\mathbf{r}_i - \mathbf{r}_j|}. \tag{6.4}$$

Similar to the atomic-scale systems, we can define the effective Bohr radius and the effective Rydberg energy[a] by

$$a_B^* = \frac{\kappa\hbar^2}{m^* e^2} = \frac{\kappa}{(m^*/m)} a_B$$

$$E_{\text{Ryd}}^* = \frac{m^* e^4}{2\kappa^2\hbar^2} = \frac{(m^*/m)}{\kappa^2} E_{\text{Ryd}}. \tag{6.5}$$

For the GaAs/AlGaAs interface, the typical values are $m^* = 0.067m$ and $\kappa \approx 12.9$. Thus, we have $a_B^* \approx 10$ nm, which is about 200 times larger than the atomic hydrogen case a_B, and $E_{\text{Ryd}} = 5.47$ meV. Using the effective-mass approximation to describe the 2DEG, the Coulomb interaction is measured by the dimensionless parameter r_s defined by $r_s \equiv 1/\sqrt{\pi n}a_B^*$, which is the interparticle spacing in units of a_B^*. We note that $r_s = 0$ corresponds to no Coulomb interaction and the Coulomb interaction becomes large as r_s becomes large, where the average electron spacing becomes large due to the strong Coulomb repulsion. The typical charge density of the 2D electron gas in GaAs $n \approx 3 \times 10^{11}$ cm^{-2} corresponds to $r_s \approx 1$, which is in a weak interaction regime (thus, we usually call 2D electron *gas*). In experiments, the electron density n is tunable by applying the gate voltage within the range of $10^{10} \sim 10^{12}$ cm^{-2}, which corresponds to the interaction parameter $r_s \approx 0 - 4$.

It should be noted that the typical interaction strengths for atomistic metallic systems are covered within this range of interaction strength, for example $r_s = 2.0$ for Al and $r_s = 3.9$ for Na. This suggests that the artificially confined electron systems such as quantum dots in 2DEG work as *artificial atoms*. We will describe their electronic and spin structures later in this chapter.

6.1.1.2 Conductance and its fluctuation from quantum level statistics

Here we describe the conductance and its fluctuation in terms of the energy level and level spacing. In Chapter 2, we see that the

---

[a] For the electron–electron interaction, screening in the 2D electron systems is expressed by the dielectric constant as $\epsilon_{2D}(q) = 1 + (q_{\text{TF}}/q)$ with $q_{\text{TF}} = 1/a_B^*$ in the Thomas–Fermi approximation. This should be compared with $\epsilon_{3D}(q) = 1 + (q_{\text{TF}}^2/q^2)$ in the 3D electron systems. More accurately it becomes $\epsilon_{2D}(q) = 1 + (q_{\text{TF}}/q)\left\{1 - \sqrt{1 - (2k_F/q)^2}\right\}$.

Table 6.1 Typical values for the 2DEG of GaAs

Typical values for GaAs 2DEG	
Effective mass	$m^* = 0.067m$
Electron density	$n = 3 \times 10^{11}$ cm^{-2}
Fermi wavelength	$\lambda_F = (2\pi/n)^{1/2} = 40$ nm
Fermi energy	$E_F = \pi\hbar^2 n/m^* = 10$ meV
Fermi velocity	$v_F = (2E_F/m^*)^{1/2}$ $= 2 \times 10^7$ cm/s
Electron mobility	$\mu = 10^6$ cm$^2/Vs$
Scattering time	$\tau = m^*\mu/e = 40$ ps
Mean free path	$\ell_{\text{mfp}} = v_F\tau = 10\ \mu$m

conductivity with impurity scatterings is given by $\sigma = e^2 D\nu = ne^2\tau/m^*$, where the scattering time τ is related to the impurity strength V by $1/\tau = 2\pi\langle V^2\rangle\nu/\hbar$.

The conductance in the d-dimensional system is defined by $G = \sigma L^{d-2}$. Thus, it is described by

$$G = e^2\left(\nu L^d\right)\frac{D}{L^2} = \frac{e^2}{\hbar}\frac{E_c}{\Delta_0}, \tag{6.6}$$

where $E_c = \hbar D/L^2 (= \hbar/t_c)$ is the Thouless energy and $\Delta_0 = \left(\nu L^d\right)^{-1}$ is the mean spacing between energy levels. t_c is the typical diffusion time for an electron in the system with the length L. Thus, we can define the average conductance as

$$\langle G\rangle = \frac{e^2}{\hbar}g \quad \text{with } g = \left\langle\frac{E_c}{\Delta_0}\right\rangle. \tag{6.7}$$

Physically, dimensionless conductance $g = N_T$ denotes the average number of levels inside an energy window of width E_c. This is called the Thouless formula for the conductance. To determine E_c (or D) from the solutions to the quantum problem, we can use the sensitivity of energy levels to a change of boundary conditions.

Now let us recall the section for the localization in Chapter 2. If we have the disorder from impurities, the wavefunctions tend to localize as $|\Psi(\mathbf{r})| \sim e^{-|r-r_0|/\xi}$ due to the interference effect of the electron phase, where ξ is the localization length. When ξ is larger than the size of the system L as $\xi \gg L$, we can say that the disorder is weak in the metallic regime and the wavefunctions are extended to show a strong dependence on the boundary conditions. This is the weak localization regime. The average conductance becomes $\langle G \rangle \gg e^2/\hbar$, since there are a number of extended states within the energy of E_c. On the other hand, when the disorder is very strong, ξ becomes smaller than L as $\xi \ll L$. The wavefunctions are localized in the insulating regime to show no dependence on the boundary conditions. Then the conductance reduces $\langle G \rangle \ll e^2/\hbar$. This is the strong localization regime. The transition, called the Anderson transition, is very important for the disorder problem.

Here we consider the weak localization regime. In experiments, presently we can reduce the disorder so small in the GaAs 2DEG system into the metallic regime $\langle G \rangle \gg e^2/\hbar$ as shown in Fig. 6.3. We can classify the regimes according to the size of the system L, the localization length ξ, and the mean free path ℓ_{mfp} as

$$\begin{aligned} \ell_{\text{mfp}} \ll L \ll \xi \quad & \text{diffusive regime} \\ L \ll \ell_{\text{mfp}} \ll \xi \quad & \text{ballistic regime.} \end{aligned} \tag{6.8}$$

In the diffusive regime, the transport is essentially independent of the form of the system, while in the ballistic regime the boundaries of the system play the role of scatters instead of impurities.

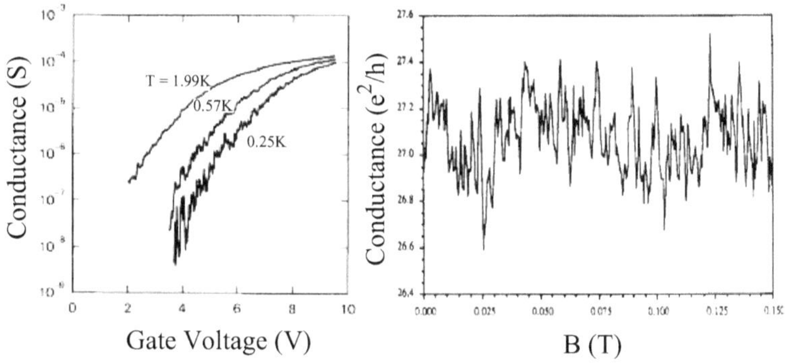

Figure 6.3 (Left) Conductance as a function of gate voltage for several temperatures. Conductance decreases as the temperature decreases with an increase of fluctuations. From A. B. Fowler, A. Hartstein, and R. A. Webb, *Phys. Rev. Lett.* **48**, 196 (1982) [212]. (Right) Conductance fluctuation in units of e^2/h as a function of applied magnetic fields measured at 0.058 K. From J. P. Bird, A. D. C. Grassie, M. Lakrimi, K. M. Hutchings, J. J. Harris, and C. T. Foxon, *J. Phys.: Cond. Mat.* **2**, 7847 (1990) [213].

The fluctuation of conductance defined by

$$\langle(\delta G)^2\rangle = \langle G^2\rangle - \langle G\rangle^2 \tag{6.9}$$

shows a property in the metallic regime, independent of microscopic details of the sample size or geometry such as

$$\langle(\delta G)^2\rangle \sim \left(\frac{e^2}{\hbar}\right)^2. \tag{6.10}$$

This shows that while the conductance is very large $\langle G\rangle \gg e^2/\hbar$, its fluctuation is always reduced to the universal value of $\langle\delta G\rangle \sim e^2/\hbar$. In view of the energy levels, this means that there are a number of states inside E_c such as $N_T \gg 1$, the number of fluctuation is $\delta N_T \sim 1$. This shows clearly that there is some correlation between the energy levels, suggesting the rigidity of the spectrum with strong repulsion between the energy levels. We note that $\delta N_T \sim \sqrt{N_T}$ for an uncorrelated spectrum case. This result is called universal conductance fluctuation (UCF) observed in the metallic regime as we see in Chapter 2 in phase coherence viewpoints.

The fluctuation of conductance can be studied to see the energy levels of the Hamiltonian matrix, whose elements take random values due to the disorder. This is called the random matrix, originally used to study the energy levels in the nuclear physics [16, 214]. Here let us see a simple case for the 2×2 Hamiltonian matrix

$$\hat{H} = \begin{pmatrix} h_{11} & h_{12} \\ h_{12} & h_{22} \end{pmatrix}, \tag{6.11}$$

whose elements h_{ij} are distributed randomly with the Gaussian distribution. The distribution function $P\left(\{h_{ij}\}\right)$ becomes

$$P\left(\{h_{ij}\}\right) = \frac{1}{Z}\exp\left(-\frac{1}{2}\left(h_{11}^2 + h_{22}^2 + 2h_{12}^2\right)\right), \tag{6.12}$$

where Z is the renormalization factor. Since the level spacing $\Delta = \epsilon_2 - \epsilon_1$ is given from the eigenvalues $\epsilon_{1,2}$ of $\hat{H}$,

$$\epsilon_{1,2} = \frac{h_{11} + h_{22} \pm \sqrt{(h_{11} - h_{22})^2 + 4h_{12}^2}}{2}, \tag{6.13}$$

we can calculate the distribution of probability that the level spacing of the random matrix $\hat{H}$ coincides to Δ as

$$P_{\mathrm{WD}}(\Delta) = \int dh_{11}\, dh_{22}\, dh_{12} P\left(\{h_{ij}\}\right) \delta(\epsilon - \Delta) = \frac{\pi}{2}\Delta e^{-(\pi/4)\Delta^2}. \tag{6.14}$$

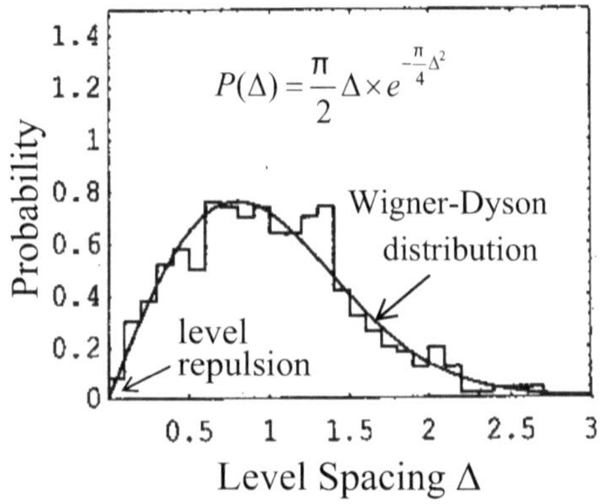

Figure 6.4 Probability distribution vs. level spacings for disordered confined systems. The Wigner–Dyson distribution is shown for reference.

This is the Wigner–Dyson distribution for the random matrix when a system has time-reversal symmetry. Here we show this from the 2×2 matrix. This is extended to larger matrix. Its important characteristic is that there is a strong level repulsion close to $\Delta = 0$, which indicates that two quantum levels repel to each other. Also, the levels are not distributed so far away from the mean level spacing $\langle \Delta \rangle$, indicating a strong correlation between quantum levels. These show that quantum levels tend to repel when distributed dense, while they tend to attract when distributed sparse. As a result, fluctuation of quantum levels is suppressed significantly.

Figure 6.4 shows numerical calculation results of the distribution of energy level spacing Δ for a simple disordered confinement model. The probability of the numerical distribution fits well to the analytical Wigner–Dyson distribution. Note that when there is no correlation between energy levels, the distribution becomes Poisson-type with no level repulsion around $\Delta \sim 0$.

6.1.1.3 Low-dimensional systems fabricated in 2DEG

Lower-dimensional systems can be fabricated to define a pattern by electron-beam lithography technique on top of the heterostructures and to use it as a gate. Applying a negative voltage to the gate, the electrons beneath it is depleted and the electron motions of 2D systems are laterally confined to the ungated region to create one- or zero-dimensional systems. In the next two sessions, we will show the 1D system—quantum wire (QW) and quantum point contact (QPC)—and the 0D system—quantum dot (QD)—and study their transport properties.

6.1.2 *Quantum Point Contact*

When we create a pattern with the width of a length scale comparable to the Fermi wavelength λ_F, we form a quasi-1D wire system called quantum wire (QW). Depending on the length of the QW and the mean free path $\ell_{\rm mfp}$, we can see diffusive or ballistic transport. When the length of the QW is longer than $\ell_{\rm mfp}$, electrons suffer from many impurity scatterings and the wire becomes diffusive. On the other hand, if the length of the QW becomes less than $\ell_{\rm mfp}$, the wire becomes ballistic. When the length of the wire is of order of λ_F, typically the same as the width, this corresponds to a QPC with a split-gate geometry and a narrow ballistic conducting channel is constructed.

In ballistic QW or QPC as shown in Fig. 6.5, by applying more and more negative bias voltage to the gate, we can change the constriction potential in the 2DEG from the defined lithographic width to zero. Since the ballistic electron transport occurs only through the point contact, we can study it as a function of the external potential barrier by changing the gate voltage. In the wire constriction, electrons are confined in the lateral y direction and the transport properties are affected by the presence of a potential barrier in the x direction. As we apply more and more negative gate voltages, the electron density is reduced and correspondingly the potential barrier height is raised. When the barrier height becomes larger than the Fermi energy ϵ_F, QPC is pinched-off and electron transport becomes zero.

6.1.2.1 Conductance quantization in QPC

As we increase the negative voltage to the gate, the electron transport through QPC is reduced. When the potential width of

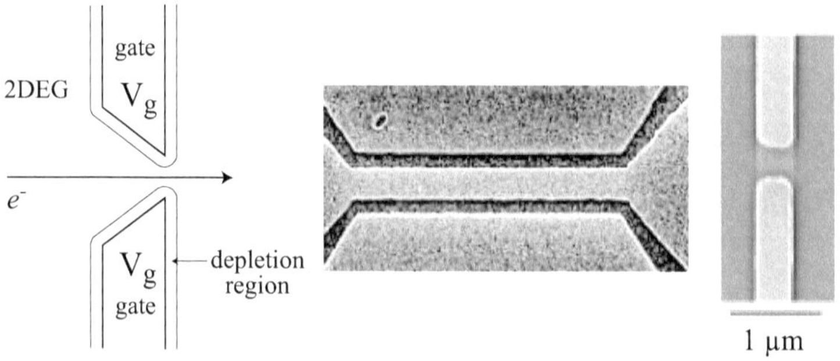

Figure 6.5 (Left) Schematic view of a split gate device. Applying a negative voltage V_g to gate depletes the underlying 2D electron gas to form a one-dimensional channel. (Right) SEM pictures of the quantum wire and QPCs. The center picture from B. E. Kane, G. R. Facer, A. S. Dzurak, N. E. Lumpkin, R. G. Clark, L. N. Pfeiffer and K. W. West, *Appl. Phys. Lett.*, **72**, 3506 (1998) [217] and the right picture from S. M. Cronenwett, H. J. Lynch, D. Goldhaber-Gordon, L. P. Kouwenhoven, C. M. Marcus, K. Hirose, N. S. Wingreen, and V. Umansky, *Phys. Rev. Lett.*, **88**, 226805 (2002) [218].

the QPC becomes narrow and comparable to the Fermi wavelength λ_F, electron wavefunctions form 1D subbands and the conductance quantization as a step of $2e^2/h$ is observed [215, 216]. In this situation, the longitudinal momentum is conserved and no dissipation takes place. The conductance of such a 1D conductor does not depend on the length of the channel and depends on the number of 1D modes in the channel. Here we see how the conductance quantization observed in QPC is explained from the transport theory.

The resistance of a point contact in the classical ballistic regime is known as the Sharvin resistance, which is due to the elastic backscattering at the geometrical narrowing of the point contact. Summing up all the electron particles coming to the point contact with various angles θ, the electric current is obtained using the width of the contact L_W, the electron number δn, and the electron velocity v_F as

$$\delta I = -e\delta n v_F L_W \int_{-\pi/2}^{\pi/2} \cos\theta \frac{d\theta}{2\pi} = -\frac{e}{\pi}\delta n v_F L_W. \tag{6.15}$$

Correspondingly the conductance becomes

$$G = \left(-\frac{dI}{dV}\right)_{V\to 0} = \frac{e^2}{\pi}\frac{dn}{dE}v_F L_W. \tag{6.16}$$

Since $v_F = \hbar k_F/m^*$ and the density of states is $dn/dE = \nu = m^*/\pi\hbar^2$, we can write the classical equation Eq. (6.16) to a semi-classical version, which includes conductance quantum $2e^2/h$ as

$$G = \frac{2e^2}{h}\frac{k_F L_W}{\pi} = \frac{2e^2}{h}\frac{2L_W}{\lambda_F}, \tag{6.17}$$

where the Fermi wavelength is $\lambda_F = \sqrt{2\pi/n}$ as shown before. Although this semi-classical form of conductance is continuous and linear in the width L_W and is not quantized, it suggests that the conductance deviates when λ_F is of order L_W due to the wave nature of electrons. When we consider that the wavefunction of an electron flowing through a narrow region with the width of L_W becomes zero at $x = 0$ and $x = a_W$, its energy is given by

$$\epsilon = \frac{p_y^2}{2m^*} + \frac{\hbar^2}{2m^*}\left(\frac{\pi N}{L_W}\right)^2 \quad (N = 1, 2, \ldots), \tag{6.18}$$

where the quantum integer number N is determined from the condition $\epsilon < \hbar^2 k_F^2/2m^*$ such as $N = \left[\frac{k_F L_W}{\pi}\right]$, where the brackets denote the integer part. This is schematically shown in Fig. 6.6. Correspondingly the conductance becomes

$$G = \frac{2e^2}{h}N \tag{6.19}$$

with the quantum number N.

This implies that with decreasing the electron density n and thus k_F by applying negative bias voltages to the constriction, the electron energy ϵ is reduced continuously and correspondingly the conductance G is also reduced. However, due to the wave nature of electrons, the electron energy for the direction that the current is flowing is discretized and the conductance G decreases in quantized steps of $2e^2/h$ as a function of the applied bias voltage. We should note that when the conductance G keeps the plateau, the reduced energy of electrons due to a negatively applied bias voltage is lost from the kinetic energy $p_y^2/2m^*$. This is the case that we assume the perfect transmission at $T = 0$ case.

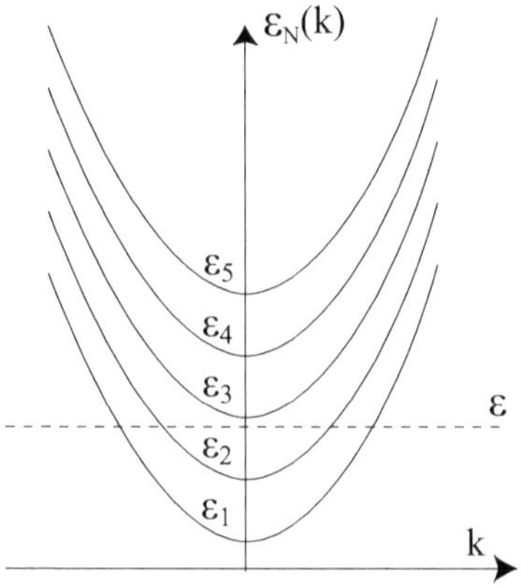

Figure 6.6 Energy E_N vs. longitudinal wave-vector k at the bottom of a QPC, assuming a parabolic confinement potential.

For more general case, we have

$$G = \frac{2e^2}{h} \sum_N \int \left(-\frac{\partial f}{\partial \epsilon} \right) \mathcal{T}_N(\epsilon) d\epsilon, \tag{6.20}$$

where $f(\epsilon)$ is the Fermi–Dirac distribution and $\mathcal{T}_N$ is the transmission coefficient. In the absence of impurities, since the potential variation is smooth as an adiabatic constriction, we expect the plateau structure of G as a function of an applied bias voltage. This is the quantization of conductance of QPC and an example of the manifestation of the Landauer formula in the ballistic electron transmission, where the cancellation of the 1D density of states $\nu \sim 1/\pi\hbar v$ and the group velocity v for the current produces the plateau structure. We call N as the 1D channel number of subband N. It should be noted, however, as the channel becomes longer and longer with more than 10 μm in length, since an ideal 1D wire with infinite length is impossible to realize in experiments, the conducting channel becomes diffusive and does not show ballistic quantization due to the scatterings from the constriction potential in addition to the usual impurity scatterings.[a]

[a]We note that the effects of Coulomb interaction on the purely one-dimensional interacting electron systems raise subtle problems as the Tomonaga–Luttinger liquid.

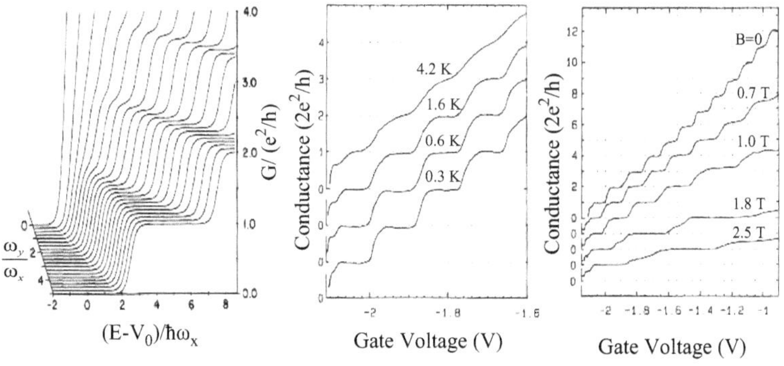

Figure 6.7 (Left) Calculated conductance as a function of gate voltage for different saddle potentials characterized by ω_y/ω_x. From M. Büttiker, *Phys. Rev. B* **41**, 7906 (1990) [219]. (Center) QPC conductance vs. gate voltage for different temperatures. Conductance quantization is broken down due to thermal averaging. (Right) QPC conductance vs. gate voltage at 0.6 K for several values of magnetic fields, showing the transition from $2e^2/h$ in zero field to e^2/h in high magnetic fields. From B. J. van Wees, L. P. Kouwenhoven, E. M. M. Willems, C. J. P. Harmans, J. E. Mooij, H. van Houten, C. W. Beenakker, J. G. Williamson, and C. T. Foxon, *Phys. Rev. B* **43**, 12431 (1991) [141].

In the QPC, the actual induced potential is not known, but self-consistent calculations indicate that it has a saddle-point shape, varies smoothly or adiabatically along the length of the QPC as

$$V(x, y) = V_0 - \frac{1}{2}m^*\omega_x^2 x^2 + \frac{1}{2}m^*\omega_y^2 y^2, \qquad (6.21)$$

where V_0 is the electrostatic potential at the center of the saddle point. In this saddle-point shape potential, the electrons are confined to y direction to form the conducting channels N, while the electrons go through the barrier of the potential in the x direction. The total conductance is expressed using the transmission coefficient $\mathcal{T}_N(\epsilon)$ of

$$\mathcal{T}_N(\epsilon) = \frac{1}{1 + \mathrm{e}^{-2\pi(\epsilon-\epsilon_N)/\hbar\omega_x}}, \qquad (6.22)$$

where $\mathcal{T}_N$ depends exponentially on the band edge energy mode $\epsilon_N = V_0 + \hbar\omega_y(N + 1/2)$. As a result, the shape of the saddle point potential plays an important role in determining the shape of the conductance quantization. In the zero-temperature limit, the conductance from Eq. (6.20) is given by

$$G = \frac{2e^2}{h}\sum_N \mathcal{T}_N(\epsilon_F) = \frac{2e^2}{h}\mathcal{T}. \quad (6.23)$$

The total transmission $\mathcal{T} = \sum_N \mathcal{T}_N(\epsilon_F)$ shows a series of well-developed steps if the transition region $\hbar\omega_x$ for the opening of a quantum channel is small compared to the channel separation $\hbar\omega_y$ such as $\omega_y \geq \omega_x$. Namely, when the quantum wire is very short ($\omega_y/\omega_x \ll 1$), conductance quantization is very hard to see. When the confinement is almost symmetric ($\omega_y/\omega_x \approx 1$), we can see clear plateaus in the conductance. As this ratio increases ($\omega_y/\omega_x \gg 1$), the plateaus become more and more flat and long. We see the conductance behavior as a function of ω_y/ω_x in Fig. 6.7 (left).

In order to observe the quantization of conductance in 1D system, we need to decrease the disorder with clean constriction potential at low temperature to reduce thermal averaging. Presently, conductance quantization is readily observable in GaAs 2DEG system due to the development of high purity heterostructure, nano-lithography techniques, and milli-Kelvin measurements, as shown in Fig. 6.7 (right).

6.1.3 *Quantum Dot*

When we confine laterally the electrons of 2DEG system furthermore to small regions by an external potential, such artificially fabricated small island is called a quantum dot with zero-dimensional system. One realization of a quantum dot in GaAs/AlGaAs heterostructure as shown in Fig. 6.8.[a]

Usually, a quantum dot is constructed of a million of atoms and most of the electrons are tightly bound to the atoms. However, a few to a few hundred electrons are free in the quantum dot. Since the Fermi wavelength of these free electrons is comparable to the size of the dot, such electrons occupy discrete quantum levels as is the similar situation to the atomic orbitals in atoms. They

[a]Quantum dots as tiny islands can be fabricated in various ways. Quantum dots created in Si inversion layers exhibit single-electron tunneling characteristics at room temperature. Self-assembled quantum dots epitaxially grown in the InAs or InP semiconductor are utilized for the applications of optoelectronic devices such as infrared photodetectors and lasers.

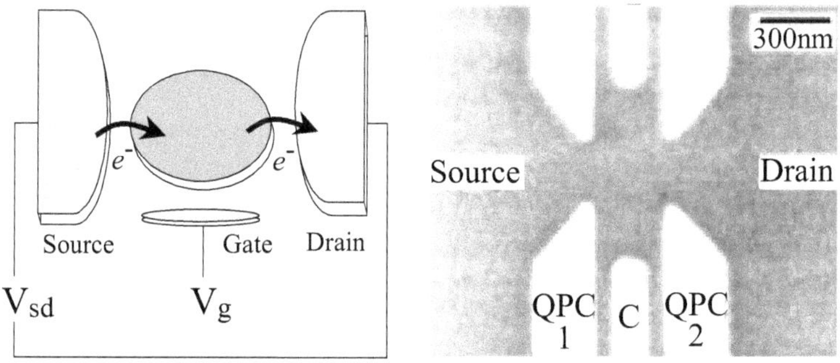

Figure 6.8 (Left) Schematic view of a quantum dot. (Right) SEM photograph for a quantum dot fabricated in GaAs/AlGaAs heterostructure. From T. H. Oosterkamp, L. P. Kouwenhoven, A. E. A. Koolen, N. C. van der Vaart, and C. J. P. M. Harmans, *Phys. Rev. Lett.* **78**, 1536 (1997) [220].

have a discrete excitation spectrum, and thus quantum dots are sometimes referred to as artificial atoms. Different from the atoms whose energy spectra are usually studied by the interaction with light or photon, the energy spectra of quantum dots are studied by measuring the transport properties with the attachment of the leads to probe the current.

Present fabrication technologies make it possible to produce quantum dots so clean that the addition spectra are reproducible from dot to dot. Especially, very well-controlled quantum dots show the atomic-like spectra in the transport measurement, which means that we can produce artificial atoms in the semiconductor. Reproducible quantum dots are considered to be important for device applications, such as high-density memory using single-electron tunneling effect, novel optical materials, quantum pumping, and realization of quantum computing using coupled quantum dots.

Here we observe the transport behaviors of quantum dots fabricated between source and drain electrodes. Experimentally, when measuring the tunnel conductance through a quantum dot or the capacitance to the dot while changing an applied voltage to the gate electrode, one observes the conductance behavior for various temperature regimes.

When the temperature is high, the conductance takes the constant value independent of the electron number of the quantum dot. Therefore, I–V characteristic shows the classical Ohmic behavior. The transport is due to the thermal excitation of electrons. As we decrease the temperature, we observe periodic oscillations of the conductance with a peak at each gate voltage. Since electrons are confined and discrete quantum levels are formed in the quantum dots, we might consider that electron transport occurs by the resonant tunneling through quantum levels ϵ_i of quantum dots over the barriers at contacts in the low-temperature regime. As we see later, electron transport through tunneling is mostly determined by the Coulomb energy of the quantum dot. The transport at low temperature is called the CB oscillation.

One example of experiments using metallic CNT devices for the conductance behaviors at different temperatures as a function of gate voltage is shown in Fig. 6.9. At very low temperature (~1.5 K), the appearance of the CB oscillations indicates that the electrons are localized over the entire length of the CNT and CNT acts as a quantum dot between electrodes. When the temperature becomes

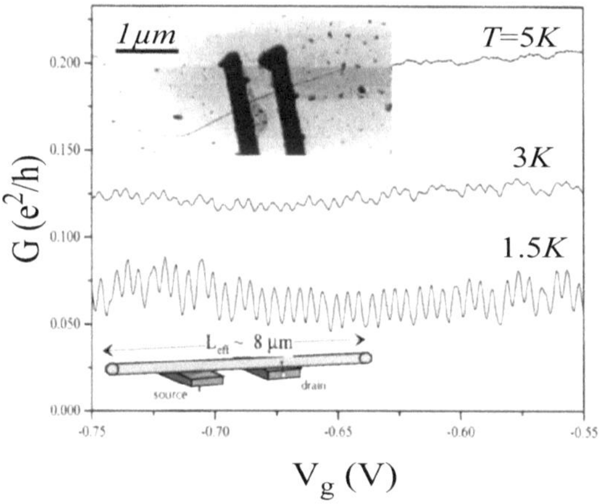

Figure 6.9 Conductance vs. gate voltage at different temperatures for the metallic CNT device. The appearance of the Coulomb blockade (CB) oscillations at low temperature (~1.5 K) indicates that the electrons are localized over the entire length of the CNT and CNT acts as a quantum dot between electrodes. From P. L. McEuen, M. Bockrath, D. H. Cobden, Y. G. Yoon, and S. G. Louie, *Phys. Rev. Lett.* **83**, 5098 (1999) [221].

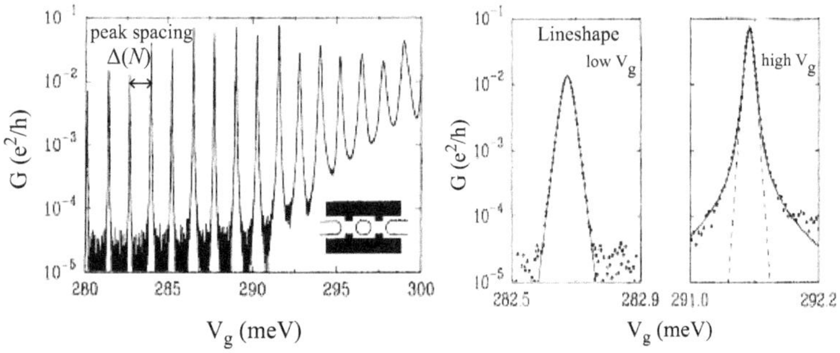

Figure 6.10 (Left) Low-bias conductance of the quantum dot vs. gate voltage V_g at low temperature $T = 60$ mK. (Right) At low V_g conductance peak lineshape is fitted well to a thermally broadened resonance $\cosh^{-2}(V_g)$ and at high V_g conductance peak lineshape is fitted to a thermally broadened Lorentzian parameters Γ. From E. B. Foxman, P. L. McEuen, U. Meirav, N. S. Wingreen, Y. Meir, P. A. Belk, N. R. Belk, M. A. Kaster, and S. J. Wind, *Phys. Rev. B* **47**, 10020 (1993) [222].

higher, conductance oscillation behaviors gradually become smaller and disappear above 5 K.

6.1.3.1 Lineshape of conductance

Let us see the lineshape of conductance of quantum dots. Figure 6.10 (left) shows typical examples at low temperature. The conductance formula due to the resonant tunneling is obtained from Chapter 2 as

$$G = \frac{2e^2}{h} \int \left(-\frac{\partial f}{\partial \epsilon} \right) \sum_i \frac{\Gamma_L \Gamma_R}{(\epsilon - \epsilon_i)^2 + (\Gamma_L + \Gamma_R)^2/4} d\epsilon, \tag{6.24}$$

where $\Gamma_{L,R}$ are the coupling constant to left (L) and right (R) electrodes. Since the gate voltage is proportional to the chemical potential, we can take it as $\mu = e\alpha V_g$, where α is the geometrical constant.

When the coupling is large as $k_B T \ll \Gamma_L + \Gamma_R$, we can approximate $-\partial f/\partial \epsilon = \delta(\epsilon - \mu)$ and obtain the conductance as

$$G \approx \frac{2e^2}{h} \sum_i \frac{\Gamma_L \Gamma_R}{\left(e\alpha V_g - \epsilon_i\right)^2 + (\Gamma_L + \Gamma_R)^2/4}. \tag{6.25}$$

This shows that the lineshape of the conductance peak is Lorentzian with the width of $\Gamma_L + \Gamma_R$ independent of the temperature. The amplitude takes the maximum of $(2e^2/h) \times 4\Gamma_L\Gamma_R/(\Gamma_L + \Gamma_R)^2$ periodically whenever the gate voltage V_g coincides with the quantum levels ϵ_i. This corresponds to the coherent transport through the quantum dots.

When the coupling is small as $\Gamma_L + \Gamma_R \ll k_B T$, the Lorentzian peak is much sharper than $-\partial f(\epsilon)/\partial\epsilon$ and thus we have

$$G \approx \frac{2e^2}{h}\sum_i \left(-\frac{\partial f}{\partial \epsilon}\right)_{\epsilon=\epsilon_i} \int \frac{\Gamma_L\Gamma_R}{(\epsilon-\mu)^2 + (\Gamma_L+\Gamma_R)^2/4} d\epsilon$$
$$= \sum_i \frac{G_0}{4k_B T} \times \cosh^{-2}\left(\frac{e\alpha V_g - \epsilon_i}{2k_B T}\right), \tag{6.26}$$

where

$$G_0 = \frac{2e^2}{\hbar}\frac{\Gamma_L\Gamma_R}{\Gamma_L+\Gamma_R}. \tag{6.27}$$

This shows that the lineshape of the conductance peak is given by $\cosh^{-2}(V_g)$ dependence on the temperature. The amplitude takes the maximum of $G_0/4k_B T$ periodically whenever gate voltage V_g coincides with the quantum levels ϵ_i. Thus, the maximum amplitude is inversely proportional to the temperature. This corresponds to the sequential transport through the quantum dots.

Now we see the experiment in Fig. 6.10 (right). The peak lineshape is fitted well to $\cosh^{-2}(V_g)$ at low V_g, while it is fitted well to the Lorentzian shape at high V_g. This shows that the coupling is smaller than $k_B T$ at low V_g regime and the quantum state of the dot is localized. On the other hand, at high V_g regime, the coupling is larger than $k_B T$ and the quantum state of the dot spreads out to the electrodes. This observation suggests that we can obtain information on the coupling Γ of the quantum dot from the fitting of the lineshape of conductance.

6.1.3.2 Charging energy

The problem is the peak spacing of the conductance. We see that conductance oscillation shows almost uniformly periodic. This is very strange if the transport is due to the resonant tunneling through the quantum levels ϵ_i. Since the shapes of confinement potentials of quantum dots are usually not symmetric, the quantum levels ϵ_i would be randomly spaced. Then obeying the UCF theory, we expect that the peak spacings of conductance show the Wigner–Dyson distribution and would not be the uniform one.

After repeating conductance measurements for various temperatures, the periodic conductance oscillation is shown not to be related to disorder. Furthermore, the conductance peaks are not split for the application of magnetic fields and spin degeneracy is already split out for each conductance peak. This shows that the transport through quantum dots, especially on the peak spacing, cannot be explained only from the single-particle quantum levels due to confinement.

In addition to the single-particle quantum levels, another important factor to describe the electronic states of quantum dots is the electron–electron interaction energy (i.e., Coulomb charging energy). Typically this is larger than the single-particle level spacing and therefore the ground-state energy is strongly dependent on the number of electrons in the quantum dot [223, 224]. We note that the spin state emerges when the Coulomb exchange energy is strong enough and plays an important role for the transport properties of quantum dots.

Let us compare the Coulomb interaction energy with the single-particle energy to understand electronic states of quantum dots in GaAs systems. Here we consider that electrons are confined within the flat disk region of the radius R. As for the Coulomb energy, it becomes

$$E_c \approx \frac{e^2}{2\kappa R/\pi} = \left(\frac{1}{\kappa^2}\frac{m^*}{m}\right)\left(\frac{me^4}{\hbar^2}\right)\frac{a_B^*}{2R/\pi} \approx \frac{a_B^*}{R} \times 17.2 \text{ meV}, \tag{6.28}$$

where we use $E_c = me^4/\hbar^2 = 27.2$ eV (=2 Ryd), $(1/\kappa^2)(m^*/m) \approx 4 \times 10^{-4}$, and $a_B^* \approx 10$ nm. As for the confinement energy, a single-particle energy spacing is approximately determined from

$$\Delta_0 \approx \frac{\hbar^2}{m^* R^2} = \left(\frac{1}{\kappa^2}\frac{m^*}{m}\right)\left(\frac{me^4}{\hbar^2}\right)\left(\frac{a_B^*}{R}\right)^2 \approx \left(\frac{a_B^*}{R}\right)^2 \times 10.9 \text{ meV}. \tag{6.29}$$

Since the Coulomb energy decreases as $E_c \sim 1/R$ and a single-particle decreases as $\Delta_0 \sim 1/R^2$, the Coulomb energy E_c dominates when the size of quantum dot is large. In the GaAs systems, we have $E_c = 0.86$ meV and $\Delta_0 = 0.0275$ meV for $R = 200$ nm, which means that the Coulomb energy is much larger than the single energy level spacing $E_c/\Delta_0 = 31.4$ and the transport properties are determined mostly by the Coulomb energy. This is the typical case shown in Fig. 6.10 and is called the CB regime for $E_c/\Delta_0 \gg 1$.

When the size of quantum dot decreases close to a_B^*, that is, several tens of nm, the single-particle energy is comparable to the Coulomb energy. The transport properties are determined with the competition of both the Coulomb energy and the single-particle energy. Since the Fermi wavelength λ_F is close to this length and the Coulomb energy is within a range of $r_s \approx 0 - 4$ in the GaAs systems, this suggests that the quantum dot confined with several tens of nm length shows similar characteristic behaviors to atoms and therefore is called an artificial atom. We note that in the hydrogen atom, $E_c/\Delta_0 = 2$ for the lowest energy level. We will show the electronic structures of quantum dots accurately using the density functional theory (DFT) later.

6.1.3.3 Single-electron transport

Here let us consider the transport behaviors of quantum dots when the Coulomb energy is dominant in the CB regime as schematically shown in Fig. 6.11. The electronic states in electrodes are filled up to μ_L and μ_R, for which source-drain voltage V_{sd} becomes $eV_{sd} = \mu_L - \mu_R$. The discrete states of a quantum dot are filled with N electrons up to $\mu_{\rm dot}$. Due to the addition of one electron, the energy of the quantum dot raises by $\Delta\epsilon + e^2/C$. In the present situation, the $(N+1)$st electron cannot tunnel on the quantum dot, since the resulting electrochemical potential $\mu(N+1)$ is higher than the potentials of electrodes. So, for $\mu(N) < \mu_L, \mu_R < \mu(N+1)$ the tunneling of electrons on and off the quantum dot is blocked at low temperature, which is the CB and N electrons are localized on the quantum dot.

The CB can be removed by changing the gate voltage $V_g(N+1)$ to align $\mu(N+1)$ as shown in Fig. 6.12 and the transport through the quantum dot become possible. This process, in which the current is carried by successive discrete charging and discharging of the quantum dot, is known as single-electron tunneling (SET). Therefore, transport at the peak is due to single-electron tunneling when the ground-state energies of states differ by one electron. Note

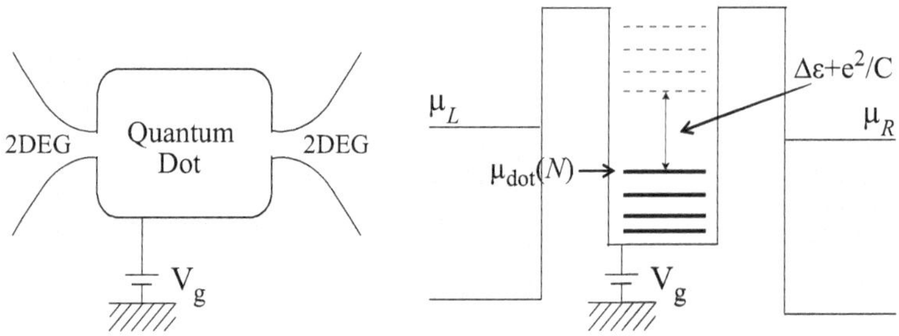

Figure 6.11 (Left) Schematic illustration of a quantum dot fabricated in the 2DEG. (Right) Energy diagram through a quantum dot. Electronic states in the electrodes are filled up to μ_L and μ_R, for which the source-drain voltage V_{sd} becomes $eV_{sd} = \mu_L - \mu_R$. The discrete states of a quantum dot are filled with N electrons up to $\mu_{\text{dot}}(N)$. Due to the addition of one electron, the energy of the quantum dot raises by $\Delta\epsilon + e^2/C$, and thus in the present situation, an electron transport is blocked at low temperature.

that the number of electrons in the quantum dot is fixed at N in the case of zero conductance. This is called the CB. The number of electrons fluctuates between N and $N+1$ in the case of non-zero conductance (no CB). Experimentally, when measuring the tunnel conductance through a quantum dot by changing an applied voltage to gate electrode, one observes a periodic peak at each gate voltage at which the average number of electrons on the dot N increases by one to $N+1$. The heights and the lineshapes of peaks as we show in the last section reflect tunneling amplitudes. The spacings of peaks

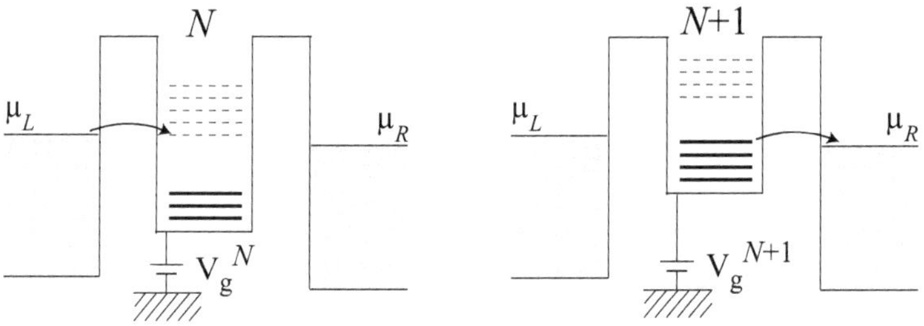

Figure 6.12 The electron addition to the quantum dot with potential $\mu_{\text{dot}}(N+1)$ aligned with potentials of μ_L and μ_R to change the gate voltage V_g. These show two parts of the sequential tunneling processes for the situation with N and with $N+1$ electrons on the quantum dot.

reflect ground-state total energy differences between states of the quantum dot with different numbers of electrons.

Here we consider the conditions for which such tunneling transport becomes possible. The first requirement is that the charging energy is larger than the thermal energy. The charging energy that one more electron is attached to quantum dots is $E_c = e^2/C$, where C is the capacitance of the dot island. So this requirement is expressed as $e^2/C \gg k_B T$. Since the temperature of order 1 K corresponds to $k_B T = 0.086$ meV, this is realized in GaAs quantum dots when the temperature is 10–50 mK range. Note that the capacitance C for a flat disk with the radius L is 1.8×10^{-16} F for $L = 200$ nm. As a result, most of the transport experiments of GaAs quantum dots have been performed at low temperature in dilution refrigerators. The second requirement is that the barriers between the quantum dot and electrodes are sufficiently large such that electrons are situated either in the source, in the drain, or on the quantum dot. In such cases, the quantum fluctuation in the number of N due to tunneling through the barriers is much less than one over the time scale of the measurement. Since the typical time to charge or discharge the quantum dots is $\Delta t = R_t C$, where R_t is the resistance of the barriers for tunneling, by using the Heisenberg uncertainty relation $\Delta E \, \Delta t = (e^2/C) R_t C > h$, we have $R_t \gg h/e^2$. This means that the coupling Γ to the quantum dot to the source and drain electrodes must be weak and the resistance R_t should be much larger than the resistance quantum $h/e^2 = 25.8$ kΩ in order for the energy uncertainty to be much smaller than the charging energy. When these two conditions are satisfied, we can observe the peak due to the SET in the transport measurements.

The present estimate is also applicable to atomic-scale transport on the same footing when the above requirements are satisfied, where the charging energy and the single-particle energy are of the order of eV. Currently there are various quantum dot systems with different materials. Therefore, we need to estimate the conditions for the electron transport. When we replace the quantum dots by the molecules, we can observe the similar transport behavior possibly in the room temperature regime. In this case, it is important to note that the contact conditions of the molecules and electrodes play the key role for the transport behaviors.

6.1.3.4 Peak spacing of conductance

Here we formulate the peak spacings, or addition energies, to add each electron to the quantum dot. The simple idea for the peak spacings is as follows. If we assume that there is no coupling to the source and drain contacts, then the quantum dot acts as an island for electrons. The number of electrons in the quantum dot is frozen at a particular integer number N and the charge is equal to Ne.

We define the energy of a quantum dot having N electrons by $E_{\text{dot}}(N)$. When we apply a bias voltage at the gate electrode, the total energy of the system becomes

$$E(N) = E_{\text{dot}}(N) - N\alpha e V_g, \tag{6.30}$$

where V_g is the gate voltage and α is a constant describing the electrostatic coupling between the dot and the gate electrode. By changing the gate voltage V_g, one controls the external potential applied to the dot and thereby controls the total number of electrons on the quantum dot. At low temperatures, electron hopping into a quantum dot containing N electrons is suppressed due to the large charging energy and becomes possible by the very small tunneling rate only when the ground-state energy $E(N)$ is equal to $E(N+1)$. This degeneracy condition determines the observed conductance oscillation peaks, which occur at the gate voltages, or electron chemical potentials $\mu(N)$ is equal to

$$\alpha e V_g^N = E_{\text{dot}}(N+1) - E_{\text{dot}}(N) \, (= \mu(N)) \, . \tag{6.31}$$

We note that the electron chemical potential $\mu(N)$ changes linearly with the gate voltage V_g with a coefficient α, which depends on the capacitance between the dot and the gate. The peak spacing $\Delta(N)$ needed to put an extra electron in the quantum dot is obtained by

$$\begin{aligned}\Delta(N) &= \alpha e(V_g^{N+1} - V_g^N) \, (= \mu(N+1) - \mu(N)) \\ &= E_{\text{dot}}(N+1) - 2E_{\text{dot}}(N) + E_{\text{dot}}(N-1). \end{aligned} \tag{6.32}$$

Here we treat the peak spacing $\Delta(N)$ using the so-called constant interaction (CI) model, which treats the Coulomb energy E_c and the single-particle level spacing Δ_0 separately and is frequently used to describe the transport of quantum dots. If we assume the charging energy E_c by the parameterized capacitance of the quantum dot keeping a constant value C for any number of electrons N in the dot, we can express the energy of the quantum dot $E_{\text{dot}}(N)$ semi-classically by

$$E_{\text{dot}}(N) = \frac{e^2}{C}\frac{N(N-1)}{2} + \sum_{i=1}^{N} \epsilon_i, \tag{6.33}$$

where ϵ_i are the single-particle energy levels of the quantum dot due to confinement. This is a reasonable assumption when the quantum dot is much larger than the screening length. In this CI model, the coupling constant α to the gate electrode is taken as $\alpha = C_g/C$, where C_g is the capacitance between the dot and the gate and $C = C_L + C_R + C_g$ is the total capacitance between the dot and all other pieces of metal around it. Here C_L and C_R are the capacitances across the barriers to the source and drain electrodes. Then the addition energy $\Delta(N)$ becomes

$$\Delta(N) = \frac{e^2}{C} + \Delta_0, \tag{6.34}$$

where $\Delta_0 = \epsilon_{N+1} - \epsilon_N$ is the single-particle level spacing. This CI model shows that the peak spacings of the SET transport are determined by the sum of the charging energy E_c and the single-particle level spacing Δ_0.[a]

When the number of electrons N in the quantum dot is very large and as a result the size of the dot is large, the charging energy is much larger than the single-particle level spacing $e^2/C \gg \Delta_0$ as we see. In this CB regime, $\Delta(N) = \mu(N+1) - \mu(N) \approx e^2/C$ displays conductance peaks with almost constant peak spacings e^2/C as a function of the gate voltage V_g. This is the periodic Coulomb oscillation phenomena. We note that each disordered quantum dot has its own, almost constant characteristic oscillation peaks. By

---

[a]In general, we can write $E(N) = E_{\text{dot}}(N) - Ne\phi_{\text{ext}}$, where $\phi_{\text{ext}} = (C_g V_g + C_L V_L + C_R V_R)/C$ is the electrostatic potential in the quantum dot.

In the linear-response transport regime $V_L = V_R = 0$ ($V_{sd} = 0$), we have $E(N) = E_{\text{dot}}(N) - N\alpha e V_g$ as in Eq. (6.30). The condition for electron transport is $E(N+1) - E(N) = 0$ and the periodic SET peaks appear at $V_g^N \propto N$. Here we neglect the single-particle level ϵ_N for simplicity.

In the non-linear transport regime ($V_{sd} \neq 0$), the condition for electron transport is $\mu_L < E(N+1) - E(N) < \mu_R$, which produces the rhombic-shaped regions in the V_g–V_{sd} plot, called the *Coulomb diamond*. Within the Coulomb diamond for each N, no current flows due to CB. For a symmetric situation $V_L = -V_R = V_{sd}/2$, the Coulomb diamond region is expressed by the condition $(N - 1/2)e < C_g V_g + (C_L + C_g/2)V_{sd} < (N + 1/2)e$ and $(N - 1/2)e < C_g V_g - (C_R + C_g/2)V_{sd} < (N + 1/2)e$.

applying a negative bias to the gate electrode, we can decrease the number of electrons N one by one and finally it becomes zero for sufficiently large negative bias voltage. When the number of electrons N is small, say 10 electrons, the single-particle level spacing Δ_0 becomes comparable to the charging energy E_c as we see, and thus we expect that the observed peak spacings $\Delta(N)$ would reflect information of the single-particle levels ϵ_N in addition to the charging energy E_c in the transport spectra.

6.1.4 *Green's Function Theory of Quantum Dots*

In the previous section, we describe the semi-classical treatment of the transport of quantum dots, where we find that both the charging energy and single-particle energy are important and must be treated on the same footing. Here we treat it purely quantum mechanically using the model that includes the Coulomb energy as an on-site energy U. It consists of two ideal electrodes coupled to a single quantum dot site where there is a Coulomb interaction of U. Here we consider only two levels ϵ_σ with a spin index σ on the site. The Hamiltonian is

$$\hat{H} = \sum_{\sigma;k\in L,R} \epsilon_{k\sigma}\hat{c}^\dagger_{k\sigma}\hat{c}_{k\sigma} + \sum_\sigma \epsilon_\sigma \hat{d}^\dagger_\sigma \hat{d}_\sigma + U\hat{n}_\uparrow \hat{n}_\downarrow + \sum_{\sigma;k\in L,R}\left[V_{k\sigma}\hat{c}^\dagger_{k\sigma}\hat{d}_\sigma + \text{h.c.}\right]. \qquad (6.35)$$

Here the states of the left (L) and right (R) electrodes have energies $\epsilon_{k\sigma}$ and are connected to the quantum dot site by the hopping elements $V_{k\sigma}$.

The linear-response conductance is calculated using the generalized transport formula described in Chapter 3,

$$G = \frac{e^2}{\hbar}\sum_\sigma \int_{-\infty}^{\infty}\left(-\frac{\partial f(\epsilon)}{\partial \epsilon}\right)\frac{\Gamma^\sigma_L(\epsilon)\Gamma^\sigma_R(\epsilon)}{\Gamma^\sigma_L(\epsilon)+\Gamma^\sigma_R(\epsilon)}\left(-\frac{1}{\pi}\text{Im}\left[G^r_\sigma(\epsilon)\right]\right)d\epsilon, \qquad (6.36)$$

as the sum of the transmission probabilities in each spin channel weighted by the derivative of the Fermi–Dirac distribution function. The transmission probabilities are obtained from the Fourier transform of the retarded Green's function of the single site of

$$g^r_\sigma(t-t') = -\frac{i}{\hbar}\theta(t-t')\langle\left[\hat{d}_\sigma(t), \hat{d}^\dagger_\sigma(t')\right]\rangle, \qquad (6.37)$$

where the brackets [ ] denote the anti-commutator. Since the Coulomb energy U is dominant and the hopping $V_{k\sigma}$ is very small in the tunneling regime, first we calculate Green's function without hopping term. Then hopping from electrodes $V_{k\sigma}$ is included perturbatively by solving the Dyson equation.

The equation of motion for Green's function $g_\sigma^r(t)$ becomes

$$i\hbar\frac{\partial}{\partial t}g_\sigma^r(t-t') = \delta(t-t') + \epsilon_\sigma g_\sigma^r(t-t') + Ug^{(2)}(t-t'), \quad (6.38)$$

where we use $i\hbar\partial\hat{d}_\sigma/\partial t = \left[\hat{d}_\sigma, \hat{H}\right] = \epsilon_\sigma\hat{d}_\sigma + U\hat{d}_\sigma\hat{n}_{\bar{\sigma}}$. Here we define the second-order Green's function $g^{(2)}$ by

$$g^{(2)}(t-t') = -\frac{i}{\hbar}\theta(t-t')\langle\left[\hat{d}_\sigma(t)\hat{n}_{\bar{\sigma}}(t), \hat{d}_\sigma^\dagger(t')\right]\rangle. \quad (6.39)$$

The equation of motion for the second-order Green's function $g^{(2)}$ becomes

$$i\hbar\frac{\partial}{\partial t}g^{(2)}(t-t') = \delta(t-t')\langle\hat{n}_{\bar{\sigma}}\rangle + \epsilon_\sigma g^{(2)}(t-t') + Ug^{(2)}(t-t'). \quad (6.40)$$

Here we use $\hat{n}_{\bar{\sigma}}\hat{n}_{\bar{\sigma}} = \hat{n}_{\bar{\sigma}}$ and $\left[\hat{n}_{\bar{\sigma}}, \hat{H}\right] = 0$. From the Fourier transform,

$$\begin{aligned}(\epsilon - \epsilon_\sigma)g_\sigma^r(\epsilon) &= 1 + Ug^{(2)}(\epsilon)\\ (\epsilon - \epsilon_\sigma - U)g^{(2)}(\epsilon) &= \langle\hat{n}_{\bar{\sigma}}\rangle,\end{aligned} \quad (6.41)$$

then we obtain $g_\sigma^r(\epsilon)$ as

$$g_\sigma^r(\epsilon) = \frac{\epsilon - \epsilon_\sigma - (1 - \langle\hat{n}_{\bar{\sigma}}\rangle)U}{(\epsilon - \epsilon_\sigma - U)(\epsilon - \epsilon_\sigma)} = \frac{1 - \langle\hat{n}_{\bar{\sigma}}\rangle}{\epsilon - \epsilon_\sigma} + \frac{\langle\hat{n}_{\bar{\sigma}}\rangle}{\epsilon - \epsilon_\sigma - U}. \quad (6.42)$$

This shows that Green's function $g_\sigma^r(\epsilon)$ has two resonances at $\epsilon = \epsilon_\sigma$ and $\epsilon = \epsilon_\sigma + U$, with the weights of $\langle\hat{n}_{\bar{\sigma}}\rangle$ and $1 - \langle\hat{n}_{\bar{\sigma}}\rangle$, respectively.

Then we add the perturbation induced by the electrodes, which would broaden the states of the quantum dots. The full Green's function $G_\sigma^r(\epsilon)$ is obtained from the Dyson equation to include the hopping term $V_{k\sigma}$

$$\begin{aligned}G_\sigma^r(\epsilon) &= g_\sigma^r(\epsilon) + g_\sigma^r(\epsilon)\Sigma_{\text{leads}}^\sigma(\epsilon)G_\sigma^r(\epsilon)\\ &= \frac{g_\sigma^r(\epsilon)}{1 - g_\sigma^r(\epsilon)\Sigma_{\text{leads}}^\sigma(\epsilon)}.\end{aligned} \quad (6.43)$$

We take the self-energy due to tunneling into the electrodes corresponding to noninteracting electrons

$$\Sigma_{\text{leads}}^\sigma(\epsilon) = \sum_{k\in L,R}|V_{k\sigma}|^2 g_{k\sigma}(\epsilon) = \sum_{k\in L,R}\frac{|V_{k\sigma}|^2}{\epsilon - \epsilon_{k\sigma}} \quad (6.44)$$

and we have

$$G^r_\sigma(\epsilon) = \frac{\epsilon - \epsilon_\sigma - (1 - \langle \hat{n}_{\bar{\sigma}} \rangle)U}{(\epsilon - \epsilon_\sigma - U)(\epsilon - \epsilon_\sigma) - \Sigma^\sigma_{\text{leads}}(\epsilon)\left(\epsilon - \epsilon_\sigma - (1 - \langle n_{\bar{\sigma}} \rangle)U\right)}$$
$$= \frac{1 - \langle \hat{n}_{\bar{\sigma}} \rangle}{\epsilon - \epsilon_\sigma - \Sigma^\sigma_{\text{leads}}(\epsilon)} + \frac{\langle \hat{n}_{\bar{\sigma}} \rangle}{\epsilon - \epsilon_\sigma - U - \Sigma^\sigma_{\text{leads}}(\epsilon)}. \quad (6.45)$$

This result indicates that the couplings to electrodes broaden the resonance from $\epsilon = \epsilon_\sigma$ to $\epsilon = \epsilon_\sigma + \Sigma^\sigma_{\text{leads}}(\epsilon)$ and from $\epsilon = \epsilon_\sigma + U$ to $\epsilon = \epsilon_\sigma + U + \Sigma^\sigma_{\text{leads}}(\epsilon)$ with the same weights and show how the tunneling broadens the isolated levels ϵ_σ. Note that $G^r_\sigma(\epsilon)$ is exact in the noninteracting limit ($U = 0$). Since the occupation is calculated from Green's function by

$$\langle \hat{n}_\sigma \rangle = \int \left(-\frac{1}{\pi} \text{Im}[G^r_\sigma(\epsilon)] \right) f(\epsilon) d\epsilon, \quad (6.46)$$

these are solved self-consistently. After we get Green's function $G^r_\sigma(\epsilon)$, we study the transport of quantum dots using the formula Eq. (6.36). We assume that the electrodes have flat density of states with the couplings

$$\Gamma^\sigma(\epsilon) = \Gamma^\sigma_L(\epsilon) + \Gamma^\sigma_R(\epsilon) = i\left(\Sigma^\sigma_{\text{leads}}(\epsilon) - \Sigma^{\sigma *}_{\text{leads}}(\epsilon)\right)$$
$$= 2\pi \sum_{k \in L, R} |V_{k\sigma}|^2 \delta(\epsilon - \epsilon_{k\sigma}) \quad (6.47)$$

to be constant for spin dependence and have the symmetric barriers as $\Gamma^\sigma_L = \Gamma^\sigma_R = \Gamma^\sigma/2$. Since the coupling to electrodes Γ^σ increases rapidly near the top of the barriers, they are very different. Examples of transport calculations in this formalism are shown in Chapter 3 for two $U/\Gamma = 5{,}100$. We note that the CB regime for quantum dots corresponds to $U/\Gamma = 100$. This formalism is generalized to multiple-level systems to mimic the quantum dot experiments in high-temperature regime by approximating the density-of-states of a level ϵ_n with a sum of delta functions $\sum_m P_n(m)\delta(\epsilon_n + mU - \epsilon)$, where the weight $P_n(m)$ is the probability that m levels are occupied for the isolated site and is given by the corresponding Boltzmann weight. Then we carry out the integral for the conductance to have

$$G = \frac{e^2}{4\hbar} \sum_n \sum_m \left(-\frac{\partial f(\epsilon_n + mU)}{\partial \epsilon} \right) \Gamma_n P_n(m). \quad (6.48)$$

The strengths of coupling to electrodes $\Gamma_n(\ll k_B T)$ are very important for the magnitude of the conductance.

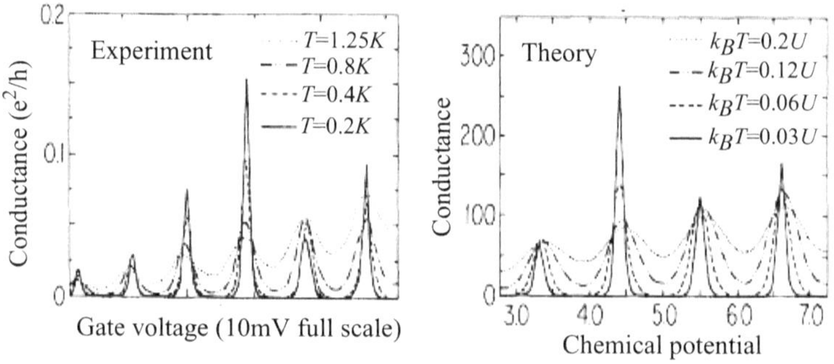

Figure 6.13 Comparison between (Left) measured and (Right) calculated Coulomb oscillations for four different temperature regimes. The level spacing is taken as $\Delta_0 = 0.1U$. For $k_B T \ll \Delta_0$, only one level contributes to each conductance peak, while for $k_B T \simeq \Delta_0$ many levels contribute and the conductance rises with temperature. From Y. Meir, N. S. Wingreen, and P. A. Lee, *Phys. Rev. Lett.* **66**, 3048 (1991) [225].

Figure 6.13 shows an example of the calculated results for the periodic conductance oscillation through the multiple levels as a function of chemical potential μ for four temperatures with $\Gamma = 0.01U$. For $k_B T \ll \Delta_0 = 0.1U$, when the chemical potential passes the lower energy level, the spin-up level fills and the weight of the spin-down level is pushed up to $\epsilon + U$ with the split by $U + \Delta_0 = 1.1U$ in the conductance, while for $k_B T \simeq \Delta_0$ many levels contribute the conductance, which rises with temperature. Thus, the phenomenology of the experiment in high-temperature regime of $k_B T \gg \Gamma$ is reproduced well with the theoretical model for $U \gg \Delta_0$ regime with large N.

6.1.5 *Kondo Effects on Electron Transport*

We have treated the electron hopping $V_{k\sigma}$ perturbatively. This is a good approximation for the high-temperature regime. However, as we decrease the temperature, higher order scattering at contacts produces anomalous conductance due to Kondo resonances.

The Kondo effect traditionally arises from the interactions between a localized single magnetic impurity atom and the many electrons in the non-magnetic metal. Due to their strong coupling with higher-order scatterings, the transport properties exhibit anomalous behaviors at low temperature [226]. In a quantum dot system, if an odd number of electrons are trapped within the dot, the total spin of the dot is necessarily non-zero and this localized spin acts as a magnetic impurity with a large number of electrons on the two electrodes between a quantum dot. The distinction between a quantum dot and a real metal is their geometries. In a metal, the electrons with plane waves are scattered from impurities to mix with different momenta, which results in the increase of resistance. In a quantum dot, since all the electrons go through the devices between two electrodes, the mixing occurs only with states belonging to the two opposite electrodes, which results in the increase of conductance or equivalently the decrease of resistance. Such an artificial quantum dot system is unique to study the Kondo effects, since it is possible to vary the parameters of the single particle level ϵ_σ, its width Γ, and the Coulomb repulsion energy U by the voltages of the gates as shown in Fig. 6.14.

Let us include such higher order contribution. We need solve Green's function directly with the hopping term $V_{k\sigma}$, instead of the second-order perturbation in Eq. (6.44). We define the following Green's functions:

$$G^r_\sigma(t-t') = -\frac{i}{\hbar}\theta(t-t')\langle[\hat{d}_\sigma(t), \hat{d}^\dagger_\sigma(t')]\rangle,$$
$$G^r_{k\sigma}(t-t') = -\frac{i}{\hbar}\theta(t-t')\langle[\hat{c}_{k\sigma}(t), \hat{d}^\dagger_\sigma(t')]\rangle,$$
$$G^{(2)}_\sigma(t-t') = -\frac{i}{\hbar}\theta(t-t')\langle[\hat{d}_\sigma(t)\hat{n}_{\bar{\sigma}}(t), \hat{d}^\dagger_\sigma(t')]\rangle. \quad (6.49)$$

The equations of motions for the first two functions become

$$i\hbar\frac{\partial}{\partial t}G^r_\sigma(t) = \delta(t) + \epsilon_\sigma G^r_\sigma(t) + U\,G^{(2)}_\sigma(t) + \sum_{k\in L,R} V^*_{k\sigma}G^r_{k\sigma}(t)$$
$$i\hbar\frac{\partial}{\partial t}G^r_{k\sigma}(t) = \epsilon_{k\sigma}G^r_{k\sigma}(t) + V_{k\sigma}G^r_\sigma(t), \quad (6.50)$$

where we use (we take $t' = 0$)

$$i\hbar\frac{\partial\hat{d}_\sigma}{\partial t} = [\hat{d}_\sigma, \hat{H}] = \epsilon_\sigma\hat{d}_\sigma + U\hat{d}_\sigma\hat{n}_{\bar{\sigma}} + \sum_{k\in L,R} V^*_{k\sigma}\hat{c}_{k\sigma}$$
$$i\hbar\frac{\partial\hat{c}_{k\sigma}}{\partial t} = [\hat{c}_{k\sigma}, \hat{H}] = \epsilon_{k\sigma}\hat{c}_{k\sigma} + V_{k\sigma}\hat{d}_\sigma. \quad (6.51)$$

From the Fourier transform,

$$(\epsilon - \epsilon_\sigma)G^r_\sigma(\epsilon) = 1 + U\,G^{(2)}_\sigma(\epsilon) + \sum_{k\in L,\,R} V^*_{k\sigma}G^r_{k\sigma}(\epsilon)$$

$$(\epsilon - \epsilon_{k\sigma})G^r_{k\sigma}(\epsilon) = V_{k\sigma}G^r_\sigma(\epsilon), \tag{6.52}$$

we have

$$(\epsilon - \epsilon_\sigma - \Sigma^\sigma_{\text{leads}}(\epsilon))G^r_\sigma(\epsilon) = 1 + U\,G^{(2)}_\sigma(\epsilon) \tag{6.53}$$

with

$$\Sigma^\sigma_{\text{leads}}(\epsilon) = \sum_{k\in L,\,R} \frac{|V_{k\sigma}|^2}{\epsilon - \epsilon_{k\sigma}} = \sum_{k\in L,\,R} |V_{k\sigma}|^2 g_{k\sigma}(\epsilon). \tag{6.54}$$

On the other hand, the equation of motion for $G^{(2)}_\sigma(\epsilon)$ becomes

$$i\hbar\frac{\partial}{\partial t}G^{(2)}_\sigma(t) = \delta(t)\langle\hat{n}_{\bar{\sigma}}\rangle + \epsilon_\sigma G^{(2)}_\sigma(t) + U\,G^{(2)}_\sigma(t)$$
$$+ \sum_{k\in L,\,R}\left[V^*_{k\sigma}\Gamma^{(2)}_{1,k\sigma}(t) + V_{k\sigma}\Gamma^{(2)}_{2,k\sigma}(t) - V^*_{k\sigma}\Gamma^{(2)}_{3,k\sigma}(t)\right],$$

where we introduce the new correlation functions defined by

$$\Gamma^{(2)}_{1,k\sigma}(t-t') = -\frac{i}{\hbar}\theta(t-t')\langle\left[\hat{c}_{k\sigma}(t)\hat{n}_{\bar{\sigma}}(t),\,\hat{d}^\dagger_\sigma(t')\right]\rangle$$
$$\Gamma^{(2)}_{2,k\sigma}(t-t') = -\frac{i}{\hbar}\theta(t-t')\langle\left[\hat{c}^\dagger_{k\bar{\sigma}}(t)\hat{d}_\sigma(t)\hat{d}_{\bar{\sigma}}(t),\,\hat{d}^\dagger_\sigma(t')\right]\rangle$$
$$\Gamma^{(2)}_{3,k\sigma}(t-t') = -\frac{i}{\hbar}\theta(t-t')\langle\left[\hat{c}_{k\bar{\sigma}}(t)\hat{d}^\dagger_{\bar{\sigma}}(t)\hat{d}_\sigma(t),\,\hat{d}^\dagger_\sigma(t')\right]\rangle.$$

Here we use the relation for $\hat{n}_{\bar{\sigma}}$ such as

$$i\hbar\frac{\partial\hat{n}_{\bar{\sigma}}}{\partial t} = \left[\hat{n}_{\bar{\sigma}},\,\hat{H}\right] = \sum_{k\in L,\,R}\left[-V_{k\sigma}\hat{c}^\dagger_{k\sigma}\hat{d}_{\bar{\sigma}} + V^*_{k\sigma}\hat{d}^\dagger_{\bar{\sigma}}\hat{c}_{k\sigma}\right]. \tag{6.55}$$

In the same way, we obtain from the Fourier transform,

$$(\epsilon - \epsilon_\sigma - U)G^{(2)}_\sigma(\epsilon) = \langle\hat{n}_{\bar{\sigma}}\rangle$$
$$+ \sum_{k\in L,\,R}\left[V^*_{k\sigma}\Gamma^{(2)}_{1,k\sigma}(\epsilon) + V_{k\sigma}\Gamma^{(2)}_{2,k\sigma}(\epsilon) - V^*_{k\sigma}\Gamma^{(2)}_{3,k\sigma}(\epsilon)\right].$$

When we use the Hartree–Fock approximation such as

$$\Gamma^{(2)}_{1,k\sigma}(\epsilon) \simeq \langle\hat{n}_{\bar{\sigma}}\rangle G^r_{k\sigma}(\epsilon),\quad \Gamma^{(2)}_{2,k\sigma}(\epsilon) = \Gamma^{(2)}_{3,k\sigma}(\epsilon) = 0, \tag{6.56}$$

then we get

$$(\epsilon - \epsilon_\sigma - U)G^{(2)}_\sigma(\epsilon) = \langle\hat{n}_{\bar{\sigma}}\rangle\left[1 + \Sigma^\sigma_{\text{leads}}G^r_\sigma(\epsilon)\right]. \tag{6.57}$$

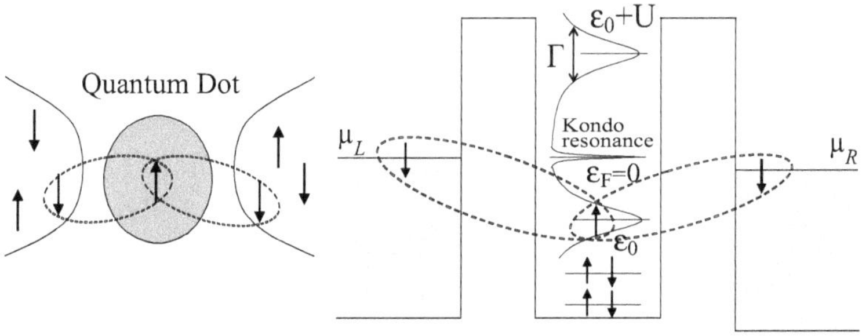

Figure 6.14 (Left) Schematic view of the formation of spin-singlet state due to the coupling of the electronic states of electrodes with a magnetic moment in the quantum dot. (Right) Schematic illustration of the energy diagram. An odd number of electrons is confined in the quantum dot, with spin-up electron occupying the level ϵ_0 below the Fermi energy. At low temperatures, this spin-up electron entangles with opposite spins in the metallic electrodes, creating a total-spin-zero ground state and a many-body Kondo resonance at the Fermi energy with the width determined by the Kondo temperature T_K. The electron tunnels through the quantum dot by using a classically forbidden virtual state.

Combining with Eq. (6.53), we obtain the following formula:

$$G_\sigma^r(\epsilon) = \frac{\epsilon - \epsilon_\sigma - (1 - \langle \hat{n}_{\bar{\sigma}} \rangle)U}{(\epsilon - \epsilon_\sigma - U)(\epsilon - \epsilon_\sigma) - \Sigma_{\text{leads}}^\sigma(\epsilon)\left(\epsilon - \epsilon_\sigma - (1 - \langle \hat{n}_{\bar{\sigma}} \rangle)U\right)}.$$

This is exactly the same result obtained in the previous session in Eq. (6.45). Therefore, we find that the previous perturbative treatment corresponds to the Hartree–Fock approximation within the mean field approach.

To obtain the higher order results on the spin-flip correlation, which leads to the Kondo effect generated by the tunneling term in Eq. (6.56), we calculate the equation-of-motion for $\Gamma_{1,2,3,k\sigma}^{(2)}(t - t')$ and their Fourier transforms. Neglecting some terms not related to spin-flip correlation, we have

$$(\epsilon - \epsilon_{k\sigma})\Gamma^{(2)}_{1,k\sigma}(\epsilon) = V_{k\sigma} G^{(2)}_{\sigma}(\epsilon)$$
$$(\epsilon + \epsilon_{k\sigma} - \epsilon_{\sigma} - \epsilon_{\bar{\sigma}} - U)\Gamma^{(2)}_{2,k\sigma}(\epsilon) = V^{\dagger}_{k\sigma} G^{(2)}_{\sigma}(\epsilon) + \sum_{k' \in L,R} V^{*}_{k'\sigma}\Gamma^{(3)}_{1,kk'\sigma}(\epsilon)$$
$$(\epsilon - \epsilon_{k\sigma} + \epsilon_{\bar{\sigma}} - \epsilon_{\sigma})\Gamma^{(2)}_{3,k\sigma}(\epsilon) = V_{k\sigma}(G^{r}_{\sigma}(\epsilon) - G^{(2)}_{\sigma}(\epsilon)) - \sum_{k' \in L,R} V_{k'\sigma}\Gamma^{(3)}_{2,kk'\sigma}(\epsilon)$$

with the third-order correlation functions approximated by

$$\Gamma^{(3)}_{1,kk'\sigma}(\epsilon) = -\delta_{kk'} f(\epsilon_k) G^{r}_{\sigma}(\epsilon)$$
$$\Gamma^{(3)}_{2,kk'\sigma}(\epsilon) = \delta_{kk'}(1 - f(\epsilon_k)) G^{r}_{\sigma}(\epsilon). \tag{6.58}$$

Substituting these results into Eq. (6.56), we obtain

$$(\epsilon - \epsilon_{\sigma} - U - \Sigma^{\sigma}_{\text{leads}}(\epsilon) - \Sigma^{\sigma}_{3}(\epsilon))G^{(2)}_{\sigma}(\epsilon) + \Sigma^{\sigma}_{1}(\epsilon)G^{r}_{\sigma}(\epsilon) = \langle \hat{n}_{\bar{\sigma}} \rangle.$$

Here we define the new self-energies as

$$\Sigma^{\sigma}_{i}(\epsilon) = \sum_{k \in L,R} A^{(i)}_{k}(\epsilon)|V_{k\sigma}|^2 \times \left[\frac{1}{\epsilon - \epsilon_{\sigma} + \epsilon_{\bar{\sigma}} - \epsilon_{k\sigma}} + \frac{1}{\epsilon - \epsilon_{\sigma} - \epsilon_{\bar{\sigma}} - U + \epsilon_{k\sigma}}\right] \tag{6.59}$$

with $A^{(1)}_{k} = f(\epsilon_k)$, $A^{(2)}_{k} = 1 - f(\epsilon_k)$, and $A^{(3)}_{k} = 1$.

Then we get Green's function of the quantum dot as

$$G^{r}_{\sigma}(\epsilon) = \frac{1 - \langle \hat{n}_{\bar{\sigma}} \rangle}{\epsilon - \epsilon_{\sigma} - \left[\Sigma^{\sigma}_{\text{leads}}(\epsilon) - U\,\Sigma^{\sigma}_{1}(\epsilon)\left(\epsilon - \epsilon_{\sigma} - U - \Sigma^{\sigma}_{\text{leads}}(\epsilon) - \Sigma^{\sigma}_{3}(\epsilon)\right)^{-1}\right]} + \frac{\langle \hat{n}_{\bar{\sigma}} \rangle}{\epsilon - \epsilon_{\sigma} - U - \left[\Sigma^{\sigma}_{\text{leads}}(\epsilon) + U\,\Sigma^{\sigma}_{2}(\epsilon)\left(\epsilon - \epsilon_{\sigma} - \Sigma^{\sigma}_{\text{leads}}(\epsilon) - \Sigma^{\sigma}_{3}(\epsilon)\right)^{-1}\right]}.$$

This expression for $G^{r}_{\sigma}(\epsilon)$ is exact in the noninteracting limit ($U = 0$) and in the isolated-site limit ($V_{k\sigma} = 0$). We see that there are two resonances, located in the vicinity of $\epsilon = \epsilon_{\sigma}$ and $\epsilon = \epsilon_{\sigma} + U$ with the weights of $1 - \langle \hat{n}_{\bar{\sigma}} \rangle$ and $\langle \hat{n}_{\bar{\sigma}} \rangle$. However, the resonances are modified and shifted by the multiple tunneling processes that electrons go back and forth between the quantum dots and the contacts, which are represented by $\Sigma^{\sigma}_{\text{leads}}$ and the new self-energies of Σ^{σ}_{i}. Generally,

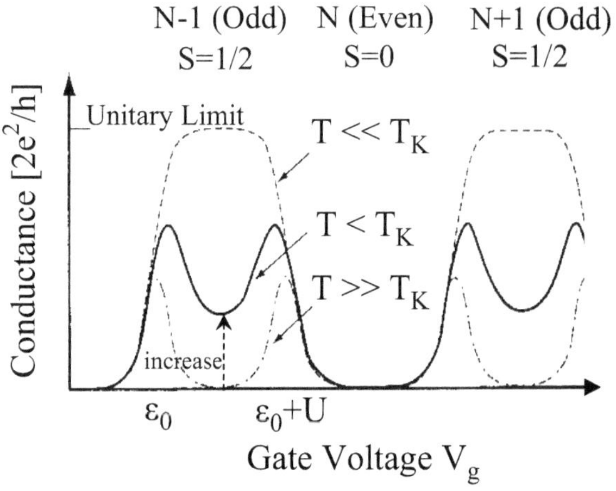

Figure 6.15 Schematic view of the conductance behavior as a function of gate voltage, which changes the number of electrons N in the quantum dot. When N is odd, the conductance increases as a decrease of temperature T. In the high-temperature regime $T \gg T_k$, the conductance behaviors are by the single-electron tunneling, while for $T \ll T_K$ the conductance approaches the quantum limit of $2e^2/h$ due to perfect conduction through a virtual Kondo state.

$G^r_\sigma(\epsilon)$ depends on the temperature, chemical potential through $\langle \hat{n}_{\bar{\sigma}} \rangle$, coupling to the electrodes $\Sigma^\sigma_{\text{leads}}$, and the new self-energies Σ^σ_i. These parameters produce the Kondo anomaly for the electron transport through a quantum dot.

Now, in order to see the characteristic features of the Kondo resonance of a quantum dot, let us consider the infinite U limit case for simplicity, where the double occupancy of electrons at the quantum dot site is prohibited. In this case, Green's functions of the dot and the self-energies become

$$G^r_\sigma(\epsilon) = \frac{1 - \langle \hat{n}_{\bar{\sigma}} \rangle}{\epsilon - \epsilon_\sigma - \Sigma^\sigma_{\text{leads}}(\epsilon) - \Sigma^\sigma_1(\epsilon)}$$

$$\Sigma^\sigma_{\text{leads}}(\epsilon) = \sum_{k \in L,R} \frac{|V_{k\sigma}|^2}{\epsilon - \epsilon_{k\sigma} + i0^+}$$

$$\Sigma^\sigma_1(\epsilon) = \sum_{k \in L,R} \frac{|V_{k\sigma}|^2 f(\epsilon_k)}{\epsilon - \epsilon_\sigma + \epsilon_{\bar{\sigma}} - \epsilon_{k\sigma} + i0^+}. \tag{6.60}$$

Since the Fermi surfaces are sharp at low temperature, $\mathrm{Re}[\Sigma_1^\sigma]$ grows logarithmically at μ and gives rise to the peak due to virtual intermediate states in the density of states. Correspondingly, the differential conductance dG/dV has a peak structure with the width determined by the Kondo temperature T_K. This signature is called the zero-bias anomaly (ZBA) of the Kondo peak and becomes a feature for the formation of anomalous Kondo state.

Let us obtain the approximate conductance formula for the Kondo transport. We write the imaginary Γ and the real part Λ of the self-energies as

$$\begin{aligned}\Gamma &= \Gamma_L + \Gamma_R = -2\mathrm{Im}\left(\Sigma^\sigma_{\mathrm{leads}}(\epsilon) + \Sigma^\sigma_1(\epsilon)\right) \approx 2\pi |V|^2 \nu \\ \Lambda &= \mathrm{Re}\left(\Sigma^\sigma_{\mathrm{leads}}(\epsilon) + \Sigma^\sigma_1(\epsilon)\right) \approx |V|^2 \mathrm{P}\left[\sum_k \frac{f(\epsilon_k)}{\epsilon - \epsilon_k}\right] \\ &\approx |V|^2 \int_D^{k_B T} \frac{\nu}{\epsilon - \epsilon'} d\epsilon' \approx \frac{\Gamma}{2\pi} \ln\left|\frac{D}{k_B T}\right|, \end{aligned} \tag{6.61}$$

where ϵ_0 is the localized state and D is the band edge energy. Contribution from Σ_1^σ to Λ is essential for the logarithmic behavior. Then we can approximate $G^r_\sigma(\epsilon)$ as

$$G^r_\sigma(\epsilon) \approx \frac{1}{\left(\epsilon - \epsilon_0 - \frac{\Gamma}{2\pi}\ln\left|\frac{D}{k_B T}\right|\right) + i\Gamma/2}. \tag{6.62}$$

Substituting these results to the conductance formula, we have

$$\begin{aligned} G &= \frac{e^2}{\hbar}\sum_\sigma \int_{-\infty}^{\infty} \left(-\frac{\partial f(\epsilon)}{\partial \epsilon}\right) \frac{\Gamma^\sigma_L(\epsilon)\Gamma^\sigma_R(\epsilon)}{\Gamma^\sigma_L(\epsilon) + \Gamma^\sigma_R(\epsilon)} \left(-\frac{1}{\pi}\mathrm{Im}\left[G^r_\sigma(\epsilon)\right]\right) d\epsilon \\ &\approx \frac{2e^2}{h} \frac{\Gamma_L \Gamma_R}{(\mu - \epsilon_0)^2 + (\Gamma/2)^2} \left(1 + \frac{\Gamma}{\pi|\mu - \epsilon_0|} \ln\left|\frac{D}{k_B T}\right| + \cdots\right). \end{aligned} \tag{6.63}$$

The first term is the usual conductance from the resonance and the second term shows that the conductance increases logarithmically when the temperature is lowered [228–230], which is the well-known signature as the Kondo effect [231].

We note that this result is obtained from the s–d Hamiltonian

$$\begin{aligned} \hat{H}_{sd} &= -\sum_{k \in L, R} J_k \left[S^+ \hat{c}^\dagger_{k\downarrow}\hat{c}_{k\uparrow} + S^- \hat{c}^\dagger_{k\uparrow}\hat{c}_{k\downarrow} + S_z(\hat{c}^\dagger_{k\uparrow}\hat{c}_{k\downarrow} - \hat{c}^\dagger_{k\downarrow}\hat{c}_{k\uparrow})\right] \\ J_k &= \frac{|V_k|^2}{\epsilon - \epsilon_k} \quad (U \to \infty) \end{aligned} \tag{6.64}$$

with

$$S^+ = d_\uparrow^\dagger d_\downarrow, \quad S^- = d_\downarrow^\dagger d_\uparrow, \quad S_z = (d_\uparrow^\dagger d_\uparrow - d_\downarrow^\dagger d_\downarrow)/2 \tag{6.65}$$

by the second-order perturbation of the spin flipping process as

$$\langle k \uparrow | \hat{H}_{sd} \frac{1}{\epsilon - \hat{H}_0 + i0} \hat{H}_{sd} | k \uparrow \rangle \approx \sum_k |J_k|^2 \frac{f(\epsilon_k)}{\epsilon - \epsilon_k + i0}. \tag{6.66}$$

Since Eq. (6.64) is written as

$$\hat{H}_{sd} = - \sum_{k\sigma\sigma' \in L, R} J_k \hat{c}_{k\sigma}^\dagger \sigma_{\sigma\sigma'} \hat{c}_{k\sigma'} \cdot S, \tag{6.67}$$

where $\sigma_{\sigma\sigma'}$ is the Pauli matrix, the Kondo state is generated by the exchange processes between localized electrons in the quantum dot and free-electron states. As many electrons are involved, the Kondo resonance is a many-body spin-singlet state.

In the artificial quantum dot systems, the Kondo effect is observed in the transport measurement through quantum dots, which act as localized spin. As the temperature becomes lower, the conductance at the chemical potential grows, even if it is between the SET peaks and off-resonance. Thus, we can observe the Kondo effect in the temperature-dependent behavior of conductance as a function of the gate voltage. When the number of electrons in the quantum dot is odd and as a result the spin remains, the conductance through quantum dots increases as a decrease of temperature as we see in Fig. 6.15.

It is important to note that the expansion in Eq. (6.63) breaks down when the second term exceeds the first term. This determines the critical temperature

$$k_B T_K \propto \exp\left[-\frac{\pi|\mu - \epsilon_0|}{\Gamma}\right] \tag{6.68}$$

called the Kondo temperature T_K.[a] Above the Kondo temperature $T > T_K$ in the high-temperature regime, the perturbative approach is correct and we see the logarithmic-dependent conductance in addition to the SET peaks. When the temperature is lower than T_K, we cannot use the perturbative approach for the transport for $T \ll T_K$. In this situation, multiple-scatterings of electrons occur between

[a] Since $\Gamma = 2\pi|V|^2\nu$, this can be written as $k_B T_K \propto \exp[-1/2\nu J]$, where $J = |V|^2/|\mu - \epsilon_0|$ is the anti-ferromagnetic exchange coupling.

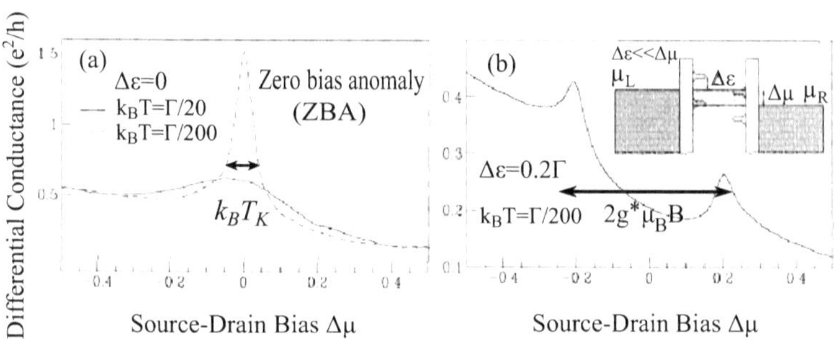

Figure 6.16 (Left) Zero bias differential conductance dI/dV as a function of the source-drain bias voltage in the nonequilibrium situation. The Kondo peak appears at the Fermi energy due to the virtual state formed between the electrodes and quantum dot, which is called the ZBA with the width determined by the Kondo temperature $k_B T_K$. (Right) Differential conductance at a finite magnetic field. The ZBA from the Kondo peaks splits due to the Zeeman energy. (Inset) Schematic views of the energy diagram. When the source-drain bias reaches the Zeeman splitting, the Kondo peaks enter the region between the chemical potentials and give rise to a peak in dI/dV. From Y. Meir, N. S. Wingreen, and P. A. Lee, *Phys. Rev. Lett.* **70**, 2601 (1993) [227].

electrodes through the quantum dot via the virtual hopping and correspondingly a strongly correlated Kondo state is formed at the Fermi energy. The magnetic moment of the quantum dot is screened entirely by the spins of mobile electrons in the electrodes. Since the Kondo state has a coherent nature and is strongly dependent on the temperature, the transport through the quantum dot via the Kondo state is also a strong function of the temperature. The ZBA, an important characteristic of the Kondo effect, is observed in the conductance at the Fermi energy when we change source-drain bias voltage. The differential conductance shows very sharp peaks with the width of $k_B T_k$. Applying a finite magnetic field to the quantum dot, the differential conductance of the ZBA from the Kondo peaks splits due to the Zeeman energy as shown in Fig. 6.16.

The behaviors of the Kondo effect at various temperature regimes are one of the central issues for the correlated electron system and have been studied intensely using various calculation techniques. It is shown that for the finite U case, the Kondo temperature T_K is expressed in terms of the two localized states at ϵ_0 and $\epsilon_0 + U$ as

$$k_B T_K = \sqrt{\Gamma U} \exp\left[\frac{\pi(\epsilon_0 - \mu)(\epsilon_0 + U - \mu)}{\Gamma U}\right]. \tag{6.69}$$

From the perturbative approach above, the temperature dependence of the conductance peak of the ZBA at the Fermi energy for $T \gg T_K$ reveals logarithmic behaviors. On the other hand, when the temperature is less than the Kondo temperature $T \ll T_K$, the conductance peak is approximated by

$$G \approx \frac{2e^2}{h}\left[1 - \frac{\pi^2 T^2}{T_K^2}\right]. \tag{6.70}$$

More detailed calculations show that the temperature dependence of the conductance peak reveals the following universal feature for all the temperature regime described as

$$G = \frac{2e^2}{h} f\left(\frac{T}{T_K}\right), \tag{6.71}$$

where the $f(x)$ is a universal function such as

$$f(x) = \left[1 + (2^{\frac{1}{s}} - 1)x^2\right]^{-s} \tag{6.72}$$

with $s = 0.23$. The conductance increases as a decrease of the temperature, and in the limiting low temperature $T \to 0$, it approaches the unitary limit[a] of $G_0 = 2e^2/h$ even if the hopping to the quantum dot from electrodes are very small in the tunneling regime due to the barrier heights. We note that the perturbative calculation in Eq. (6.63) is divergent as $T \to 0$.

Here we see two experimental observations of the Kondo effects in the transport measurements of the quantum dots. The top panels of Fig. 6.17 show the conductance behavior as a function of the gate voltage for various temperatures from 100 mK up to 3.8 K, which corresponds to Fig. 6.15. The vertical dashed lines mark gate voltages at which two charge states are degenerate. Between the dashed lines, the charge state of the quantum dot site is odd and we see that the Kondo effect enhances conductance. The temperature

[a] Here we assume the symmetric barriers $\Gamma_L = \Gamma_R = \Gamma/2$ for $G_0 = 2e^2/h$. In case of asymmetric barriers, $G_0 = (2e^2/h) \times 4\Gamma_L\Gamma_R/(\Gamma_L + \Gamma_R)^2$.

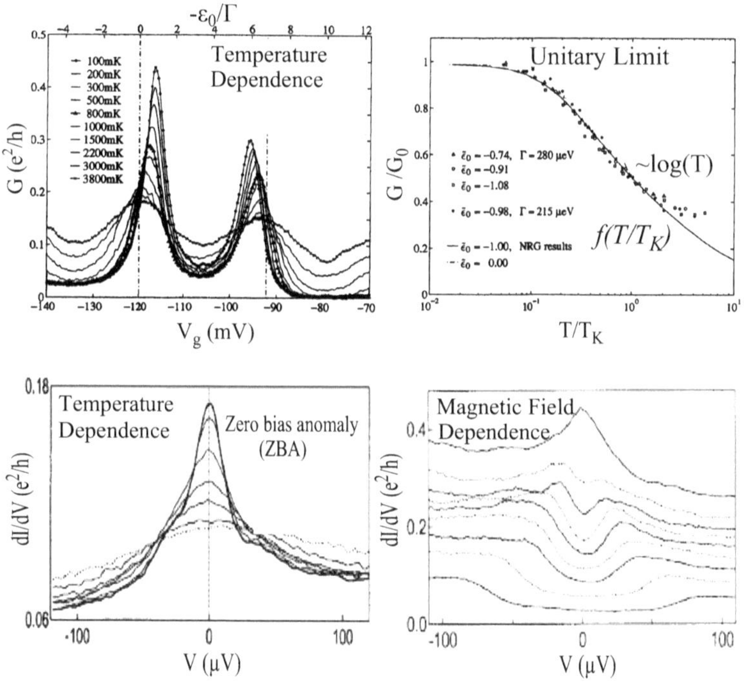

Figure 6.17 (Top) (Left) Conductance vs. gate voltage V_g at various temperatures. (Right) Normalized conductance $\tilde{G} = G/G_0$ as a function of $\tilde{T} = T/T_K$ with the universal function. From D. Goldhaber-Gordon, J. Göres, M. A. Kastner, H. Shtrikman, D. Mahalu, and U. Meirav, *Phys. Rev. Lett.* **81**, 5225 (1998) [232]. (Bottom) (Left) Differential conductance dI/dV vs. source-drain bias. (Right) The splitting of the ZBA peak in dI/dV with a magnetic field. From S. M. Cronenwett, T. H. Oosterkamp, and L. P. Kouwenhoven, *Science* **281**, 540 (1998) [233].

dependence of the conductance is shown to fit the universal function $f(T/T_K)$ very well.

The bottom panels of Fig. 6.17 show the differential conductance as a function of the source-drain voltage for various temperatures (left) and the splitting of the ZBA peak by the magnetic field $B_{||}$ in the plane of the 2DEG (right), which corresponds to Fig. 6.16. The temperature for the left panel is $T = 45$, 50, 75, 100, 130, 200, and 270 mK, respectively. We see the evolution of the ZBA as the temperature T decreases. The peak maximum shows logarithmic in

T and width is linear in T. The magnetic field for the right panel is $B_{||} = 0.10, 0.43, 0.56, 0.80, 0.98, 1.28, 1.48, 2.49$, and 3.49 T. We resolve a splitting that increases linearly with $B_{||}$. These experiments of the transport through quantum dots show the clear evidence of the Kondo resonance.

6.2 Density Functional Theory of Quantum Dots

In the previous section, we see that electronic states of quantum dots in 2DEG systems are determined not only by their single-particle energy ϵ_i, but also by the Coulomb charging energy. Recent experiments make it possible to fabricate the quantum dots whose structures are controlled very well. Such quantum dots show atomic-like spectra in the SET transport measurements, reflecting internal electronic states and thereby are called artificial atoms. Since a number of useful applications might be possible using the well-controlled quantum dots, the advent of such atomic-like spectra in quantum dots calls for appropriately quantitative theoretical tools for calculating $E_{\mathrm{dot}}(N)$, including charging energy, single-particle energy, and spin states, precisely from the accurate electronic-state calculations to compare to experiment [234, 235].

We present the electronic states of quantum dots based on the density functional formalism. This method treats a large number of inhomogeneous interacting electron systems for various confining external potentials and provides accurate ground-state energies $E_{\mathrm{dot}}(N)$, including the effects of spin, magnetic fields, as well as the deformations of the potentials and also the disorder. We note that other methods such as exact diagonalization, Hartree–Fock, and Quantum Monte Carlo (QMC), have merits and demerits in accuracies, the tractable sizes, computational times, and so on [236].

Beginning from the classical capacitance model, which includes Coulomb interaction but no quantum states, we compare $E_{\mathrm{dot}}(N)$ with SDFT and other methods to show that SDFT provides very accurate results. Then electronic and spin states of quantum dots are shown in the framework of the DFT, which includes the magnetic fields effect by CDFT and excitation energies by TDDFT. In the artificial quantum dot systems, different from real atoms, disorder and deformation cannot be avoided, which induce localized nature of wavefunctions and thus affect the transport properties. Since the Coulomb interaction is much larger than the quantum-state energy spacing, it is interesting to ask which of them determines their transport behaviors. Finally, we treat conductance fluctuations of quantum dots due to disorder.

6.2.1 *Spin-DFT for Single-Electron Transport*

Quantum dots are tiny islands artificially created variously, and thus their actual induced potentials are different from sample to sample. In the well-controlled circular-shape quantum dots, the external potential is expressed approximately by the parabolic confining potential $V_{\text{ext}}(\mathbf{r}) = (1/2)m^*\omega_0^2 r^2$. As we show in the previous section, the Coulomb charging energy is larger than the energy level spacing $E_c/\Delta_0 \gg 1$ for $N \gg 1$ in the quantum dots. Thus, we consider first the classical Coulomb energy contribution to the addition energy $\Delta(N)$, neglecting the quantum effects of exchange and correlation energy and the kinetic energy of electrons.

6.2.1.1 Classical capacitance model of quantum dots

First, we view the Coulomb energy effect as a classical electrostatic capacitance C problem and see how many electrons N a quantum dot can occupy for repulsive Coulomb interaction. The addition energy $\Delta(N)(= \mu(N+1) - \mu(N))$ is obtained from

$$\mu(N) = V_{\text{ext}}(r) + \frac{e^2}{\kappa}\int_0^{2\pi} d\theta \int_0^R \frac{n(r')}{|\mathbf{r}-\mathbf{r}'|} r' dr' \tag{6.73}$$

with the restriction of

$$2\pi \int_0^R n(r) r dr = N. \tag{6.74}$$

Here R is the radius of the 2D disk-type circular quantum dot in Fig. 6.8. The solutions for the electron density $n(r)$ and $\mu(N)$ are obtained by

$$n(r) = \frac{3N}{2\pi R^2}\sqrt{1 - \frac{r^2}{R^2}}$$

$$\mu(N) = N^{2/3}\left[\left(\frac{3\pi e^2}{4\kappa}\right)^2 m^*\omega_0^2\right]^{1/3} \tag{6.75}$$

with $R^3 = (3\pi e^2/4\kappa(N/m^*\omega_0^2)$ [237]. This shows that the addition energy decreases as $\Delta(N) \sim d\mu(N)/dN \sim N^{-1/3}$ for an increase of N. Alternatively, the size of the quantum dot R increases as $N^{1/3}$ due to the repulsion of electrons. Then the capacitance $C \propto R$ of the quantum dot increases as $N^{1/3}$ and the addition energy $\Delta(N) = e^2/C$ decreases as $N^{-1/3}$.

The quantum correction is estimated from the Thomas–Fermi approximation to include the kinetic energy of electrons as

$$\mu(N) = \frac{\pi\hbar^2}{2m^*}n(r) + V_{\rm ext}(r) + \frac{e^2}{\kappa}\int_0^{2\pi} d\theta \int_0^R \frac{n(r')}{|\mathbf{r}-\mathbf{r}'|}r'dr'. \quad (6.76)$$

This is written in terms of dimensionless variables as

$$\text{const} = \frac{\pi}{2}\frac{a_B^*}{R}\tilde{n}(\tilde{r}) + V_{\rm ext}(r)\frac{\kappa R}{e^2 N} + \int_0^1 \frac{\tilde{n}(\tilde{\mathbf{r}}')}{|\tilde{\mathbf{r}}-\tilde{\mathbf{r}}'|}d\tilde{\mathbf{r}}' \quad (6.77)$$

with the scaling of $\tilde{r} = r/R$ and $\tilde{n} = nR^2/N$ and the effective Bohr radius $a_B^* = \kappa\hbar^2/m^*e^2$. Thus, the quantum correction is small if $a_B^*/R \ll 1$. Since $a_B^* \sim 10$ nm for GaAs, the quantum correction is negligible for large quantum dots with $R \gg 200$ nm and $N \gg 200$. As the number of electrons in the dots decreases by applying a negative bias to the gate, we can reach the situation where only a few electrons exist in the quantum dots.

Here let us estimate the Coulomb energy for a small quantum dot with a parabolic potential. An approximate size of a quantum dot is obtained from the minimization of $\Delta E = \hbar^2/2m^*\Delta r^2 + (1/2)m^*\omega_0^2\Delta r^2$ to give $\ell_0 = \sqrt{\hbar/m^*\omega_0}$. When we take the typical confinement parameter of $\hbar\omega_0 = 3.0$ meV and the material parameters of GaAs $m^* = 0.067m$ and $\kappa = 12.9$, the corresponding interaction strength becomes $(e^2/\kappa\ell_0)/\hbar\omega_0 = 1.91$ with $\ell_0 \simeq 19.5$ nm. This shows that the Coulomb energy $E_c \sim e^2/\kappa\ell_0$ still dominates the single-particle spectrum $\Delta_0 \sim \hbar\omega_0$, but we expect the manifestations of the quantum phenomena as artificial atoms.

6.2.1.2 Single-particle states and degeneracy of 2D circular quantum dots

Second, we consider the case where one electron ($N = 1$) is occupied in the 2D quantum dot with a parabolic potential. In this case, there is no capacitance C effect with no electron–electron interaction. Since this corresponds to the noninteracting electron problem, we solve the 2D Schrödinger equation

$$\left[-\frac{\hbar^2}{2m^*}\nabla^2 + \frac{1}{2}m^*\omega_0^2 r^2\right]\varphi_{n,k}(r) = \epsilon_{n,k}\varphi_{n,k}(r) \tag{6.78}$$

to have the eigenstates of

$$\varphi_{n,k}(r,\theta) = |n,k\rangle = \sqrt{\frac{n!}{\pi(n+|k|)!}}\frac{1}{\ell_0}e^{-\frac{r^2}{2\ell_0^2}}\left(\frac{r^2}{\ell_0^2}\right)^{\frac{|k|}{2}}\mathrm{L}_n^{|k|}\left(\frac{r^2}{\ell_0^2}\right)e^{-ik\theta}. \tag{6.79}$$

Here $\mathrm{L}_n^{|k|}(x)$ is a Laguerre polynomial and $\ell_0 = \sqrt{\hbar/m^*\omega_0}$. This is called the Fock–Darwin states.

The electronic states are labeled by two quantum numbers (n, k), where $n(= 0, 1, 2, \cdots)$ corresponds to the principal value and $k(= \cdots, -2, -1, 0, 1, 2, \cdots)$ corresponds to the angular momentum. The single-particle energy levels form a ladder,

$$\epsilon_{n,k} = (2n + |k| + 1)\hbar\omega_0 = M\hbar\omega_0 \tag{6.80}$$

with rung degeneracy M. These are $|0, 0\rangle$ for $M = 1$; $|0, \pm 1\rangle$ for $M = 2$; $|1, 0\rangle$ and $|0, \pm 2\rangle$ for $M = 3$; $|1, \pm 1\rangle$ and $|0, \pm 3\rangle$ for $M = 4$; $|2, 0\rangle$, $|1, \pm 2\rangle$, $|0, \pm 4\rangle$ for $M = 5$; and so on. The total number of the degeneracy of the states is equal to $\sum_{M=1}^{n} M = n(n + 1)$ and there are energy gaps of $\hbar\omega_0$ between these numbers. Thus, the electronic states form closed-shell structures at the electron numbers of $N = 2n(n + 1) = 2, 6, 12, 20, 30, 42, 56, \cdots$, where we include the spin degeneracy. We will see the closed-shell structures in the transport spectrum measurements of the quantum dots later.

Here we note the similarity to and the difference from the real atom. In the hydrogen atom case ($N = 1$), the confining potential is constructed from the nucleus of a proton $V_{\text{ext}}(\mathbf{r}) = -e^2/r$. The electronic states are expressed by three quantum numbers (n, l, m), where $l = 0, 1, 2, \cdots, n-1$ and $m = -l, -l+1, \cdots, l-1, l$, and the single-particle energy levels are $\epsilon_{n,l,m} = -1/2n^2$ in atomic units of $me^4/\hbar^2 \simeq 27.2$ eV. In the spherically symmetric Coulomb field, each state with a principal quantum number n is degenerate. In this case, the total number of the degeneracy of the states is equal to $\sum_{l=0}^{n-1}(2l + 1) = n^2$. The electronic states thus form closed-shell structures at the electron numbers of $N = 2n^2 = 2, 8, 18, 32, \cdots$, where we include the spin degeneracy. The closed-shell structures in the real atomic systems are observed in the optical spectrum measurements.

6.2.1.3 Formalism of SDFT for 2D quantum dots

Finally, we consider the many-electron cases ($N > 1$) with electron–electron interactions, which break the degeneracy M. We apply the spin-density functional method and solve the Kohn–Sham equations to obtain the total energy $E_{\text{dot}}(N)$. It is reasonable to expand the basis function

$$\psi_{i\sigma}(\mathbf{r}) = \sum_{nk} C^{\sigma}_{nk}\varphi_{nk}(r, \theta)\chi_{\sigma} \tag{6.81}$$

by the Fock–Darwin states and a spin function χ_{σ}.

In the spin-density functional theory (SDFT), the ground-state energy of a quantum dot with N electrons is obtained from

$$E_{\text{dot}}(N) = \sum_{i,\sigma}^{N} \epsilon_i^{\sigma} - \frac{e^2}{2\kappa}\int\int \frac{n(\mathbf{r})n(\mathbf{r}')}{|\mathbf{r}-\mathbf{r}'|}d\mathbf{r}d\mathbf{r}' + E_{\text{xc}} - \sum_{\sigma}\int n^{\sigma}(\mathbf{r})\frac{\delta E_{\text{xc}}}{\delta n^{\sigma}(\mathbf{r})}d\mathbf{r}, \tag{6.82}$$

where σ denotes the spin index. The Kohn–Sham eigenvalues ϵ_i^{σ} and the electron density $n(\mathbf{r})$ are obtained numerically from the self-consistent solution to the Kohn–Sham equations,

$$\left[-\frac{\hbar^2}{2m^*}\nabla^2 + V_{\text{eff}}(\mathbf{r})\right]\psi_{i\sigma}(\mathbf{r}) = \epsilon_i^{\sigma}\psi_{i\sigma}(\mathbf{r}) \tag{6.83}$$

$$n(\mathbf{r}) = \sum_{\sigma} n^{\sigma}(\mathbf{r}) = \sum_{\sigma}\sum_{i}|\psi_{i\sigma}(\mathbf{r})|^2, \tag{6.84}$$

where the effective potential is constructed from

$$V_{\text{eff}}(\mathbf{r}) = V_{\text{ext}}(\mathbf{r}) + \frac{e^2}{\kappa}\int\frac{n(\mathbf{r}')}{|\mathbf{r}-\mathbf{r}'|}d\mathbf{r}' + \frac{\delta E_{\text{xc}}}{\delta n^{\sigma}(\mathbf{r})}. \tag{6.85}$$

The first term is an external potential for the confinement and impurities, the second term is the classical Coulomb interaction (Hartree) potential, and the third term is the exchange-correlation energy E_{xc}, which is treated by the local-density approximation (LDA)

$$E_{\text{xc}} = \int n(\mathbf{r})\epsilon_{\text{xc}}[n(\mathbf{r}), \zeta(\mathbf{r})]d\mathbf{r}$$

$$\epsilon_{\text{xc}}[n(\mathbf{r}), \zeta(\mathbf{r})] = \epsilon_{\text{x}}[n(\mathbf{r}), \zeta(\mathbf{r})] + \epsilon_{\text{c}}[n(\mathbf{r}), \zeta(\mathbf{r})] \tag{6.86}$$

with the spin polarization $\zeta(\mathbf{r}) = [n^{\uparrow}(\mathbf{r}) - n^{\downarrow}(\mathbf{r})]/n(\mathbf{r})$. The exchange-correlation potential is obtained from

$$\frac{\delta E_{xc}}{\delta n^{\sigma}(\mathbf{r})} = \epsilon_{xc}(\mathbf{r}) - \frac{r_s}{2}\frac{\partial \epsilon_{xc}(\mathbf{r})}{\partial r_s} - (\zeta - \mathrm{sgn}\sigma)\frac{\partial \epsilon_{xc}(\mathbf{r})}{\partial \zeta}, \tag{6.87}$$

where

$$\epsilon_x[r_s, \zeta] = -\frac{2\sqrt{2}}{3\pi r_s}\left[(1+\zeta)^{3/2} + (1-\zeta)^{3/2}\right] \tag{6.88}$$

is the exchange energy for 2D electrons and $r_s = 1/\sqrt{\pi n} a_B^*$ is the local interatomic distance in units of the effective Bohr radius $a_B^* = \kappa\hbar^2/m^* e^2$.

As for the correlation energy ϵ_c, which reflects the correlated quantum effects for the 2D interacting electrons, there are several proposed approximate functionals. Here we show two of them. One is given by Tanatar and Ceperley [238] as

$$\epsilon_c[r_s, \zeta] = \epsilon_c[r_s, 0] + \frac{(1+\zeta)^{3/2} + (1-\zeta)^{3/2} - 2}{2(\sqrt{2}-1)}(\epsilon_c[r_s, 1] - \epsilon_c[r_s, 0]) \tag{6.89}$$

with

$$\epsilon_c[r_s, 0/1] = a_0^{0/1}\frac{1 + a_1^{0/1} r_s^{1/2}}{1 + a_1^{0/1} r_s^{1/2} + a_2^{0/1} r_s + a_3^{0/1} r_s^{3/2}}, \tag{6.90}$$

and the other is given by Attaccalite *et al.* [239]

$$\epsilon_c[r_s, \zeta] = (e^{-\beta r_s} - 1)\left\{\epsilon_x[r_s, \zeta] - \left(1 + \frac{3}{8}\zeta^2 + \frac{3}{128}\zeta^4\right)\epsilon_x[r_s, 0]\right\}$$
$$+ \alpha_0(r_s) + \alpha_1(r_s)\zeta^2 + \alpha_2(r_s)\zeta^4 \tag{6.91}$$

with

$$\alpha_i(r_s) = A_i + (B_i r_s + C_i r_s^2 + D_i r_s^3)$$
$$\times \ln\left[1 + \frac{1}{E_i r_s + F_i r_s^{3/2} + G_i r_2^2 + H_i r_s^3}\right]. \tag{6.92}$$

The parameters $a_{0-3}^{0/1}$ and A_{0-2}–H_{0-2} are given in the references [238, 239]. The total energy difference is almost negligible in the interaction regime in the present quantum dot systems.

To check the accuracies of the SDFT for the interacting 2D electron systems of quantum dots, we compare the total energies

Table 6.2 Comparison of ground-state total energies $E_{\text{dot}}(N)$ for a quantum dot with $2 \leq N \leq 13$ electrons by Hartree–Fock (HF), spin-density functional (SDFT), and diffusion Monte Carlo (DMC) methods

N	E_{HF}	E_{SDFT}	E_{DMC}	ΔE_{HF}	ΔE_{SDFT}
2	1.1420	1.04684	1.0216	0.1204	0.0252
3	2.4048	2.26308	2.2339	0.1709	0.0292
4	3.9033	3.74632	3.7145	0.1889	0.0318
5	5.8700	5.56919	5.5338	0.3362	0.0354
6	8.0359	7.63500	7.6001	0.4358	0.0349
7	10.5085	10.07176	10.0342	0.4743	0.0376
8	13.1887	12.72691	12.6900	0.4987	0.0369
9	16.1544	15.61889	15.5801	0.5743	0.0388
10	19.4243	18.76357	18.7232	0.7011	0.0404
11	22.8733	22.11130	22.0738	0.7995	0.0375
12	26.5490	25.67597	25.6356	0.9134	0.0404
13	30.4648	29.53617	29.4938	0.9710	0.0424

Note: The parameters are $\kappa = 12.4$, $m^* = 0.067m$, and $\hbar\omega_0 = 3.32$ meV. Energies are in units of $E^* = 11.86$ meV$(= 2E^*_{\text{Ryd}})$. Note that the errors by SDFT $\Delta E_{\text{SDFT}} = E_{\text{LSDA}} - E_{\text{DMC}}$ are approximately one order of magnitude smaller than those by HF method $\Delta E_{\text{HF}} = E_{\text{HF}} - E_{\text{DMC}}$ for all the electron numbers.

Source: From F. Pederiva, C. J. Umrigar, and E. Lipparini, *Phys. Rev. B* **62**, 8120 (2000); **68**, 089901 (2003) [240].

with those obtained from other calculations. Table 6.2 gives the ground-state total energies for the parabolic quantum dot for N from 2 to 13 by the Hartree–Fock (HF) (left), the SDFT (center), and diffusion Monte Carlo (DMC) (right) methods. The DMC method is known to be the most accurate here and gives energies close to the exact-diagonalization method for a small number of electrons [240]. Thus, we assume the DMC method reliable at least up to $N = 13$ electrons. We see that the errors by SDFT are approximately one order of magnitude smaller than those by HF. Moreover, in HF the error increases linearly with the number of electrons, while in SDFT the error remains almost constant.

It is interesting to note that the HF method often gives better total energies than SDFT in atoms or molecules, while in 2D quantum dots SDFT reliably gives better energies. The reason for this difference might be that in a 2D electron system the correlation energy E_c, which is missing in HF calculations, is much more important.

Another reason is that electron densities in quantum dots tend to become more uniform than in atoms or molecules, so the local-density approximation works better. These results motivate us to use SDFT to calculate electronic states for larger numbers of electrons in quantum dots.[a]

6.2.1.4 Electronic states of circular and elliptical quantum dots

Now we see the SDFT results for the full quantum calculations of GaAs quantum dots. Figure 6.18(a) (left) shows the addition energy $\Delta(N)$ as a function of electron number from $N = 1$ up to $N = 58$. The dotted line indicates $\Delta(N)$ obtained from a classical electrostatic analysis with no kinetic energy in Eq. (6.75). Overall, $\Delta(N)$ decreases with $N^{1/3}$, as the dot and its capacitance grow. On average, the addition energy obtained from the density functional calculation is close to the classical electrostatic result e^2/C even for small N.

However, we see small zigzag structures, and unusually large addition energies at electron numbers $N = 2, 6, 12, 20, 30, 42$, and 56. This shows apparently the manifestation of the quantum phenomena. As we see, these numbers correspond to the formation of closed-shell structures. Even in the presence of electron–electron interaction, extra energy is required to add one more electron to a closed shell and the addition energies show peaks at these numbers. The visibility of the shell structure in transport becomes difficult as the number N increases, consistent with experiments on parabolic quantum dots, in which the shell structure in the addition energy was observed up to 12 electrons as shown in Fig. 6.19.

By analogy to atoms, one might expect that Hund's rule for total spin would apply in the present situation. According to Hund's rule, as degenerate states are filled, the total spin S takes the maximum value allowed by the exclusion principle and becomes zero for closed shells. In the right panel of Fig. 6.18(a), we show the spin configuration as a function of the electron number N for the circular potential. The dotted line represents the spin configuration when

[a]The method of direct total-energy minimization by the conjugate-gradient method in a plane-wave basis can also solve the Kohn–Sham equation [59]. Both methods produce the same results.

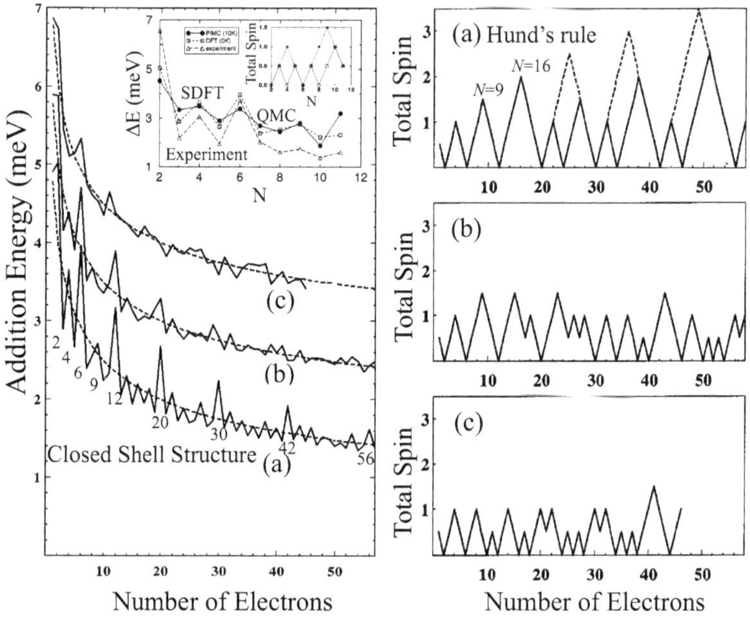

Figure 6.18 (Left) Addition energy $\Delta(N)$ as a function of electron number N in the quantum dot (a) for a confining parabolic potential $V(\mathbf{r}) = m^*\omega_0^2 r^2/2$ and (b, c) for the elliptical potentials $V(\mathbf{r}) = \frac{1}{2}m^*(\omega_x^2 x^2 + \omega_y^2 y^2)$. The dotted lines indicate the addition energy according to a classical electrostatic analysis in Eq. (6.75). (Inset) Comparison of the DFT calculation with experiments and QMC calculation. From J. Harting, O. Mülken, and P. Borrmann, *Phys. Rev. B* **62**, 10207 (2002) [241]. (Right) Ground-state spin as a function of electron number N (a) for the parabolic confining potential and (b-c) for the elliptical potentials. The dotted line in (a) indicates the spin configuration when Hund's rule is satisfied. From K. Hirose and N. S. Wingreen, *Phys. Rev. B* **59**, 4604 (1999) [242].

Hund's rule is satisfied. We can see that for up to 22 electrons filling the dot, the spin configurations obey Hund's rule. For larger dots, Hund's rule is violated and the high spin states are suppressed. In particular, the total spin becomes zero at electron numbers $N = 24,\ 34,\ 46$ instead of the expected $S = 2$.

The breakdown of Hund's rule is due to the non-parabolic effective potential caused by Coulomb interactions. For example, without interactions $N = 24$ corresponds to 20 electrons in filled

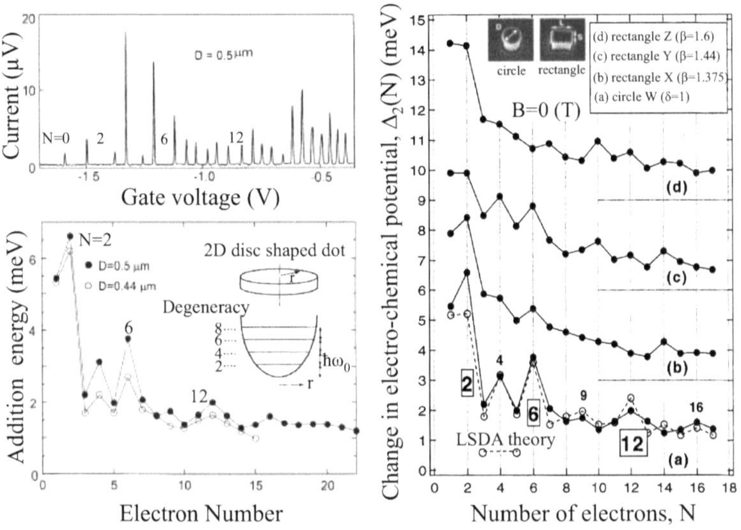

Figure 6.19 (Left) Experimental observations of Coulomb oscillations in the current vs. gate voltage at $B = 0$ T observed for a 0.5 μm quantum dot and addition energy vs. electron number. The inset shows a schematic diagram of the vertical quantum dot device and the model with a harmonic lateral potential. The single-particle states are confined into discrete 0D levels whose degeneracies are 2, 4, 6, 8, $\cdots$, including spin degeneracy from the lowest level. From S. Tarucha, D. G. Austing, T. Honda, R. J. van der Hage, and L. P. Kouwenhoven, *Phys. Rev. Lett.* **77**, 3613 (1996) [243]. (Right) Change in the electrochemical potential as a function of electron number N up to 17 at $B = 0$ T for a circular quantum dot with (a) $D = 0.5$ μm ($\delta = \omega_x/\omega_y = 1$) and (b–d) three kinds of rectangular dots ($\beta = L/S$), respectively. From D. G. Austing, S. Sasaki, S. Tarucha, S. M. Reimann, M. Koskinen, and M. Manninen, *Phys. Rev. B* **60**, 11514 (1999) [244].

inner shells and four "valence" electrons distributed among 10 degenerate states: $|n, k\rangle = |0, \pm 4\rangle$, $|1, \pm 2\rangle$, and $|2, 0\rangle$, spin-up and spin-down. Coulomb interactions deform the radial potential and lower the energy of the single-particle states with larger angular momentum $|k|$. The system could minimize its exchange energy by creating an $S = 2$ state, i.e., putting all four valence electrons into spin-up states. Instead, for $N = 24$, to minimize total energy, the system minimizes its single-particle energy by putting all four electrons into $k = \pm 4$ states, giving a total spin $S = 0$, and breaking Hund's rule.

Next, we consider elliptically deformed potentials to investigate the effect of removing circular symmetry

$$V_{\text{ext}}(\mathbf{r}) = \frac{1}{2} m^* (\omega_x^2 x^2 + \omega_y^2 y^2), \tag{6.93}$$

with $\omega_0^2 = (\omega_x^2 + \omega_y^2)/2$, which lifts the large degeneracies of the electronic shell structure. Figure 6.18(b, c) (left) shows that as the deformation grows the regular zigzag pattern found for the circular potential becomes irregular and the large peaks at large N disappear. One can still see large peaks at $N = 2, 6, 12$, and 20 electrons in curve (b), which are the remnants of the closed-shell structures. However, with increasing ellipticity, such large peaks are missing and in curve (c) large peaks disappear completely.

Figure 6.18 (right) shows the spin configurations for the deformed external potentials. We can see that Hund's rule is satisfied up to $N = 15$ in (b). In (c), we do not see any remnants of Hund's rule. The high spin states are suppressed as the deformation is increased: The loss of the closed-shell structures results in a more Pauli-like behavior of the total spin.

6.2.1.5 Charge-density wave states

For a quantum dot with circular symmetry, the eigenstates can always be chosen to have definite angular momentum L_z, and hence circularly symmetric charge density. Nevertheless, for certain N, SDFT calculations predict a rotational charge-density wave (CDW) near the edge of the dot. Figure 6.20 shows an example of such a CDW state for $N = 31$. The spin-up density $n^{\uparrow}(\mathbf{r})$ indicates a strong spin-density modulation, while a weaker modulation occurs in the spin-down density. The number $N = 31$ corresponds to a closed shell plus one electron, indicating that the extra electron is added to the lowest orbital in the next shell, namely the one with highest angular momentum. By circular symmetry, this orbital is doubly degenerate. Hence the ground state of the entire dot is doubly degenerate with total angular momentum $\pm L_z$. In contrast to atoms, the spin-orbit interaction in GaAs dots is too small to split this degeneracy. The SDFT result is a mixture of these two

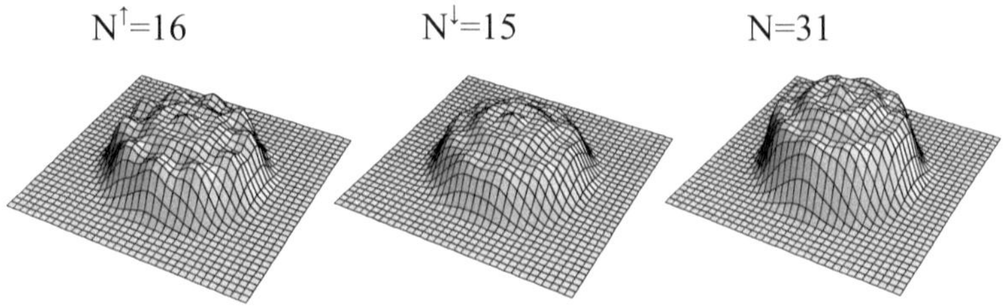

Figure 6.20 Charge-density distributions at $N = 31$: (Left) spin-up $N^{\uparrow} = 16$, (center) spin-down $N^{\downarrow} = 15$, and (right) total charge-density distribution for a parabolic potential. From K. Hirose and N. S. Wingreen, *Phys. Rev. B* **59**, 4604 (1999) [242].

degenerate ground states. For example, for $N = 31$, the total angular momentum will be $L_z = \pm 5$, giving a charge-density modulation $\sim |\exp(i5\theta) + \exp(-i5\theta)|^2 \sim \cos 2(5\theta)$ as observed in Fig. 6.20.

The electron density

$$n^{\sigma}(\mathbf{r}) = N \int \cdots \int |\Psi(\mathbf{r}, \mathbf{r}_2, \cdots, \mathbf{r}_N)|^2 d\mathbf{r}_2 \cdots d\mathbf{r}_N \tag{6.94}$$

for $N = 3$ by exact diagonalization confirms the above interpretation. As expected, there are two degenerate ground states, with $L_z = \pm 1$, and that a coherent mixture of these states produces almost exactly the same electron density obtained in the density functional calculation. The degeneracy of positive and negative angular-momentum ground states for certain numbers of electrons in a circular dot implies a high sensitivity to deformation of the external potential. Indeed for these electron numbers the true CDW ground states are induced by the increasing elliptical deformation of external potential.

6.2.2 *Current-DFT for Magnetic-Field Effects*

6.2.2.1 Single-particle states of 2D dots in magnetic fields

Here we examine the electronic states of a 2D-quantum dot when magnetic fields are applied perpendicular to the plane of the quantum dot. Magnetic fields break the degeneracy of opposite angular-momentum states $\pm k$ and induce spin polarization. The

Hamiltonian under the magnetic fields becomes

$$H = \sum_i \left\{ \frac{1}{2m^*} \left[-i\hbar\nabla_i + \frac{e}{c}\mathbf{A}(\mathbf{r}) \right]^2 + V_{\text{ext}}(\mathbf{r}) + g^*\mu_B B S_z \right\} + \frac{e^2}{\kappa} \sum_{i<j} \frac{1}{|\mathbf{r}_i - \mathbf{r}_j|^2}. \tag{6.95}$$

Here $g^*\mu_B B S_z$ is the Zeeman energy with the Bohr magnetron $\mu_B = \hbar e/2mc$ and the effective g factor of $g^* = 0.44$ for GaAs. Taking the symmetric gauge for $\mathbf{A}(\mathbf{r}) = (B/2)(-y, x, 0)$, we can change the first single-particle term H_0 as

$$H_0 = \sum_i \left\{ -\frac{\hbar^2}{2m^*}\nabla_i^2 + \frac{1}{2}\omega_c L_z + \frac{1}{2}m^* \left(\omega_0^2 + \frac{1}{4}\omega_c^2 \right) + g^*\mu_B B S_z \right\}, \tag{6.96}$$

where $\omega_c = eB/m^*c$ is the cyclotron frequency and $L_z = -i\hbar(x\partial/\partial y - y\partial/\partial x) = -i\hbar\partial/\partial\theta$ is the angular momentum. Note that $\mathbf{A} = (\mathbf{B} \times \mathbf{r})/2$, so that $\nabla \cdot \mathbf{A} = 0$.

In this gauge, since $[H_0, L_z] = 0$ and $[H_0, S_z] = 0$, the single-particle state can be written again with the quantum number (n, k) in Fock–Darwin state

$$\varphi_{n,k}(r, \theta) = |n, k\rangle = \sqrt{\frac{n!}{\pi(n + |k|)!}\frac{1}{\ell}} e^{-\frac{r^2}{2\ell^2}} \left(\frac{r^2}{\ell^2}\right)^{\frac{|k|}{2}} \mathrm{L}_n^{|k|}\left(\frac{r^2}{\ell^2}\right) e^{-ik\theta} \tag{6.97}$$

with the modified length of $\ell = \sqrt{\hbar/m^*\sqrt{\omega_0^2 + \omega_c^2/4}}$. The single-particle levels become

$$\epsilon_{n,k} = (2n + |k| + 1)\hbar\sqrt{(\omega_0^2 + \frac{1}{4}\omega_c^2)} - \frac{1}{2}\hbar\omega_c k + g^*\mu_B B S_z. \tag{6.98}$$

This shows that in the magnetic field, the electron motion is determined by the cyclotron frequency ω_c in addition to the confinement frequency ω_0. The breaking of the degeneracy of orbitals occurs with angular momentum index $\pm k$ with the energies of electrons split by $\Delta\epsilon_{n,\pm k} = -\hbar\omega_c k$.

6.2.2.2 Current-density functional theory

To include the electron–electron interaction for many-electron systems in magnetic fields, we use the current-DFT formalism. Since the ground-state total energy of electronic structures in the presence of magnetic fields is determined not only by the electron density $n(\mathbf{r})$ but also by the paramagnetic orbital current

$$\mathbf{j}_p(\mathbf{r}) = \frac{e\hbar}{2m^* i} \sum_{i,\sigma} [\psi_{i\sigma}^*(\mathbf{r}) \nabla \psi_{i\sigma}(\mathbf{r}) - \psi_{i\sigma}(\mathbf{r}) \nabla \psi_{i\sigma}^*] \tag{6.99}$$

due to the angular momentum L_z, we need to take the paramagnetic orbital current effect into account within the DFT formalism. Such theory is called the current-density functional theory (CDFT) [245], which provides a way for describing a system of interacting electrons in a gauge field.

The important development of CDFT is to show that due to the gauge invariance, the exchange-correlation energy E_{xc} is a functional of vorticity $\upsilon(\mathbf{r})$ defined by

$$\upsilon(\mathbf{r}) = \nabla \times \frac{\mathbf{j}_p(\mathbf{r})}{n(\mathbf{r})} \tag{6.100}$$

and

$$\frac{1}{c}\mathbf{A}_{\text{xc}}(\mathbf{r}) = -\frac{1}{n(\mathbf{r})} \nabla \times \frac{\delta E_{\text{xc}}}{\delta \upsilon(\mathbf{r})}. \tag{6.101}$$

This becomes the source of $\mathbf{j}_p(\mathbf{r})$ and we can use the local approximation

$$E_{\text{xc}} = \int n(\mathbf{r}) \epsilon_{\text{xc}}[n(\mathbf{r}), \zeta(\mathbf{r}), \upsilon(\mathbf{r})] d\mathbf{r}. \tag{6.102}$$

Specifically, we solved the following Kohn–Sham equations:

$$\begin{aligned}
&\left[-\frac{\hbar^2}{2m^*}\nabla^2 - i\frac{e\hbar}{2m^* c} \{\nabla \cdot [\mathbf{A}(\mathbf{r}) + \mathbf{A}_{\text{xc}}(\mathbf{r})] + [\mathbf{A}(\mathbf{r}) + \mathbf{A}_{\text{xc}}(\mathbf{r})] \cdot \nabla\} \right. \\
&+ \frac{e^2}{2m^* c^2}\mathbf{A}(\mathbf{r})^2 + V_{\text{ext}}(\mathbf{r}) + g^* \mu_B B S_z + \frac{e^2}{\kappa} \int \frac{n(\mathbf{r}')}{|\mathbf{r} - \mathbf{r}'|} d\mathbf{r}' \\
&\left. + \frac{\delta E_{\text{xc}}}{\delta n^\sigma(\mathbf{r})} - \frac{1}{c}\mathbf{A}_{\text{xc}}(\mathbf{r}) \frac{\mathbf{j}_p(\mathbf{r})}{n(\mathbf{r})} \right] \psi_{i\sigma}(\mathbf{r}) = \epsilon_i^\sigma \psi_{i\sigma}(\mathbf{r})
\end{aligned} \tag{6.103}$$

with

$$n(\mathbf{r}) = \sum_\sigma n^\sigma(\mathbf{r}) = \sum_\sigma \sum_i |\psi_{i\sigma}(\mathbf{r})|^2. \tag{6.104}$$

After the self-consistent solutions are obtained, the ground-state total energy of a quantum dot with N electrons in a magnetic field is obtained from

$$E_{\rm dot}(N) = \sum_{i,\sigma}^{N} \epsilon_i^\sigma - \frac{e^2}{2\kappa} \int\int \frac{n(\mathbf{r})n(\mathbf{r}')}{|\mathbf{r}-\mathbf{r}'|} d\mathbf{r}d\mathbf{r}' + E_{\rm xc}$$
$$- \sum_{\sigma} \int n^\sigma(\mathbf{r}) \frac{\delta E_{\rm xc}}{\delta n^\sigma(\mathbf{r})} d\mathbf{r} - \frac{1}{c} \int \mathbf{j}_p(\mathbf{r})\mathbf{A}_{\rm xc}(\mathbf{r}) d\mathbf{r}. \quad (6.105)$$

For practical use of the local-density approximation, we consider the homogeneous electron gas in the presence of magnetic field. Due to the gauge invariance of the physical current

$$\mathbf{j}(\mathbf{r}) = \mathbf{j}_p(\mathbf{r}) - \frac{e^2}{m^* c} n(\mathbf{r})\mathbf{A}(\mathbf{r}) = 0, \quad (6.106)$$

the correlation effects from the vorticity can be expressed in terms of the local effective magnetic field as

$$\upsilon(\mathbf{r}) = \nabla \times \frac{\mathbf{j}_p(\mathbf{r})}{n(\mathbf{r})} = \frac{e^2}{m^* c} B_{\rm eff}(\mathbf{r}). \quad (6.107)$$

Following reference [246], we use the Pade approximation

$$\epsilon_{xc}[n, \zeta, \nu] = \frac{\epsilon_{xc}^{\rm LWM}[\nu] + \nu^4 \epsilon_{xc}[n, \zeta]}{1+\nu^4}, \quad (6.108)$$

where the local filling factor $\nu(\mathbf{r}) = 2\pi\ell_B^2 n(\mathbf{r})$ is

$$\nu(\mathbf{r}) = \frac{2\pi\hbar e n(\mathbf{r})}{m^*} \left[\nabla \times \frac{\mathbf{j}_p(\mathbf{r})}{n(\mathbf{r})}\right]^{-1} \quad (6.109)$$

and $\epsilon_{xc}^{\rm LWM}[\nu]$ is the exchange-correlation energy in high magnetic field [247, 248]

$$\epsilon_{xc}^{\rm LWM}[\nu] = -0.782133\sqrt{\nu}(1 - 0.211\nu^{0.74} + 0.012\nu^{1.7})(e^2/\ell_B). \quad (6.110)$$

6.2.2.3 Electronic structures of QDs in magnetic fields

In the quantum dots in magnetic fields, electrons are confined by the potential with $\hbar\omega_0$ and also by the magnetic field characterized by $\hbar\omega_c$. So as the magnetic field increases, we expect different regimes of conductance according to the strengths of the two parameters ω_0 and ω_c. Here we use $\hbar\omega_0 = 3.0$ meV and $\hbar\omega_c$ is obtained from

$\hbar\omega_c = 2\mu_B B(m/m^*)$, where $\mu_B = e\hbar/2mc$ is the Bohr magnetron. Due to the small effective mass for GaAs systems, $\omega_0 \sim \omega_c$ is satisfied around $B = 1.8$ T. This situation is difficult to realize for the real atoms, where we need too strong magnetic fields to confine electrons. The strength of the magnetic field is generally described by the filling factor $\nu = 2\pi\ell_B^2 n$ with the magnetic length $\ell_B = \sqrt{c\hbar/eB}$. In an infinite 2D system, this corresponds to the number of filled Landau levels or equivalently the ratio of electrons to magnetic flux quantum hc/e with the Hall conductivity of $\sigma_{xy} = \nu(e^2/h)$. For example, the states are doubly occupied by spin-up and spin-down electrons for $\nu = 2$ and singly occupied by spin-polarized electrons for $\nu = 1$.

Figure 6.21 (left) [249] shows the electron chemical potential $\mu(N)$ as a function of magnetic field up to $B = 4.5$ T for electron numbers up to $N = 35$. At $B = 0$, we can see that there are large

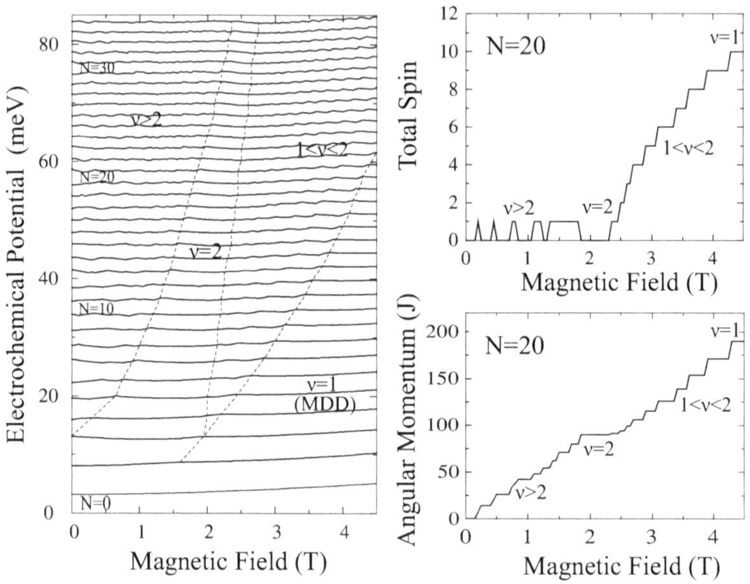

Figure 6.21 (Left) Electrochemical potential $\mu(N)$ as a function of magnetic field B. The dotted lines indicate the regions having different filling factor ν. (Right) Total spin and angular momentum J for a quantum dot with $N = 20$ electrons as a function of magnetic field B.

gaps at $N = 2, 6, 12, 20$, and so on, corresponding to the shell structure shown in Fig. 6.18. These gaps collapse even for weak magnetic fields of 0.3 T. As we increase the magnetic fields, there appear four distinct regimes:

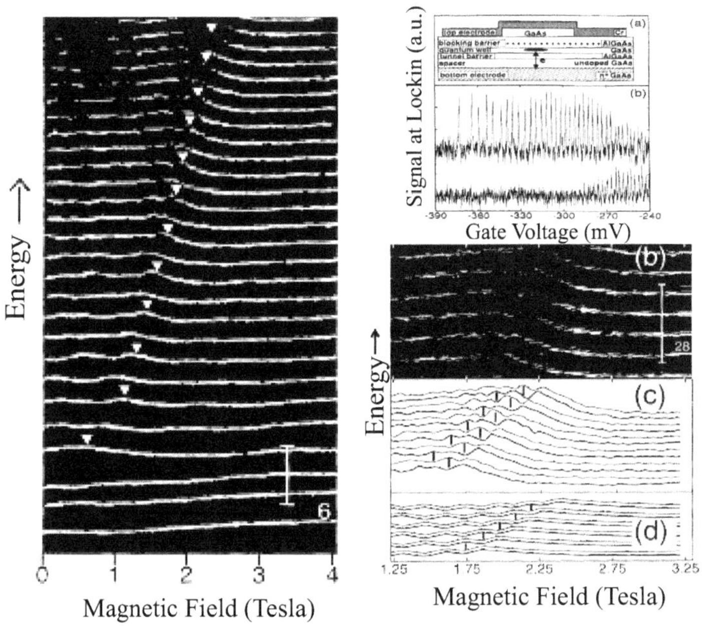

Figure 6.22 (Left) Conductance as a function of magnetic field and gate bias. The vertical bars represent an energy of 5 meV. (Right) (top) Schematic view of the sample and conductance vs. gate bias at $B = 0$ T. (Bottom) Measured conductance for $N = 27 - 32$ and N-electron ground-state energies as a function of magnetic field extracted from the observed data. The behaviors of zigzag structures of conductance correspond to spin flipping due to magnetic fields and the regime with no spin flipping correspond to the filling factor of $\nu = 2$. From R. C. Ashoori, H. L. Stormer, J. S. Weiner, L. N. Pfeiffer, K. W. Baldwin, and K. W. West, *Phys. Rev. Lett.* **71**, 613 (1993) [250].

(1) In weak magnetic fields ($0\text{ T} - 1.8\text{ T}$, $\hbar\omega_c \ll \hbar\omega_0$), the single-particle levels become

$$\epsilon_{n,k} \sim (2n + |k| + 1)\hbar\omega_0 - \frac{1}{2}\hbar\omega_c k + g^* \mu_B B S_z. \qquad (6.111)$$

The k and S_z degeneracies are lifted by the B-linear terms, which push upward the levels with $k < 0$ and $S_z > 0$ and push downward the levels with $k > 0$ and $S_z < 0$. Thus, as we increase the magnetic field B, electron chemical potentials $\mu(N)$ reveal zigzag structures, due to crossings of the single-particle levels $\epsilon_{n,k}$ and the corresponding changes of the spin state. Therefore, the result shows a quite different behavior from the shell structure at $B = 0$.

(2) In intermediate fields (1.8 T – 2.5 T, $\hbar\omega_c \sim \hbar\omega_0$), the quantum dot has a region of constancy. The spin states and angular momentum do not change even if we increase B (Fig. 6.21 (right)). The filling factor stays close to $\nu = 2$. In this regime, the electrons fully occupy the lowest Landau level orbitals with both up and down spins, forming a stable structure.

(3) In large magnetic fields (2.5 T – 4.0 T, $\hbar\omega_c > \hbar\omega_0$), the single-particle levels become

$$\epsilon_{n,k} \sim (\bar{n} + \frac{1}{2})\hbar\omega_c + (2n + |k| + 1)\frac{\hbar\omega_0^2}{\omega_c} + g^*\mu_B B S_z, \quad (6.112)$$

where $\bar{n} = n + (|k| - k)/2 = 0, 1, 2\cdots$ is Landau level index. In this regime, the confinement due to the magnetic fields becomes so strong that the quantum dot shrinks in size with increasing B. In order to reduce the Coulomb energy, the quantum dot transfers electrons from spin-up orbitals near the center of the dot to spin-down orbitals near the edge, thereby flipping spins and rearranging the charge density. The angular momentum increases each time that an electron is transferred into a lowest-Landau-level orbital as shown in Fig. 6.21 (right). It is important to note that the level structure from the parabolic confinement and the Landau level structure are very different as regards the single-particle angular momentum.

(4) In strong magnetic fields (more than 4.0 T, $\hbar\omega_c \gg \hbar\omega_0$), the quantum dot becomes fully spin polarized with filling factor $\nu = 1$, that is, all the electrons fall into the lowest energy states of the lowest Landau level, giving a total angular momentum of $J = 0 + 1 + 2 + \cdots + (N - 1) = N(N - 1)/2$. This state is called the maximum density droplet (MDD) state. The electron density $n(\mathbf{r})$ is flat except at the edge of the quantum dot and forms a stable incompressible state, corresponding to the

extreme integer quantum Hall regime in a 2D electron system. Further increase of the magnetic field leads to the strongly correlated regime of the fractional quantum Hall effect, where the kinetic energy is quenched and the electronic states are determined from the Coulomb interaction. Therefore, the mean field approach is not applicable.

These conductance behaviors of the quantum dots, zigzag structures due to spin flipping in the weak fields, constant in the intermediate fields, rearrangement of the charge density and spin in the strong fields, are observed in experiments as shown in Fig. 6.22.

6.2.3 *TDDFT for Excited-State Energy*

Up to now, we showed the ground-state electronic structures of quantum dots, which are reflected to the peaks of the single-electron transport measurements as a function of gate voltages. Here we consider the excited states of quantum dots, which would be detected in the transport measurements when the finite source-drain voltages are applied, thus in nonequilibrium situations.

A knowledge of electronic excited states of quantum dots is important for many applications, such as optical spectroscopy, quantum dot lasers, and coherent control of electron and spin dynamics of quantum dots. Also, quantum dots are expected to become fundamental units for constructing quantum information devices as quantum bits (qubits), where coherent manipulations of the dynamics of electrons between the ground and excited states of coupled quantum dots are needed. Therefore, the dynamics of electrons in the excited states of quantum dots have been a subject of intense study and efficient theoretical methods to provide accurate excitation energies of quantum dots are needed [251].

6.2.3.1 Excitation energy problem of DFT

The excitation energy ΔE, or equivalently bandgap problem, is one of the fundamental issues of the DFT. We note that the Kohn–Sham equation maps an interacting N electron system to a noninteracting *fictitious* N electron system and gives us the ground-state energy, not the excited-state energy as we see in Chapter 4.

The excitation energy is defined as the difference between the electron affinity $A = E(N) - E(N+1) = -\epsilon_{\text{LUMO}}$ and the ionization energy $I = E(N-1) - E(N) = -\epsilon_{\text{HOMO}}$ such as $\Delta E = I - A = \epsilon_{\text{LUMO}} - \epsilon_{\text{HOMO}}$. The first approximation often used to get to the excitation energy is to take the difference between the ground-state Kohn–Sham eigenvalues;

$$\Delta\epsilon = \epsilon_{N+1} - \epsilon_N, \tag{6.113}$$

where we represent ϵ_n as the n-th Kohn–Sham eigenvalue. This is true for a *true* noninteracting electron system with no Coulomb interaction. However, for an interacting electron system, there is no physical interpretation for the Kohn–Sham eigenvalues ϵ and orbitals $\psi(\mathbf{r})$ except for the roles of auxiliary variables to obtain the ground-state electron density $n(\mathbf{r}) = \sum_n^N |\psi_i(\mathbf{r})|^2$ and the total energy of *fictitious* N electron system. Thus, this approximation is not true for ΔE of an interacting system.

The excitation energy can be obtained from $\Delta E = \epsilon_{N+1}(N+1) - \epsilon_N$, where $\epsilon_{N+1}(N+1)$ is the highest energy of the occupied Kohn–Sham state of the $N+1$ electron system and ϵ_N is the HOMO energy of the Kohn–Sham N electron system. Using Eq. (6.113), we have the excitation energy as

$$\Delta E = \Delta\epsilon + \Big(\epsilon_{N+1}(N+1) - \epsilon_{N+1}\Big). \tag{6.114}$$

The second term is the difference between the $(N+1)$st Kohn–Sham eigenvalues that correspond to *neutral* and *ionized* electron systems, and becomes the central issue for the excitation energy problem. For $N \gg 1$ in solids, the addition of an extra electron induces an infinitesimal change of the electron density. Then the two corresponding Kohn–Sham potentials are practically the same up to a constant energy shift and thus the electron density do not change. Since the Hartree term depends explicitly on the electron density, the energy difference is approximately expressed by a rigid shift plus the variation of the exchange-correlation functional. We derive the formula ΔE explicitly using the TDDFT formalism in the next sections.

6.2.3.2 Time-dependent density functional theory

Time-dependent density functional theory (TDDFT), which is a rigorous extension of the ground-state SDFT into the time domain, allows a number of dynamic and excited-state properties to be efficiently calculated [103, 252]. Here, we have applied TDDFT to the excitation energies of quantum dots with a finite number of electrons and compare the results with exact diagonalization studies.

The calculations are carried out as follows. We use the time-dependent Kohn–Sham equations to obtain the excitation energies,

$$\begin{aligned} i\hbar\frac{\partial\phi_{i\sigma}(\mathbf{r},t)}{\partial t} &= \left(-\frac{\hbar^2}{2m^*}\nabla^2 + V_{\text{ext}}(\mathbf{r},t) + \frac{e^2}{\kappa}\int\frac{n(\mathbf{r}',t)}{|\mathbf{r}-\mathbf{r}'|}d\mathbf{r}' \right. \\ &\quad \left. + \frac{\delta E_{\text{xc}}[n(\mathbf{r},t),\zeta(\mathbf{r},t)]}{\delta n^{\sigma}(\mathbf{r},t)}\right)\phi_{i\sigma}(\mathbf{r},t) \\ &= \left[H_{0\sigma}(\mathbf{r}) + \delta V_{\sigma}(\mathbf{r},t)\right]\phi_{i\sigma}(\mathbf{r},t) \\ n(\mathbf{r},t) &= \sum_{\sigma} n_{\sigma}(\mathbf{r},t) = \sum_{\sigma}\sum_{i}|\phi_{i\sigma}(\mathbf{r},t)|^2, \end{aligned} \tag{6.115}$$

where $H_{0\sigma}(\mathbf{r})$ is the Kohn–Sham Hamiltonian for the ground-state energy and $\delta V_{\sigma}(\mathbf{r},t)$ is the time-dependent Coulomb interaction contribution:

$$\begin{aligned} \delta V_{\sigma}(\mathbf{r},t) &= \delta V_{\text{ext}}(\mathbf{r},t) + \frac{e^2}{\kappa}\int\frac{n(\mathbf{r}',t)-n(\mathbf{r}')}{|\mathbf{r}-\mathbf{r}'|}d\mathbf{r}' \\ &\quad + \frac{\delta E_{xc}[n(\mathbf{r},t),\zeta(\mathbf{r},t)]}{\delta n^{\sigma}(\mathbf{r},t)} - \frac{\delta E_{xc}[n(\mathbf{r}),\zeta(\mathbf{r})]}{\delta n^{\sigma}(\mathbf{r})}. \end{aligned} \tag{6.116}$$

We use two different methods to obtain excitation energies. One uses the linear-response perturbation theory and the other is a direct integration of the time-dependent Kohn–Sham equation of Eq. (6.115) for the response to an external perturbation.

6.2.3.3 Linear-response approach for excitations

In a circumstance where an external time-dependent potential is small, we can use the perturbation theory, which will be sufficient to determine the low excitation energies. We focus on the linear change of the electron density, which enables us to obtain the optical absorption spectrum. We expand the electron density in perturbative series

$$n(\mathbf{r}, t) = n^{(0)}(\mathbf{r}) + n^{(1)}(\mathbf{r}, t) + n^{(2)}(\mathbf{r}, t) + \cdots, \tag{6.117}$$

where $n^{(1)}(\mathbf{r}, t)$ depends linearly on the external potential $v^{(1)}$ in the frequency space

$$n^{(1)}(\mathbf{r}, \omega) = \int \chi(\mathbf{r}, \mathbf{r}', \omega) v^{(1)}(\mathbf{r}', \omega) d\mathbf{r}'. \tag{6.118}$$

Here $\chi(\mathbf{r}, \mathbf{r}', \omega)$ is the linear density-density response function of the system. Since the density in the *interacting* electron systems is mapped to a fictitious system of *noninteracting* electron systems in DFT, we can calculate the linear change of the electron density from

$$n^{(1)}(\mathrm{r}, \omega) = \int \chi_{KS}(\mathbf{r}, \mathbf{r}', \omega) v^{(1)}_{KS}(\mathbf{r}', \omega) d\mathbf{r}'. \tag{6.119}$$

In TDDFT, the time-dependent effective potential $v^{(1)}_{KS}(\mathbf{r}, t)$ is

$$\begin{aligned} v^{(1)}_{KS}(\mathbf{r}, t) = v^{(1)}(\mathbf{r}, t) + \frac{e^2}{\kappa} \int \frac{n^{(1)}(\mathbf{r}', t)}{|\mathbf{r} - \mathbf{r}'|} d\mathbf{r}' \\ + \int\int f_{xc}(\mathbf{r}t, \mathbf{r}'t') n^{(1)}(\mathbf{r}', t') d\mathbf{r}' dt', \end{aligned} \tag{6.120}$$

where the exchange-kernel $f_{xc}(\mathbf{r}t, \mathbf{r}'t')$ is defined by

$$f_{xc}(\mathbf{r}t, \mathbf{r}'t') = \frac{\delta^2 E_{xc}}{\delta n^{\sigma}(\mathbf{r}, t) \delta n^{\tau}(\mathbf{r}', t')}. \tag{6.121}$$

Combining Eqs. (6.118), (6.119), and (6.120), we have

$$\begin{aligned} \chi(\mathbf{r}, \mathbf{r}', \omega) &= \chi_{KS}(\mathbf{r}, \mathbf{r}', \omega) \\ &+ \int\int \chi_{KS}(\mathbf{r}, \mathbf{r}'', \omega) \left[\frac{e^2}{\kappa} \frac{1}{|\mathbf{r}'' - \mathbf{r}'''|} + f_{xc}(\mathbf{r}'', \mathbf{r}''', \omega) \right] \\ &\times \chi(\mathbf{r}''', \mathbf{r}', \omega) d\mathbf{r}'' d\mathbf{r}'''. \end{aligned} \tag{6.122}$$

This method relies on the theorem that the frequency-dependent response function of a finite interacting system has discrete poles at the excited-state energies in the limit $\hbar\omega \to \Omega$, where $\Omega = E_m - E_0$ is the excitation energy of the interacting system. This implies that the function Ξ

$$\Xi(\mathbf{r}, \mathbf{r}', \omega) = \int \chi_{KS}(\mathbf{r}, \mathbf{r}'', \omega) \left[\frac{e^2}{\kappa} \frac{1}{|\mathbf{r}'' - \mathbf{r}'|} + f_{xc}(\mathbf{r}'', \mathbf{r}', \omega) \right] d\mathbf{r}'' \tag{6.123}$$

has eigenvalues at $\hbar\omega = \Omega$. Since the noninteracting density-density response function is given by

$$\chi_{KS}(\mathbf{r}, \mathbf{r}', \omega) = \sum_{ij}^{\infty}(f_j - f_i)\frac{\psi_i(\mathbf{r})\psi_i^*(\mathbf{r}')\psi_j(\mathbf{r}')\psi_j^*(\mathbf{r})}{\hbar\omega - (\epsilon_i - \epsilon_j) + i0}, \qquad (6.124)$$

the following eigenvalue equation is derived for the excitation energies ΔE:

$$(\Delta E)^2 = (\epsilon_{i\sigma} - \epsilon_{j\sigma})^2\delta_{ik}\delta_{jl}\delta_{\sigma\tau} + 2\sqrt{(f_{i\sigma} - f_{j\sigma})(\epsilon_{j\sigma} - \epsilon_{i\sigma})}$$
$$\times K_{ij\sigma,kl\sigma}\sqrt{(f_{k\tau} - f_{l\tau})(\epsilon_{l\tau} - \epsilon_{k\tau})},$$
$$K_{ij\sigma,kl\tau} = \int\int \psi_{i\sigma}^*(\mathbf{r})\psi_{j\sigma}(\mathbf{r})\left\{\frac{e^2}{\kappa}\frac{1}{|\mathbf{r} - \mathbf{r}'|} + f_{\text{xc}}(\mathbf{r}, \mathbf{r}', \omega)\right\}$$
$$\times\psi_{k\tau}(\mathbf{r}')\psi_{l\tau}^*(\mathbf{r}')d\mathbf{r}d\mathbf{r}'. \qquad (6.125)$$

In the above formulas, $\psi_{i\sigma}(\mathbf{r})$, $\psi_{k\tau}(\mathbf{r})$ are occupied and $\psi_{j\sigma}(\mathbf{r})$, $\psi_{l\tau}(\mathbf{r})$ are unoccupied Kohn–Sham orbitals in the ground state, σ and τ are spin index and $f_{i\sigma}$ is the orbital occupation number. $K_{ij\sigma,kl\sigma}$ gives the shift of excitation energies due to interactions [253].

We note that by neglecting the higher-order terms, manipulation of Eq. (6.125) yields the lowest-order excitation energy as

$$\Delta E \simeq \Delta\epsilon + K_{ij\sigma,ij\sigma}, \qquad (6.126)$$

where $\Delta\epsilon = \epsilon_{j\sigma} - \epsilon_{i\sigma}$ is the difference between the Kohn–Sham eigenvalues of the unoccupied and occupied orbitals [254]. This formula is an expression based on the linear-response TDDFT formalism corresponding to the excitation energy in Eq. (6.114). As for the exchange-correlation kernel, we use simply the adiabatic local-density approximation (ALDA) $f_{\text{xc}}(\text{r}t, \text{r}'t')$ as

$$f_{\text{xc}}^{ALDA}(\mathbf{r}t, \mathbf{r}'t') = \delta(\mathbf{r} - \mathbf{r}')\delta(t - t')\frac{\delta^2 E_{xc}}{\delta n^\sigma(\mathbf{r}, t)\delta n^\tau(\mathbf{r}', t')}$$
$$= \frac{1}{n(\mathbf{r})}\left[-\frac{r_s}{4}\frac{\partial\epsilon_{xc}(\mathbf{r})}{\partial r_s} + \frac{r_s^2}{4}\frac{\partial^2\epsilon_{xc}(\mathbf{r})}{\partial r_s^2} + \left(\zeta - \frac{\text{sgn}\sigma + \text{sgn}\tau}{2}\right)\right.$$
$$\left.\times r_s\frac{\partial^2\epsilon_{xc}(\mathbf{r})}{\partial r_s\partial\zeta} + (\zeta - \text{sgn}\sigma)(\zeta - \text{sgn}\tau)\frac{\partial^2\epsilon_{xc}(\mathbf{r})}{\partial\zeta^2}\right]. \qquad (6.127)$$

6.2.3.4 Direct time integration of TDDFT Hamiltonian

The second method we use to obtain excited-state energies from TDDFT is to directly time-integrate the time-dependent Kohn–Sham equations. This method can be applied to the many-electron systems where any time-dependent perturbative field is imposed. The time integration for the i-th Kohn–Sham orbital is implemented by the split-operator method

$$\begin{aligned}\phi_{i\sigma}(\mathbf{r}, t+\Delta t) &= e^{-i\int_t^{t+\Delta t}(H_{0\sigma}(\mathbf{r})+\delta V_\sigma(\mathbf{r},t))dt/\hbar}\phi_{i\sigma}(\mathbf{r}, t) \\ &\approx e^{-iH_{0\sigma}(\mathbf{r})\Delta t/2\hbar}e^{-i\delta V_\sigma(\mathbf{r},t+\Delta t/2)\Delta t/\hbar} \\ &\quad \times e^{-iH_{0\sigma}(\mathbf{r})\Delta t/2\hbar}\phi_{i\sigma}(\mathbf{r}, t). \qquad (6.128)\end{aligned}$$

Here, to obtain the excitation energy spectrum, we use the following perturbation for an initial perturbation at $t = 0$:

$$\phi_{i\sigma}(\mathbf{r}, +0) = e^{ik^*\{\sin(qx)+\sin(qy)\}/q}\psi_{i\sigma}(\mathbf{r}), \qquad (6.129)$$

which promotes transitions to all the excited states, not just dipole excitations [255]. The strength of the perturbation from the linear-response regime to the non-linear-response regime is controlled by the magnitude of k^*. Here we use $q = 0.45\ \text{nm}^{-1}$ and a small wave number $k^* = 0.005\ \text{nm}^{-1}$ to restrict the perturbation to the linear-response regime.

The excitation energies are obtained from the frequency spectrum of the projection of $\phi_{i\sigma}(\mathbf{r}, t)$ onto the various ground-state Kohn–Sham orbitals $\psi_{i\sigma}(\mathbf{r})$,

$$P(\omega) = \int_0^{t_{\max}} \langle\phi_{i\sigma}(\mathbf{r}, t)|\psi_{i\sigma}(\mathbf{r})\rangle e^{-i\omega t}dt. \qquad (6.130)$$

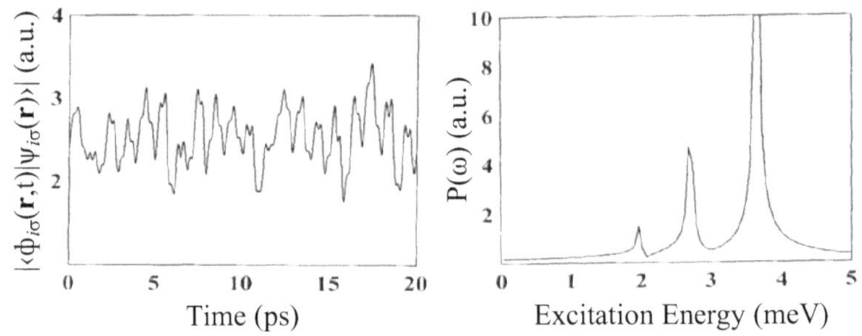

Figure 6.23 (Left) An example of the time-dependent correlation $|\langle\phi_{i\sigma}(\mathbf{r}, t)|\phi_{i\sigma}(\mathbf{r}, t = 0)\rangle|$. (Right) The Fourier transform of the correlation function $P(\omega)$ for the spectra. The excitation energies are obtained using the Hanning method.

Accurate peak positions of the frequency spectrum are obtained by the Hanning method [256]. The time integration is performed by expanding the wavefunctions with the ground-state Kohn–Sham orbitals,

$$\phi_{i\sigma}(\mathbf{r},t) = \sum_j C_{ij\sigma}(t)\psi_{j\sigma}(\mathbf{r}), \tag{6.131}$$

where

$$C_{ij\sigma}(t+\Delta t) = \sum_{j'} \exp\left[-i\frac{\epsilon_{j\sigma}+\epsilon_{j'\sigma}}{2}\frac{\Delta t}{\hbar}\right] \times \left[\int \psi^*_{j\sigma}(\mathbf{r})e^{-i\delta V_\sigma(\mathbf{r},t+\Delta t/2)\frac{\Delta t}{\hbar}}\psi_{j'\sigma}(\mathbf{r})d\mathbf{r}\right] C_{ij'\sigma}(t).$$

Figure 6.23 shows an example for the direct calculation of the time-dependent Kohn–Sham equation (left) and the Fourier transform of the correlation function $P(\omega)$.

6.2.3.5 Excitation energies of quantum dots

Let us apply the TDDFT for the excitation energies of quantum dots. For the direct integration method, we use $\Delta t = 0.01$ ps and integrated up to $t_{\rm max} = 200$ ps. Both our methods for obtaining excited-state energies produced almost the same results for low-energy excitations, though the direct integration method is numerically difficult. Specifically, we show the result for the case of elliptical quantum dots of Eq. (6.93) with $\omega_x^2/\omega_y^2 = 3/5$ in closed-shell structures and compared the low-energy excited states with those obtained by exact diagonalization method.

Table 6.3 shows the excitation energies obtained by exact diagonalization and by TDDFT for the spin-polarized case ($\zeta = 1$) with $N = 3$ electrons and spin $S_z = 3/2$ up to fifth excited level. We see that TDDFT produces fairly accurate results. The deviations from the exact diagonalization results are less than 0.20 meV.

Using the excitation energies obtained from TDDFT, we calculated the photoabsorption spectra. Within linear-response theory, optical susceptibility is expressed in the dipole approximation as

$$\chi(\omega) = -\frac{e^2}{m^*}\sum_i \frac{f_i}{2\omega_i}\left[\frac{1}{\omega-\omega_i+i\delta} - \frac{1}{\omega+\omega_i+i\delta}\right], \tag{6.132}$$

Table 6.3 Excitation energies by exact diagonalization and TDDFT up to the fifth excited state for a spin-polarized dot with $N = 3$ electrons ($S_z = 3/2$)

Level	E_{exact} (meV)	E_{TDDFT} (meV)	Error (meV)
1st	1.296	1.183	0.113
2nd	1.980	1.868	0.112
3rd	2.598	2.597	0.001
4th	2.911	2.714	0.197
5th	3.354	3.353	0.001

where the oscillator strength f_i is

$$f_i = \frac{2m^*}{\hbar}|\langle\Psi_i|x|\Psi_0\rangle|^2(\omega - \omega_i). \tag{6.133}$$

Here Ψ_0 and Ψ_i are the ground-state and i-th excited-state wavefunctions and $\hbar\omega$ is the photon energy. δ gives the peak width due to inelastic scatterings from other degrees of an environment. Here we arbitrarily set $\delta = 0.02$ meV. The photoabsorption cross section is obtained from

$$\sigma(\omega) = \frac{4\pi\omega}{c}\text{Im}\{\chi(\omega)\}. \tag{6.134}$$

Figure 6.24 shows the photoabsorption cross sections by (a) exact diagonalization, (b) naive SDFT, and (c) TDDFT. The SDFT spectra in (b) are obtained using the occupied and unoccupied Kohn–Sham orbitals $\psi_i(\mathbf{r})$ and eigenvalues ϵ_i from the Kohn–Sham ground state [257]. We use the formula in Eq. (6.125) for the TDDFT photoabsorption spectra in (c). We see that two peaks appear in the photoabsorption spectra for TDDFT, in excellent agreement with the exact-diagonalization results. For the confinement potential of elliptical quantum dots, the center-of-mass coordinates and the relative coordinates are separable and thus, according to the generalized Kohn's theorem, Coulomb interaction has no effect on the photoabsorption spectra. Figure 6.24 shows that Kohn's theorem is satisfied in TDDFT and the two peaks correspond to the bare confinement energies $\hbar\omega_x$ and $\hbar\omega_y$. On the other hand, in the naive SDFT case the peak positions are incorrectly shifted to a lower energy and spurious additional peaks appear.

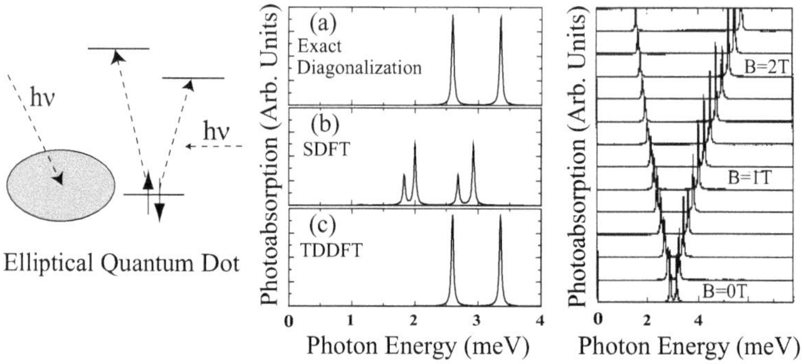

Figure 6.24 (Left) Schematic view of the electron excitation by the photon energy of $h\nu$. (Center) Photoabsorption spectra for an elliptical quantum dot with $\omega_y^2/\omega_x^2 = 3/5$ obtained by (a) exact diagonalization, (b) SDFT, and (c) TDDFT. (Right) Magnetic field dependence of the excitation spectra due to the Zeeman splitting.

The TDDFT method is also applicable to the spin-unpolarized case, where we see that singlet-triplet transitions due to exchange interaction are well reproduced and that errors with respect to exact diagonalization are less than 0.20 meV for the case of $N = 2$ electrons and spin $S = 0$ [257].

6.2.4 *Conductance Fluctuation of Quantum Dots*

6.2.4.1 Quantum level vs. Coulomb energy

In the previous sessions, we showed the electronic states of quantum dots by using the DFT formalism, which produces accurate total energies, and compare the calculated results of quantum dots with the transport experiments of SETs for the well-controlled circular quantum dots. Here we consider the effects of disorder on the SET peak spacings. As we see in the first overview, the conductance fluctuation due to the disorder in 2DEG systems shows a fascinating feature in the weak localization regime. That is, the magnitude of fluctuation is of the order $2e^2/h$, which is attributed to the interference effects of the quantum states. On the other hand, according to the random-matrix theory (RMT), the distribution of fluctuations of energy level spacing $\delta\Delta_0$ obeys Wigner–Dyson statistics, where the energy levels become repulsive when they are close in energy, while they become attractive when they are apart in energy. This results in the UCF observed in the conductance measurements of the electron devices.

In the quantum dots, however, electronic states are determined not only by their single-particle energies, but the electron–electron interactions play very important roles. The conductances are observed as SET peaks at low temperatures and their peak spacings reflect the total energy of quantum dots. Furthermore, the peak spacings $\Delta(N)$ in the quantum dots reflect differences between ground-state energies of interacting electrons *for different numbers of electrons*, one cannot directly apply the RMT to evaluate the distribution of peak spacings. In such situations, it would be interesting to ask how about the fluctuation of conductance in the SET of quantum dots where both the quantum levels and Coulomb energies are important [258]. We might ask what the main contribution to the fluctuation of peak spacings is, how about the magnitude of fluctuations, and what will become of spin behaviors in disordered quantum dots. The key concepts for the disorder quantum dots are shown in Fig. 6.25. In these respects, the interplay of disorder and electron–electron interactions in quantum dots has attracted much attention.

Experiments on disordered quantum dots have measured the spacings between conductance peaks as a function of gate voltage in the CB region. The shell structure found for clean circular dots disappears completely in these disordered dots. In such systems, single-particle levels ϵ_i are spaced randomly. In the CB regime, the charging energy e^2/C dominates the average single-particle level spacing

$$\frac{e^2}{C} \gg \Delta_0 \;\; (= \langle \epsilon_{N+1} - \epsilon_N \rangle), \tag{6.135}$$

so that peak spacings, i.e., addition energies, are almost equal but have measurable fluctuations. Experiments find a more symmetric distribution of peak spacings than the Wigner–Dyson form as shown in Fig. 6.26. Experiments, however, disagree on whether the fluctuations are several times larger, scaling as $10 - 15\%$ of the total charging energy e^2/C [259] in Fig. 6.26 (Top), or are comparable to the mean level spacing $\langle \Delta_0 \rangle$ [260] in Fig. 6.26 (Bottom).

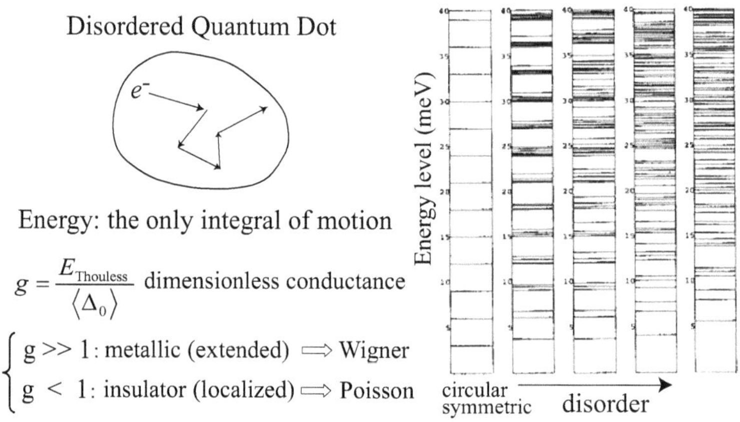

Figure 6.25 (Left) Schematic views of the electron motion in the disordered quantum dots. Since there is no symmetry, the energy is the only integral of motions. The distribution for dimensionless conductance g behaves as the Wigner–Dyson type for the metallic case due to the level repulsions. On the other hand, the distribution of g behaves as the Poisson type for the insulating case where all the states are strongly localized and behave independently without level repulsions. (Right) Energy levels as the strength of disorder becomes large from left (circular symmetric) to right. Quantum levels of a ladder with degeneracy are lifted due to disorder and are randomly spaced in the spectra.

Here we apply the SDFT to clarify the origin, the magnitude, the distribution of CB peak-spacing *fluctuations*, and spin behaviors. We consider fluctuations of the addition energy for $N = 10$ and $N = 11$ electrons. Since the properties of quantum dots would change with different material parameters m^* and κ, in order to study their electronic states systematically, we measure the dimensionless interaction strength by $(e^2/\kappa\ell_0)/\hbar\omega_0$, or approximately by $r_s(= 1/\sqrt{\pi n_0}a_B^*)$, where n_0 is the electron density at the center of the quantum dot. First, we treat the spin-polarized case to investigate the magnitude and distribution of fluctuation of CB peak spacings. Then we consider the spin-unpolarized case to see the behavior of spin and the spin-related even–odd alternations.

The external potential for the present disordered dots is the sum of a confining parabola and multiple "impurity" potentials each with

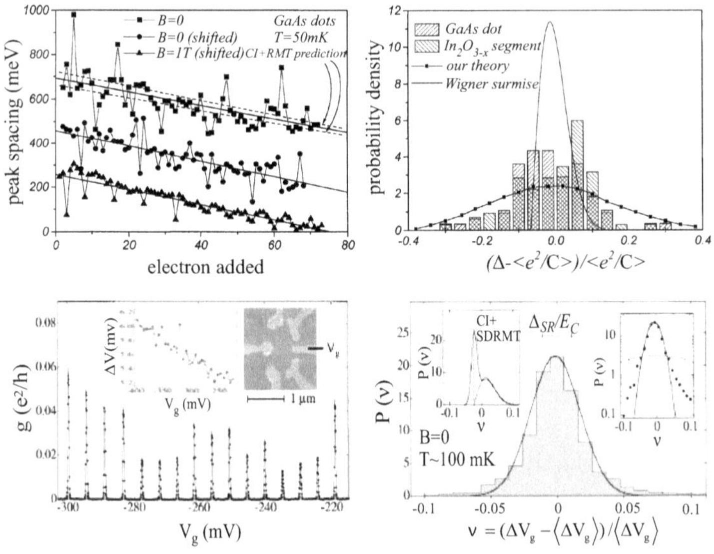

Figure 6.26 (Top) (Left) Peak spacing vs. number of electrons added to quantum dots. (Right) Histogram of the normalized fluctuations of quantum dots. From U. Sivan, R. Berkovits, Y. Aloni, O. Prus, A. Auerbach, and G. Ben-Yoseph, *Phys. Rev. Lett.* **77**, 1123 (1996) [259]. (Bottom) (Left) Coulomb blockade peaks as a function of gate voltage V_g. Solid lines show fits to $\cosh^{-2}$ lineshape to find the peaks accurately and peak spacings extracted from the data. (Right) Histograms of normalized peak spacing ν with the mean level spacing Δ_{SR}/E_C. Solid curve shows best fit to normalized Gaussian for $B = 0$ T. From S. R. Patel, S. M. Cronenwett, D. R. Stewart, C. M. Marcus, C. I. Duruöz, J. S. Harris. Jr., K. Campman, and A. C. Gossard, *Phys. Rev. Lett.* **80**, 4522 (1998) [260].

a Gaussian profile:

$$V_{\text{ext}}(\mathbf{r}) = \frac{1}{2}m^*\omega_0^2 r^2 + \frac{\gamma}{2\pi\lambda^2}\sum_i^{N_{\text{imp}}} \exp\left(-\frac{|\mathbf{r}-\mathbf{r}_i|^2}{2\lambda^2}\right). \tag{6.136}$$

The impurity potentials are randomly distributed with density $n_i = 1.03 \times 10^{-3}$ nm^{-2} and strength γ uniformly distributed on $[-W/2, W/2]$ with $W = 10\hbar^2/m^*$. The width of each impurity is taken as $\lambda = \ell_0/(2\sqrt{2})$, where $\ell_0 = \sqrt{\hbar/m^*\omega_0} \simeq 19.5$ nm. We use more than 1,000 different impurity configurations for the disorder average at each Coulomb interaction strength. We treat interaction strengths of $0 \leq r_s \leq 4.85$ in the spin-polarized case and $0 \leq r_s \leq 3.90$ in the spin-unpolarized case, by varying the dielectric constant κ.

The strength of disorder is measured by the dimensionless conductance g, which is evaluated in disordered quantum dots as follows. From a scattering phase-shift analysis, the mean free path of electrons $\ell_{\mathrm{mfp}} = v_F \tau \simeq 170$ nm to be slightly larger than the dot diameter $L = 120 - 160$ nm, where the quantum dot diameter increases with r_s. The dimensionless conductance is given by $g = \hbar/(t_c\langle\Delta_0\rangle)$ where the t_c is the time an electron takes to cross the quantum dot and $\langle\Delta_0\rangle$ is the mean level spacing. For quantum dots in the ballistic regime, such as those studied here, $g \sim k_F L/\pi$. Therefore, the disordered quantum dots we study here are marginally in the ballistic regime and have a dimensionless conductance $g = 2 - 4$, in the metallic regime where SDFT works well. We note that the experiments on GaAs have $r_s \sim 1$ and $g > 1$ and thus corresponds closely to the range of interaction strengths and dimensionless conductance treated here.

6.2.4.2 Addition energy and its fluctuation

Figure 6.27 shows the distribution of addition energies Δ (right) for interacting quantum dots with $(e^2/\kappa\ell_0)/\hbar\omega_0 = 2.39$, and the distribution of Δ_0 for noninteracting quantum dots of the same size (left). In the inset, we show the electron density $n(\mathbf{r})$ for one realization of disorder. While interactions considerably enhance the average addition energy

$$\langle\Delta\rangle \simeq 6.50\langle\Delta_0\rangle, \tag{6.137}$$

the fluctuation in the interacting case is only 13% larger than the noninteracting fluctuation:

$$\frac{\delta\Delta}{\delta\Delta_0} \equiv \frac{\sqrt{\langle\Delta^2\rangle - \langle\Delta\rangle^2}}{\sqrt{\langle\Delta_0^2\rangle - \langle\Delta_0\rangle^2}} \simeq 1.13. \tag{6.138}$$

The distribution of level spacings Δ_0 in the noninteracting dots has the expected Wigner–Dyson form, while that in the interacting dots is somewhat more symmetrical. The symmetry continues to increase with interaction strength.

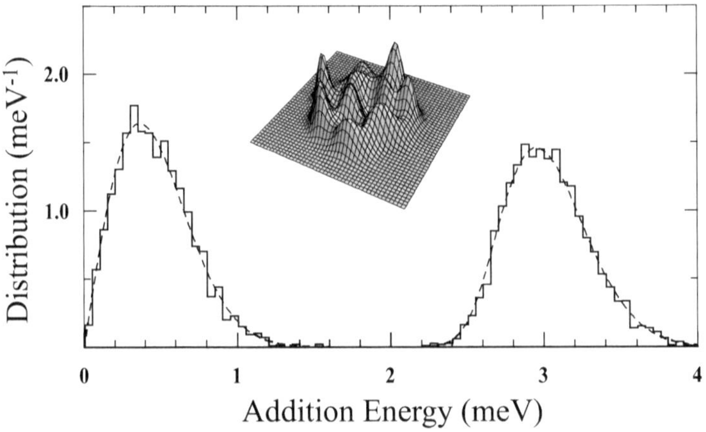

Figure 6.27 Distribution of addition energies Δ to add the 11th electron for interacting quantum dots with $(e^2/\kappa\ell_0)/\hbar\omega_0 = 2.39$ (right), and for noninteracting quantum dots of the same size (left). The energy bandwidth is 0.05 meV. The dashed lines show the distribution function obtained from Eq. (6.156). (Inset) The charge density profile $n(\mathbf{r})$ for $N = 10$ electrons for one configuration of impurities. From K. Hirose, F. Zhou, and N. S. Wingreen, *Phys. Rev. B* **63**, 075301 (2001) [261].

Figure 6.28 shows the average addition energy $\langle\Delta\rangle$ and its rms fluctuations $\delta\Delta$ for disordered quantum dots as a function of the Coulomb-interaction strength $(e^2/\kappa\ell_0)/\hbar\omega_0$. For comparison, we also plot results for disordered, noninteracting dots of the same size, which we obtain as follows. First, we find the effective potential for the clean dot, without impurities. Then we solve for the single-particle level energies ϵ_0^i for this effective potential plus the random impurity potentials in Eq. (6.136). The addition energy is simply given by

$$\Delta_0 = \epsilon_{N+1} - \epsilon_N \tag{6.139}$$

with $N = 10$. All quantum dot sizes satisfy the relation

$$\delta\Delta_0 = \sqrt{\frac{4}{\pi} - 1}\langle\Delta_0\rangle \simeq 0.52\langle\Delta_0\rangle \tag{6.140}$$

predicted by the RMT for the Gaussian orthogonal ensemble.

In Fig. 6.28(a), the average addition energy in the noninteracting case is seen to decrease with increasing Coulomb interaction.

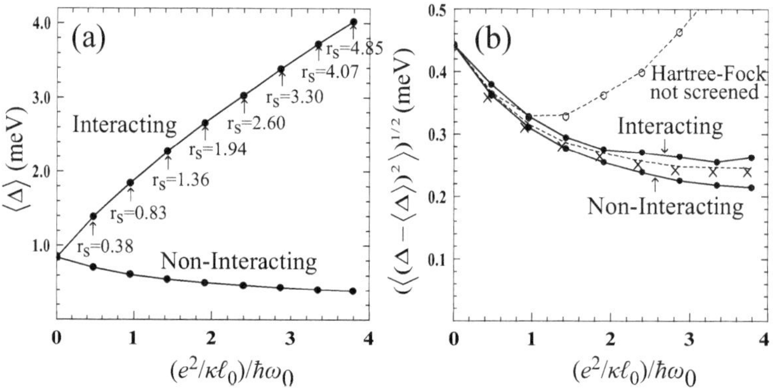

Figure 6.28 (a) Average addition energy $\langle\Delta\rangle$ and (b) fluctuation $\delta\Delta$ as a function of electron–electron interaction strength $(e^2/\kappa\ell_0)/\hbar\omega_0$. The measure of interaction strength $r_s(= 1/\sqrt{\pi\rho_0}a_B^*)$ indicated by arrows in (a) can be applied to the data in all panels. For each data point, the disorder average is taken over more than 1000 different impurity configurations. At each r_s, the noninteracting data are taken for dots of the same size as the interacting dots, and the relation $\delta\Delta_0 \simeq 0.52\langle\Delta_0\rangle$ expected for noninteracting level-spacing statistics is always satisfied. Also plotted in (b) are the fluctuations due to noninteracting level spacings plus the *screened* Coulomb interaction between two electrons at the Fermi surface (crosses), or the *unscreened* exchange interaction between the two electrons (open circles). From K. Hirose, F. Zhou, and N. S. Wingreen, *Phys. Rev. B* **63**, 075301 (2001) [261].

This is because the increasing Coulomb repulsion among electrons causes the quantum dot to grow and hence the level spacing to shrink. The average addition energy in the interacting case increases considerably with the Coulomb interaction strength, as expected from the classical electrostatics relation

$$\langle\Delta\rangle \simeq \frac{e^2}{C}, \tag{6.141}$$

where C is the capacitance of the quantum dot. However, Fig. 6.28(b) shows that increasing interaction strength only slightly increases the addition-energy *fluctuations*. For GaAs quantum dots with $r_s \simeq 2$, the enhancement is only about 10 %, in rough agreement with the experiment [260].

6.2.4.3 Magnitude of fluctuation

To understand the magnitude of addition-energy fluctuations, we use the phenomenological framework for the regime $r_s < 1$, where the random-phase approximation (RPA) is valid, and show that it applies to the SDFT results at least up to $r_s \simeq 5$. Consider first a quantum dot containing $N - 1$ electrons in the ground state. The addition of the N-th electron to form the N electron ground state requires an electron chemical potential $\mu(N)$. To form instead the first excited state of N electrons requires the higher chemical potential $\mu(N) + \Delta\epsilon$. For an ensemble of disordered metallic quantum dots, $\Delta\epsilon$ will have Wigner–Dyson statistics, with $\langle\Delta\epsilon\rangle$ equal to the mean noninteracting level spacing $\langle\Delta_0\rangle$, since the lowest excitation of a Fermi liquid is a single electron promoted across Fermi surface.

The addition energy Δ is the *increase* in chemical potential from $\mu(N)$ required to add one more electron to the quantum dot and thus form the $N + 1$ electron ground state. This $(N + 1)$st electron must have an extra energy $\Delta\epsilon$ to occupy the lowest empty level *plus* an extra energy $U_{N,N+1}$ due to its Coulomb interaction with the N-th electron. The total addition energy Δ will be approximately given by the sum of these two contributions:

$$\Delta \simeq \Delta\epsilon + U_{N,N+1}. \tag{6.142}$$

The distribution of $\Delta\epsilon$ is given by the Wigner–Dyson distribution of level spacings for a noninteracting quantum dot of the same size. The average interaction energy $\langle U_{N,N+1}\rangle$ is the capacitive charging energy e^2/C, which is much larger than Δ_0 for $r_s > 1$

$$\langle U_{N,N+1}\rangle \gg \langle\Delta_0\rangle. \tag{6.143}$$

We estimate the *fluctuations* in $U_{N,N+1}$ by calculating the screened Coulomb interaction between two electrons at the Fermi surface. Specifically, we treat the screening effect as

$$\tilde{U}_{N,N+1} = e\int \tilde{\varphi}_{N,N}(\mathbf{r})n_{N+1,N+1}(\mathbf{r})d\mathbf{r}. \tag{6.144}$$

Here $\tilde{\varphi}_{i,j}(\mathbf{r})$ is the screened potential due to an electron, which is evaluated in Fourier representation as

$$\tilde{\varphi}_{i,j}(\mathbf{q}) = \frac{2\pi e}{\kappa|\mathbf{q}|}\left[\frac{n_{i,j}(\mathbf{q})}{\epsilon(\mathbf{q})}\right], \tag{6.145}$$

where

$$n_{i,j}(\mathbf{r}) = \phi_i(\mathbf{r})\phi_j(\mathbf{r}) \qquad (6.146)$$

is the density of the single-particle wavefunction $\phi_n(\mathbf{r})$ of a noninteracting disordered quantum dot. The dielectric function $\epsilon(\mathbf{q})$ is approximated as

$$\epsilon(\mathbf{q}) = 1 - v_q\chi(q) = 1 - \frac{2\pi}{|\mathbf{q}|}\chi(q) \qquad (6.147)$$

with the susceptibility

$$\chi(q) = -\frac{e^2}{\kappa}\nu F\left(\frac{q}{2k_F}\right). \qquad (6.148)$$

We use the Thomas–Fermi approximation $F(0) = 1$, which gives

$$\tilde{\varphi}_{N,N}(\mathbf{q}) = \frac{2\pi e}{\kappa}\left[\frac{n_{N,N}(\mathbf{q})}{|\mathbf{q}| + q_{TF}}\right] \qquad (6.149)$$

with the Thomas–Fermi wavevector

$$q_{TF} = \frac{2\pi e^2}{\kappa}\nu = \frac{1}{a_B^*}. \qquad (6.150)$$

This implies that the wavelengths larger than $2\pi/q_0$ ($\approx$ 60 nm in GaAs case) are screened out and do not contribute to the fluctuation of Coulomb interaction. We note that a correct estimate for the average interaction energy $\langle U_{N,N+1}\rangle$ needs to a finite geometry of the dot. However, the fluctuations in $U_{N,N+1}$ can be estimated using the bulk screening expression for $\varphi(\mathbf{q})$. The final results do not depend on the forms of screening such as Thomas-Fermi or RPA in the metallic regime.

From the above, we find that the fluctuations of the screened Coulomb interaction are always considerably *smaller* than the noninteracting level-spacing fluctuations

$$\delta\tilde{U} \equiv \sqrt{\langle(\tilde{U}_{N,N+1})^2\rangle - \langle\tilde{U}_{N,N+1}\rangle^2} \ll \delta\Delta_0 \qquad (6.151)$$

up to at least $r_s \simeq 5$. The total fluctuation estimated as

$$\delta\tilde{\Delta} = \sqrt{(\delta\Delta_0)^2 + (\delta\tilde{U})^2} \qquad (6.152)$$

is shown in Fig. 6.28(b) by crosses. We see that the fluctuations in the Thomas–Fermi screening model agree well with the SDFT results with *no* free parameters. This supports the picture that the addition-energy fluctuation arises from two quasi-particles above a filled Fermi sea interacting via a *screened* Coulomb potential.

This is also the same for fluctuations due to the exchange potential. If it is treated as unscreened, it will produce large fluctuations. To see this, we calculate the unscreened exchange interaction between two electrons near the Fermi surface in the quantum dots

$$U^{\text{exch}}_{N,N+1} = \frac{e^2}{\kappa} \int\int \frac{\phi^*_N(\mathbf{r})\phi_N(\mathbf{r}')\phi^*_{N+1}(\mathbf{r}')\phi_{N+1}(\mathbf{r})}{|\mathbf{r}-\mathbf{r}'|} d\mathbf{r}d\mathbf{r}'. \quad (6.153)$$

We note that the fluctuations in the unscreened direct Coulomb interaction are $\sim$10% larger than the fluctuations in the unscreened exchange interaction over the range of r_s considered. In Fig. 6.28, we have plotted as open circles the fluctuations taken by summing the unscreened exchange interaction given above with the noninteracting level spacing. It is clear that for $r_s > 1$, the unscreened exchange interaction noticeably overestimates the addition-energy fluctuations. In contrast, the DFT correctly accounts for screening within the electron gas, including exchange interactions.

6.2.4.4 Distribution function

Within this picture, the increase of the fluctuation of $U_{N,N+1}$ with increasing interaction strength leads naturally to greater symmetry of the distribution of addition energies. Numerically, the distribution of $\tilde{U}_{N,N+1}$ has a symmetric Gaussian form in agreement with Eq. (6.142), and we observe that the addition-energy distribution function $P(\Delta)$ is always very well described by the convolution of a Wigner–Dyson distribution for $\Delta\epsilon$,

$$P_{\text{WD}}(\Delta\epsilon) = \frac{\pi}{2}\frac{\Delta\epsilon}{\langle\Delta_0\rangle^2}\exp\left[-\frac{\pi}{4}\frac{\Delta\epsilon^2}{\langle\Delta_0\rangle^2}\right], \quad (6.154)$$

with a Gaussian distribution for interaction energies $U_{N,N+1}$,

$$P_{\text{Gauss}}(U_{N,N+1}) = \frac{1}{\sqrt{2\pi}\delta U}\exp\left[-\frac{(U_{N,N+1} - \langle\Delta\rangle + \langle\Delta_0\rangle)^2}{2(\delta U)^2}\right] \quad (6.155)$$

to become

$$P(\Delta) = \int\int d\Delta\epsilon \, dU_{N,N+1} \, P_{\rm WD}(\Delta\epsilon) P_{\rm Gauss}(U_{N,N+1}) \times \delta(\Delta\epsilon + U_{N,N+1} - \Delta)$$
$$= \frac{1}{2}\sqrt{\frac{\pi}{2}\frac{\delta U}{\alpha\langle\Delta_0\rangle^2}} e^{-\frac{\pi}{4}\frac{\tilde{\Delta}^2}{\alpha\langle\Delta_0\rangle^2}} \times \left\{ e^{-\frac{\tilde{\Delta}^2}{2\alpha(\delta U)^2}} + \sqrt{\frac{\pi}{2\alpha}}\frac{\tilde{\Delta}}{\delta U}\left[1 + \mathrm{erf}\left(\frac{\tilde{\Delta}}{2\alpha(\delta U)^2}\right)\right]\right\}.$$

Here

$$\alpha = \frac{\pi \delta U^2}{2\langle\Delta_0\rangle^2} + 1 \quad \text{and} \quad \tilde{\Delta} = \Delta - \langle\Delta\rangle + \langle\Delta_0\rangle, \tag{6.156}$$

where $\langle\Delta\rangle$ is the center of the distribution and δU is a fitting parameter giving the width of the fluctuations of $U_{N,N+1}$. In the noninteracting case, $\delta U = 0$ so that $P(\Delta)$ reduces to a Wigner–Dyson distribution as expected. In the other limit, $P(\Delta)$ becomes nearly symmetric for sufficiently large δU. In Fig. 6.27, we show $P(\Delta)$ given by Eq. (6.156) as a dashed line. It is seen that the SDFT distribution is described very well by Eq. (6.156) with the best fit value of $\delta U = 0.11$ meV very close to the value $\delta\tilde{U} = 0.09$ meV estimated from the Thomas–Fermi screened Coulomb interaction between two electrons at the Fermi surface.

To test whether the distribution of addition energies is well described by the sum of noninteracting level spacings and a symmetric distribution due to interactions, we propose to compare the third moment of the distribution $P(\Delta)$ with and without a magnetic field $B_\perp$ normal to the plane of the quantum dot. Since the interaction part, coming from the screened Coulomb interaction in our picture, is symmetric it does not contribute to the third moment of $P(\Delta)$. Therefore, we expect

$$\frac{\langle(\Delta - \langle\Delta\rangle)^3\rangle_{B_\perp \neq 0}}{\langle(\Delta - \langle\Delta\rangle)^3\rangle_{B_\perp = 0}} \sim (2 - 5\pi/8)/(2 - 6/\pi) \simeq 0.405, \tag{6.157}$$

which applies to level spacings taken from a Gaussian orthogonal ensemble ($B_\perp = 0$) and a Gaussian unitary ensemble ($B_\perp \neq 0$). Since our results apply only to the case of spin-polarized electrons, it is necessary to apply a large magnetic field in the plane of the quantum dots. The result can also be tested numerically, e.g., by exact diagonalization studies as in Ref. [259].

6.2.4.5 Absence of even–odd alternation

Here we consider the spin states in the disordered quantum dots. The control of spin is essential for a number of applications such as nanometer-scale spintronics, quantum bits, and for fundamental studies of the Kondo effect in quantum dots. In disordered or chaotic quantum dots, high spin states are suppressed by the rarity of degenerate or nearly degenerate levels. However, the observed absence of even–odd alternation of addition energies in quantum dots is consistent with ground states of nonminimal spin. Evidence for such high-spin states has been obtained from studies of the magnetic dispersion of CB peaks. Here we apply SDFT to study the ground-state spin of small, disordered quantum dots for which $g > 1$, over a wide range of interaction strengths.

Let us first recall that in the noninteracting case $(e^2/\kappa\ell_0)/\hbar\omega_0 = 0$, the distribution of $N = 10$ (even) has the Wigner–Dyson form, while the distribution of $N = 11$ (odd) is a delta function since no additional energy is needed to add another electron with opposite spin to the same spatial orbital,

$$\delta\Delta(\text{noninteracting}) = \begin{cases} \delta\Delta_0 & \text{for } N = \text{even} \\ 0 & \text{for } N = \text{odd}. \end{cases} \tag{6.158}$$

This is schematically shown in the top of Fig. 6.29.

Figure 6.29 (left) shows (a) the average addition energy $\langle\Delta\rangle$ and (b) its fluctuation $\delta\Delta \equiv [\langle(\Delta - \langle\Delta\rangle)^2\rangle]^{1/2}$, as functions of electron–electron interaction strength $(e^2/\kappa\ell_0)/\hbar\omega_0$ for $N = 10$ and $N = 11$ electrons. Dashed curves are for total spin fixed at its minimum $S = 0$ or $S = 1/2$. Solid curves are for unrestricted spin; i.e., the lowest energy $E(N)$ is found among all possible spin values. Also shown in (a) is the average noninteracting level spacing $\langle\Delta_0\rangle$ and in (b) its fluctuations $\langle\delta\Delta_0\rangle$, for quantum dots of the same size as in the interacting case.

Such even–odd alternation and the bimodal structure of the distribution disappear gradually with increasing interaction strength. In Fig. 6.29(a), there is an even–odd alternation in addition energies in the weak-interaction regime $(e^2/\kappa\ell_0)/\hbar\omega_0 \leq 0.95$ ($r_s \leq 0.76$), which reflects the energy cost of adding every odd electron to a new orbital level. No alternation is observed in the regime of strong interactions ($r_s > 1$) where $\langle\Delta\rangle$ increases roughly linearly as a

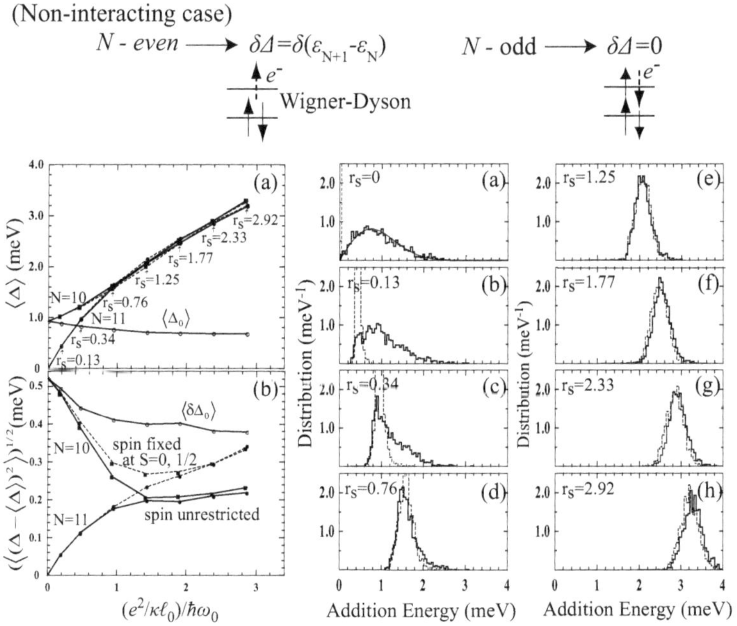

Figure 6.29 (Top) Schematic representation of the level occupation. (Left) (a) Average addition energy $\langle\Delta\rangle$ and (b) fluctuation $\delta\Delta = \langle(\Delta - \langle\Delta\rangle)^2\rangle^{1/2}$, as functions of electron–electron interaction strength $(e^2/\kappa\ell_0)/\hbar\omega_0$ for $N = 10$ and $N = 11$ electrons. Solid curves are for unrestricted spin; dashed curves are for total spin fixed at $S = 0$ or $1/2$. For each data point, the disorder average is taken over more than 1000 different impurity configurations. The average noninteracting level spacing and its fluctuations are shown for the same size dots as in the interacting case. (Right) Distribution of addition energies Δ for $N = 10$ (solid line) and for $N = 11$ (dotted line) for interaction strength $(e^2/\kappa\ell_0)$ and r_s respectively equal to (a) 0.0, 0.0, (b) 0.20, 0.13, (c) 0.47, 0.34, (d) 0.95, 0.76, (e) 1.43, 1.25, (f) 1.91, 1.77, (g) 2.40, 2.33, (h) 2.86, 2.92. From K. Hirose and N. S. Wingreen, *Phys. Rev. B* **65**, 193305 (2002) [262].

function of interaction. In Fig. 6.29 (left) (b), the magnitudes of the addition-energy fluctuations for $N = 10$ and $N = 11$ electrons are seen to merge around $r_s \sim 1.25$. The interacting fluctuations $\delta\Delta$ are always *smaller* than the fluctuations $\delta\Delta_0$ of the noninteracting level spacing, where the latter satisfy the RMT relation $\delta\Delta_0 \simeq \sqrt{4/\pi - 1}\langle\Delta_0\rangle \simeq 0.52\langle\Delta_0\rangle$. The interacting fluctuations are roughly

$\delta\Delta \approx 0.28\langle\Delta_0\rangle$, and clearly do not scale with the average addition energy $\langle\Delta\rangle$. In fact, interacting fluctuations are slightly smaller than those for the same number of spin-polarized electrons in the same potential. Comparing Figs. 6.29 (left) (a) and Fig. 6.29 (left) (b), one sees that in the range $r_s = 0.76 - 1.25$, the average addition energy is the same for $N = 10$ and $N = 11$, but the $N = 10$ distribution has a significantly larger width. For $r_s = 1.25$ and below, the shapes of the $N = 10$ and $N = 11$ distributions are qualitatively different, consistent with experiments for a high-electron-density GaAs quantum dot with $r_s = 0.72$.

For the stronger-interaction regime $r_s > 1$, fluctuations in addition energy are reduced when the spin states are unrestricted. In Fig. 6.29 (right), we show the distributions of $\Delta(N)$ for $N = 10$ (solid line) and for $N = 11$ (dotted line) for several interaction strengths $(e^2/\kappa\ell_0)/\hbar\omega_0$ and r_s. The distributions of Δ are roughly Gaussian, which agrees with experiments at $r_s > 1$. For the strength at $(e^2/\kappa\ell_0)/\hbar\omega_0 = 1.91$ ($r_s = 1.77$), the distributions of Δ become narrower in the spin unrestricted case: the low-energy tail of the $N = 10$ distribution is reduced, while the high-energy tail is reduced for the $N = 11$ distribution. We trace this behavior to the appearance of $S = 1$ ground states for $N = 10$ electrons. Since $\Delta(N = 10) = E(11) - 2E(10) + E(9)$, the appearance of new low values for $E(10)$ removes some of the *lowest* values of $\Delta(N = 10)$. In a similar way, since $\Delta(N = 11) = E(12) - 2E(11) + E(10)$, new low values of $E(10)$ removes some of the *highest* values of $\Delta(N = 11)$.

6.2.4.6 Behavior of spin

Figure 6.30 (left) shows the probabilities of the different ground-state spins S versus electron–electron interaction strength. Solid curves are for $N = 10$ electrons (integer spin) and dashed curves are for $N = 11$ electrons (half-integer spin). With increasing interaction strength, the probabilities of ground states of $S = 1, 3/2$, and 2 increase, while the probabilities of even higher spin states ($S = 5/2$, etc.) remain negligible. The probability of $S = 1$ is always higher than that of $S = 3/2$. The probability of an $S = 1$ ground state increases rapidly up to $(e^2/\kappa\ell_0)/\hbar\omega_0 \simeq 1.91$ ($r_s \simeq 1.77$) and slows down to almost saturate.

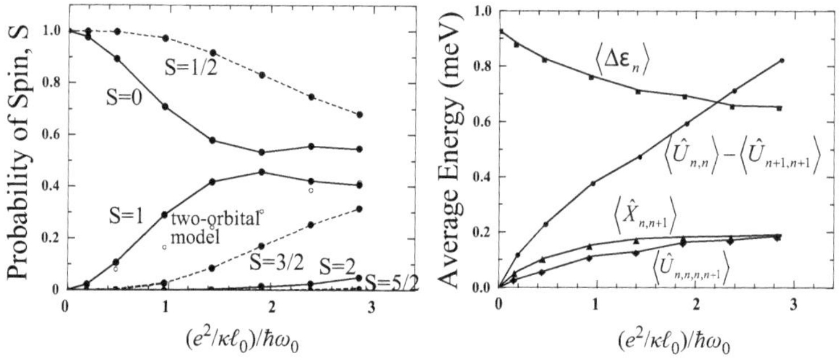

Figure 6.30 (Left) Probability of a spin S ground state as a function of electron–electron interaction strength $(e^2/\kappa\ell_0)/\hbar\omega_0$. Solid curves are for $N = 10$ electrons (integer spin) and dashed curves are for $N = 11$ electrons (half-integer spin). The circles give the probability of a spin $S = 1$ ground state for the full two-orbital model. (Right) Average level spacing and screened Coulomb interactions for two-orbital model for various parameters. From K. Hirose and N. S. Wingreen, *Phys. Rev. B* **65**, 193305 (2002) [262].

To understand the saturation behavior of the probability of high-spin ground states at larger r_s and the stabilization of minimal-spin states, we consider the possible states of two electrons occupying two spin-degenerate orbitals near the Fermi energy. (The other electrons in the quantum dot are assumed to pair up in lower-energy orbitals.) For this two-orbital model, there are three degenerate $S = 1$ states consisting of one electron in each of the two orbitals, and three nondegenerate $S = 0$ states. To evaluate the energies of the competing $S = 0$ and $S = 1$ two-electron states, we use the basis of single-particle eigenstates $\phi_n(\mathbf{r})$ and $\phi_{n+1}(\mathbf{r})$ with energies ϵ_n and ϵ_{n+1} of noninteracting electrons. The various diagonal and off-diagonal Coulomb matrix elements are evaluated in the RPA, which accounts for the screening effect of the other electrons (more accurate than the Thomas–Fermi approximation) in quantum dots.

The energy of the three degenerate $S = 1$ states is

$$\tilde{E}(S = 1) = \epsilon_n + \epsilon_{n+1} + \tilde{U}_{n,n+1} - \tilde{X}_{n,n+1}, \tag{6.159}$$

where

$$\tilde{U}_{i,j} = e \int \tilde{\varphi}_{i,i}(\mathbf{r}) n_{j,j}(\mathbf{r}) d\mathbf{r} \tag{6.160}$$

is the screened Coulomb interaction between two electrons in orbital i and j, and

$$\tilde{X}_{i,j} = e \int \tilde{\varphi}_{i,j}(\mathbf{r}) n_{j,i}(\mathbf{r}) d\mathbf{r} \tag{6.161}$$

is the screened exchange interaction, where $\tilde{\varphi}_{i,j}(\mathbf{r})$ and $n_{i,j}(\mathbf{r})$ are defined in Eqs. (6.145) and (6.146). Here we use the Lindhard polarizability for 2D in Eq. (6.148):

$$F(x) = \begin{cases} 1 & x \leq 1 \\ 1 - \sqrt{1 - \left(\dfrac{1}{x}\right)^2} & x \geq 1. \end{cases} \tag{6.162}$$

The energy $\tilde{E}(S = 0)$ of the lowest $S = 0$ state is obtained by diagonalizing the following 3×3 matrix:

$$\begin{bmatrix} 2\epsilon_n + \tilde{U}_{n,n} & \sqrt{2}\tilde{V}_{n,n+1} & \tilde{X}_{n,n+1} \\ \sqrt{2}\tilde{V}_{n,n+1} & \epsilon_n + \epsilon_{n+1} + \tilde{U}_{n,n+1} + \tilde{X}_{n,n+1} & \sqrt{2}\tilde{V}_{n+1,n} \\ \tilde{X}_{n,n+1} & \sqrt{2}\tilde{V}_{n+1,n} & 2\epsilon_{n+1} + \tilde{U}_{n+1,n+1} \end{bmatrix},$$

where the off-diagonal Coulomb matrix elements are

$$\tilde{V}_{i,j} = e \int \tilde{\varphi}_{i,i}(\mathbf{r}) n_{i,j}(\mathbf{r}) d\mathbf{r}. \tag{6.163}$$

We see that the magnitudes of the $\tilde{V}_{n,n+1}$ are comparable to the exchange energy $\tilde{X}_{n,n+1}$, as shown in Fig. 6.30 (right).

The average total energy difference between states with $S = 0$ and $S = 1$ obtained from the diagonalization of the matrix

$$\Delta\tilde{E} = \tilde{E}(S = 0) - \tilde{E}(S = 1). \tag{6.164}$$

shows that for the two-orbital model, $\Delta\tilde{E}$ agrees reasonably well with our SDFT results for all strengths of interaction, as shown in Fig. 6.31. In contrast, placing two electrons in the lowest single-particle orbital $\phi_n(\mathbf{r})$ gives an energy

$$\tilde{E}(S = 0) = 2\epsilon_n + \tilde{U}_{n,n} \tag{6.165}$$

and the average energy difference of

$$\Delta\tilde{E} = \xi - \Delta_0, \tag{6.166}$$

where

$$\xi = \tilde{U}_{n,n} - (\tilde{U}_{n,n+1} - \tilde{X}_{n,n+1}), \tag{6.167}$$

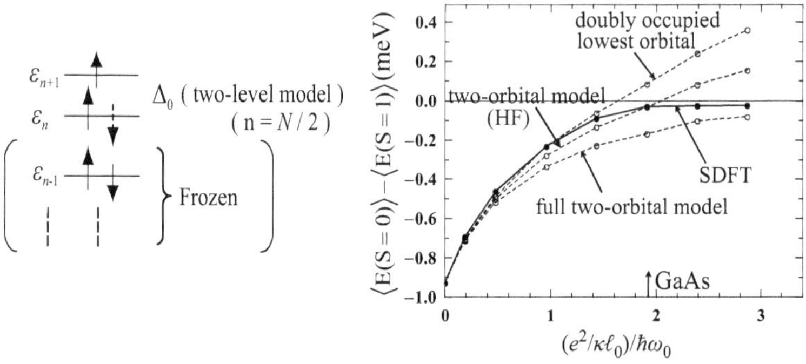

Figure 6.31 (Left) Schematic picture of the two-orbital model for the saturation behavior of high-spin ground states at larger r_s and the stabilization of minimal-spin states. The possible states of two electrons occupying two spin-degenerate orbitals near the Fermi energy, while the other electrons are assumed to pair up in lower-energy orbitals. For the two-orbital model, there are three degenerate $S = 1$ states consisting of one electron in each of the two orbitals, and three nondegenerate $S = 0$ states. (Right) Average total energy difference ΔE between states with $S = 0$ and $S = 1$ as a function of electron–electron interaction. The solid curve shows the SDFT results. Also shown are results of the two-orbital model: exact (lower dashed curve), Hartree–Fock (upper dashed curve), with the parameters evaluated from the noninteracting orbitals. From K. Hirose and N. S. Wingreen, *Phys. Rev. B* **65**, 193305 (2002) [262].

which is significantly larger at large r_s. Here we note that if we assume $\tilde{U}_{n,n+1} \approx \tilde{U}_{n,n}$, then the criterion for the emergence of an $S = 1$ ground state is that the exchange energy is larger than the single-particle level spacing $\tilde{X}_{n,n+1} > \Delta_0$. This corresponds to the Stoner instability for quantum dots. The Hartree–Fock (HF) approximation for Eq. (6.163), i.e., the doubly occupied single orbital, overestimates the energy of the lowest $S = 0$ state and fails to account for the saturation of the $S = 1$ probability at large r_s. It is evident that for the two-orbital model, the off-diagonal Coulomb matrix elements help stabilize the $S = 0$ basis states.

These calculations show that the fluctuations of conductance of quantum dots as SET peak spacings do not scale with the average addition energy but remain proportional to the single-particle level spacing, influenced by the fluctuations in the screened Coulomb interaction between a pair of electrons at the Fermi energy.

6.3 Spin-Dependent Transport of QPCs

In the previous section, we describe the quantization of conductance in QPCs. Due to the confinement to one direction by the split-gate electrode, the transmission of electron is restricted and the conductance takes integer values in steps of $2e^2/h$. In this context, we consider the transport as a noninteracting electron transmission. However, as we see in the quantum dot case, interaction effects become very important and correspondingly the spin effects emerge in the low-dimensional structures. Recent experiments of QPC transport show the spin-related behaviors in the lowest channel, indicating the importance of the Coulomb interaction for the spin-dependent transport through QPCs.

6.3.1 *Instability for Spin Flipping in Low-dimensional Systems*

Here we show that the 2DEG electron system becomes unstable to the spin polarization when it is confined to a narrow region to create low-dimensional systems. We estimate the interaction strength r_s where such spin instability appears in the quasi-1D electron gas system. Due to confinement, the kinetic energy in the quasi-1D electron system with the width of L_y becomes

$$\begin{aligned} E_{kin} &= \frac{1}{2\pi}\int_{-(1+\zeta)k_f}^{(1+\zeta)k_f}\frac{\hbar^2k^2}{2m^*} + \frac{1}{2\pi}\int_{-(1-\zeta)k_f}^{(1-\zeta)k_f}\frac{\hbar^2k^2}{2m^*} \\ &= \frac{1+3\zeta^2}{3}\frac{\hbar^2k_f^2}{2m^*}n_{1d} \\ &= \frac{1+3\zeta^2}{12}\left(\frac{L_y}{a_B^*}\right)^2\frac{1}{r_s^4}. \end{aligned} \tag{6.168}$$

Since the exchange-correlation energy for the 2DEG system due to the electron interaction is given in Eqs. (6.88–6.92), the total energy becomes

$$E = \frac{1+3\zeta^2}{12}\left(\frac{L_y}{a_B^*}\right)^2\frac{1}{r_s^4} - \frac{4\sqrt{2}}{3\pi r_s}\left[(1+\zeta)^{3/2}+(1-\zeta)^{3/2}\right] + E_c(r_s,\zeta). \tag{6.169}$$

When we use the numerical data for $E_c(r_s, \zeta)$ in Eq. (6.90) from the Quantum Monte Carlo calculations, used in the previous quantum dot case, we find that the spin-polarization occurs at $r_s = 3.8$ for $L_y = 60$ nm. This value of r_s is very small compared with $r_s = 37$, which corresponds to very low density as $n_{2d} = 2.8 \times 10^8$ cm^{-2} for 2D electron systems and much smaller as $r_s = 75$ for 3D electron systems. This estimation suggests that as we confine the electrons to a narrow region, the system becomes unstable to the spin flipping.

We confirm this estimation by the numerical calculations. Let us treat the uniform quantum wire, a quasi-1D system, which has a parabolic confinement potential only in one direction and solve the electronic states for the channel of the band using the SDFT. In Fig. 6.32 (right), we show the band edge energies at $T = 0$ relative to the Fermi energy as a function of side gate energy. There is a region in the lowest subband where the band edge energies show numerical instability. In this regime, a spin-polarized state with lower density has the same chemical potential as a spin-unpolarized state with higher density. Since we assume the translational invariance of the density, two coexisting phases cannot be obtained, instead we find the competing states in this regime producing an instability. The onset of spin polarization in

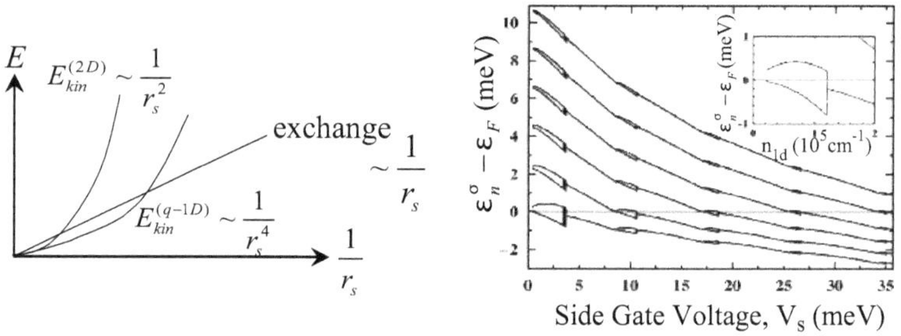

Figure 6.32 (Left) Schematic view of the energy for electron–electron interaction strength $1/r_s$. The exchange energy, which is proportional to $1/r_s$ dominates for strong electron–electron interaction ($1/r_s \ll 1$) over the kinetic energy of $1/r_s^2$ for the 2D case and $1/r_s^4$ for the quasi-1D case. Note that electron–electron interaction becomes larger as $1/r_s$ becomes smaller. (Right) Band edge energies at $T = 0$ allowing for spontaneous spin polarization. When the band edge energies separate, electrons are partially or completely spin polarized. The inset is the lowest band edge energies as a function of electron density n_{1d}. From K. Hirose, S. S. Li, and N. S. Wingreen, *Phys. Rev. B* **63**, 033315 (2001) [264].

the lowest subband appears at an interaction strength $r_s = 3.3$, which is close to the estimated value above for $L_y = 60$ nm. The essential difference between a quasi-1D wire and a uniform 2DEG is the quenching of part of the kinetic energy by confinement. The dependence $E_{kin} \sim 1/r_s^4$ for the wire due to the confinement implies that kinetic energy is relatively unimportant at low density, as we see in the left panel of Figure 6.32.

However, we need to take care of this estimation for the spin polarization of the ground state in the low density limit, since it is strictly proven that spin polarization does not occur in a purely one-dimensional electron system (Lieb's theorem) [263], for which the lowest channel state might apply. It should be noted that the polarized and unpolarized states in the lowest subband are close in energy, and thus the correlated state might appear for those states [264]. Such spin effects may be observable to the transport behaviors. We note that one such example is the transport via the Kondo effect in the quantum dots, which is the zero-dimensional system. The Kondo effect appears as an interaction of the conduction electron with the localized spin, where the quantum dot acts as localized spin. The conductance is significantly temperature dependent.

6.3.2 *Anomalous Transport of 0.7 Structure*

In the transport measurement of the QPC, we see that the conductance increases in units of $2e^2/h$ through the n-th transverse channel and spin σ. Surprising experimental observation is that in addition to these usual integer conductance steps, there exists an extra conductance plateau another around $0.7(2e^2/h)$. This feature is now well known and is called the *0.7 structure*.

In the pioneering experiment [265], the conductance shows up to 26 quantized plateau with applying negative biases to a side-gate voltages of the QPC systems and in addition an extra plateau appears close to $0.7(2e^2/h)$. This 0.7 structure is reproducible on thermal cycling and thus rules out the possibility of UCF due to impurity scatterings. Figure 6.33 (left) shows the temperature dependence of the 0.7 structure from 0.07 K to 1.5 K. We see that as the temperature decreases, the 0.7 structure gradually

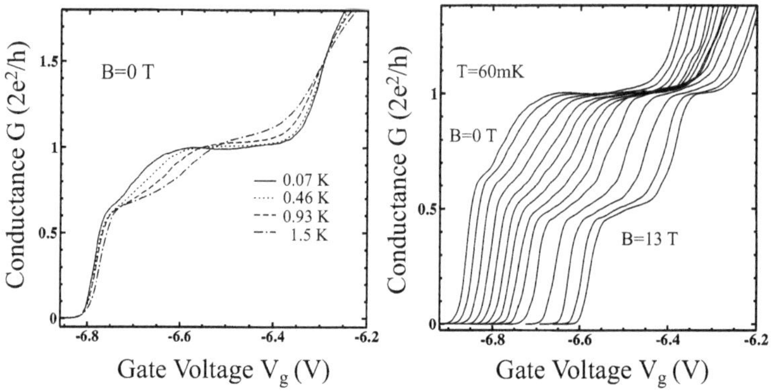

Figure 6.33 (Left) Temperature dependence of the 0.7 structure compared to the quantized plateau at $2e^2/h$. (Right) The evolution of the structure at $0.7(2e^2/h)$ into a step at e^2/h in a parallel magnetic field $B_{||} = 0$–13 T, in steps of 1 T. From K. J. Thomas, J. T. Nicholls, M. Y. Simmons, M. Pepper, D. R. Mace, and D. A. Ritchie, *Phys. Rev. Lett.* **77**, 135 (1996) [265].

increases toward the $2e^2/h$ plateau. On the other hand, the 0.7 structure becomes evident as the temperature increases, while the usual plateau at $2e^2/h$ disappears due to the thermal averaging. The magnetic field dependence in Fig. 6.33 (right) shows that with increasing in-plane magnetic field and the breaking of spin degeneracy, plateaus in units of e^2/h are observed and the 0.7 structure moves down to $0.5(2e^2/h)$. Interestingly, the transition from the 0.7 to the 0.5 plateau is continuous. The gradual evolution into the spin-split plateau at e^2/h means that the 0.7 structure is related to spin effect in the channel, and thus the transport is related to the spin-dependent phenomena. This experimental observation attracted much attention and subsequent experiments [266, 267] were performed. As for the theoretical works, various mechanisms to account for this phenomenon were proposed, including spontaneous spin polarization [268], ferromagnetism [269], spin singlet-triplet bound states [270], electron–phonon scattering [271], and Wigner crystal [272].

Here we present a mechanism by the Kondo effect from the binding of an electron in the vicinity of the point contact. Experimental observations in the transport of QPC suggestive of the Kondo effect [218] are shown such as evolution of an ZBA peak in the differential conductance, which splits in a magnetic field and a perfect transmission below a characteristic Kondo temperature T_K consistent with the peak width. The SDFT results predict a bound state in the vicinity of the QPC in the lowest channel [273], which leads to the relevance of an Anderson model [274] to explain the anomalous transport phenomena. We present the SDFT and QMC calculations for the QPC and show the detailed experimental observations for the 0.7 anomaly. Then we introduce the effective model and a perturbative approach to solve the model for anomalous transport of QPCs.

6.3.3 *Electronic Structure in Lowest Channel*

6.3.3.1 Spin-DFT calculations of QPCs

First, we provide the calculation method of SDFT for QPCs. We note that, as we showed in the previous section for the quantum dots, computations based on the SDFT provide us fairly accurate electronic and spin states of quantum dots for the interacting 2D confined systems. Thus, we expect the applications of SDFT to the 2D open systems give us clues for the mechanisms of unusual transport behaviors of QPCs. We present the formation of a quasi-bound state with a local moment and see the instability of the spin flipping at the center of a QPC. We provide also the QMC calculations later.

We begin to apply SDFT to the electronic states of a uniform quantum wire. This method allows us to treat both classical electrostatics from the confinement and electron–electron interaction in a unified framework. We take the x direction for the quantum wire where the current is flowing and the electrons are confined to the y direction with a parabolic potential. Since there is no scattering in the x direction, the wavefunctions are expanded by

$$\Psi^{\sigma}_{n,k_n}(\mathbf{r}) = e^{ik_n x} \sum_m c^{\sigma}_{n,m} \varphi_m(y), \tag{6.170}$$

where σ is the spin index, n is the subband number, and $\varphi_m(y)$ is the m-th eigenmode of the bare parabolic potential given by

$$\varphi_m(y) = \frac{1}{\sqrt{2^n n!}} \left(\frac{m^*\omega_y}{\pi\hbar}\right)^{1/4} H_n\left(\sqrt{\frac{m^*\omega}{\hbar}}y\right) \exp\left[-\frac{m^*\omega_y}{2\hbar}y^2\right]. \tag{6.171}$$

Here H_n is the n-th Hermite polynomials. We solve the following Kohn–Sham equation numerically:

$$\left[-\frac{\hbar^2}{2m^*}\nabla^2 + \frac{1}{2}m^*\omega_y^2 y^2 + V_H(y) + \frac{\delta E_{xc}[n,\zeta]}{\delta\rho^\sigma(\mathbf{r})} + g\mu_B B\sigma\right]\Psi^\sigma_{n,k_n}(\mathbf{r})$$
$$= \left(\epsilon_n^\sigma + \frac{\hbar^2 k_n^2}{2m^*}\right)\Psi^\sigma_{n,k_n}(\mathbf{r})$$
$$n(\mathbf{r}) = \sum_\sigma \rho^\sigma(\mathbf{r}) = \frac{1}{2\pi}\sum_{\sigma,n}\int_{-\infty}^{\infty} f_n^\sigma |\Psi^\sigma_{n,k_n}(\mathbf{r})|^2 dk_n \tag{6.172}$$

and iterate until self-consistent solutions are obtained for each electron density

$$n_{1d} = \int n(\mathbf{r})dy = \frac{1}{2\pi}\sum_{\sigma,n}\int_{-\infty}^{\infty} f_n^\sigma dk_n. \tag{6.173}$$

Here ϵ_n^σ is the band edge energy for subband n with spin σ and

$$f_n^\sigma = f\left(\epsilon_n^\sigma + \frac{\hbar^2 k_n^2}{2m^*}\right). \tag{6.174}$$

The Fermi–Dirac distribution function is $f(\epsilon) = 1/(e^{(\epsilon-\mu_{\text{wire}})/k_B T} + 1)$, where the chemical potential of electrons in the wire μ_{wire} is determined self-consistently for each density n_{1D}. The term $g\mu_B B\sigma$ is due to the Zeeman effect with $g = 0.44$ and the exchange-correlation energy $E_{xc}[n,\zeta]$ is treated in the local-density approximation with the local spin polarization $\zeta(\mathbf{r}) = [n^\uparrow(\mathbf{r}) - n^\downarrow(\mathbf{r})]/n(\mathbf{r})$. The Hartree potential is given by

$$V_H(\mathbf{r}) = -\frac{e^2}{\kappa}\int_{-\infty}^{\infty} n(y')\ln\left[\frac{(y-y')^2}{(y-y')^2+a^2}\right]dy', \tag{6.175}$$

where we introduce an image-charge plane at a distance $a = 100$ nm from the wire to model the experimental geometry. We use the constant parameters for GaAs and an external potential with $\hbar\omega_y = 2.0$ meV in the y direction. From the solution to the Kohn–Sham equations, the total free energy per unit length L is obtained from

$$\frac{E}{L} = \frac{1}{2\pi}\sum_{\sigma,n}\int_{-\infty}^{\infty}\left(\epsilon_n^\sigma + \frac{\hbar^2 k_n^2}{2m^*}\right) f_n^\sigma dk_n - \frac{1}{2}\int_{-\infty}^{\infty} V_H(y)n(\mathbf{r})dy$$
$$-\sum_\sigma\int_{-\infty}^{\infty} n^\sigma(\mathbf{r})\frac{\delta E_{xc}[n,\zeta]}{\delta n^\sigma(\mathbf{r})}dy + E_{xc}$$
$$+\frac{k_B T}{2\pi}\sum_{\sigma,n}\int_{-\infty}^{\infty}\left[f_n^\sigma \ln f_n^\sigma + (1-f_n^\sigma)\ln(1-f_n^\sigma)\right]dk_n, \tag{6.176}$$

where E_{xc} is the exchange-correlation energy per unit length L.

In experiments, the wire is coupled to 2D leads and the side gate voltage V_s is used to change the electron density n_{1D} in the wire. The absolute chemical potential of the electrons in the wire is the sum of the internal chemical potential $\mu_{\text{wire}} = d(E/L)/dn_{1D}$ and the external potential energy. The latter varies linearly with side gate voltage as $\propto -\alpha e V_s$. The capacitive lever arm α depends on the detailed gate geometry, which is the same as the quantum dot cases. Here, we take $\alpha = 1$ for simplicity. To model the fixed leads, we set the absolute chemical potential to zero, so that $\mu = \mu_{\text{wire}} - eV_s = 0$. From these procedures, we have the electronic states of quantum wires n^0 and the wavefunction

$$\psi_n^\sigma(y) = \sum_m c_{n,m}^\sigma \varphi_m(y). \tag{6.177}$$

Next, we introduce the potential barrier for the x direction to describe the realistic QPC situations as

$$V_{\text{QPC}}(x, y) = V(x) + \frac{1}{2}m^* \left[\omega_y + \frac{V(x)}{\hbar}\right]^2 y^2 - \frac{1}{2}m^*\omega_y^2 y^2, \tag{6.178}$$

where

$$V(x) = \frac{V_0}{\cosh^2(\omega_x\sqrt{m^*/2V_0}x)} = \frac{V_0}{\cosh^2(x/d)} \tag{6.179}$$

with the decay length of $d = \sqrt{2V_0/m^*}/\omega_x$. Note that $V(x)$ approaches the saddle potential $V(x) \approx V_0 - 1/2m^* w_x^2 x^2$ around $x \approx 0$. Then we solve the following Kohn–Sham equation for wavefunctions with energy ϵ:

$$\left[-\frac{\hbar^2}{2m^*}\frac{\partial^2}{\partial x^2} + \delta V_{\text{QPC}}(x, y) + \delta V_H(x, y) + g\mu_B B\sigma + \epsilon_n^\sigma \right.$$
$$\left. + \frac{\delta E_{xc}[n, \zeta]}{\delta n^\sigma(\mathbf{r})} - \frac{\delta E_{xc}[n^0, \zeta^0]}{\delta n^{0\sigma}(y)}\right] \Psi_{n,k_n}^\sigma(x, y) = \epsilon \Psi_{n,k_n}^\sigma(x, y). \tag{6.180}$$

Electrons incident from the two leads ($x \to \pm\infty$) are scattered elastically by the effective QPC potential $\delta V^\sigma(x, y) = V_{QPC}(x, y) + \delta V_H(x, y) + \delta V_{xc}^\sigma(x, y)$, which is the difference between the QPC self-consistent potential and that of the clean wire we obtain above. The Hartree and exchange-correlation parts of the potentials are written, respectively, as

$$\delta V_H(x, y) = \frac{e^2}{\kappa} \int\int dx'dy' \left\{n(x', y') - n^0(y')\right\}$$
$$\times \left[\frac{1}{\sqrt{(x-x')^2 + (y-y')^2}} - \frac{1}{\sqrt{(x-x')^2 + (y-y')^2 + a^2}}\right], \tag{6.181}$$

and

$$\delta V_{xc} = \frac{\delta E_{xc}[n, \zeta]}{\delta n^\sigma(\mathbf{r})} - \frac{\delta E_{xc}[n^0, \zeta^0]}{\delta n^{0\sigma}(y)}. \tag{6.182}$$

The wavefunctions of scattering states can be characterized as waves incident from the left $\Psi^L_{n,k^\sigma_n}(x, y)$ and from the right $\Psi^R_{n,k^\sigma_n}(x, y)$. Expanding these wavefunctions using $\psi^\sigma_n(y)$ as

$$\Psi^{L/R}_{n,k^\sigma_n}(x, y) = \sum_m u^{L/R}_{n,m,k^\sigma_n}(x)\psi^\sigma_m(y), \tag{6.183}$$

$u^{L/R}_{n,m,k^\sigma_n}(x)$ have the forms of plane waves with wavevectors $k^\sigma_n = \sqrt{2m^*(\epsilon - \epsilon^\sigma_n)}/\hbar$ far from the QPC region

$$u^L_{n,m,k^\sigma_n}(x) = \begin{cases} e^{ik^\sigma_n x}\delta_{n,m} + r^\sigma_{n,m}e^{-ik^\sigma_m x} & (x \le -x_0) \\ t^\sigma_{n,m}e^{ik^\sigma_m x} & (x \ge x_0), \end{cases} \tag{6.184}$$

$$u^R_{n,m,k^\sigma_n}(x) = \begin{cases} t^\sigma_{n,m}e^{-ik^\sigma_m x} & (x \le -x_0) \\ e^{-ik^\sigma_n x}\delta_{n,m} + r^\sigma_{n,m}e^{ik^\sigma_m x} & (x \ge x_0), \end{cases} \tag{6.185}$$

where $r^\sigma_{n,m}$ and $t^\sigma_{n,m}$ are the elements of unknown reflection and transmission matrices. $|x_0| = 500$ nm is sufficiently far from the QPC that $\delta V(x, y)$ is negligible. Now we obtain the scattering states by the recursion-transfer-matrix (RTM) method from the solution to the Kohn–Sham equation with the above boundary conditions. The different point described precisely in Chapter 4 is that we use the localized basis sets instead the plane wave for the expansion of y direction. Therefore, instead of the Fourier transform of the effective potential to obtain the matrix elements, we need to calculate the integral of

$$V^\sigma_{i,j}(x) = \int \psi^{\sigma *}_i(y)\delta V^\sigma(x, y)\psi^\sigma_j(y)dy$$
$$= \sum_{m,n} c^{\sigma *}_{i,m}c^\sigma_{j,n} \int \varphi_m(y)\delta V^\sigma(x, y)\varphi_n(y)dy \tag{6.186}$$

for each x point. Then we use the RTM method in the same procedure as we showed. From the resulting wavefunctions, electron density $n(x, y)$ is constructed as a sum over occupied states,

$$n(x, y) = \frac{1}{2\pi} \sum_{n,\sigma,L/R} \int_0^{\infty} f\left(\epsilon_n^\sigma + \frac{\hbar^2 k_n^{\sigma 2}}{2m^*}\right) |\Psi_{n,k_n^\sigma}^{L/R}(x, y)|^2 dk_n^\sigma . \tag{6.187}$$

These procedures are iterated until self-consistent solutions are obtained for the electron density n_{1D}.

6.3.3.2 Local moment formation in QPCs

Figure 6.34 shows the electronic properties of QPCs with $V_0 = 3.0$ meV at $T = 0.1$ K for three different lengths: (a) $d = 82.6$ nm ($\hbar\omega_x = 1.0$ meV); (b) $d = 55.0$ nm ($\hbar\omega_x = 1.5$ meV); (c) $d = 41.3$ nm ($\hbar\omega_x = 2.0$ meV). The electron density far from the QPC is taken as $n_{1D} = 2.80 \times 10^{-2}$ nm^{-1}, where the electrons far into the wire are unpolarized and only the lowest two spin subbands contribute to transport. In this low-density regime, we find that a solution with broken spin symmetry coexists with an unpolarized solution. To obtain the broken-symmetry solution, we first apply an in-plane magnetic field up to $B = 6$ T and then solve the Kohn–Sham equation self-consistently while reducing the magnetic field to zero. We show the self-consistent QPC barrier as a function of position x in the direction of current flow. Specifically, we plot the energy of the bottom of the lowest 1D subband $\epsilon_0^\sigma(x)$ relative to the band edge ϵ_0^σ far into the wire, for both spin-up (solid lines) and spin-down (dashed lines) electrons in Figs. 6.34(a–c). We see small Friedel oscillations with a period of $2\pi/2k_f \approx 72$ nm are present far into the wire. The self-consistent QPC barrier is strongly spin dependent in all these cases, with the chemical potential μ lying above the spin-up barrier but below the spin-down barrier, so between two barriers.

In the panels of the insets of Figs. 6.34(a-c), we plot the transmission coefficient $\mathcal{T}_\sigma(\epsilon)$ (left) and the local 1D density of states $\nu_\sigma(\epsilon)$ at the center of the QPC (right) for the respective physical parameters. The solid lines give the spin-up and the dashed lines give the spin-down $\mathcal{T}_\sigma(\epsilon)$ and $\nu_\sigma(\epsilon)$. The 1D electron density in the vicinity of a QPC is obtained from $n_{1D}^\sigma = \int \nu_\sigma(\epsilon) d\epsilon$, which approach the spin-unpolarized states of $n_{1D}^\uparrow = n_{1D}^\downarrow = 1.40 \times$

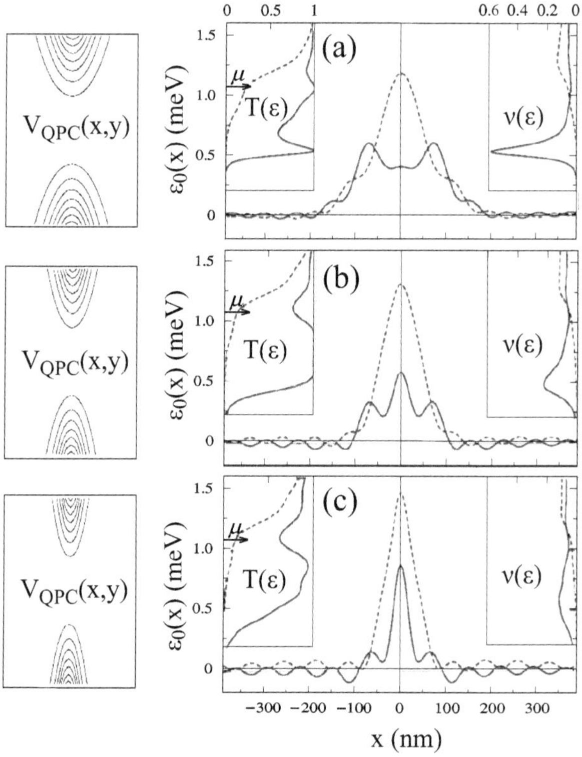

Figure 6.34 Self-consistent barrier, i.e., energy of the bottom of the lowest 1D subband at temperature $T = 0.1$ K as a function of position x in the direction of current flow through QPC. The sharpness of potentials is measured as (a) $\hbar\omega_x = 1.0$ meV, (b) $\hbar\omega_x = 1.5$ meV, and (c) $\hbar\omega_x = 2.0$ meV. The chemical potential μ is indicated by an arrow on the left. Solid lines are for spin-up and dashed lines are for spin-down electrons. Contour plot of the QPC potential $V_{\mathrm{QPC}}(x, y)$ is shown in the left side. (Inset) (Left) $\mathcal{T}_\sigma(\epsilon)$ as a function of energy. (Right) Local density of states $\nu_\sigma(\epsilon)$ at the center of the density. From K. Hirose, Y. Meir, and N. S. Wingreen, *Phys. Rev. Lett.* **90**, 026804 (2003) [273].

10^{-2} nm^{-1} far into the wire. Below the chemical potential μ, the local density of states (LDOS) $\nu_\uparrow(\epsilon)$ for spin-up electrons shows a resonance that broadens as the QPC is shortened. Correspondingly, for all these QPC lengths, there is an excess spin-up density in the vicinity of the barrier. In each case, the integrated net spin-up

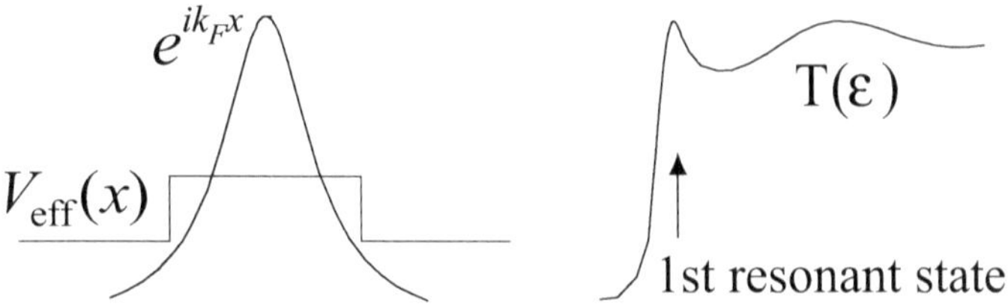

Figure 6.35 (Left) Schematic illustration of the multiple electron scattering for the repulsive potential forming the local moment. (Right) Transmission as a function of energy.

density $n_{1D}^{\uparrow}$ is close to 1 spin: 0.85 for (a), 0.93 for (b), and 0.90 for (c). Thus, there is a local moment with a spin of 1/2 formed at the QPC. The data from SDFT gives strong evidence for a quasi-bound state centered at the QPC, that is, there is a resonance in the LDOS $\nu(\epsilon)$ for spin-up with a net of one spin bound in the vicinity of the QPC. The transmission coefficient $\mathcal{T}(\epsilon)$ for electrons in the lowest channel is approximately 1 over a range of energies above the spin-up resonance. This implies an onset of strong hybridization at energies above quasi-bound state.

It is important to note that this local moment found within SDFT results from the self-consistent flattening of the QPC barrier, such as a series of quasi-bound states resulting from multiple reflections from the edges of the barrier, as shown schematically in Fig. 6.35. While the bare QPC potential does not give rise to resonances, the self-consistent potential is flattened for spin-up, particularly for the longer QPCs and the first resonance in the spin-up LDOS is clearly resolved. Since this resonance lies below the chemical potential, it is fully occupied. Then the result is a localized spin of 1/2 at the QPC.

Figure 6.36 shows the dependence of the Coulomb energy U and the hybridization Γ on the length of the QPC. U is obtained from the energy difference between the resonance center of $\nu_{\uparrow}(\epsilon)$ and the energy at which the derivative of $\nu_{\downarrow}(\epsilon)$ is a maximum. Γ is obtained from the full width at half maximum (FWHM) of a Lorentzian fit to the resonance in $\nu_{\uparrow}(\epsilon)$. The hybridization energy Γ increases sharply from $\Gamma \approx 0.1$ meV up to ≈ 0.6 meV as the QPC is shortened from $d = 80$ nm to $d = 40$ nm. On the other hand, the Coulomb energy U stays nearly constant at ≈ 0.6 meV over this range. We note that the

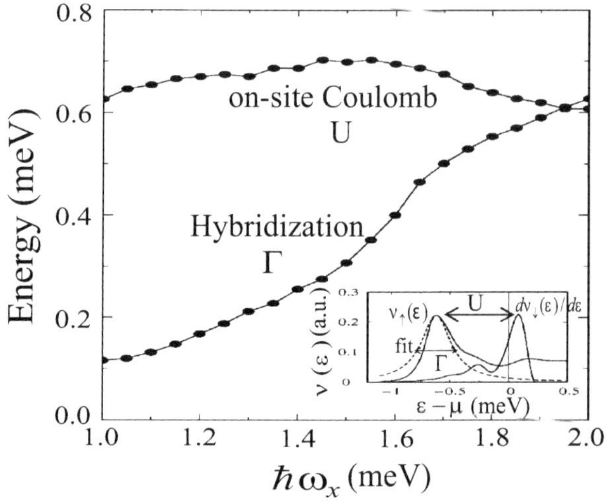

Figure 6.36 Hybridization energy Γ and on-site Coulomb energy U as a function of $\hbar\omega_x$, the sharpness of the external QPC potential. Γ is obtained from the FWHM of the Lorentzian fit to the resonance. U is obtained from the energy difference between the resonance center and the maximum of $d\nu(\epsilon)/d\epsilon$ for the high-energy spin. (Inset) $\nu_\uparrow(\epsilon)$ and $d\nu_\downarrow(\epsilon)/d\epsilon$ at the center of QPC for $\hbar\omega_x = 1.5$ meV. The dashed line is the Lorentzian fit to the resonance of $\nu_\uparrow(\epsilon)$. Hybridization energy Γ and Coulomb energy U are indicated. From K. Hirose, Y. Meir, and N. S. Wingreen, *Phys. Rev. Lett.* **90**, 026804 (2003) [273].

Coulomb energy $U \approx 0.6$ meV is quite reasonable for the charging energy as we estimated before, while the significant change of the hybridization energy $\Gamma \approx 0.1 - 0.6$ meV, which is larger than that of the quantum dot tunneling case, implies an onset of strong coupling at energies above the quasi-bound state. As the external QPC barrier becomes flatter in the x direction, the width of the quasi-bound state decreases.

These results show that a localized spin of 1/2 might be formed at the QPC as a result of multiple reflections at the edges of the barrier and corresponding self-consistent flattening of the QPC barrier. These also show that the quasi-bound electron at the QPC strongly hybridizes the electrons with energies above the quasi-bound state. We realize that this situation is very close to the Kondo model situation where a localized spin with the energy below

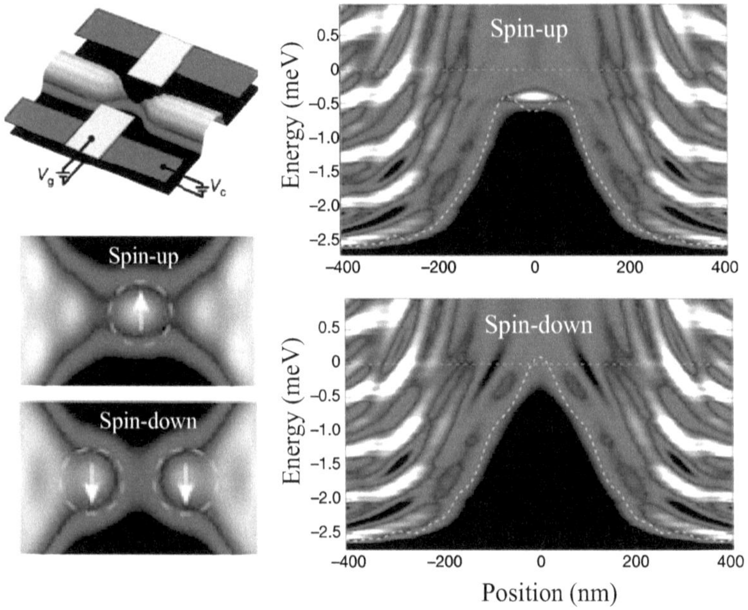

Figure 6.37 (Left) (top) The device configuration used in the calculation. Gate voltage V_g is applied to split gates and forms a QPC between two electrodes, which is negatively biased by V_c. (Bottom) Electron density for spin-up and spin-down. (Right) Spin-up and spin-down LDOS and the effective potential for electrons. The stripes on both sides of the QPC correspond to the transverse modes of the electrodes. The effective potential for spin-up electrons has a double-barrier structure in the vicinity of the QPC, and forms a quasi-bound state about 0.5 meV below the Fermi level in the lowest mode. From T. Rejec and Y. Meir, *Nature* **442**, 900 (2006) [275].

the Fermi level couples strongly with the transmitted electrons, producing a correlated state. As we show in the previous section, the transport through the Kondo resonance is characterized by the ZBA observed in the differential conductance dI/dV, whose width is determined not by the hybridization energy Γ but by the Kondo temperature T_K and whose temperature dependence of the peak shows a universal feature. Such a strong Kondo peak in the differential conductance is expected when

$$k_B T < k_B T_K < \Gamma \ll U. \tag{6.188}$$

From the present calculations, we see that the resonance width Γ will be too broad for a Kondo effect to develop for very short QPC ($\Gamma \approx U$), while Γ, and thus T_K might be too small for very long QPC. There is an optimal range of QPC lengths for observing the Kondo effect. We will see the results of experiments showing the formation of a Kondo-like correlated spin state in QPCs later.

Next, we show the formation of local moment in QPCs appears in more realistic device configurations. Figure 6.37 shows the electron densities (left) and the potential profiles (right) of the realistic QPC configurations, as described in the left top panel, for spin-up and spin-down LDOS and the effective potential for electrons in the lowest transverse mode at $V_g = -0.125$ V. The stripes on both sides of the QPC correspond to the transverse modes of the electrodes. The effective potential for spin-up electrons has a double-barrier structure in the vicinity of the QPC, and forms a quasi-bound state about 0.5 meV below the Fermi level in the lowest mode. This result is close to Fig. 6.34(b) and shows that the formation of local moment appears in the realistic device configurations. Furthermore spin-down electrons, which is absent in the center of the QPC, form quasi-bound states in the shoulders of the potential on both sides of the QPC.

6.3.3.3 Quantum Monte Carlo calculations of QPCs

Here we see the formation of local moment in QPC constriction obtained with the many-body variational and diffusion QMC methods, which include the correlation effects beyond the mean-field approximation. The electron density in the electrodes is $r_s = 2.6$ and that in the constriction is controlled by the gate voltage V_g with $H^* = 11.9$ meV.

Figure 6.38 (left) shows the electron density of the QPC constriction with the saddle potential. A single electron is localized in the low-density region with substantial barriers to the high-density leads, corresponding to the SDFT calculations. The spin of the localized electron is "not" static but dynamical, and it changes as the electron tunnels to the leads, yielding a spin density of zero. This result shows that it is possible to have localized electrons in a low-density region separated from the leads.

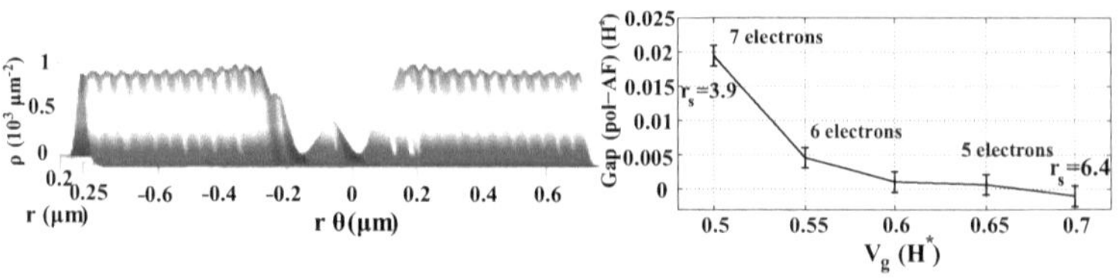

Figure 6.38 (Left) Two-dimensional ground-state density for a short constriction of $V_g = 0.5\ H^*$ with a length of 0.34 μm, showing a single localized electron. (Right) Difference of total energy when electrons in the constriction are fully spin polarized or anti-ferromagnetically ordered as a function of V_g. From A. D. Güclü, C. J. Umrigar, H. Jiang, and H. U. Baranger *Phys. Rev. B.* **80**, 201302 (2009) [276].

The ground state cannot be ferromagnetically spin polarized for strictly 1D system (Lieb's theorem) [263], but the situation might change for quasi-1D system of the QPC constriction due to the transverse degree as we argue. Figure 6.38 (right) shows the difference of total energy when electrons in the constriction are fully polarized or anti-ferromagnetically ordered. We see that at $V_g = 0.5\ H^*$ with $r_s = 3.7$, close to the QPC situation in Fig. 6.34, the low-density part is not spin-polarized with the spin gap as large as $\sim$3 K. As V_g increases, the overlap with leads becomes small resulting in having larger r_s and the energy gap becomes small.

6.3.4 *Kondo-like Correlated State in Experiment*

Here we see the experiment on the anomalous transport behavior of QPC by Cronenwett *et al.* [218], which shows the quantum transport through QPC via the Kondo effects. Using the surface gate depletion method to construct a 1D channel in a 2DEG, the experiments were performed for source-drain bias measurements.

The linear-response conductance around $V_{sd} \sim 0$ in the top panels of Fig. 6.39 exhibits a characteristic evolution from spin-degenerate plateaus at $B = 0$ at integer multiples of $2e^2/h$, into spin-resolved plateaus at integer multiples of e^2/h in high magnetic field. A remnant of the spin-resolved plateau remains at $B = 0$ and $T = 80$ mK as a barely visible shoulder below the $2e^2/h$ plateau. As the temperature is increased, conductance at this shoulder decreases and a plateau near $0.7(2e^2/h)$ forms. It is noted that while

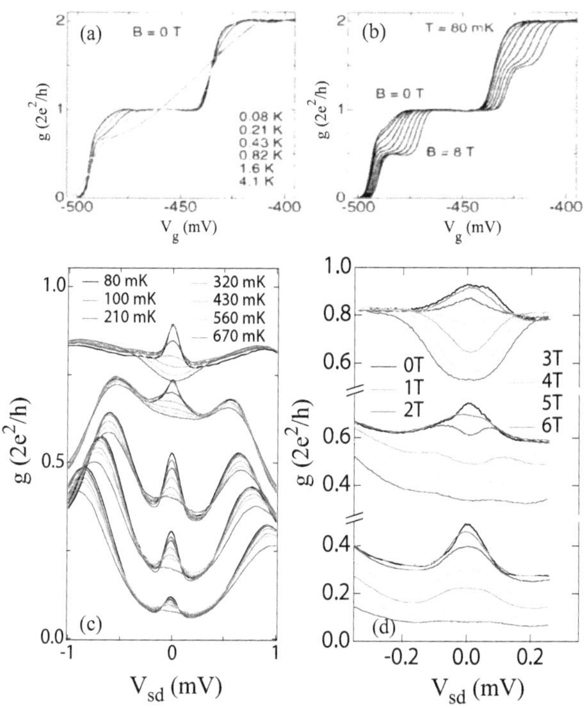

Figure 6.39 (Top) (a) Linear conductance vs. gate voltage V_g at several temperatures. The extra plateau at $\sim 0.7(2e^2/h)$ appears with increasing temperature, while the plateaus at multiples of $2e^2/h$ become less visible due to thermal smearing. (b) Linear g versus V_g for in-plane field B from 0 to 8 T in 1 T steps, showing spin-resolved plateaus at odd multiples of e^2/h at high fields. (Bottom) (c) Temperature dependence of the ZBA for different gate voltages, at temperatures from 80 to 670 mK. (d) Evolution of the ZBA with in-plane B, at V_g corresponding to high, intermediate, and low conductance. From S. M. Cronenwett, H. J. Lynch, D. Goldhaber-Gordon, L. P. Kouwenhoven, C. M. Marcus, K. Hirose, N. S. Wingreen, and V. Umansky, *Phys. Rev. Lett.* **88**, 226805 (2002) [218].

the 0.7 structure becomes stronger at elevated temperatures, the plateaus at multiples of $2e^2/h$ become more washed out. That is, as the temperature is lowered, the plateaus at multiples of $2e^2/h$ become sharper, while the plateau at $0.7(2e^2/h)$ rises to the unitary limit of $2e^2/h$ and thus disappears.

The bottom left panel of Fig. 6.39 shows the non-linear differential conductance as a function of source-drain bias at several temperatures at $B = 0$. We see a narrow peak in conductance around $V_{sd} = 0$ for the whole range $0 < G < 2e^2/h$. This ZBA forms as the temperature is lowered and is closely linked to the disappearance of the 0.7 structure at low temperature. The formation of a zero-bias conductance peak, and the associated enhancement of the linear conductance up to the unitary limit $2e^2/h$ at low temperature are reminiscent of the Kondo effect seen in quantum dots containing an odd number of electrons. A characteristic feature of the Kondo regime ($T < T_K$) in quantum dots is that the ZBA peak is split by $2g^*\mu_B B$ upon application of an in-plane magnetic field when $g^*\mu_B B >\sim T_K$. In the QPC, we see that the ZBA peak splits clearly near $g \sim 0.7(2e^2/h)$ in Fig. 6.37(d), consistent with $2g^*\mu_B B$. That is, splitting roughly linear in field for $B <\sim 3$ T. At higher conductances the ZBA peak does not split but merely collapses with B. This is because $2g^*\mu_B B < T_K$ in this regime.

Figure 6.40 shows the non-linear transport data. In this representation, plateaus in $G(V_g)$ appear as accumulations of traces, for instance in the half-plateaus at higher bias ($V_{sd} >\sim 0.5$ mV) at $G \sim 1/2,\ 3/2$, and $5/2$ in units of $2e^2/h$. The linear-response plateaus appear as accumulated traces around zero bias at multiples of $2e^2/h$ and, in the $B = 8$ T data, at odd multiples of e^2/h. Within the first subband ($G < 2e^2/h$), the non-linear data for $B = 0$ look strikingly similar to the $B = 8$ T data, including the wing shape of the extra plateau that extends out from the 0.7 feature. Higher subbands at $B = 0$ do not show extra plateaus. Comparing these figures, the zero-field plateau at $\sim 0.7(2e^2/h)$, which extends to $\sim 0.8(2e^2/h)$ at high bias, results from a splitting of spin bands. This leads to transport signatures in the lowest mode, which greatly resemble the situation at $B = 8$ T, where spin degeneracy is explicitly lifted in all modes by the applied field.

We see that the temperature- and magnetic field dependence of conductance in QPC is similar to that in the quantum dot. Guided by this similarity, we consider a scaling of the temperature dependence of the conductance using a single scaling parameter that we indicate as the Kondo temperature T_K. Experimentally, this single parameter

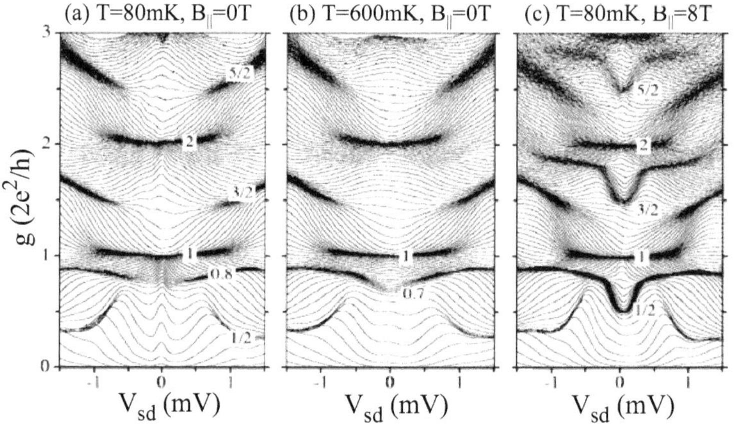

Figure 6.40 (Left) Non-linear G at 80 mK, $B = 0$ at V_g intervals of 1.25 mV. Plateaus at multiples of $2e^2/h$ around $V_{sd} \sim 0$ and half-plateaus at odd multiples of e^2/h at high bias are visible. An ZBA is present only at low magnetic field and low temperatures. At high bias, an extra plateau appears at $G \sim 0.8(2e^2/h)$. (Center) Non-linear G at 600 mK, $B = 0$ at V_g intervals of 1.0 mV. Note absence of an ZBA and accumulation of traces at $G \sim 0.7(2e^2/h)$ around $V_{sd} \sim 0$ that merges with the high-bias plateau at $0.8(2e^2/h)$. (Right) Non-linear G at 80 mK, $B = 8$ T at V_g intervals of 1.2 mV. Spin-resolved plateaus at odd multiples of e^2/h around $V_{sd} \sim 0$ merge with high-bias plateaus at $0.8(2e^2/h)$, and $2.8(2e^2/h)$. The high-bias feature at $0.8(2e^2/h)$ looks similar to that in the $B = 0$ data. From S. M. Cronenwett, H. J. Lynch, D. Goldhaber-Gordon, L. P. Kouwenhoven, C. M. Marcus, K. Hirose, N. S. Wingreen, and V. Umansky, *Phys. Rev. Lett.* **88**, 226805 (2002) [218].

allows data from a broad range of gate voltages to be scaled onto a single curve as a function of scaled temperature T/T_K. The left panel of Fig. 6.41 shows the linear g as a function of scaled temperature T/T_K where T_K is the single fit parameter described from the *modified* expression for the Kondo conductance,

$$G = \frac{e^2}{h}\left[f\left(\frac{T}{T_K}\right) + 1\right], \tag{6.189}$$

where $f(T/T_K)$ is a universal function for the Kondo conductance in Eq.(6.72) well approximated by

$$f\left(\frac{T}{T_K}\right) = \left[1 + (2^{1/s} - 1)\left(\frac{T}{T_K}\right)^2\right]^{-s} \tag{6.190}$$

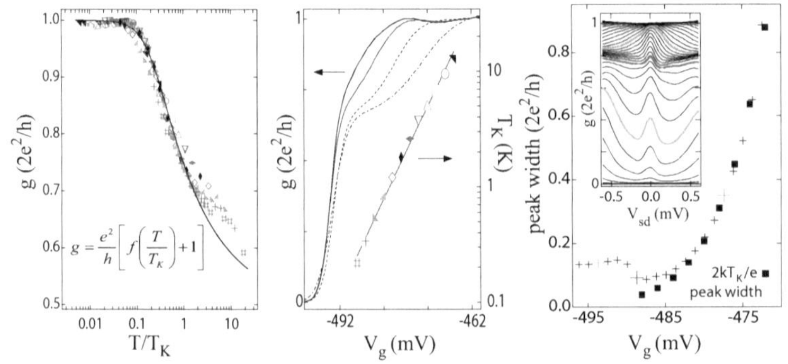

Figure 6.41 (Left) Linear G as a function of scaled temperature T/T_K where T_K is the single fit parameter. (Center) T_K (right axis) obtained from the fits of $G(T/T_K, V_g)$ along with the conductance (left axis) at temperatures of 80 mK (solid line), 210 mK (dotted), 560 mK (dashed), and 1.6 K (dot-dashed). (Right) Widths of the ZBA peak (crosses) defined as the FWHM taken from non-linear G for $G < 2e^2/h$ in the inset and values for $2k_BT_K/e$ (squares) for the range of V_g. From S. M. Cronenwett, H. J. Lynch, D. Goldhaber-Gordon, L. P. Kouwenhoven, C. M. Marcus, K. Hirose, N. S. Wingreen, and V. Umansky, *Phys. Rev. Lett.* **88**, 226805 (2002) [218].

with $s = 0.23$. Eq. (6.189) differs from the usual form for the quantum dots of $G = (2e^2/h)f(T/T_K)$ in Eq. (6.71) by the addition of a constant e^2/h term that sets the high-temperature limit to e^2/h rather than zero. The mechanism for this additional constant term of e^2/h to the usual dot form will be shown in the next section.

The one-parameter fit to the $G(T)$ data using Eqs. (6.189) and (6.190) yields values of T_K that increase exponentially with the gate voltage, $\ln(T_K) \sim a(V_g - V_g^0)$, with $a = 0.18$ from a best fit line to $\ln(T_K)$ as shown in the center panel of Fig. 6.41. The exponential dependence of T_K on V_g is also observed for quantum dots, since $T_K \sim \exp[\pi\epsilon_0(\epsilon_0 + U)/\Gamma U]$ depends on ϵ_0, the energy of the bound spin relative to the Fermi energy of the leads, $(\epsilon_0 + U)$, the energy to the next available state, and Γ, the energy broadening due to coupling to the reservoirs. Another feature of the Kondo effect as it appears in quantum dots is that the width of the ZBA is set by T_K rather than the larger level-broadening scale Γ. In the QPC, the width of the ZBA in the inset of the right panel of Fig. 6.40 is roughly constant for $G < 0.7$. At $G \sim 0.7$, the ZBA width first decreases significantly then increases as G approaches $2e^2/h$. The ZBA peak width in the right panel of Fig. 6.41 is very close to $2kT_K/e$ for $G > 0.7$.

6.3.5 *Kondo Model for Anomalous Transport*

Now we see that using the SDFT, the local moment with spin 1/2 is formed at the QPC as a result of multiple reflections at the edges of the barrier and the quasi-bound electron at the QPC strongly hybridizes the electrons as we sweep the gate voltage. Also, we see a number of experimental data that show the similarities of anomalous transport behaviors of QPC at the 0.7 structure with the Kondo physics in the quantum dots cases. In order for a Kondo-like effect to appear in a QPC, the splitting between spin bands must be *dynamic* rather than a frozen spin polarization at zero field. The universal feature above suggests the high-temperature limit is $G = 0.5(2e^2/h)$ and G decays logarithmically very slowly with increasing temperature, leaving a significant residual enhancement around $G = 0.7(2e^2/h)$ for a wide range of temperature. The recovery of the conductance to the unitary limit, $2e^2/h$, with decreasing temperature is a natural feature of the Kondo effect. These are schematically shown in Fig. 6.42.

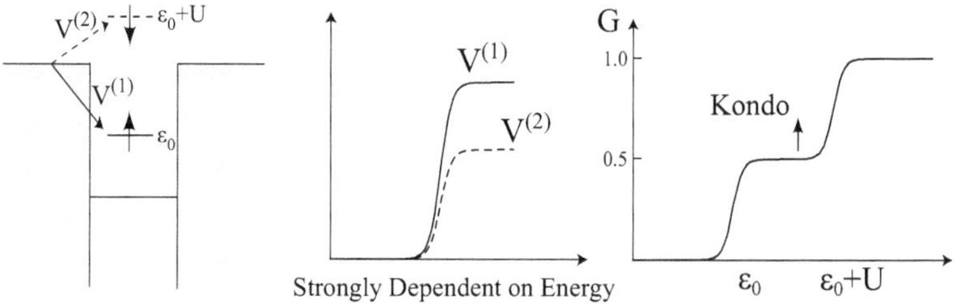

Figure 6.42 (Left) Schematic view of an electron transport through a QPC. (Center) Behaviors of the hybridization energy $V^{(1,2)}$ as a function of the energy, or equivalently the gate voltage V_g, assuming $V^{(1)} > V^{(2)}$. (Right) Conductance behavior as a function of energy. Conductance increases rapidly up to $0.5(2e^2/h)$ with large $V^{(1)}$. The temperature dependence of G between 0.5 and 1.0 is due to the Kondo effect.

Here we present calculations based on the Kondo model for the anomalous transport behaviors at QPCs. Since the Kondo physics is beyond the single-particle picture, we write down an effective Hamiltonian based on the SDFT results and demonstrate the applicability of the effective model to transport through a QPC to compare the results to experimental data in the previous section. A novel feature in the model distinguishes the transport through a QPC from transport through quantum dots and explains the large residual conductance from the hybridization to the band, which is a strong function of gate voltage energy. Since we show a quasi-bound state is formed at the center of QPCs, we model the QPC and leads by the following Hamiltonian:

$$\begin{aligned}\hat{H} = & \sum_{\sigma;k\in L,R} \epsilon_{k\sigma}\hat{c}^{\dagger}_{k\sigma}\hat{c}_{k\sigma} + \sum_{\sigma}\epsilon_{\sigma}\hat{d}^{\dagger}_{\sigma}\hat{d}_{\sigma} + U\hat{n}_{\uparrow}\hat{n}_{\downarrow} \\ & + \sum_{\sigma;k\in L,R}\left[V^{(1)}_{k\sigma}(1-\hat{n}_{\bar{\sigma}})\hat{c}^{\dagger}_{k\sigma}\hat{d}_{\sigma} + \text{h.c.}\right] \\ & + \sum_{\sigma;k\in L,R}\left[V^{(2)}_{k\sigma}\hat{n}_{\bar{\sigma}}\hat{c}^{\dagger}_{k\sigma}\hat{d}_{\sigma} + \text{h.c.}\right], \end{aligned} \tag{6.191}$$

where $\hat{c}^{\dagger}_{k\sigma}$ ($\hat{c}_{k\sigma}$) creates (destroys) an electron with momentum k and spin σ in lead L or R, $\hat{d}^{\dagger}_{\sigma}$ ($\hat{d}_{\sigma}$) creates (destroys) a spin-σ electron on the quasi-bound state at the center of the QPC and $\hat{n}_{\sigma} = \hat{d}^{\dagger}_{\sigma}\hat{d}_{\sigma}$. As for the hybridization matrix elements, $V^{(1)}_{k\sigma}$ is for transitions between 0 and 1 electrons on the site and $V^{(2)}_{k\sigma}$ is for transitions between 1 and 2 electrons, which are strongly dependent on gate voltage energy and are taken as step-like functions, mimicking the exponentially increasing transparency. From the SDFT calculations, we consider $V^{(2)}_{k\sigma} < V^{(1)}_{k\sigma}$, since the Coulomb potential of an electron already occupying the QPC will reduce the tunneling rate of a second electron through the bound state. In the absence of magnetic field, the two spin directions are degenerate, so $\epsilon_{\uparrow} = \epsilon_{\downarrow} = \epsilon_0$.

For the present QPC system, we expect the *high-temperature* contribution from $0 \leftrightarrow 1$ valence fluctuations to the conductance G to saturate at $0.5(2e^2/h)$ for $\epsilon_0 < \epsilon_F$, because the probability of an opposite spin electron occupying the site is $\sim$0.5. Since $V^{(2)}_{k\sigma}$ is significantly smaller than $V^{(1)}_{k\sigma}$, the contribution to the conductance

G from $1 \leftrightarrow 2$ valence fluctuations is small until $\epsilon_F \approx \epsilon_0 + U$

$$G = \frac{2e^2}{h}\left[(1 - \langle \hat{n}_{\bar{\sigma}} \rangle)\frac{(\Gamma_1/2)^2}{(\mu - \epsilon_0)^2 + (\Gamma_1/2)^2} + \langle \hat{n}_\sigma \rangle \frac{(\Gamma_2/2)^2}{(\mu - \epsilon_0 - U)^2 + (\Gamma_2/2)^2}\right] \tag{6.192}$$

for $T \gg T_K$, where the coupling constants Γ are

$$\Gamma_{1,2} = 2\pi \sum_{k\sigma} |V_{k\sigma}^{(1,2)}|^2 \delta(\epsilon - \epsilon_{k\sigma}). \tag{6.193}$$

On the other hand, the Kondo effect enhances this contribution from $V_{k\sigma}^{(2)}$ with decreasing temperature, resulting in the conductance of $2e^2/h$ at zero temperature due to Friedel sum rule:

$$G = \frac{2e^2}{h}\sin^2(\pi\langle \hat{n}_\sigma \rangle) \rightarrow \frac{2e^2}{h} \quad \text{for } T \rightarrow 0. \tag{6.194}$$

The first term of Eq. (6.192) saturates at $0.5(2e^2/h)$ due to the rapid increase of ϵ_F by the gate voltage, which is a different point from the quantum dot case. The second term contributes to the temperature-dependent transport due to the Kondo effect. Then the generalized temperature dependence of the conductance is expected to behave as

$$G = \frac{2e^2}{h}\left[\frac{1}{2} + \frac{1}{2} f\left(\frac{T}{T_K}\right)\right] = \frac{e^2}{h}\left[1 + f\left(\frac{T}{T_K}\right)\right]. \tag{6.195}$$

This expression is consistent with the experimental observations.

Since the gate voltage range corresponds to the singly occupied site regime, we change the Anderson Hamiltonian Eq. (6.191) into the Kondo limit, for a quantitative estimate of the conductance. The Hamiltonian corresponding to the singly occupied site becomes

$$\hat{H} = \sum_{\sigma;k\in L,R} \epsilon_{k\sigma}\hat{c}_{k\sigma}^\dagger \hat{c}_{k\sigma} + \sum_{\sigma\sigma';kk'\in L,R} J_{kk'\sigma\sigma'}\hat{c}_{k\sigma}^\dagger \sigma_{\sigma\sigma'}\hat{c}_{k'\sigma'} \cdot S + \sum_{\sigma\sigma';kk'\in L,R} W_{kk'\sigma\sigma'}\hat{c}_{k\sigma}^\dagger \hat{c}_{k'\sigma'}, \tag{6.196}$$

where $J_{kk'\sigma\sigma'}$ and $W_{kk'\sigma\sigma'}$ are

$$\begin{aligned}
J_{kk'\sigma\sigma'} &= 2(J^{(1)}_{kk'\sigma\sigma'} + J^{(2)}_{kk'\sigma\sigma'}) \\
&= \frac{1}{2}\left(\frac{V^{(1)}_{k\sigma}V^{(1)*}_{k'\sigma'}}{\epsilon_{k\sigma} - \epsilon_{\sigma}} - \frac{V^{(1)}_{k\sigma}V^{(1)*}_{k'\sigma'}}{\epsilon_{\sigma} + U - \epsilon_{k\sigma}} - \frac{V^{(2)}_{k\sigma}V^{(2)*}_{k'\sigma'}}{\epsilon_{k\sigma} - \epsilon_{\sigma}} + \frac{V^{(2)}_{k\sigma}V^{(2)*}_{k'\sigma'}}{\epsilon_{\sigma} + U - \epsilon_{k\sigma}}\right) \\
W_{kk'\sigma\sigma'} &= J^{(1)}_{kk'\sigma\sigma'} - J^{(2)}_{kk'\sigma\sigma'} \\
&= \frac{1}{4}\left(\frac{V^{(1)}_{k\sigma}V^{(1)*}_{k'\sigma'}}{\epsilon_{k\sigma} - \epsilon_{\sigma}} - \frac{V^{(1)}_{k\sigma}V^{(1)*}_{k'\sigma'}}{\epsilon_{\sigma} + U - \epsilon_{k\sigma}} + \frac{V^{(2)}_{k\sigma}V^{(2)*}_{k'\sigma'}}{\epsilon_{k\sigma} - \epsilon_{\sigma}} - \frac{V^{(2)}_{k\sigma}V^{(2)*}_{k'\sigma'}}{\epsilon_{\sigma} + U - \epsilon_{k\sigma}}\right).
\end{aligned} \tag{6.197}$$

The Pauli spin matrix is indicated by σ, and the local spin due to the bound state is $S = \frac{1}{2}\hat{d}^{\dagger}_{\sigma}\sigma_{\sigma\sigma'}\hat{d}_{\sigma'}$. Here we note that the potential term $W_{kk'\sigma\sigma'}$, usually ignored in Kondo problem, is essential to produce the large background conductance at high temperature.

We treat the above Kondo Hamiltonian perturbatively in the couplings J. The differential conductance up to the third-order J^3 is given by

$$G \approx G_2 + G_3 \times f(T), \tag{6.198}$$

where conductance from the second-order element G_2 is given by

$$\begin{aligned}
G_2 = \frac{4\pi e^2}{\hbar}\nu_L(\mu)\nu_R(\mu) \times \Big\{(J^{(-)})^2 + (J^{(+)})^2 \\
\times \left[3 + 2\langle M\rangle\left(\tanh\frac{\Delta + eV}{2k_B T} + \tanh\frac{\Delta - eV}{2k_B T}\right)\right]\Big\}.
\end{aligned} \tag{6.199}$$

Here we include the magnetic field B contribution through the Zeeman splitting $\Delta = g^*\mu_B B$ as the magnetization for the uncoupled site $\langle M\rangle = -(1/2)\tanh(\Delta/2k_B T)$ and $\nu_{L/R}(\epsilon) = \sum_{k\in L/R}\delta(\epsilon - \epsilon_k)$ is the single-electron density of states in the leads. The coupling energies $J_{kk'}$ and $W_{kk'}$ are approximated and replaced by the values at the Fermi energy

$$\begin{aligned}
J^{(+)} = J^{(1)} + J^{(2)} = \frac{1}{2}\left(\frac{|V^{(1)}|^2}{\mu - \epsilon_0} + \frac{|V^{(2)}|^2}{\epsilon_0 + U - \mu}\right) f_{FD}\left(-\frac{\mu}{\delta}\right) \\
J^{(-)} = J^{(1)} - J^{(2)} = \frac{1}{2}\left(\frac{|V^{(1)}|^2}{\mu - \epsilon_0} - \frac{|V^{(2)}|^2}{\epsilon_0 + U - \mu}\right) f_{FD}\left(-\frac{\mu}{\delta}\right),
\end{aligned}$$

where the Fermi–Dirac form $f_{FD}(x) = 1/[1+\exp(x)]$ is included for a steplike increase of J due to the gate voltage and δ. $V^{(1)}$ and $V^{(2)}$ are taken constants for simplicity.

The third-order element G_3 is

$$G_3 = -\frac{96\pi e^2}{\hbar}\nu_L(\mu)\nu_R(\mu)J^2 \tag{6.200}$$

and the divergent term in Eq. (6.198) is

$$f(T) = \frac{J^{(2)}}{6}\left\{\left[3 + 2\langle M\rangle\left(\tanh\frac{\Delta + eV}{2k_B T} + \tanh\frac{\Delta - eV}{2k_B T}\right)\right]F(eV)\right.$$
$$\left. + \frac{1}{2}\sum_{\pm}\left[3 + 4\langle M\rangle\tanh\frac{\Delta \pm eV}{2k_B T}\right]F(eV \pm \Delta)\right\}, \tag{6.201}$$

where the function $F(x)$ is

$$F(x) = -\int_{-\infty}^{\infty} g(\epsilon)\frac{\partial f(\epsilon - x)}{\partial \epsilon}d\epsilon \tag{6.202}$$

with

$$g(\epsilon) = \sum_k \frac{f(\epsilon_k)}{\epsilon_k - \epsilon} = \mathrm{P}\int_D^{k_B T}\frac{f(\epsilon')}{\epsilon' - \epsilon}d\epsilon'. \tag{6.203}$$

Here we set the band edge energy D approximate the diverging integral $F(x)$ as

$$F(x) \approx \ln\left[\frac{|x| + k_B T}{D}\right]\left(\approx \ln\left[\frac{\sqrt{x^2 + (k_B T)^2}}{D}\right]\right). \tag{6.204}$$

At low temperatures, the Kondo effect leads to a logarithmically diverging contribution G_3 to the differential conductance at order J^3 due to integrals running from the Fermi energy to either band edge. In the present case where $V^{(1)}$ increases rapidly by the gate voltage, the band integral for $V^{(2)}$ runs from μ up to $\epsilon_0 + U$ and the logarithmic contribution from $V^{(2)}$ dominates G_3, producing the temperature-dependent behavior of the conductance. In the linear-response conductance in the absence of magnetic field case ($eV = 0$, $B = 0$), $F(0) \approx -\ln[(\epsilon_0 + U - \mu)/k_B T]$ and the differential conductance shows a logarithmic behavior as

$$G \approx G_2 + |G_3| \times J^{(2)}\ln\left[\frac{\epsilon_0 + U - \mu}{k_B T}\right]. \tag{6.205}$$

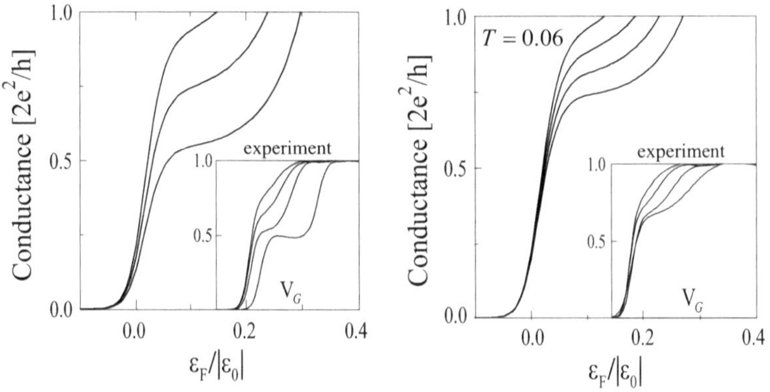

Figure 6.43 (Left) Conductance at temperatures of $T = 0.05$, 0.1, 0.2 as a function of Fermi energy in all energies in units of $|\epsilon_0|$. The parameters are $U = 1.45$, $\nu V_1^2 = 0.12$, $\nu V_2^2 = 0.015$, and $\delta = 0.02$. The inset is the experimental conductance of QPC at four different temperatures. (Right) Conductance in a magnetic field, for Zeeman splitting $\Delta = 0, 0.07,\ 0.12,\ 0.4$ at $T = 0.06$. The inset is the experimental conductance of QPC at different magnetic fields. From Y. Meir, K. Hirose, and N. S. Wingreen, *Phys. Rev. Lett.* **89**, 196802 (2002) [274].

The left panel of Fig. 6.43 shows the linear-response conductance of the first subband as a function of the Fermi energy μ relative to the energy of the quasi-bound state ϵ_0 at three different temperatures obtained from Eq. (6.205). As the temperature increases, a distinct plateau forms at $0.7(2e^2/h)$ in excellent agreement with the experimental data shown in the inset. Since G_2 depends only on the values of $V^{(1)}$ and $V^{(2)}$ at μ, it is dominated by $V^{(1)}$, while the Kondo enhancement is dominated by $V^{(2)}$. The contribution due to $V^{(1)}$ is set around $0.5(2e^2/h)$ by construction. On the other hand, the contribution due to $V^{(2)}$, resulting from the $1 \leftrightarrow 2$ valence fluctuations, is small at high temperature but grows with decreasing temperature in a form following the Kondo scaling function $F(T/T_K)$. Here $k_B T_K \approx U \exp[-1/2\nu J^{(2)}] = U \exp[(\mu - \epsilon_0 - U)/\nu|V^{(2)}|^2]$ is in agreement with the experimental observation of a Kondo temperature, which increases exponentially with gate voltage $V_G(\propto \mu)$ such as $\ln(T_K) \propto V_G$. Note that the conductance is not bound by its physical limit of $2e^2/h$ in the present perturbation theory. The calculations also show the formation of a strong plateau

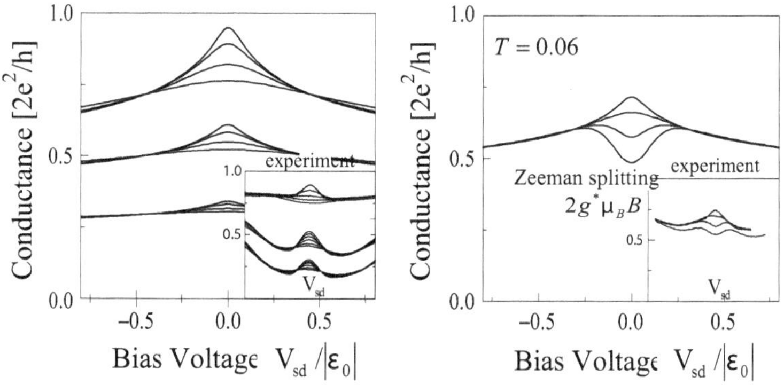

Figure 6.44 (Left) Differential conductance dI/dV for the Kondo model vs. bias voltage V_{sd} at Fermi energies of $\epsilon_F = 0.1,\ 0.03,\ 0.01$ from top group to bottom. For each curve, the temperatures are $T = 0.06,\ 0.1,\ 0.2,\ 0.4$, respectively. All the other parameters are the same as in Fig. 6.43. The inset is the experimental differential conductance data. (Right) dI/dV in magnetic field with Zeeman splitting $\Delta = 0,\ 0.04,\ 0.07,\ 0.1$ at $T = 0.06$ and $\mu = 0.04$. The inset is the experimental differential conductance data at different magnetic fields. From Y. Meir, K. Hirose, and N. S. Wingreen, *Phys. Rev. Lett.* **89**, 196802 (2002) [274].

with increasing parallel magnetic field as shown in the right panel of Fig. 6.43. The Kondo logarithms in G_3 are suppressed and the term in G_2 that depends on $\langle M \rangle$ gives negative contribution, leading to the evolution of the 0.7 plateau toward 0.5. In agreement with experiment, the conductance is no longer monotonically increasing with gate voltage.

The left panel of Fig. 6.44 shows the differential conductance as a function of bias voltage for several values of μ and temperatures. Even at the lowest conductances for small μ, there is a clear Kondo peak as is seen in experiment (inset). Because of the suppression of the Kondo effect by bias voltage, the large voltage accumulation is independent of temperature in agreement with experiment [218]. Magnetic field splits the Kondo peak as shown in the right panel of Fig. 6.44 in agreement with experiment (inset). Here we note that the width of source-drain bias of a transition from $g(2e^2/h) \sim 0.5$ to $g(2e^2/h) \sim 0.8$ with $|V_{sd}| \sim 0.2$ mV shown in distinctive wing shape of the spin-resolved plateaus of the non-linear differential conductance at $B = 8$ T in Fig. 6.39 (right) is consistent with the Zeeman splitting as $\delta = g^*\mu_B B/e = 25\mu\text{V/T} \times 8\text{T} = 0.2$ mV. An important prediction of the Kondo model is that the current through a QPC will be spin-polarized if the Zeeman splitting is larger than both $k_B T$ and $k_B T_K$. Therefore, at low temperature and in the vicinity of the $0.7(2e^2/h)$ plateau where T_K is small, a QPC may become an effective spin filter.

The 0.7 anomaly has been already observed in the first experiment of the quantization of conductance in the QPCs [215, 216] and has been a longstanding puzzle. Here we have presented a local moment formation in the electronic states of QPC by the SDFT calculations and the many-body QMC calculations, experimental observations to display the Kondo physics in the ZBA, which show the 0.7 structure is related to the *dynamical* spin, and a microscopic model for the anomalous transport through a point contact. As we noted, there are other mechanisms proposed for the 0.7 anomaly other than the Kondo effect, spontaneous *static* spin polarization, ferromagnetism, spin singlet-triplet bound states, electron–phonon scattering, Wigner crystal, and others. The electronic states of quasi-1D systems, especially in the low-density regime where the electron–electron interaction becomes strong, are still not understood sufficiently. Various situations might be realized for different geometrical constrictions. The 0.7 structure problem for the conductance anomaly in QPCs has been an active and challenging topic of transport in the low-dimensional systems [277].

Chapter 7

Epilogue

In this book, we present the descriptions of electron transport, starting with the Drude model, Langevin and master equation, Liouville equation, and the semi-classical Boltzmann equation. These are based mostly on the classical mechanics. Then we present the quantum mechanical approaches, which include the linear-response Kubo formula and the scattering-wave Landauer formula.

Green's function method and its extension to the nonequilibrium Green's function (NEGF) provide powerful ways for the mathematical treatments of these formulas not only in the single-particle but also in the many-electron cases. The Kubo formula is written using the particle-hole Green's function and the Landauer formula, extended to the nonequilibrium situations with finite bias voltages, is expressed by the NEGF. These basic formulas are then used to implement the calculation methods for realistic materials from atomistic points of views.

Recent rapid progress in nanoscience and nanotechnology enables us to construct nanometer-scale devices. Correspondingly, much attention has been paid to the transport properties from atomic-scale viewpoints. We describe several computational approaches for the quantum transport of nanosystems. Here let us review the characteristics of these calculation methods again.

Quantum Transport Calculations for Nanosystems
Kenji Hirose and Nobuhiko Kobayashi

ISBN 978-981-4267-32-8 (Hardcover), 978-981-4267-59-5 (eBook)
www.panstanford.com

7.1 Brief Summary of Quantum Transport Calculations

7.1.1 *Formulation by the Difference of Bias Voltages*

First, we recall the Landauer formula based on scattering approach

$$G = \frac{2e^2}{h} \sum_{\alpha} \mathcal{T}_{\alpha}(\epsilon_F), \tag{7.1}$$

where the transmission $\mathcal{T}_{\alpha}(\epsilon)$, which includes the elastic scatterings from impurities and phonons as well as the effects from geometrical constrictions, determines the electron transport for conducting channels α. In case of no scattering as $\mathcal{T}_{\alpha} = 1$ for each channel α, the conductance becomes simply $G = (2e^2/h)N_{\alpha}$, where N_{α} is the number of open channels.

To extend to treat a nonequilibrium situation with a finite bias voltage V, we assign different chemical potentials to the left (L) and right (R) electrodes and compute the electric current flowing within the energy window of these chemical potentials as

$$I_{\text{elastic}} = \frac{2e}{h} \sum_{\alpha} \int_{\mu_R}^{\mu_L} \mathcal{T}_{\alpha}(\epsilon) \left[f_L(\epsilon) - f_R(\epsilon) \right] d\epsilon, \tag{7.2}$$

where $f_{L/R}(\epsilon)$ are the Fermi–Dirac distributions of electrodes.

The transmission function $\mathcal{T}_{\alpha}(\epsilon)$ is obtained directly from the scattering amplitudes of the wavefunctions using effectively by the recursion-transfer-matrix (RTM) method or by the Lippmann–Schwinger (LS) method, or equivalently we can use the NEGF method to obtain the total transmission as $\sum_{\alpha} \mathcal{T}_{\alpha}(\epsilon) = \text{Tr}\left[G^r \Sigma^L G^a \Sigma^R\right]$.[a]

When we consider inelastic scatterings of electrons with lattice vibrations as phonons, which accompany the energy transfer to phonon emission and induce local heating, an additional term is needed for this formula to represent inelastic phonon emissions for energy dissipation:

$$I = \frac{2e}{h} \sum_{\alpha} \int_{\mu_R}^{\mu_L} \mathcal{T}_{\alpha}(\epsilon) \left[f_L(\epsilon) - f_R(\epsilon) \right] d\epsilon + I_{\text{inelastic}}. \tag{7.3}$$

Here inelastic term $I_{\text{inelastic}}$ is obtained based on the NEGF formalism with an inclusion of electron–phonon interactions as self-energy term.

---

[a]Note that this formulation is effective in the noninteracting case within the framework of the mean field theory.

7.1.2 Formulation by the Response to Electric Fields

Next, we describe the time-dependent wave-packet diffusion (TD-WPD) method, which enables us to treat huge number of atoms up to more than millions using the localized basis sets. This method is based on the linear-response approach of the Kubo formula

$$G = \frac{S}{L} e^2 \nu(\epsilon_F) \lim_{t \to \infty} D(t, \epsilon_F), \tag{7.4}$$

for which electron transport is described as a response to an external electric field E in the lowest approximation. Here we note that the diffusion term $D(t, \epsilon) = \langle \hat{x}(t) - \hat{x}(0))^2 \rangle_{eq} / t$ is written by the current–current correlation function $\int_{-\infty}^{\infty} \langle \hat{\mathbf{j}}(t) \cdot \hat{\mathbf{j}}(0) \rangle_{eq} dt$. It looks that this approach is different from the scattering approach above. However, we note that when there is no scatterings, using the relation $\nu(\epsilon_F) = 2/(2\pi \hbar v_F / N_\alpha) S$ and $D(\epsilon_F) = v_F L$, we see that the conductance becomes the same formula as the scattering approach by $G = (2e^2/h) N_\alpha$.

In this formulation, the effects of scatterings with impurities and phonons are included in the calculations of diffusion constant $D(\epsilon_F)$. However, it should be noted that within the linear-response theory in the zero-bias limit, inelastic scatterings accompanied by energy dissipation processes are not included.

Recall that the Kubo formula is derived from the linear term of the time evolution of the density matrix as

$$\delta \hat{\rho}(t) = -\frac{i}{\hbar} \int_{-\infty}^{t} e^{-i \hat{H}_0 (t - t')/\hbar} \left[\hat{H}_1(t'), \hat{\rho}(t') \right] e^{i \hat{H}_0 (t - t')/\hbar} dt', \tag{7.5}$$

where an electric field $\mathbf{E}(t) = -(1/c) \partial \mathbf{A}(t) / \partial t$ is included in $\hat{H}_1(t)$ through the vector potential $\mathbf{A}(t)$. In order to extend this method to include inelastic scatterings for finite bias cases, we need to integrate the equation of the density matrix equation directly

$$i\hbar \frac{\partial \hat{\rho}(t)}{\partial t} = \left[\hat{H}_{\text{tot}}(t), \hat{\rho}(t) \right], \tag{7.6}$$

which leads us to the time-dependent quantum master equation (QME) method.

To derive general computational methods for the electron plus phonon reservoir system with energy dissipation, however, is very difficult. Here we present the Lindblad formula for the reduced density matrix of the system based on the weak (second-order) coupling for electron–phonon interaction within the Markov approximation, which takes the form of

$$\frac{\partial}{\partial t}\hat{\rho}(t) = -\frac{i}{\hbar}\left[\hat{H}(t), \hat{\rho}(t)\right] + \hat{C}\left[\hat{\rho}(t)\right]. \tag{7.7}$$

Here the inelastic effects from phonon scatterings are included in the last term $\hat{C}\left[\hat{\rho}(t)\right]$. In the numerical calculations, the relation of detailed balance has to be checked to reach the equilibrium state, with inelastic scattering effects to balance with the electron acceleration energy due to an electric field $\mathbf{E}$ *via* $\mathbf{A}(t) = -c\mathbf{E}(t)t$. We note that in this formalism electronic state calculations are performed with periodic boundary conditions.

Let us consider the analogy between quantum and classical formula for transport. We can see that the formula of Eq. (7.7) is an analog of the semi-classical Boltzmann equation

$$\frac{\partial}{\partial t}f(\mathbf{r}, \mathbf{p}, t) = \left\{H(t), f(\mathbf{r}, \mathbf{p}, t)\right\} + \left[\frac{\partial f(\mathbf{r}, \mathbf{p}, t)}{\partial t}\right]_{\text{coll}} \tag{7.8}$$

to the quantum version with the classical distribution function $f(\mathbf{r}, \mathbf{p}, t)$ changing into the quantum density matrix $\rho(t) = \sum_i |\psi_i\rangle\langle\psi_i|$. Here we use the relations of the classical mechanics $d\mathbf{r}/dt = \partial H/\partial \mathbf{p}$ and $d\mathbf{p}/dt = -\partial H/\partial \mathbf{r}$ and the Poisson bracket $\{A, B\} = (\partial A/\partial \mathbf{p} \cdot \partial B/\partial \mathbf{r} - \partial A/\partial \mathbf{r} \cdot \partial B/\partial \mathbf{p})$, instead of the anti-commutator relation $[\hat{A}, \hat{B}] = \hat{A}\hat{B} - \hat{B}\hat{A}$. While the electron density and electrical current are given by

$$n(\mathbf{r}, t) = \int f(\mathbf{r}, \mathbf{p}, t)d\mathbf{p}, \quad \mathbf{j}(\mathbf{r}, t) = e\int \frac{\mathbf{p}}{m} f(\mathbf{r}, \mathbf{p}, t)d\mathbf{p}, \tag{7.9}$$

in the Boltzmann equation, these are given in the QME method by

$$n(\mathbf{r}, t) = \mathrm{Tr}\left[\hat{\rho}(t)\hat{n}(\mathbf{r})\right], \quad \mathbf{j}(\mathbf{r}, t) = \mathrm{Tr}\left[\hat{\rho}(t)\hat{\mathbf{j}}(\mathbf{r})\right]. \tag{7.10}$$

This shows that if we neglect the off-diagonal elements of the density matrix, this method results in the semi-classical Boltzmann equation. Thus, we can see the correspondence for classical and quantum versions of the kinetic equations for the transport.

We note that as the dynamics of an electron becomes irreversible due to the scattering term $[\partial f(\mathbf{r}, \mathbf{p}, t)/\partial t]_{\text{coll}}$ in the semi-classical Boltzmann equation, the last term $\hat{C}\left[\hat{\rho}(t)\right]$, which includes the dissipation due to phonon emission drives the electron dynamics irreversible in the QME method. This enables us to treat the nonequilibrium quantum transport problems in open systems.

7.2 Future Direction of Transport through Nanosystem

7.2.1 *Multiscale Transport Calculations*

In this book, we describe various calculation methods. Let us summarize in view of the multiscale transport calculations.

The most accurate method is to use the plane-wave basis sets. We describe the RTM method and the LS method in the Laue representation. These methods enable us to treat electronic states for various geometrical structures at the atomic scale, which are important since we cannot control the atomic-scale structures completely. The contact problem of single molecules to electrodes is an example. However, these methods are computationally very demanding and only several tens atoms can be treated presently. In this direction, the NEGF method with numerically localized atomic basis sets is applicable to larger number of atoms up to several thousands. For that purpose, the inverse of the matrix to obtain Green's function, which needs maximally $O(N^3)$ computation, should be reduced.

On the other hand, for the transport calculations by the response to electric fields, such as the Kubo formula based on the linear-response theory and master equation method, we can utilize efficient computational techniques for the time integration. For example, the TD-WPD method enables us to perform an $O(N)$ computation and to treat up to approximately 10^8 electrons. As for the QME approach, we might use the standard electronic calculation methods for the density matrix, since we set the periodic boundary conditions where the motions of electrons are controlled by the electromagnetic fields of $A(\mathbf{r})$ and $E(\mathbf{r})$. However, we need to treat a huge number of atoms, including atomic structures deep inside electrodes, which are not taken into account in the scattering approaches through the boundary conditions.

Although each calculation method has its merits and demerits in views of computational cost, stability, and accuracy, we expect that these methods contribute to the multiscale calculations for the transport properties to cover in various scales in future.

7.2.2 Characteristic Length for Quantum Transport

Transport behaviors change with the size of the system. In this book, we present two important length scales.

The first is the mean free path ℓ_{mfp}, which is an average length that electrons are scattered from impurities, phonons, etc. For the system with $L \gg \ell_{\mathrm{mfp}}$, electron motions are determined from scatterings, leading to Ohm's law in the diffusive regime. On the other hand, for $L \ll \ell_{\mathrm{mfp}}$ in the ballistic regime, electron motions are determined from the geometrical structures described by the number of quantum channels. To obtain ℓ_{mfp} from atomistic viewpoints, we need to treat a huge number of atoms. Here we calculate the mean free paths of CNTs and organic semiconductors using the TD-WPD method. It is an interesting and important problem to find how the nonequilibrium situation changes ℓ_{mfp} where phonon emission due to electron–phonon coupling is treated explicitly and to relate ℓ_{mfp} to energy dissipation as heat resistance. The resulting ℓ_{mfp} might be smaller than that from Fermi golden rule.

The second is the phase coherence length L_{ϕ}, which separates the system from the classical to quantum one. Usually to determine the sources of phase breaking is very difficult. At low temperature, the electron–electron scattering is the main source, while the effects of phonon scatterings become significant at high temperature. In the weak localization regime, the formulas of phase coherence are derived for various dimensions. Here we calculate the temperature dependence of L_{ϕ} for CNTs from phonon scatterings in the weak localization regime. In organic semiconductors, due to the rapid progress of fabricating single-crystal structures, the various transport behaviors from hopping to band-like natures are observed. This suggests that the localization behavior of carriers and correspondingly the phase coherence changes significantly in the strong and weak localization regimes.[a] Different from the inorganic materials with rigid crystal or amorphous structures, organic materials change their structures flexibly for various environments. The understanding of phase coherence problems for various organic materials is an important problem.

[a] In experiments, such behaviors are observed in the Hall effect measurements.

7.2.3 *Coulomb Interaction Effects on Electron Transport*

The Coulomb interactions become important as the system size becomes small at the nanometer scale. Its classical energy is

$$E_c = \int n(\mathbf{r}) V_H(\mathbf{r}) d\mathbf{r} = e^2 \int\int \frac{n(\mathbf{r})n(\mathbf{r}')}{|\mathbf{r}-\mathbf{r}'|} d\mathbf{r} d\mathbf{r}' \simeq \frac{e^2}{C}\frac{N(N-1)}{2}, \tag{7.11}$$

where N is the number of electrons, C is the classical capacitance, and V_H is obtained from the solution to the Poisson equation. When the nanosystems are well separated from electrodes and are confined, the total number of electrons are almost conserved as N, approaching the particle description.[a]

On the other hand, when the nanosystems have good contacts with electrodes or the temperature is high enough, the wavefunctions of electrons spread out through the whole system, approaching the wave description.[b]

To treat electron transport between the confined and open systems in the presence of strong Coulomb interaction needs special care. In this book, we treat the electron transport through the Kondo effect, which occurs due to the quantum exchange process of conducting electrons coupled with the localized spin states, resulting in the strongly temperature- and magnetic field–dependent properties. The transport behavior at low-density regime of QPC is another example. The interplay of the Coulomb energy and the confinement produces anomalous transport behaviors.

Although the mean field approach based on the DFT, including the exchange and correlation energy, gives clues to describe the situations, the many-body treatment is needed for such transport calculations. We give the ways to use Green's function method for the higher-order expansion. To reproduce the experimental situations, the combination of electronic-state calculations based on the DFT and Green's function method is a powerful way. Also the numerical calculations for Green's function using the quantum Monte Carlo (QMC) are of importance. The QMC method is extended to treat

[a]The number of electrons N is fixed in the Coulomb blockade regime, where the classical treatment of master equation for the transport becomes possible.

[b]We can describe electron transport using the scattering approaches.

nonequilibrium situations, which enables us to treat nonequilibrium transport of correlated electron systems.

7.2.4 *Thermal Transport and Thermoelectric Effects*

The understanding of the thermal transport or heat conduction becomes important to reduce the heating problem of nanometer-scale electric devices. The thermal transport is attributed to two contributions from electrons and lattice phonons. In the metallic systems where large electric current flows, electron transport is dominant for the thermal transport, which obeys the Wiedemann–Franz law. On the other hand, in the insulating systems phonon transport plays the major role. In the semiconductor systems, both contributions from carrier transport by electrons and holes and from phonon transport have to be taken into account.

Phonon transport occurs by the difference of temperatures. Since phonons are generated by the emission due to the electron–phonon interaction, we need to treat electron and phonon transport simultaneously. In the macroscopic materials, the heat conduction problem is analyzed using the familiar heat diffusion equation based on Fourier's law. In the nanometer-scale materials, it is not well understood how the heat flows and much attention has been paid to the thermal transport in the nanosystems.

Here we give the formulas based on the Boltzmann equation, molecular dynamics with linear-response theory, nonequilibrium molecular dynamics, and the NEGF method for phonons. These are categorized into the frameworks of particle description and wave description. The behaviors of phonon transport would change with the size from diffusive to ballistic natures where the mean free path of phonon separates these regimes. Therefore, it is important to identify the mean free path not only for electrons but also for phonons in the functional devices with nanometer-scale sizes.

The problem of energy conversion from the thermal energy to electric one using the thermoelectric relations becomes important for the creation of efficient thermoelectric device. Its efficiency is expressed by the dimensionless parameter of the figure of merit, proportional to the ratio of carrier transport and thermal transport.

Control of the nanometer-scale interfaces would be effective to reduce the thermal conductance and thus to increase the efficiency of thermoelectric devices.

7.2.5 *Transport Based on the Density Functional Theory*

Finally, we consider the transport calculation methods in view of the density functional theory (DFT). There have been a number of works for the transport calculations in the scattering approach combined with DFT within the local-density approximation.

On the basis of the DFT, the application of the conventional Kohn–Sham equation to the transport problem is not sufficient. This is apparent since the Kohn–Sham equation is derived for the ground-energy electronic states, while the transport problem in itself needs to include excited states in the formalism. Since the bandgap energies obtained from the Kohn–Sham eigenvalues are known usually to take only half of the true excitation energies, this suggests that the energies of resonant states within the finite bias voltage between two Fermi levels are different from the true ones, which results in the errors of the obtained electric current.

One of the calculation methods to overcome this difficulty is to use the formulation and numerical computation based on the time-dependent density functional theory (TDDFT). In the transport problems, we solve the time-dependent Kohn–Sham equation as

$$\frac{\partial}{\partial t}\psi_i(\mathbf{r}, t) = -\frac{i}{\hbar} H_{KS}(t)\psi_i(\mathbf{r}, t) \tag{7.12}$$

in the scattering formalism with open boundary conditions, or alternatively we solve the time-dependent QME

$$\frac{\partial}{\partial t}\hat{\rho}(t) = -\frac{i}{\hbar}\left[\hat{H}_{KS}(t), \hat{\rho}(t)\right] + \hat{C}\left[\hat{\rho}(t)\right] \tag{7.13}$$

in the density matrix formalism. Here the time-dependent Kohn–Sham Hamiltonian is given by

$$\hat{H}_{KS}(t) = -\frac{\hbar^2}{2m}\nabla^2 + V_{\text{ext}}(\mathbf{r}, t) + e^2\int \frac{n(\mathbf{r}', t)}{|\mathbf{r} - \mathbf{r}'|}d\mathbf{r}' + V_{\text{xc}}\left[n(\mathbf{r}, t)\right]. \tag{7.14}$$

To apply the TDDFT to transport problems, there are a number of problems to overcome. These include the exchange-correlation energy problem for TDDFT,[a] treatment of memory effects in the scattering approach, stable and fast TD-calculation methods, etc. The development of the practical simulation method by TDDFT for electron transport of the realistic systems is expected.

[a]We note that to include the retardation effects in time associated with the non-local electric current, the time-dependent current-DFT (TD-CDFT) formalism is proposed and has been intensively studied [104]

References

1. Ashcroft, N. W., and Mermin, N. D. (1976). *Solid State Physics* (Thomson Learning, Philadelphia).
2. Kubo, R., Toda, M., and Hashitume, N. (1991). *Statistical Physics II: Nonequilibrium Statistical Mechanics* (Springer, Berlin).
3. Zwanzig, R. (2001). *Nonequilibrium Statistical Mechanics* (Oxford University Press).
4. Gershenfeld, N. (2000). *The Physics of Information Technology* (Cambridge University Press).
5. Breuer, H.-P., and Petruccione, F. (2006). *The Theory of Open Quantum Systems* (Oxford University Press).
6. Jacoboni, C., and Reggiani, L. (1983). The Monte Carlo Method for the Solution of Charge Transport in Semiconductors with Applications to Covalent Materials, *Rev. Mod. Phys.* **55**, 645.
7. Chaikin, P. M., and Lubensky, T. C. (2000). *Principles of Condensed Matter Physics* (Cambridge University Press).
8. Green, M. S. (1954). Markov Random Processes and the Statistical Mechanics of Time-Dependent Phenomena, II. Irreversible Processes in Fluids, *J. Chem. Phys.* **22**, 398.
9. Kubo, R. (1957). Statistical-Mechanical Theory of Irreversible Processes. I. General Theory and Simple Applications to Magnetic and Conduction Problems, *J. Phys. Soc. Jpn.* **12**, 570.
10. Dugdale, J. S. (1977). *The Electrical Properties of Metals and Alloys* (Edward Arnold Publishers, London).
11. Chen, G. (2005). *Nanoscale Energy Transport and Conversion* (Oxford University Press).
12. Landauer, R. (1957). Spatial Variation of Currents and Fields due to Localized Scatterers in Metallic Conduction, *IBM J. Res. Dev.* **1**, 223.
13. Büttiker, M., Imry, Y., Landauer, R., and Pinhas, S. (1985). Generalized Many-Channel Conductance Formula with Application to Small Rings, *Phys. Rev. B* **31**, 6207.

14. Anderson, P. W. (1958). Absence of Diffusion in Certain Random Lattices, *Phys. Rev.* **109**, 1492.
15. Abraham, E., Anderson, P. W., Licciardello, D. C., and Ramakrishnan, T. V. (1979). Scaling Theory of Localization: Absence of Quantum Diffusion in Two Dimensions, *Phys. Rev. B* **22**, 3519.
16. Altshuler, B. L., Lee, P. A., and Webb, R. A. (ed.) (1991). *Mesoscopic Phenomena in Solids* (Elsevier Science, Netherlands).
17. Abrikosov, A. A., Gorkov, L. P., and Dzyaloshinski, I. E. (1963). *Methods of Quantum Field Theory in Statistical Physics* (Dover Publications, Inc., New York).
18. Fetter, A. L., and Walecka, J. D. (1971). *Quantum Theory of Many-Particle Systems* (McGraw-Hill, Inc.).
19. Doniach, S., and Sondheimer, E. H. (1974). *Green's Functions for Solid State Physicists* (Imperial College Press, Inc.).
20. Negele, J. W., and Orland, H. (1988). *Quantum Many-Particle Systems* (Addison-Wesley Publishing, New York).
21. Mahan, G. D. (1990). *Many-Particle Physics* (Plenum, New York).
22. Economou, E. N. (1990). *Green's Functions in Quantum Physics* (Springer-Verlag, Berlin).
23. Datta, S. (1995). *Electronic Transport in Mesoscopic Systems* (Cambridge University Press, New York).
24. Haug, H., and Jauho, A.-P. (1996). *Quantum Kinetics in Transport and Optics of Semiconductors* (Springer-Verlag, Berlin).
25. Ferry, D. K., and Goodnick, S. M. (1997). *Transport in Nanostructures* (Cambridge University Press, UK).
26. Datta, S. (2005). *Quantum Transport: Atom to Transistor* (Cambridge University Press, New York).
27. Ventra, M. Di (2008). *Electrical Transport in Nanoscale Systems* (Cambridge University Press, UK).
28. Buot, F. A. (2009). *Nonequilibrium Quantum Transport Physics in Nanosystems: Foundation of Computational Nonequilibrium Physics in Nanoscience and Nanotechnology* (World Scientific, Singapore).
29. Lee, P. A., and Ramakrishnan, T. V. (1985). Disordered Electronic Systems, *Rev. Mod. Phys.* **57**, 287.
30. Kadanoff, L. P., and Baym, G. (1962). *Quantum Statistical Mechanics* (Benjamin, New York).
31. Keldysh, L. V. (1965). Diagram Technique for Nonequilibrium Processes, *Sov. Phys. JETP* **20**, 1018.

32. Langreth, D. C. (1976). *Linear and Nonlinear Electron Transport in Solids*, vol. 17, NATO Advanced Study Institute (Plenum, New York).
33. Meir, Y., and Wingreen, N. S. (1992). Landauer Formula for the Current through an Interacting Electron Region, *Phys. Rev. Lett.* **68**, 2512.
34. Fisher, D. S., and Lee, P. A. (1981). Relation between Conductivity and Transmission Matrix, *Phys. Rev. B* **23**, 6851.
35. Wingreen, N. S., Jacobsen, K. W., and Wilkins, J. W. (1988). Resonant Tunneling with Electron–Phonon Interaction: An Exactly Solvable Model, *Phys. Rev. Lett.* **61**, 1396.
36. Jauho, A.-P., Meir, Y., and Wingreen, N. S. (1994). Time-Dependent Transport in Interacting and Noninteracting Resonant-Tunneling Systems, *Phys. Rev. B* **50**, 5528.
37. Fröhlich, H. (1954). Electrons in Lattice Fields, *Adv. Phys.* **3**, 325.
38. Matsubara, T. (1955). A New Approach to Quantum-Statistical Mechanics, *Prog. Theor. Phys.* **14**, 351.
39. Hirsch, J. E., and Fye, R. M. (1986). Monte Carlo Method for Magnetic Impurities in Metal, *Phys. Rev. Lett.* **56**, 2521.
40. Gull, E., Millis, A. J., Lichtenstein, A. I., Rubtsov, A. N., Troyer, M., and Werner, P. (2011). Continuous-Time Monte Carlo Methods for Quantum Impurity Models, *Rev. Mod. Phys.* **83**, 349.
41. Lundqvist, S., and March, N. H. (ed.) (1983). *Theory of the Inhomogeneous Electron Gas* (Plenum, New York).
42. Parr, R. G., and Yang, W. (1989). *Density-Functional Theory of Atoms and Molecules* (Oxford University Press, New York).
43. Gross, E. K. U and Dreizler, R. M. (ed.) (1995). *Density-Functional Theory* (Plenum, New York).
44. Joubert, D. (ed.) (1998). *Density Functionals: Theory and Applications* (Springer-Verlag, Berlin).
45. Fiolhais, D., Nogueira, F., and Marques, M. (ed.) (2003). *A Primer in Density Functional Theory* (Springer-Verlag, Berlin).
46. Martin, R. M. (2004). *Electronic Structure: Basis Theory and Practical Methods* (Cambridge University Press, UK).
47. Kotliar, R. M., Savrasov, S. Y., Haule, K., Oudovenko, U. S., Parcollet, O., and Marianetti, C. A. (2006). Electronic Structure Calculations with Dynamical Mean-Field Theory, *Rev. Mod. Phys.*, **78**, 865.
48. Hohenberg, P., and Kohn, W. (1964). Inhomogeneous Electron Gas, *Phys. Rev.* **136**, B864.

49. Kohn, W., and Sham, L. J. (1965). Self-Consistent Equations Including Exchange and Correlation Effects, *Phys. Rev.* **140**, A1133.
50. Ceperley, D. M., and Alder, B. J. (1980). Ground State of the Electron Gas by a Stochastic Method, *Phys. Rev. Lett.* **45**, 566.
51. Perdew, J. P., and Zunger, A. (1981). Self-Interaction Correction to Density-Functional Approximations for Many-Electron Systems, *Phys. Rev. B* **23**, 5048.
52. Perdew, J. P., Burke, K., and Ernzerhof, M. (1996). Generalized Gradient Approximation Made Simple, *Phys. Rev. Lett.* **77**, 3865.
53. Ashcroft, N. W. (1966). Electron–Ion Pseudopotentials in Metals, *Phys. Lett.* **23**, 48.
54. Schlüter, M., Chelikowsky, J. R., Louic, S. G., and Cohen, M. L. (1975). Self-Consistent Pseudopotential Calculations for Si(111) Surfaces: Unreconstructed (1×1) and Reconstructed (2×1) Model Structures, *Phys. Rev. B* **12**, 4200.
55. Kleinman, L., and Bylander, D. M. (1982). Efficient Form for Model Pseudopotentials, *Phys. Rev. Lett.* **48**, 1425.
56. Bachelet, G. B., Hamann, D. R., and Schlüter, M. (1982). Pseudopotentials that Work: From H to Pu, *Phys. Rev. B* **26**, 4199.
57. Troullier, N., and Martins, J. L. (1991). Efficient Pseudopotentials for Plane-Wave Calculations. II. Operators for Fast Iterative Diagonalization, *Phys. Rev. B* **43**, 8861.
58. Car, R., and Parrinello, M. (1985). Unified Approach for Molecular Dynamics and Density-Functional Theory, *Phys. Rev. Lett.* **55**, 2471.
59. Payne, M. C., Teter, M. P., Allan, D. C., Arias, T. A., and Joannopoulos, J. D. (1992). Iterative Minimization Techniques for ab initio Total-Energy Calculations: Molecular Dynamics and Conjugate Gradients, *Rev. Mod. Phys.* **64**, 1045.
60. Hirose, K., and Tsukada, M. (1995). First-Principles Calculation of the Electronic Structure for a Bielectrode Junction System under Strong Field and Current, *Phys. Rev. B* **51**, 5278.
61. Kobayashi, K. (1999). Tunneling into Bloch States from a Tip in Scanning Tunneling Microscopy, *Phys. Rev. B* **59**, 13251.
62. Hirose, K., Kobayashi, N., and Tsukada, M. (2004). Ab initio Calculations for Quantum Transport through Atomic Bridges by the Recursion Transfer-Matrix Method, *Phys. Rev. B* **69**, 245412.
63. Kobayashi, N., and Tsukada, M. (1999). Numerical Method for Local Density of States and Current Density Decomposed into Eigenchannels in Multichannel System, *Jpn. J. Appl. Phys.* **38**, 3805.

64. Lang, N. D. (1995). Resistance of Atomic Wires, *Phys. Rev. B* **52**, 5335.
65. Kobayashi, N., Aono, M., and Tsukada, M. (2001). Conductance Channels of Al Wires at Finite Bias, *Phys. Rev. B* **64**, 121402(R).
66. Taylor, J., Guo, H., and Wang, J. (2001). Ab initio Modeling of Quantum Transport Properties of Molecular Electronic Devices, *Phys. Rev. B* **63**, 245407.
67. Xue, Y., Datta, S., and Ratner, M. A. (2002). First-Principles Based Matrix Green's Function Approach to Molecular Electronic Devices: General Formalism, *Chem. Phys.* **281**, 151.
68. Brandbyge, M., Mozos, J.-L., Ordejón, P., Taylor, J., and Stokbro, K. (2002). Density-Functional Method for Nonequilibrium Electron Transport, *Phys. Rev. B* **65**, 165401.
69. Soler, J. M., Artacho, E., Gale, J. D., García, A., Junquera, J., Ordejón, P., and Sánchez-Portal, D. (2002). The SIESTA Method for ab initio Order-N Materials Simulation, *J. Phys. Condens. Matter* **14**, 2745.
70. Ozaki, T. (2003). Variationally Optimized Atomic Orbitals for Large-Scale Electronic Structures, *Phys. Rev. B* **67**, 155108.
71. Sancho, M. P. L., Sancho, J. M. L, and Rubio, J. (1984). Quick Iterative Scheme for the Calculation of Transfer Matrices: Application to Mo(100), *J. Phys. F: Matter Phys.* **14**, 1205.
72. Rego, L. G. C., and Kirczenow, G. (1999). Fractional Exclusion Statistics and the Universal Quantum of Thermal Conductance: A Unifying Approach, *Phys. Rev. B* **59**, 13080.
73. Yamamoto, T., Watanabe, S., and Watanabe, K. (2004). Universal Features of Quantized Thermal Conductance of Carbon Nanotubes, *Phys. Rev. Lett.* **92**, 075502.
74. Dubi, Y., and Ventra, M. Di. (2011). Colloquium: Heat Flow and Thermoelectricity in Atomic and Molecular Junctions, *Rev. Mod. Phys.* **83**, 131.
75. Zhang, W., Mingo, N., and Fisher, T. S. (2007). Simulation of Phonon Transport across a Non-Polar Nanowire Junction Using an Atomistic Green's Function Method, *Phys. Rev. B* **76**, 195429.
76. Markussen, T., Jauho, A.-P., and Brandbyge, M. (2009). Electron and Phonon Transport in Silicon Nanowire: Atomistic Approach to Thermoelectric Properties, *Phys. Rev. B* **79**, 035415.
77. Callen, H. B. (1985). *Thermodynamics and an Introduction to Thermo-statistics* (Wiley).
78. Onsager, L. (1931). Reciprocal Relations in Irreversible Processes I. *Phys. Rev.* **37**, 405.

79. Goldsmid, H. J. (2009). *Introduction to Thermoelectricity* (Springer Series in Materials Science).

80. Yamamoto, K., Ishii, H., Kobayashi, N., and Hirose, K. (2011). Effects of Vacancy Defects on Thermal Conduction of Silicon Nanowire: Nonequilibrium Green's Function Approach, *Appl. Phys. Express* **4**, 085001.

81. Ishii, H., Kobayashi, N., and Hirose, K. (2008). Time-Dependent Wave-Packet Diffusion Method for Quantum Transport Calculation: From Diffusive to Ballistic Regimes, *Appl. Phys. Express* **1**, 123002.

82. Ishii, H., Kobayashi, N., and Hirose, K. (2010). Order-N Electron Transport Calculations from Ballistic to Diffusive Regimes by a Time-Dependent Wave-Packet Diffusion Method: Application to Transport Properties of Carbon Nanotubes, *Phys. Rev. B* **82**, 085435.

83. Haydock, R. (1984). *Solid State Physics*, vol. 58 (Turnbull, D., and Ehrenreich, H., eds.) (Springer, Berlin).

84. Ishii, H., Kobayashi, N., and Hirose, K. (2011). Hall Conductivity Calculations by the Time-Dependent Wave-Packet Diffusion Method, *Phys. Rev. B* **83**, 233403.

85. Watanabe, N., and Tsukada, M. (2000). Fast and Stable Method for Simulating Quantum Electron Dynamics, *Phys. Rev. E* **62**, 2914.

86. Sugino, O., and Miyamoto, Y. (1999). Density-Functional Approach to Electron Dynamics: Stable Simulation under a Self-Consistent Field, *Phys. Rev. B* **59**, 2579.

87. Roche, S., and Mayou, D. (1997). Conductivity of Quasiperiodic Systems: A Numerical Study, *Phys. Rev. Lett.* **79**, 2518.

88. Griebel, M., Knapek, S., and Zumbusch, G. (2007). *Numerical Simulation in Molecular Dynamics: Numerics, Algorithms, Parallelization, Applications* (Springer-Verlag, Berlin).

89. Daw, O., and Baskes, M. I. (1984). Embedded-Atom Method: Derivation and Application to Impurities, Surfaces, and Other Defects in Metals, *Phys. Rev. B* **29**, 6443.

90. Stillinger, F. H., and Weber, T. (1985). Computer Simulation of Local Order in Condensed Phases of Silicon, *Phys. Rev. B* **31**, 5262.

91. Tersoff, J. (1989). Modeling Solid-State Chemistry: Interatomic Potentials for Multicomponent Systems, *Phys. Rev. B* **39**, 5566.

92. Brenner, D. W. (1990). Empirical Potential for Hydrocarbons for Use in Simulating the Chemical Vapor Deposition of Diamond Films, *Phys. Rev. B* **42**, 9458.

93. Esfarjani, K., and Stokes, H. T. (2008). Method to Extract Anharmonic Force Constants from First Principles Calculations, *Phys. Rev. B* **77**, 144112.

94. Nosé, S. (1984). A Molecular Dynamics Method for Simulations in the Canonical Ensemble, *Molecular Physics* **52**, 255.

95. Hoover, W. G. (1985). Canonical Dynamics: Equilibrium Phase-Space Distribution, *Phys. Rev. A* **31**, 1695.

96. Weiss, U. (2008). *Quantum Dissipative Systems* (World Scientific Publishing, Singapore).

97. Nakajima, S. (1958). On Quantum Theory of Transport Phenomena, *Prog. Theor. Phys.* **20**, 948.

98. Zwanzig, R. (1960). Ensemble Method in the Theory of Irreversibility, *J. Chem. Phys.* **33**, 1338.

99. Lindblad, G. (1976). On the Generators of Quantum Dynamical Semigroups, *Commun. Math. Phys.* **48**, 119.

100. Cohen-Tannoudji, C., Dupont-Roc, J., and Grynberg, G. (1992). *Atom-Photon Interactions: Basic Processes and Applications* (John Wiley & Sons, Inc., New York).

101. Louisell, W. H. (1973). *Quantum Statistical Properties of Radiation* (John Wiley & Sons, Inc., New York).

102. Gebauer, R., and Car, R. (2004). Kinetic Theory of Quantum Transport at the Nanoscale, *Phys. Rev. B.* **70**, 125324.

103. Runge, E., and Gross, E. K. U. (1984). Density-Functional Theory for Time-Dependent Systems, *Phys. Rev. Lett.* **52**, 997.

104. Marques, M. A., Ullrich, C. A., Noqueira, F., Rubio, A., Burke, K., Gross, E. K. U. (ed.) (2006). *Time-Dependent Density Functional Theory (Lecture Notes in Physics* (Springer-Verlag, Berlin).

105. Kurth, S., Stefanucci, C., Almbladh, C.-O., Rubio, A., and Gross, E. K. U (2005). Time-Dependent Quantum Transport: A Practical Scheme Using Density Functional Theory, *Phys. Rev. B* **72**, 035308.

106. Gebauer, R., and Car, R. (2004). Current in Open Quantum Systems, *Phys. Rev. Lett.* **93**, 160404.

107. Nielsen, M. A., and Chuang, I. L. (2001). *Quantum Computation and Quantum Information* (Cambridge University Press).

108. Ishii, H., Tomita, Y., Shigeno, Y., and Nakayama, T. (2008). Relaxation Process of Transient Current through Nanoscale Systems; Density Matrix Calculations, *e-J. Surf. Sci. Nanotech* **6**, 213.

109. Yamada, K., Yamamoto, T., and Watanabe, K. (2009). Transient Current Behavior of Nanoscale Objects: Role of Displacement Current and Polaron Effects, *Jpn. J. Appl. Phys.* **48**, 075001.

110. Binnig, G., Rohrer, H., Gerber, C., and Weibel, E. (1982). Surface Studies by Scanning Tunneling Microscopy, *Phys. Rev. Lett.* **57**, 49.

111. Chen, C. J. (1993). *Introduction to Scanning Tunneling Microscopy* (Oxford University Press, New York).

112. Wiesendanger, R. Güntherrodt (ed.) (1993). *Scanning Tunneling Microscopy I-III* (Springer-Verlag, Berlin).

113. Bardeen, J. (1961). Tunneling from a Many-Particle Point of View, *Phys. Rev. Lett.* **6**, 57.

114. Tersoff, J., and Hamman, D. R. (1985). Theory of the scanning tunneling microscope, *Phys. Rev. B* **31**, 805.

115. Takayanagi, K., Tanishiro, Y., Takahashi, S., and Takahashi, M. (1985). Structure Analysis of Si(111)-7 $\times$ 7 Reconstructed Surface by Transmission Electron Diffraction, *Surf. Sci.* **164**, 367.

116. Hamers, R. J., Tromp, R. M., and Demuth, J. E. (1987). Surface Electronic Structure of Si (111)-(7 $\times$ 7) Resolved in Real Space, *Phys. Rev. Lett.* **56**, 1972.

117. Brommer, K. D., Needels, M., Larson, B. E., and Joannopoulos, J. D. (1992). Ab initio Theory of the Si(111)-(7 $\times$ 7) Surface Reconstruction: A Challenge for Massively Parallel Computation, *Phys. Rev. Lett.* **68**, 1355.

118. Tsukada, M., and Shima, N. (1987). Theory of Electronic Processes of Scanning Tunneling Microscopy, *J. Phys. Soc. Jpn.* **56**, 2875.

119. Eigler, D. M., Lutz, C. P., and Rudge, W. E. (1991). An Atomic Switch Realized with the Scanning Tunneling Microscope, *Nature (London)* **352**, 600;

120. Lyo, I.-W., and Avouris, P. (1991). Field-Induced Nanometer- to Atomic-Scale Manipulation of Silicon Surfaces with the STM, *Science* **253**, 173.

121. Hirose, K., and Tsukada, M. (1994). First-Principles Theory of Atom Extraction by Scanning Tunneling Microscopy, *Phys. Rev. Lett.* **73**, 150.

122. Kobayashi, N., Hirose, K., and Tsukada, M. (1996). Theoretical Study of Current and Barrier Height between Aluminum Tip and Silicon Surface in Scanning Tunneling Microscopy, *Jpn. J. Appl. Phys.* **35**, 3710.

123. Kobayashi, N., Hirose, K., and Tsukada, M. (1997). Theoretical Study of Silicon Adatom Transfer from the Silicon Surface in Scanning Tunneling Microscopy, *Jpn. J. Appl. Phys.* **36**, 3791.

124. Uchida, H., Huang, D., Grey, F., and Aono, M. (1993). Site-Specific Measurement of Adatom Binding Energy Differences by Atom Extraction with the STM, *Phys. Rev. Lett.* **70**, 2040.

125. Kobayashi, A., Grey, F., Williams, R. S., and Aono, M. (1993). Formation of Nanometer-Scale Grooves in Silicon with a Scanning Tunneling Microscope, *Science* **259**, 1724.

126. Kobayashi, N., Hirose, K., and Tsukada, M. (2004). First-Principles Calculation of Field Emission from Adsorbed Atom on Metallic Electrode, *Appl. Surf. Sci.* **237**, 568.

127. Lang, N. D., Yacoby, A., and Imry, Y. (1989). Theory of a Single-Atom Point Source for Electrons, *Phys. Rev. Lett.* **63**, 1499.

128. Gimzewski, J. K., and Möller, R. (1987). Transition from the Tunneling Regime to Point Contact Studied Using Scanning Tunneling Microscope, *Phys. Rev. B* **36**, 1284.

129. Dürig, U. Züger, O., and Pohl, D. W. (1990). Observation of Metallic Adhesion Using the Scanning Tunneling Microscopy, *Phys. Rev. Lett.* **65**, 349.

130. Muller, C. J., van Ruitenbeek, J. M., and de Jongh, L. J. (1992). Conductance and Supercurrent Discontinuities in Atomic-Scale Metallic Constrictions of Variable Width, *Phys. Rev. Lett.* **69**, 140.

131. Scheer, E., Joyes, P., Esteve, D., Urbina, C., and Devoret, M. H. (1997). Conduction Channel Transmissions of Atomic-Size Aluminum Contacts, *Phys. Rev. Lett.* **78**, 3535.

132. Ohnishi, H., Kondo, Y., and Takayanagi, K. (1998). Quantized Conductance through Individual Rows of Suspended Gold Atoms, *Nature* **395**, 780.

133. Olesen, L. Lægsgaard, E., Stensgaard, I., Besenbacher, F., Schiøtz, J., Stoltze, P., Jacobsen, K. W., and Nørskov J. K. (1994). Quantized Conductance in an Atom-Sized Point Contact, *Phys. Rev. Lett.* **72**, 2251.

134. Kobayashi, N., Brandbyge, M., and Tsukada, M. (2000). First-Principles Study of Electron Transport through Monatomic Al and Na Wires, *Phys. Rev. B* **62**, 8430.

135. Kusaka, H., and Kobayashi, N. (2012). First-Principles Study of Electron Transport in Si Atom Wires under Finite Bias Voltage, *Appl. Surf. Sci.* **258**, 1985.

136. Li, S. S., Chang, K., Xia, J. B., and Hirose, K. (2003). Spin-Dependent Transport through $Cd_{1-x}Mn_xTe$ Diluted Magnetic Semiconductor Quantum Dots, *Phys. Rev. B* **67**, 1160.

137. Garcia, N., Munoz, M., and Zhao, Y. -W (1999). Magnetoresistance in Excess of 200 % in Ballistic Ni Nanocontacts at Room Temperature and 100 Oe, *Phys. Rev. Lett.* **82**, 2923.

138. Oshima, H., and Miyano, K. (1998). Spin-Dependent Conductance Quantization in Nickel Point Contacts, *Appl. Phys. Lett.* **73**, 2203.

139. Ono, T., Ooka, Y., Miyajima, H., and Otani, Y. (1999). $2e^2/h$ to e^2/h Switching of Quantum Conductance Associated with a Change in Nanoscale Ferromagnetic Domain Structure, *Appl. Phys. Lett.* **75**, 1622.

140. Imamura, H., Kobayashi, N., Takahashi, S., and Maekawa, S. (2000). Conductance Quantization and Magnetoresistance in Magnetic Point Contacts, *Phys. Rev. Lett.* **84**, 1003.

141. van Wees, B. J., Kouwenhoven, L. P., Willems, E. M. M., Harmans, C. J. P., Mooij, J. E., van Houten, H., Beenakker, C. W., Williamson, J. G., and Foxon, C. T. (1991). Quantum Ballistic and Adiabatic Electron Transport Studied with Quantum Point Contacts, *Phys. Rev. B* **43**, 12431.

142. Imamura, H., Kobayashi, N., Takahashi, S., and Maekawa, S. (2001). Effect of the Quantum Domain Wall on Conductance Quantization and Magnetoresistance in Magnetic Point Contacts, *Mat. Sci. Eng. B* **84**, 107.

143. Kobayashi, N., Ozaki, T., and Hirose, K. (2006). First-Principles Calculation of Spin Transport in Magnetic Nanowire Using Green's Function Method with Localized Basis Set, *J. Phys. Conf. Ser.* **38**, 95.

144. Iijima, S. (1991). Helical Microtubules of Graphitic Carbon, *Nature* **354**, 56.

145. Novoselov, K. S., Geim, A. K., Morozov, S. V., Jiang, D., Zhang, Y., Dubonos, S. V., Grigorieva, I. V., and Firsov, A. A. (2004). Electric Field Effect in Atomically Thin Carbon Films, *Science* **306**, 666.

146. Javey, A., Guo, J., Wang, Q., Lundstrom, M., and Dai, H. (2004). Ballistic Carbon-Nanotube Field-Effect Transistors, *Nature* **424**, 654.

147. Miyazaki, H., Okada, S., Sato, T., Tanaka, S., Goto, H., Kanda, A., Tsukagoshi, K., Ootuka, Y., and Aoyagi, Y. (2008). Inter-Layer Screening Length to Electric Field in Thin Graphite Film *Appl. Phys. Express.* **1**, 034007.

148. Saito, R., Dresselhaus, G., and Dresselhaus, D. S. (1998). *Physical Properties of Carbon Nanotubes* (Imperial College Press, London).

149. Ando, T. (2005). Theory of Electronic States and Transport in Carbon Nanotubes, *J. Phys. Soc. Jpn.* **74**, 777.

150. Charlier, J. C., Blase, X., and Roche, S. (2007). Electronic and Transport Properties of Nanotubes, *Rev. Mod. Phys.* **79**, 677.

151. Ajiki, H., and Ando, T. (1993). Electronic States of Carbon Nanotubes, *J. Phys. Soc. Jpn.* **62**, 1255.

152. Harrison, W. A. (1980). *Electronic Structure and the Properties of Solids: The Physics of the Chemical Bond* (W. H. Freeman and Company, San Francisco).

153. Park, J. Y., Rosenblatt, S., Yaish, Y., Sazonova, V., Üstunel, H., Braig, S., Arias, T. A., Brouwer, P. W., and McEuen, P. L. (2004). Electron–Phonon Scattering in Metallic Single-Walled Carbon Nanotubes, *Nano Lett.* **4**, 517.

154. Purewal, M. S., Hong, B. H., Ravi, A., Chandra, B., Hone, J., and Kim, P. (2007). Scaling of Resistance and Electron Mean Free Path of Single-Walled Carbon Nanotubes, *Phys. Rev. Lett.* **98**, 186808.

155. Suzuura, and Ando, T. (2002). Phonons and Electron–Phonon Scattering in Carbon Nanotubes, *Phys. Rev. B* **65**, 235412.

156. Stojetz, B., Miko, C., Forró, L., and Strunk, C. (2005). Effect of Band Structure on Quantum Interference in Multiwall Carbon Nanotubes, *Phys. Rev. Lett.* **94**, 186802.

157. Abrikosov, A. A. (1988). *Fundamentals of the Theory of Metals* (Elsevier Science, Netherlands).

158. Akkermans, E., Montambaux, G., Pichard, J.-L., and Zinn-Justin, J. (ed.) (1995). *Mesoscopic Quantum Physics: Les Houches Session LXI 1994)* (Elsevier Science, Netherlands).

159. Ishii, H., Roche, S., Kobayashi, N., and Hirose, K. (2010). Inelastic Transport in Vibrating Disordered Carbon Nanotubes: Scattering Times and Temperature-Dependent Decoherence Effects, *Phys. Rev. Lett.* **104**, 116801.

160. Stern, A., Aharanov, Y., and Imry, Y. (1990). Phase Uncertainty and Loss of Interference: A General Picture, *Phys. Rev. A.* **41**, 3436.

161. Tans, S. J., Verschueren, A. R. M., and Dekker, C. (1998). Room-Temperature Transistor Based on a Single Carbon Nanotubes, *Nature* **393**, 49.

162. Martel, R., Schmidt, T., Shea, H. R., and Avouris, P. (1998). Single- and Multi-Wall Carbon Nanotube Field-Effect Transistors, *Appl. Phys. Lett.* **73**, 2447.

163. Radosavljević, M., Heinze, S., Tersoff, J., and Avouris, Ph. (2003). Drain Voltage Scaling in Carbon Nanotube Transistors, *Appl. Phys. Lett.* **83**, 2435.

164. Nakanishi, T., Bachtold, A., and Dekker, C. (2002). Transport through the Interface between a Semiconducting Carbon Nanotube and a Metal Electrode, *Phys. Rev. B* **66**, 073307.

165. Guo, J., Datta, S., and Lundstrom, M. (2004). A Numerical Study of Scaling Issues for Schottky-Barrier Carbon Nanotube Transistors, *IEEE Trans. Electron Devices* **51**, 172.

166. Ishii, H., Kobayashi, N., and Hirose, K. (2007). Quantum Transport Properties of Carbon Nanotube Field-Effect Transistors with Electron–Phonon Coupling, *Phys. Rev. B* **76**, 205432.

167. Dürkop, T., Getty, S. A., Cobas, E., and Fuhrer, M. S. (2004). Extraordinary Mobility in Semiconducting Carbon Nanotubes, *Nano Lett.* **4**, 35.

168. Kobayashi, N., Ozaki, T., and Hirose, K. (2007). Length Dependence of Effect of Electrode Contact on Transport Properties of Semiconducting Carbon Nanotubes, *Surf. Sci.* **601**, 4136.

169. Novoselov, K. S., Geim, A. K., Morozov, S. V., Jiang, D., Katsnelson, M. I., Grigorieva, I. V., Dubonos, S. V., and Firsov, A. A. (2005). Two-Dimensional Gas of Massless Dirac Fermions in Graphene, *Nature* **438**, 197.

170. Bolotin, K. I., Sikes, K. J., Jiang, Z., Klima, M., Fudenberg, G., Hone, J., Kim, P., and Stormer, H. L. (2008). Ultrahigh Electron Mobility in Suspended Graphene, *Solid State Commu.* **146**, 351.

171. Han, M. Y. Özyilmaz, B., Zhang, Y., and Kim P. (2007). Energy Band-Gap Engineering of Graphene Nanoribbons, *Phys. Rev. Lett.* **98**, 206805.

172. Ishii, Kobayashi, N., and Hirose, K. (2010). Edge-Phonon Scattering Effects on Electron Transport of Graphene Nanoribbons, *Appl. Phys. Express* **10**, 283.

173. Wang, X., Ouyang, Y., Li, X., Wang, H., Guo, J., and Dai H. (2008). Room-Temperature All-Semiconducting Sub-10-nm Graphene Nanoribbon Field-Effect Transistors, *Phys. Rev. Lett.* **100**, 206803.

174. Nakada, K., Fujita, M., Dresselhaus, G., and Dresselhaus, M. S. (1996). Edge State in Graphene Ribbons: Nanometer Size Effect and Edge Shape Dependence, *Phys. Rev. B* **54**, 1996.

175. Reed, M. A., Zhou, C., Muller, C. J., Burgin, T. P., and Tour, J. M. (1997). Conductance of a Molecular Junction, *Science* **278**, 252.

176. Zhou, C., Muller, C. J., Deshpande, M. R., Sleight, J. W., and Reed, M. A. (1995). Microfabrication of a Mechanically Controllable Break Junction in Silicon, *Appl. Phys. Lett.* **67**, 1160.

177. Park, H., Lim, A. K. L., Alivisatos, A. P., Park, J., and McEuen, P. L. (1999). Fabrication of Metallic Electrodes with Nanometer Separation by Electromigration, *Appl. Phys. Lett.* **75**, 301.

178. Park, H., Park, J., Lim, A. K. L., Anderson, E. H., Alivisatos, A. P., and McEuen, P. L. (2000). Nanomechanical Oscillations in a Single-C_{60} Transistor, *Nature* **407**, 57.

179. Kushmerick, J. G., Naciri, J., Yang, J. C., and Shashidhar, R. (2003). Conductance Scaling of Molecular Wires in Parallel, *Nano Lett.* **3**, 897.

180. Araki, K., Endo, H., Tanaka, H., and Ogawa, T. (2004). Multi-Curve Fitting Analysis of Temperature-Dependent I–V Curves of Poly-Hexathienyl phenanthroline-Bridged Nanogap Electrodes, *Jpn. J. Appl. Phys.* **43**, L634.

181. Kobayashi, N., Ozaki, T., and Hirose, K. (2006). First-Principles Calculations of Quantum Transport in Single Molecule, *Jpn. J. Appl. Phys.* **45**, 2151.

182. Prodan, E., and Car, R. (2009). Theory of Tunneling Transport in Periodic Chains, *Phys. Rev. B* **80**, 035124.

183. Venkataraman, L., Klare, J. E., Nuckolls, C., Hybertsen, M. S., and Steigerwald, M. L (2006). Dependence of Single-Molecule Junction Conductance on Molecular Conformation, *Nature* **442**, 904.

184. Tsukada, M., Tagami, K., Hirose, K., and Kobayashi, N. (2005). Theory of Quantum Conductance of Atomic and Molecular Bridges, *J. Phys. Soc. Jpn.* **74**, 1079.

185. Hirose, K., and Kobayashi, N. (2005). Theory for Electron Hopping through Nanometer-Scale Contacts: From Tunneling Regime to Ballistic Regime, *Physica E* **29**, 515.

186. Ventra, M. D., Pantelides, S. T., and Lang, N. D. (2000). First-Principles Calculation of Transport Properties of a Molecular Device, *Phys. Rev. Lett.* **84**, 979.

187. Hirose, K., and Kobayashi, N. (2007). Effects of Atomic-Scale Contacts on Transport Properties through Single Molecules: ab initio Study, *Surf. Sci.* **601**, 4113.

188. Avouris, P., Lyo, I. W., Bozso, F., and Kaxiras, E. (1990). Adsorption of Boron on Si(111): Physics, Chemistry, and Atomic-Scale Electronic Devices, *J. Vac. Sci. Technol. A* **8**, 3405.

189. Amlani, I., Rawlett, A. M., Nagahara, L. A., and Tsui, R. K. (2002). An Approach to Transport Measurements of Electronic Molecules, *Appl. Phys. Lett.* **80**, 2761.

190. Ho, W. (2002). Single-Molecule Chemistry, *J. Chem. Phys.* **117**, 11033.

191. Smit, R. H. M., Hoat, Y., Untiedt, C., Lang, N. D., van Hemert, M. C., and Ruitenbeek, J. M. van (2002). Measurement of the Conductance of a Hydrogen Molecule, *Nature* **419**, 906.

192. Asai, Y. (2004). Theory of Inelastic Electric Current through Single Molecules, *Phys. Rev. Lett.* **93**, 246102.

193. Frederiksen, T., Paulsson, M., Brandbyge, M., and Jauho, A.-P. (2007). Inelastic Transport Theory from First-Principles: Methodology and Application to Nanoscale Devices, *Phys. Rev. B* **75**, 205413.

194. Gershenson, M. E., Podzorov, V., and Morpurgo, A. F. (2006). Colloquim: Electronic Transport in Single-Crystal Organic Transistors, *Rev. Mod. Phys.* **78**, 973.

195. Hasegawa, and Takeya, J. (2009). Organic Field-Effect Transistors Using Single Crystals, *Sci. Technol. Adv. Mater.* **10**, 024314.

196. Nelson, S. F., Lin, Y.-Y., Gundlach, D. J., and Jackson, T. N. (1998). Temperature-Independent Transport in High-Mobility Pentacene Transistors, *Appl. Phys. Lett.* **72**, 1854.

197. Takeya, J., Goldmann, C., Haas, S., Pernstich, K. P., Ketterer, B., and Batlogg, B. (2003). Field-Induced Charge Transport at the Surface of Pentacene Single Crystals: A Method to Study Charge Dynamics of Two-Dimensional Electron Systems in Organic Crystals, *J. Appl. Phys.* **94**, 5800.

198. Podzorov, V., Menard, E., Borissov, A., Kiryukhin, V., Rogers, J. A., and Gershenson, M. E. (2004). Intrinsic Charge Transport on the Surface or Organic Semiconductors, *Phys. Rev. Lett.* **93**, 086602.

199. Takeya, J., Tsukagoshi, K., Aoyagi, Y., Takenobu, T., and Iwasa, Y. (2005). Hall Effect of Quasi-Hole Gas in Organic Single-Crystal Transistors, *Jpn. J. Appl. Phys.* **44**, L1393.

200. Troisi, A., and Orlandi, G. (2006). Charge-Transport Regime of Crystalline Organic Semiconductors: Diffusion Limited by Thermal Off-Diagonal Electronic Disorder, *Phys. Rev. Lett.* **96**, 086601.

201. Fratini, S., and Ciuchi, S. (2009). Bandlike Motion and Mobility Saturation in Organic Molecular Semiconductors, *Phys. Rev. Lett.* **103**, 266601.

202. Böhlin, J., Linares, M., and Stafström, S. (2011). Effect of Dynamic Disorder on Charge Transport along a Pentacene Chain, *Phys. Rev. B* **83**, 085209.

203. Ortmann, F., Bechstedt, F., and Hannewald, K. (2009). Theory of Charge Transport in Organic Crystals: Beyond Holstein's Small-Polaron Model, *Phys. Rev. B* **79**, 235206.

204. Grimme, S. (2004). Accurate Description of van der Waals Complexes by Density Functional Including Empirical Correlations, *J. Comp Chem.* **25**, 1463.

205. Ishii, H., Honma, K., Kobayashi, N., and Hirose, K. (2012). Wave-Packet Approach to Transport Properties of Carrier Coupled with Intermolecular and Intramolecular Vibrations of Organic Semiconductors, *Phys. Rev. B* **85**, 245206.

206. Tamura, H., Tsukada, M., Ishii, H., Kobayashi, N., and Hirose, K. (2012). Roles of Intramolecular and Intermolecular Electron–Phonon Coupling on the Formation and Transport of Large Polarons in Organic Semiconductors, *Phys. Rev. B* **86**, 035208.

207. Ando, T., Fowler, A. B., and Stern, F. (1982). Electronic Properties of Two-Dimensional Systems, *Mod. Rev. Phys.* **54**, 437.

208. Sze, S. M. (1985). *Semiconductor Devices: Physics and Technology* (John Wiley & Sons, Inc., New York).

209. Sohn, L. L., Kouwenhoven, L. P., and Schön, G. (ed.) (1997). *Mesoscopic Electron Transport* (Kluwer, Dordrecht).

210. Lin, B. J. F., Tsui, D. C., Paalanen, M. A., and Gossard, A. C. (1984). Mobility of the two-dimensional electron gas in GaAs-$Al_xGa_{1-x}As$ Heterostructures, *Appl. Phys. Lett.* **45**, 695.

211. Pfeiffer, L., West, K. W., Stormer, H. L., and Baldwin, K. W. (1989). Electron Mobilities Exceeding $10^7\mathrm{cm}^2$/Vs in Modulation-Doped GaAs, *Appl. Phys. Lett.* **55**, 1888.

212. Fowler, A. B., Hartstein, A., and Webb, R. A. (1982). Conductance in Restricted-Dimensionality Accumulation Layers, *Phys. Rev. Lett.* **48**, 196.

213. Bird, J. P., Grassie, A. D. C., Lakrimi, M., Hutchings, K. M., Harris, J. J., and Foxon, C. T. (1990). Conductance Fluctuations and Non-Diffusive Motion in GaAs/AlGaAs Heterojunction Wires, *J. Phys.: Cond. Mat.* **2**, 7847.

214. Mehta, M. L. (1991). *Random Matrices* (Academic Press, Boston).

215. van Wees, B. J., van Houten, H., Beenakker, C. W. J., Williamson, J. G., Kouwenhoven, L. P., van der Marel, D., and Foxon, C. T. (1988). Quantized Conductance of Point Contacts in a Two-Dimensional Electron Gas, *Phys. Rev. Lett.* **60**, 848.

216. Wharam, D. A., Thorntom, T. J., Newbury, R., Pepper, M., Ahmed, H., Frost, J. E. F Hasko, D. G., Peacock, D. C., Ritchie, D. A., and Jones, G. A. C. (1988). One-Dimensional Transport and the Quantisation of the Ballistic Resistance, *J. Phys. C* **21**, L209.

217. Kane, B. E., Facer, G. R., Dzurak, A. S., Lumpkin, N. E., Clark, R. G., Pfeiffer, L. N., and West, K. W. (1998). Quantized Conductance in Quantum Wires with Gate-Controlled Width and Electron Density, *Appl. Phys. Lett.* **48**, 196.

218. Cronenwett, S. M., Lynch, H. J., Goldhaber-Gordon, D., Kouwenhoven, L. P., Marcus, C. M., Hirose, K., Wingreen, N. S., and Umansky, V. (2002). Low-Temperature Fate of the 0.7 Structure in a Point Contact: A Kondo-like Correlated State in an Open System, *Phys. Rev. Lett.* **88**, 226805.

219. Büttiker, M. B. (1990). Quantized Transmission of a Saddle-Point Constriction, *Phys. Rev. B* **41**, 7906.

220. Oosterkamp, T. H., Kouwenhoven, L. P., Koolen, A. E. A., van der Vaart, N. C., and Harmans, C. J. P. M. (1997). Photon Sidebands of the Ground State and First Excited State of a Quantum Dot, *Phys. Rev. Lett.* **78**, 1536.

221. McEuen, P. L., Bockrath, M., Cobden, D. H., Yoon, Y. G., and Louie, S. G. (1999). Photon Sidebands of the Ground State and First Excited State of a Quantum Dot, *Phys. Rev. Lett.* **83**, 5098.

222. Foxman, E. B., McEuen, P. L., Meirav, U., Wingreen, N. S., Meir, Y., Belk, P. A., Belk, N. R., Kastner, M. A., and Wind, S. J. (1993). Effects of Quantum Levels on Transport through a Coulomb Island, *Phys. Rev. B* **47**, 10020.

223. Grabert, H., and Devoret, M. H. (ed.) (1991). *Single Charge Tunneling in Nanostructures* (Plenum Press, New York).

224. Kastner, M. A. (1992). The Single Electron Transistor, *Rev. Mod. Phys.* **64**, 849.

225. Meir, Y., Wingreen, N. S., and Lee, P. A. (1991). Transport through a Strongly Interacting Electron System: Theory of Periodic Conductance Oscillations, *Phys. Rev. Lett.* **66**, 3048.

226. Hewson, A. C. (1993). *The Kondo Problem to Heavy Fermions* (Cambridge University Press, UK).

227. Meir, Y., Wingreen, N. S., and Lee, P. A. (1993). Low-Temperature Transport through a Quantum Dot: The Anderson Model Out of Equilibrium, *Phys. Rev. Lett.* **70**, 2601.

228. Ng, T. K., and Lee, P. A. (1988). On-Site Coulomb Repulsion and Resonant Tunneling, *Phys. Rev. Lett.* **61**, 1768.

229. Glazman, L. I., and Raikh, M. E. (1988). From the Kondo Regime to the Mixed-Valence Regime in a Single-Electron Transistor, *JETP Lett.* **47**, 452.

230. Kawabata, A. (1991). On the Electron Transport through a Quantum Dot, *J. Phys. Soc. Jpn.* **60**, 3222.

231. Kondo, J. (1964). Resistance Minimum in Dilute Magnetic Alloys, *Prog. Theor. Phys.* **32**, 37.

232. Goldhaber-Gordon, D. Göres, J., Kastner, M. A., Shtrikman, H., Mahalu, D., and Meirav, U. (1998). From the Kondo Regime to the Mixed-Valence Regime in a Single-Electron Transistor, *Phys. Rev. Lett.* **81**, 5225.

233. Cronenwett, S. M., Oosterkamp, T. H., and Kouwenhoven, L. P. (1998). A Tunable Kondo Effect in Quantum Dots, *Science* **281**, 540.

234. Jacak, L., Hawrylak P., and Wójs, A. (1998). *Quantum Dots* (Springer, Berlin).

235. Chakraborty T. (1999). *Quantum Dots: A Survey of the Properties of Artificial Atoms* (North Holland, Amsterdam).

236. Reimann S., and Manninen M. (2002). Electronic Structure of Quantum Dots, *Rev. Mod. Phys.* **74**, 1283.

237. Shikin, V., Nazin, S., Heitmann, D., and Demel, T. (1991). Dynamic Response of Quantum Dots, *Phys. Rev. B* **43**, 11903.

238. Tanatar, B., and Ceperley, D. M. (1989). Ground State of the Two-Dimensional Electron Gas, *Phys. Rev. B* **39**, 5005.

239. Attaccalite, C., Moroni, S., Gori-Giorgi, P., and Bachelet, G. B. (2002). Correlation Energy and Spin Polarization in the 2D Electron Gas, *Phys. Rev. Lett.* **88**, 256601; **91**, 109902.

240. Pederiva, F., Umrigar, C. J., and Lipparini, E. (2000). Diffusion Monte Carlo Study of Circular Quantum Dots, *Phys. Rev. B* **62**, 8120; **68**, 089901.

241. Harting, J. Mülken, and Borrmann, P. (2002). Interplay between Shell Effects and Electron Correlations in Quantum Dots, *Phys. Rev. B* **62**, 10207.

242. Hirose, K., and Wingreen, N. S. (1999). Spin-Density-Functional Theory of Circular and Elliptical Quantum Dots, *Phys. Rev. B* **59**, 4604.

243. Tarucha, S., Austing, D. G., Honda, T., van der Hage, R. J., and Kouwenhoven, L. P. (1996). Shell Filling and Spin Effects in a Few Electron Quantum Dot, *Phys. Rev. Lett.* **77**, 3613.

244. Austing, D. G., Sasaki, S., Tarucha, S., Reimann, S. M., Koskinen, M., and Manninen, M. (1999). Ellipsoidal Deformation of Vertical Quantum Dots, *Phys. Rev. B* **60**, 11514.

245. Vignale, G., and Rasolt, M. (1987). Density-Functional Theory in Strong Magnetic Fields, *Phys. Rev. Lett.* **59**, 2360; Current- and Spin-Density-Functional Theory for Inhomogeneous Electronic Systems in Strong Magnetic Fields, *Phys. Rev. B* **37**, 10685.

246. Rasolt, M., and Perrot, F. (1992). Spontaneously Symmetry-Broken-Current-Carrying States in a Strongly Inhomogeneous Interacting Two-Dimensional Fermi Liquid, *Phys. Rev. Lett.* **69**, 2563.

247. Levesque, D., Weiss, J. J., and MacDonald, A. H. (1984). Crystallization of the Incompressible Quantum-Fluid State of a Two-Dimensional Electron Gas in a Strong Magnetic Field, *Phys. Rev. B* **30**, 1056.

248. Wensauer, A., and Rössler, U. (2004). Exchange-Correlation Vector Potentials and Vorticity-Dependent Exchange-Correlation Energy Densities in Two-Dimensional Systems, *Phys. Rev. B* **69** 155302.

249. Hirose, K., and Wingreen, N. S. (2005). Electronic States and Spin in Clean and Disordered Quantum Dots, in *Quantum Dots: Research Developments* (Ling, P. A., ed.) (Nova Science Publisher, New York).

250. Ashoori, R. C., Stormer, H. L., Weiner, J. S., Pfeiffer, L. N., Baldwin, K. W., and West, K. W. (1993). N-Electron Ground State Energies of a Quantum Dot in Magnetic Field, *Phys. Rev. Lett.* **71**, 613.

251. Haug, H., and Koch, S. W. (1990) *Quantum Theory of the Optical and Electronic Properties of Semiconductors,*, 3rd Ed. (World Scientific, Singapore).

252. Burke, K., and Gross, E. K. U. (1998). A Guided Tour of Time-Dependent Density Functional Theory, in *Density Functionals: Theory and Applications* (Joubert, D., ed.) (Springer-Verlag, Berlin).

253. Casida, M. E. (1996). Time-Dependent Density Functional Response Theory of Molecular Systems: Theory, Computational Methods, and Functionals, in *Recent Developments and Application of Modern Density Functional Theory* (Seminario, J. M., ed.) (Elsevier Science, Amsterdam).

254. Petersilka, M., Gossmann, U. J., and Gross, E. K. U. (1996). Excitation Energies from Time-Dependent Density-Functional Theory, *Phys. Rev. Lett.* **76**, 1212.

255. Yabana, K., and Bertsch, G. F. (1996). Time-Dependent Local-Density Approximation in Real Time, *Phys. Rev. B.* **54**, 4484.

256. Feit, M. D., Fleck. Jr., J. A., and Steiger, A. (1982). Solution of the Schrödinger Equation by a Spectral Method, *J. Comp. Phys.* **47**, 412.
257. Hirose, K., Meir, Y., and Wingreen, N. S. (2004). Time-Dependent Density-Functional Theory of Excitation Energies of Closed-Shell Quantum Dots, *Physica E* **22**, 486.
258. Alhassid, Y. (2000). The Statistical Theory of Quantum Dots, *Rev. Mod. Phys.* **72**, 895.
259. Sivan, U., Berkovits, R., Aloni, Y., Prus, O., Auerbach, A., and Ben-Yoseph, G. (1996). Mesoscopic Fluctuations in the Ground State Energy of Disordered Quantum Dots, *Phys. Rev. Lett.* **77**, 1123.
260. Patel, S. R. Cronenwett, S. M. Stewart, D. R. Marcus, C. M. Duruöz, C. I., Harris. Jr., J. S. Campman, K., and Gossard, A. C. (1998). Statistics of Coulomb Blockade Peak Spacings, *Phys. Rev. Lett.* **80**, 4522.
261. Hirose, K. Zhou, F., and Wingreen, N. S. (2001). Density-Functional Theory of Spin-Polarized Disordered Quantum Dots, *Phys. Rev. B* **63**, 075301.
262. Hirose, K., and Wingreen, N. S. (2002). Ground-State Energy and Spin in Disordered Quantum Dots, *Phys. Rev. B* **65**, 193305.
263. Lieb, E., and Mattis, D. (1962). Theory of Ferromagnetism and the Ordering of Electronic Energy Levels, *Phys. Rev.* **125**, 164.
264. Hirose, K., Si, S. S., and Wingreen, N. S. (2001). Mechanisms for Extra Conductance Plateaus in Quantum Wires, *Phys. Rev. B* **63**, 033315.
265. Thomas, K. J., Nicholls, J. T., Simmons, M. Y., Pepper, M., Mace, D. R., and Ritchie, D. A. (1996). Possible Spin Polarization in a One-Dimensional Electron Gas, *Phys. Rev. Lett.* **77**, 135.
266. Kristensen, A., Bruus, H., Hansen, A. E., Jensen, J. B., Lindelof, P. E., Marckmann, C. J., Nygård, J., Sørensen, C. B., Beuscher, F., Forchel, A., and Michel, M. (2000). Bias and Temperature Dependence of the 0.7 Conductance Anomaly in Quantum Point Contacts, *Phys. Rev. B* **62**, 10950.
267. Reilly, D. J., Facer, G. R., Dzurak, A. S., Kane, B. E., Clark, R. G., Stiles, P. J., Hamilton, A. R., O'Brien, J. L., Lumpkin, N. E., Pfeiffer, L. N., and West, K. W. (2001). Many-Body Spin-Related Phenomena in Ultra Low-Disorder Quantum Wires, *Phys. Rev. B* **63**, 121311.
268. Wang, C.-K., and Berggren, K.-F. (1998). Local Spin Polarization in Ballistic Quantum Point Contacts, *Phys. Rev. B* **57**, 4552.
269. Spivak, B., and Zhou, F. (2000). Ferromagnetic Correlations in Quasi-One-Dimensional Conducting Channels, *Phys. Rev. B* **61**, 16730.

270. Flambaum, V. V., and Kuchiev, M. Y (2000). Possible Mechanism of the Fractional Conductance Quantization in a One-Dimensional Constriction, *Phys. Rev. B* **61**, 7869.

271. Seelig, G., and Matveev, K. A (2003). Electron–Phonon Scattering in Quantum Point Contacts, *Phys. Rev. Lett.* **90**, 176804.

272. Matveev, K. A (2004). Conductance of a Quantum Wire in the Wigner-Crystal Regime, *Phys. Rev. Lett.* **92**, 106801.

273. Hirose, K., Meir, Y., and Wingreen, N. S. (2003). Local Moment Formation in Quantum Point Contacts, *Phys. Rev. Lett.* **90**, 026804.

274. Meir, Y., Hirose, K., and Wingreen, N. S. (2002). Kondo Model for the "0.7 Anomaly" in Transport through a Quantum Point Contact, *Phys. Rev. Lett.* **89**, 196802.

275. Rejec, T., and Meir, Y. (2006). Magnetic Impurity Formation in Quantum Point Contacts, *Nature* **442**, 900.

276. Güclü, A. D., Umrigar, C. J., Jiang, H., and Baranger, H. U. (2009). Localization in an Inhomogeneous Quantum Wire, *Phys. Rev. B* **80**, 201302.

277. Special issue (2008). *J. Phys: Condens. Matter.* **20**, 164201–164217.

Index